Signals and Systems

Signals and Systems

Bernd Girod
Universität Erlangen-Nürnberg

JOHN WILEY & SONS, LTD
Chichester • Weinheim • New York • Brisbane • Singapore • Toronto

Other Wiley Editorial Offices

John Wiley & Sons, Inc., 605 Third Avenue,
New York, NY 10158-0012, USA

Wiley-VCH Verlag GmbH
Pappelallee 3, D-69469 Weinheim, Germany

John Wiley & Sons (Australia) Ltd, 33 Park Road, Milton,
Queensland 4064, Australia

John Wiley & Sons (Canada) Ltd, 22 Worcester Road
Rexdale, Ontario, M9W 1L1, Canada

John Wiley & Sons (Asia) Pte Ltd, 2 Clementi Loop #02-01,
Jin Xing Distripark, Singapore 129809

Library of Congress Cataloging-in-Publication Data

Girod, Bernd
   [Einführung in der Systemtheorie, English]
   Systems and Signals / Bernd Girod, Rudolf Rabenstein, Alexander Stenger.
      p. cm.
   Includes bibliographical references and index.
   ISBN 0 471 98800 6
      1. System theory. I. Rabenstein, Rudolf. II. Stenger, Alexander. III Title.
   Series

   Q295. G57 2001
   003–dc21   2001024349

British Library Cataloguing in Publication Data

A catalogue record for this book is available from the British Library

ISBN 0 471 98800 6

Produced from LaTeX files supplied by the translator

Signals and Systems

B. Girod, R. Rabenstein, A. Stenger

March 29, 2001

Contents

Preface

Analysing and designing systems with the help of suitable mathematical tools is extraordinarily important for engineers. Accordingly, systems theory is a part of the core curriculum of modern electrical engineering and serves as the foundation of a large number of subdisciplines. Indeed, access to specialised areas of electrical engineering demands a mastery of systems theory.

An introduction to systems theory logically begins with the simplest abstraction: linear, time-invariant systems. We find applications of such systems everywhere, and their theory has attained advanced maturity and elegance. For students who are confronted with the theory of linear, time-invariant systems for the first time, the subject unfortunately can prove difficult, and, if the required and deserved academic progress does not materialise, the subject might be downright unpopular. This could be due to the abstract nature of the subject area coupled with the deductive and unclear presentation in some lectures. However, since failure to learn the fundamentals of systems theory would have catastrophic repercussions for many subsequent subjects, the student must persevere.

We have written this book as an easily accessible introduction to systems theory for students of electrical engineering. The content itself is nothing new; the theory has already been described in other books. What is new is how we deliver the material. By means of small, clear explanatory steps, we aim to present the abstract concepts and interconnections of systems theory so simply as to make learning easy and fun. Naturally, only the reader can assess whether we have achieved our goal.

To aid understanding, we generally use an inductive approach, starting with an example and then generalising from it. Additional examples then illustrate further aspects of an idea. Wherever a picture or a figure can enrich the text, we provide one. Furthermore, as the text progresses, we continuously order the statements of systems theory in their overall context. Accordingly, in this book a discussion of the importance of a mathematical formula or a theorem takes precedence over its proof. While we might omit the derivation of an equation, we never neglect a discussion of its applications and consequences! The numerous exercises at the end of each chapter (with detailed solutions in the appendix) help to reinforce the reader's knowledge.

Although we have written this book primarily for students, we are convinced that it will also be useful for practitioners. An engineer who wants to brush up quickly on some subject will appreciate the easy readability of this text, its practice-oriented presentation, and its many examples.

This book evolved out of a course on systems theory and the corresponding laboratory exercises at the Friedrich Alexander University in Erlangen-Nürnberg. The course is compulsory for students of electrical engineering in the fifth semester. As such, the material in this book can be worked through completely in about 50 hours of lectures and 25 hours of exercises. We do assume knowledge of the fundamentals of engineering mathematics (differential and integral calculus, linear algebra) and basic knowledge of electrical circuits. Assuming that this mathematical knowledge has been acquired earlier, the material is also suitable for use in the third or fourth semester. An engineering curriculum often encompasses complex function theory and probability theory as well; although these fields are helpful, we do not assume familiarity with them.

This book is also suitable for self-study. Assuming full-time, concentrated work, the material can be covered in four to six weeks.

Our presentation begins with continuous signals and systems. Contrary to some other books that first introduce detailed forms of description for signals and only much later add systems, we treat signals and systems in parallel. The purpose of describing signals by means of their Laplace or Fourier transformations becomes evident only through the characteristics of linear, time-invariant systems. In our presentation we emphasise the clear concept of Eigen functions, whose form is not changed by systems. To take into account initial states, we use state space descriptions, which elegantly allow us to couple an external and an internal component of the system response. After covering sampling, we introduce time-discrete signals and systems and so extend the concepts familiar from the continuous case. Thereafter discrete and continuous signals and systems are treated together. Finally, we discuss random signals, which are very important today.

To avoid the arduous and seldom perfect step of correcting camera-ready copy, we handled the layout of the book ourselves at the university. All formulas and most of the figures were typeset in LaTeX and then transferred onto overhead slides that were used for two years in the systems theory lectures. We are most grateful to some 200 registered students whose attentive and astute criticism helped us to debug the presentation and the typeset equations. In addition, one year's students read the first version of the manuscript and suggested diverse improvements. Finally numerous readers of the German version reported typographic errors and sent comments by e-mail.

Our student assistants Lutz and Alexander Lampe, Stephan Gödde, Marion Schabert, Stefan von der Mark and Hubert Rubenbauer demonstrated tremendous commitment in typesetting and correcting the book as well as the solutions to the exercises. We thank Ingrid Bärtsch, who typed and corrected a large portion of the text, as well as Susi Koschny, who produced many figures.

For their attentive and tireless proof-reading, we especially thank Peter Eisert, Achim Hummel, Wolfgang Sörgel, Gerhard Runze and Reinhard Bernstein. For their generous availability for discussions about tricky mathematical questions, we sincerely thank Peter Steffen and Ulrich Forster. Edward Kimber has mastered the ambitious task of translating the German manuscript into English. Finally, we express our gratitude to John Wiley & Sons for their uncomplicated co-operation and their support of this project.

When the second edition of this book appears, we would like to extend our list of acknowledgements. Therefore we have the following request to our readers. Please send us your comments and suggestions. The simplest route is per e-mail to `stbuch@LNT.de`. Whatever error you might detect and however small it may be, please do not keep it to yourself. We promise that we will take to heart all serious comments.

Erlangen, Germany, October 2000

Bernd Girod   Rudolf Rabenstein   Alexander Stenger

1 Introduction

Systems theory concerns signals and systems. What are signals? What are systems? Before defining these terms, let us first examine some examples.

1.1 Signals

Signals describe quantities that change. Figure 1.1 depicts the electrical voltage that a microphone produces in response to the spoken word 'car'. This voltage corresponds largely to the acoustic pressure on our ear, which reacts to the changes in this pressure over time. The curve in Figure 1.1 shows the value of microphone voltage in relation to time. Since there is a voltage value for every point in time, we term this a *continuous-time* signal. We call time the *independent* variable and the voltage changing over time the dependent variable or signal amplitude. We usually represent the independent variable horizontally (x-axis) and the dependent variable vertically (y-axis).

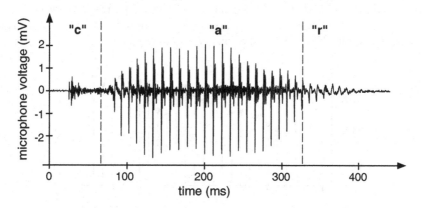

Figure 1.1: Example of a continuous-time signal: voice signal for the syllable 'car'

Figure 1.2 depicts another continuous signal. The diagram shows the temperature curves for a house wall, not over time, but in relation to the location. The curves show the temperature profile inside a 15 cm thick brick wall where the air temperature at the right side suddenly rose by 10 K. One hour later the local temperature follows the curve represented by the thick line. At another time we would have a different temperature curve. In contrast to Figure 1.1, time here is a parameter of a family of curves; the independent continuous variable is the location in the wall.

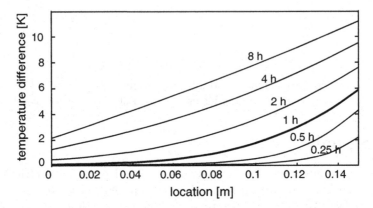

Figure 1.2: Temperature curve for a house wall

Figure 1.3 shows another kind of variable quantity, the stock market index over time. Although this index changes all the while the stock market is open, the diagram shows only the weekly average. Thus the depicted value does not change continuously, but only once a week. When the signal amplitude occurs only at certain fixed points in time (discrete times), but not for points in between, we call the signal *discrete* or, more precisely, *discrete-time*. In our example, however, the signal amplitude itself is not discrete but continuous.

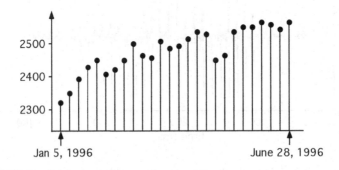

Figure 1.3: The weekly German stock market index between January 5, 1996, and June 28, 1996

In Figure 1.4 we have entered the frequency of earned marks for a test in system theory at the University of Erlangen–Nürnberg in April, 1996. The individual marks assume only discrete values (1.0 – 5.0); the frequencies (in contrast to the average stock index) are whole numbers and so likewise discrete. In this case both the independent and the dependent variables are discrete.

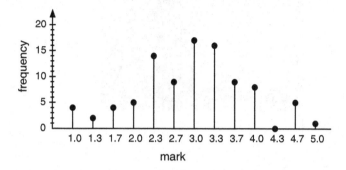

Figure 1.4: Frequency of earned marks for a test in systems theory

The signals we have considered thus far have been quantities that depend on a *single* independent variable. However, there are quantities with dependencies on two or more variables. The greyscales of Figure 1.5 depend on both the x and the y co-ordinates. Here both axes represent independent variables. The dependent variable $s(x, y)$ is entered along one axis, but is a greyscale value between the extreme values black and white.

When we add motion to pictures, we have a dependency on three independent variables (Figure 1.6): two co-ordinates and time. We call these two- or three-dimensional (or generally multidimensional) signals. When greyscale values change continuously over space or over space and time, these are continuous signals.

All our examples have shown parameters (voltage, temperature, stock index, frequencies, greyscale) that change in relation to values of the independent variables. Thereby they transmit certain information. In this book we define a signal as follows:

Definition 1: Signal

A signal is a function or sequence of values that represents information.

The preceding examples have shown that signals can assume different forms. Signals can be classified according to various criteria, the most important of which are summarised in Table 1.1.

Figure 1.5: A picture as a continuous two-dimensional signal

Figure 1.6: Moving picture as an example of a continuous three-dimensional signal

Table 1.1: Criteria for classifying signals

continuous(-time)	-	discrete(-time)
amplitude-continuous	-	amplitude-discrete
analogue	-	digital
real-valued	-	complex-valued
unidimensional	-	multidimensional
finite domain	-	infinite domain
deterministic	-	stochastic

We have already discussed the difference between continuous and discrete signals on the basis of Figures 1.1 and 1.3. Discrete signals are also termed discontinuous. Most of the preceding signals have been amplitude-continuous, because their dependent variable can take on any value. However, the signal in Figure 1.4 is amplitude-discrete, for the dependent variable (number of examinees) can assume only integer values. Taken precisely, the stock index in Figure 1.3 is likewise amplitude-discrete, since the stock index is specified to only a certain number of decimal places. Signals whose dependent and independent variables are continuous are called *analogue* signals. If both variables are discrete, we call the signal *digital*. The output voltage of a microphone is an analogue signal, for at any given time amplitude values can be read with any desired precision. Sequences of values stored in a computer are always digital, since the amplitude values can be stored only with finite word length in distinct (discrete) storage cells.

All of the signals we have considered so far had real amplitudes and so are classified as *real-valued*. Signals whose dependent variable assumes complex values are called *complex-valued*.

The signals in Figures 1.1 to 1.4 are unidimensional, while those in Figures 1.5 and 1.6 are multidimensional. For reasons of graphic representation, all the signals in the previous examples had finite domains of their independent variables and so are classified as *finite-domain* signals. However, if we consider the signal in Figure 1.6 as the picture of a television camera, then the domain of the location variable becomes finite again due to the restricted picture excerpt, but the domain of the time variable is infinite (neglecting the finite lifetime of the camera).

Signals are termed *deterministic* if their behaviour is known and can be represented, e.g., by a formula. The deflection voltage of an oscilloscope is a deterministic signal, for its behaviour is known and can be represented as a sawtooth wave. By contrast, we cannot define the amplitude values of a voice signal (see Figure 1.1) by means of formulae or graphical elements; furthermore, their continued behaviour is not known. Such signals are termed *stochastic*. Since it is impossible to specify their behaviour in terms of functions, such signals are described by expected values (mean, variance and many others).

1.2 Systems

1.2.1 What is a System?

We have seen that signals represent information. In many technical applications we want to do more than just view information; we want to store, transfer, or couple it with other information. This requires establishing and describing relationships between signals. This leads us to the definition of a *system*:

Definition 2:   System

A system is the abstraction of a process or object that puts a number of signals into some relationship.

In this general form we can imagine a system as a black box that communicates with the outside world via various signals. Figure 1.7 depicts such a system that establishes a relationship among the signals x_1 to x_n.

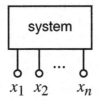

Figure 1.7: General system

In many cases we can classify a system's signals as input and output signals. Input signals exist independently of the system and are not affected by the system; instead, the system reacts to these signals. Output signals bear information generated by the system, often in response to input signals. The simple system in Figure 1.8 has one input signal x and one output signal y. We also term y the *system response* to x.

Naturally a system might contain multiple inputs and outputs. The system determines the influence of individual inputs on the output signals. In general, each output depends on all inputs. To simplify the notation, we combine input and output signals in vectors (Figure 1.8).

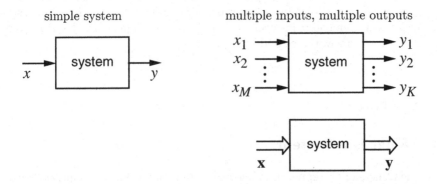

Figure 1.8: Input/output systems

1.2.2 The Domain of Systems Theory

Systems theory does not encompass the implementation of a system from given components, but with the relationships that the system imposes between its signals. Systems theory represents a powerful mathematical tool for the study and design of systems, because omitting the implementation details helps to maintain an overview of the overall system. In systems theory the focus is on the formal nature of the interconnections rather than any specialisation for specific applications. This allows systems theory a uniform representation of processes from different application domains (e.g., physics, engineering, economics, biology) and supports an interdisciplinary view.

The high degree of abstraction brings the advantages of learning economy and clarity. Learning economy ensues because the regularities of one field are easier to transfer to other fields if they are formulated in general form. Clarity results because separating the detail problems from the general relationships is elevated to a principle. However, this is countered by the drawback of a certain amount of unclearness that encumbers initial learning in systems theory.

1.2.3 Linear, Time-Invariant Systems

An important subfield of systems theory is the theory of linear, time-invariant systems. This represents the classical core domain of systems theory and is well developed, elegant and clear. This theory also proves suitable for describing nonlinear systems that can be linearised for small signal amplitudes. Systems theory for linear, time-invariant systems evolved from the practical problems of electrical engineering over more than a century [24]. Important application domains for the theory of linear, time-invariant systems in electrical engineering today include:

- Analysis and design of electrical circuits

- Digital signal processing

- Communications

- Control engineering

- Measurement engineering

This book covers only linear, time-invariant systems. First, however, we need to explain the terms linearity and time invariance.

1.2.3.1 Linearity of a System

To define the term linearity, let us consider the system in Figure 1.9. It responds to an input signal $x_1(t)$ with the output signal $y_1(t)$ and to the input signal $x_2(t)$

with the output signal $y_2(t)$. Can we deduce the output signal $y(t)$ associated with the following input signal?

$$x(t) = Ax_1(t) + Bx_2(t) \tag{1.1}$$

In general we cannot make this step, but for many relationships between input and output parameters, from (1.1) the output signal follows as

$$y(t) = Ay_1(t) + By_2(t) . \tag{1.2}$$

Examples include the relationship between current and voltage on a resistor as given by Ohm's Law, between charge and voltage in a capacitor, and between force and stretching of a spring according to Hook's Law.

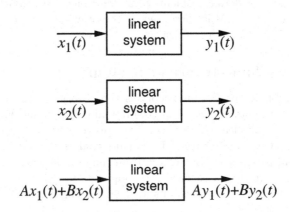

Figure 1.9: Definition of a linear system (A, B are arbitrary complex constants)

The relationship expressed in (1.1), (1.2) is called the *superposition principle*. It can be defined more generally as follows:

Definition 3:   Superposition principle

If the response of a system to a linear combination of input signals always consists of the corresponding combination of the individual output signals, then for this system the superposition principle, applies.

Due to the great importance of such systems, we also use a more tangible term:

Definition 4:   Linear systems

Systems for which the superposition principle applies are called linear *systems.*

For the system in Figure 1.9 the superposition principle is

$$x(t) = Ax_1(t) + Bx_2(t) \quad \rightarrow \quad y(t) = Ay_1(t) + By_2(t). \tag{1.3}$$

Here A and B can be any complex constants. The superposition principle can be extended directly to linear combinations of more than two signals [19] and formulated for systems with multiple inputs and outputs [23]. One important special case derives directly from (1.3) with $A = B = 0$:

$$x(t) = 0 \quad \rightarrow \quad y(t) = 0, \qquad \forall\, t \in \mathbb{R}. \tag{1.4}$$

If we apply to the input of a linear system a signal $x(t)$ that is *always* null, then the output signal must also always be null; otherwise it is not a linear system. However, equation (1.4) must not be misinterpreted as indicating that at every point in time at which the input signal passes through zero, the output signal must also pass through zero.

To be precise, the superposition principle applies for real systems only for a restricted range of the constants A and B. For the voltage drop at a real resistor, Ohm's Law applies only within limits where the resistor is not destroyed by the heat produced in it. Even within these limits, we observe small deviations from Ohm's Law caused by the temperature dependency of the material parameter. On the other hand, resistors are normally operated only within the currency and voltage ranges in which deviations from Ohm's Law are negligible. The same applies analogously for all other applications of the superposition principle. Assuming the linearity of a system is thus always an idealisation that applies at acceptable precision only within certain limits. However, since this assumption very significantly simplifies the analysis of systems, it is employed as much as possible.

1.2.3.2   Time Invariance of a System

The second important system characteristic that we generally assume in this book is time invariance. If we know, as in Figure 1.10, the response $y(t)$ of a system to the input signal $x(t)$, can we then conclude the response to the input signal $x(t - \tau)$, delayed by time τ? Certainly this is possible if the characteristics of the system **S** do not change over time. The response of the system would be the same, so that we can count on a correspondingly delayed output signal for a delayed input signal. This consideration leads us directly to the definition of a time-invariant system:

Definition 5:   Time-invariant system

A system that responds to a delayed input signal with a correspondingly delayed output signal is called a time-invariant *system.*

This definition can be generalised for systems with multiple inputs and outputs

[19]. In the definition we could exchange a delayed and a nondelayed signal and recognise that the delay in Figure 1.10 could assume positive or negative values.

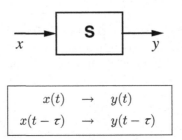

Figure 1.10: Definition of a time-invariant system

1.2.3.3 LTI-Systems

Linear systems are not generally time-invariant. Likewise, time-invariant systems need not be linear. However, as already mentioned, systems that are both linear and time-invariant play a particularly important role in systems theory. They have been assigned the acronym LTI.

Definition 6: LTI-system

A system that is both time-invariant and linear is termed an LTI-system (Linear Time-Invariant system).

The characteristics of LTI-systems and the tools for their analysis are the subjects of subsequent chapters.

1.2.4 Examples of Systems

1.2.4.1 Electrical Circuits

As an example of the description of an electrical circuit as a system, we employ the branching circuit in Figure 1.11. The time-dependent voltages $u_1(t)$ and $u_2(t)$ represent continuous signals, such as the voice signal in Figure 1.1. The circuit establishes a relationship between two signals and is thus a system. To abstract away from the electrical nature of the inner workings and the enclosed components with their regularities, we move to the representation as an input/output system in Figure 1.8. As long as we have no further information on the origin of these signals, the assignment of input signals to output signals is random. The laws of circuit theory for the idealised components (ideal resistors and capacitors) allow us to represent this circuit as a linear and time-invariant system (LTI-system).

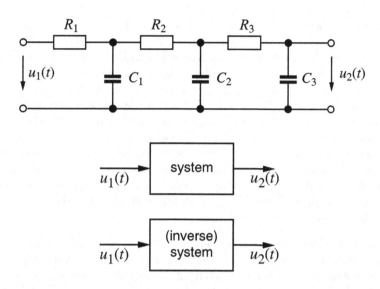

Figure 1.11: Electrical circuit as a system

1.2.4.2 Further Examples of Systems

Figure 1.12 shows additional examples of well-known relationships drawn using system theory.

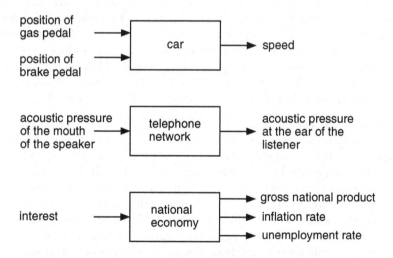

Figure 1.12: Examples of systems

The speed of a car depends on the positions of the gas and brake pedals. This relationship can be represented by a system with two inputs (the positions of the respective pedals) and one output (the speed). The system is certainly not linear. For example, the relationship between responses to the positions of the gas pedal and brake pedal is not simple, as would be the case in a linear system; instead, the braking effect depends on the speed and thus on the preceding position of the gas pedal. Also, the system is not time-invariant, since the gas pedal (input parameter) affects the speed (output parameter) differently depending on the position of the gear shift lever and the inclination of the street. Under ideal conditions, for example, with an automatic transmission on a flat, straight test route, the system would approach time invariance. Over the same time, given angles of the gas pedal and brake pedal would result in the same speed at a later time. In practice, time invariance is a frequent goal in technical systems in order to make the system response predictable.

The second system describes a telephone network with the acoustic pressure from the mouth of a person (speaker) as input signal and the acoustic pressure at the ear of another person (listener) as output signal. At first glance, in the range of acoustic amplitudes that occur while telephoning, the system can be viewed as linear and time-invariant. However, if we do not wish to neglect distortion and other disturbances, then this idealisation as a linear system no longer applies. Likewise, the system is time-variant (i.e., not time-invariant), because a later connection might yield different transmission quality.

The third example shows the application of systems theory to a nontechnical domain. The interest rate established by the national bank influences the economy, which can be assessed via various parameters such as the gross national product, the inflation rate, and the unemployment rate. In contrast to the automobile and telephone examples above, here our system analysis cannot establish relationships between input and output parameters based on recognised laws of nature. Instead, we have economic models that more or less describe economic phenomena. Here a system description is based on the assumed validity of the underlying model.

In all cases the system description is based on simplified assumptions, for it is never possible to incorporate all parameters into a model. Thus the speed of a car also depends on a number of other parameters, such as the road condition, the wind direction and velocity, the fuel quality, and the vehicle's condition. For a single vehicle and with painstaking effort, it might be possible to accommodate all parameters correctly. On the other hand, to describe the functioning of a worldwide telephone network, we would never succeed if we began with the Maxwell equation for each electrical component. As a major strength, systems theory permits describing input/output relationships on different levels of abstraction. It makes sense to begin a system analysis with few parameters and simple models. As necessary, individual systems can be modelled more precisely with more detailed subsystems.

Table 1.2: Criteria for classifying systems

continuous	-	discrete
analogue	-	digital
real-valued	-	complex-valued
unidimensional	-	multidimensional
deterministic	-	stochastic
causal	-	noncausal
with memory	-	memoryless
linear	-	nonlinear
time-invariant	-	time-variant
translation-invariant	-	translation-variant

1.2.5 Classification of Systems

As with signals, we can classify systems according to various criteria. The most important criteria are summarised in Table 1.2.

A number of these criteria are familiar from signals. Thus continuous or discrete, analogue or digital, real-valued or complex-valued, unidimensional or multidimensional systems are systems that establish relationships between signals with these characteristics. A digital system is thus one that processes digital signals. A system is causal if its response to the arrival of a time signal does not begin before this arrival. This sounds trivial, for all systems described by the laws of nature are necessarily causal. However, some important idealisations result in noncausal systems. In some cases it is easier to work with idealised, noncausal systems than with real, causal ones. Furthermore, there are systems whose independent variable is not time. For memoryless systems, the response to a time signal at a certain time depends only on the value of the input signal at the same time. By contrast, for memory systems the values of input signals of other times also play a role; naturally, for causal memory systems these other values must be those of previous times. Linear and time-invariant systems were discussed in Sections 1.2.3.1 and 1.2.3.2. For systems whose input parameters depend not on time but on other independent variables, we can define a characteristic corresponding to time invariance. Thus the temporal delay in a location-independent signal (e.g., a picture) is a translation. In more general terms, this is a translation-invariant system.

1.3 Overview of the Book

Chapter 2 shows several possibilities for describing continuous-time LTI-systems over time. We base the material on known methods for solving linear differential equations with constant coefficients.

Studying LTI-systems over a frequency range (Chapter 3) leads us to the representation of continuous-time signals with the help of the Laplace transform, which we discuss in detail in Chapter 4. Chapter 5 presents the inverse formula of the Laplace transform and its fundamentals in complex function theory. The analysis of LTI-systems with the Laplace transform and its characterisation via the system function is the subject of Chapter 6. Although linear differential equations with constant coefficients and specified start values are no longer LTI-systems, LTI methods can be elegantly extended for this important class of problems (Chapter 7); this occurs primarily via system description in state space.

Another kind of characterisation of LTI-systems in time by convolution with the impulse response is discussed in Chapter 8. In order to be able to describe the impulse response mathematically, we introduce generalised functions in this context.

An integral transformation equal in importance to the Laplace transform is the Fourier transform, whose characteristics and laws are discussed in Chapter 9. The graphical analysis of the frequency response of systems by means of Bode diagrams is the subject of Chapter 10.

Chapter 11 concerns sampled and periodic signals as well as the sampling theorem and leads us to discrete-time signals and their Fourier spectrum (Chapter 12). In Chapter 13 we handle discrete signals with the z-transform, the discrete counterpart of the Laplace transformation, and in Chapter 14 we use it to analyse discrete-time LTI-systems.

In the subsequent chapters continuous and discrete systems and signals are treated in combination. The characteristics of causal systems and signals and their description with the Hilbert transformation is the subject of Chapter 15, and Chapter 16 presents stability characteristics of systems.

In Chapter 17 we introduce random signals and their description via expected values; in addition, we discuss a frequency response representation of random signals via power density spectra. Finally, Chapter 18 is dedicated to the question of how expected values and power density spectra are modified by LTI-systems.

1.4   Exercises

Exercise 1.1

Are the following signals amplitude-discrete, discrete-time and/or digital?

a) number of days of rain per month

b) average high temperature per month

c) current temperature

d) momentary population of China

e) daily milk production of a cow

f) brightness of a pixel on a television screen

Exercise 1.2

Are the signals stored on a computer hard disk analogue or digital signals? Answer the question from different viewpoints.

Exercise 1.3

An ideal A/D converter could be constructed as follows:

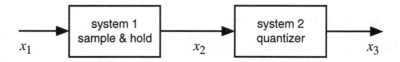

a) Which of the signals $x_1 \ldots x_3$ are analogue, amplitude-discrete, discrete-time, digital?

b) For both systems, specify whether they are linear, time-invariant, analogue, with memory, causal.

Exercise 1.4

Which of the following system descriptions designate linear, time-invariant, memory, or causal systems?

a) $y(t) = x(t)$

b) $y(t) = x^2(t)$

c) $y(t) = x(t - T), \quad T > 0$ (delay component)

d) $y(t) = x(t + T), \quad T > 0$ (accelerator)

e) $y(t) = \dfrac{dx}{dt}$

f) $y(t) = \frac{1}{T} \int\limits_{t-T}^{t} x(t')\, dt' \quad T > 0$ (moving average)

g) $\dfrac{dy}{dt} + ay(t) = x(t)$ (electrical circuit)

h) $y(t) = x(t - T(t)), \quad T(t) \geq 0$ (phase modulation)

i) $y(t) = x(t - T(t))$, $T(t)$, arbitrary

Exercise 1.5

Two systems $\mathcal{S}_1$ and $\mathcal{S}_2$ respond to an input signal $x(t)$ with the output signals

$$
\begin{array}{lll}
y_1(t) & = & \mathcal{S}_1\{x(t)\} & = & m \cdot x(t) \cdot \cos(\omega_T t) \\
y_2(t) & = & \mathcal{S}_2\{x(t)\} & = & [1 + m \cdot x(t)] \cdot \cos(\omega_T t) & m \in \mathbb{R}.
\end{array}
$$

Are the systems a) linear, b) time-invariant, c) real-valued, d) memoryless?

Exercise 1.6

A black-box system is to be examined to determine whether it is time-invariant and linear. Three measurements of the system yielded the following data:

1. Input of $x_1(t)$ produces the output signal $y_1(t)$.

2. Input of $x_2(t)$ produces the output signal $y_2(t)$.

3. Input of $x_3(t) = x_1(t - T) + x_2(t - T)$ produces the output signal $y_3(t) \neq y_1(t - T) + y_2(t - T)$.

Can you make an unambiguous statement about the above system characteristics? Defend your answer.

2 Time-Domain Models of Continuous LTI-Systems

In the last chapter we introduced the concept of systems and some of their general properties. In this and some of the following chapters we will get to know some of the different methods that can be used to model these systems. We will begin by looking at continuous-time systems and restrict ourselves to LTI (linear time-invariant)-systems.

In Chapter 1 we established a requirement for a system model in which we are not concerned about the details of the individual components. Instead, we are looking for a standardised form of system model that represents the input–output characteristics of a system by mathematical equations, independent from the implementation of the system.

This chapter deals with the following three modelling techniques for continuous-time systems:

- differential equations as a mathematical representation of the input–output relationship,

- block diagrams as a graphical representation of the relationship between input, output and internal states,

- state models that are the equivalent of the block diagrams.

Common to all three of these modelling techniques is the use of time-dependent signals, in which the derivative and the integral with respect to time plays an important role. Therefore these types of system model can be classified as 'time-domain models'. Their complements are 'frequency-domain models', which will be examined in the next chapter.

2.1 Differential Equations

2.1.1 System Analysis

Our goal is to find a system model without details of the system implementation. How can this be achieved? We will use the analysis of an electrical circuit to show the essential steps.

In many cases it is possible to ignore the spatial expansion of electrical components on a circuit board, and instead work with equivalent circuits consisting of concentrated elements. Semiconductor devices are an example of this because their complicated internal behaviour can only be accurately modelled using solid-state physics. Their effects within an electrical circuit are, however, often linear enough to be adequately modelled by simple components like resistances and ideal sources.

The second step is the replacement of the physical components with their ideal equivalents, for example, real resistances become ohmic, wires become ideal conductors, capacitors have ideal capacitance, etc.

The resulting electrical networks (for example, Figure 1.11) can be analysed using standard methods, for example, mesh or nodal analysis [18, 22]. This results in ordinary differential equations with constant coefficients, in which only the input and output signals and their derivatives occur.

This process can also be applied to other physical arrangements which, like electrical circuits, can be described by potential (e.g. electrical voltage) and flow quantities (e.g. electrical current). The analysis accordingly simplifies mechanical, pneumatic, hydraulic and thermal systems to differential equations. The same applies to other kinds of system, for example, from chemistry, biology or economics.

The simplifications mentioned are of course not always permissible. Ordinary differential equations are, for example, unsuitable for the field of fluid dynamics problems. In many other uses, however, they are of great importance, and we will therefore examine them in more depth.

2.1.2 Linear Differential Equations with Constant Coefficients

Differential equations establish relationships between derivatives of dependent quantities with respect to independent variables. They are called *ordinary differential equations* if the derivatives only occur with respect to *one* of the independent variables (e.g. time). Differential equations with derivatives with respect to more than one independent variable (e.g. time and three spatial coordinates) are called *partial differential equations*. A differential equation is said to be *linear* if the individual derivatives are only multiplied by factors and combined by addition. Additionally, if the factors of the derivatives do not depend on the independent variables, the term 'differential equation with constant coefficients' is used.

For modelling of continuous-time systems, we just need ordinary differential equations with time as the only independent variable. In such equations the input and output signals of the system must occur as dependent variables. We will soon discover that linear, and time-invariant systems can be modelled by linear differential equations with constant coefficients, so we will restrict ourselves to this type of equation.

Simple examples for linear differential equations with coefficients are

$$\ddot{y} + 2\dot{y} = x \tag{2.1}$$

$$\ddot{y} + 3\dot{y} + 2y = 2\dot{x} - x \ . \tag{2.2}$$

The general form of an ordinary linear differential equation with constant coefficients is

$$\sum_{i=0}^{N} \alpha_i \frac{d^i y}{dt^i} = \sum_{k=0}^{M} \beta_k \frac{d^k x}{dt^k} \ . \tag{2.3}$$

The greatest index N of a non-zero coefficient α_N determines what is called the *order of the differential equation*. In order to simplify this discussion, we let $M = N$ and allow some, but not all of the coefficients β_k to be equal to zero.

For a given function $x(t)$ there are up to N different *linearly independent* solutions $y(t)$ to (2.3). For a particular solution, we need to give N conditions. For initial condition problems, these would be N initial conditions $y(0), \dot{y}(0), \ddot{y}(0), \ldots$.

The differential equation (2.3) describes a continuous-time system, if $x(t)$ is the input signal, and $y(t)$ is the output signal. In order to characterise this system, we refer back to Definitions 4 and 5 and also Figs. 1.9 and 1.10. For now, we ignore possibly given initial conditions; their influence will be discussed in depth in Chapter 7.

Let us now show that (2.3) represents a time-invariant system. Through substitution of variables $t' = t - \tau$ in (2.3), it follows immediately that $x(t - \tau)$ leads to the solution $y(t - \tau)$. To show linearity we consider the two different input signals $x_1(t)$ and $x_2(t)$ and the corresponding solutions $y_1(t)$ and $y_2(t)$. Plugging the linear equation $x_3(t) = Ax_1(t) + Bx_2(t)$ into (2.3) verifies that $y_3(t) = Ay_1(t) + By_2(t)$ is a solution of the differential equation, and therefore the output signal of the system.

Every system that can be modelled using linear differential equations with constant coefficients (2.3) is thus an LTI system. This means we have found our first method for modelling such systems in the form of a differential equation. This method fulfills our initial requirements:

- Modelling of an LTI-system independent from its realisation.

- Representation of the input–output relationship, without details of the system's internal behaviour.

2.2 Block Diagrams

Block diagrams can represent more information than differential equations as they show not only the input and output signals but also internal states of a system. If only the input-output relationship is of interest then the choice of internal states is not relevant, and there are many block diagrams that correspond to the same differential equation. In case additional specifications of the internal structure are available, the block diagram can be constructed in a way that represents the internal states of the system, for example, the energy stored in the components of an electrical network.

From the many possible structures of block diagrams, there are some which are particularly suitable if the differential equation of a system is already known. We will examine three such structures in more detail.

2.2.1 Direct Form I

Our starting point is an LTI-system that is modelled by differential equation (2.3), with $M = N$. We integrate both sides N times,

$$\sum_{i=0}^{N} a_i \int_{(i)} y\,dt = \sum_{k=0}^{N} b_k \int_{(k)} x\,dt \qquad (2.4)$$

with $a_i = \alpha_{N-i}$ and $b_k = \beta_{N-k}$. We write $\int_{(i)} y\,dt$ for the i-times integral

$$\int_{(i)} y\,dt = \int_{-\infty}^{t}\left[\int_{-\infty}^{\tau_i}\cdots\left[\int_{-\infty}^{\tau_2} y(\tau_1)\,d\tau_1\right]\cdots d\tau_{i-1}\right]d\tau_i . \qquad (2.5)$$

The rearrangement

$$y = \int_{(0)} y\,dt = \frac{1}{a_0}\left[\sum_{k=0}^{N} b_k \int_{(k)} x\,dt - \sum_{i=1}^{N} a_i \int_{(i)} y\,dt\right] \qquad (2.6)$$

leads immediately to the block diagram of an LTI-system in 'Direct form I' (Figure 2.1). The rectangular boxes represent multiplication by the factor shown within, or alternatively a single integration, and the circle with a summation sign stands for addition of the input signals.

The advantage of block diagrams with this structure is that the coefficients of the differential equation appear directly as the values of the multipliers. The disadvantage is that for a N-order differential equation, $2N$ integrations must be performed.

2.2.2 Direct Form II

We will now construct a new form of block diagram with a different structure that only uses N integrators. To create the new structure, we manipulate direct form I, interchanging the first and second stages (Figure 2.2). This is permissible, as cascaded LTI-systems can be interchanged without affecting their overall transfer

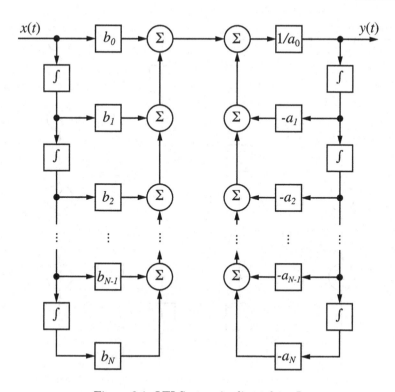

Figure 2.1: LTI-System in direct form I

function. We will show this general concept more elegantly in Chapter 6.6.1 with the help of a frequency-domain model. At the moment, however, we will just view this property as a useful assumption. Both cascades of integrators in Figure 2.2 run in parallel since the input signals of the integrators from time $t = -\infty$ are equal, and so are their output signals. We can therefore unite the two cascades, arriving at direct form II shown in Figure 2.3.

As with direct form I, the multiplier coefficients of direct form II are the coefficients of the differential equation. More importantly, this form requires the only N integrators, the minimum number for an N-order differential equation. Block diagrams that use the minimum number of energy stores (integrators) for the realisation of an N-order differential equation are also called *canonical forms*.

The signals z_i, $i = 1, \ldots, N$ at the integrator outputs describe the internal state of a system, that is not only modelled by the corresponding differential equation (2.3), but is further given an internal structure by direct form II. Without knowledge of the actual realisation of the system, this assignment of states is, of course, entirely arbitrary.

We constructed direct form I directly from the differential equation, and it is

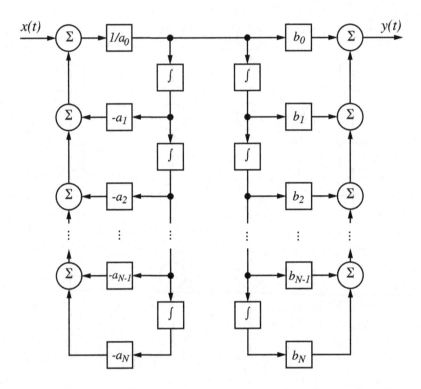

Figure 2.2: Derivation of the block diagram for direct form II from direct form I

clear that systems with this structure satisfy the differential equation (2.3). We still need to ensure that direct form II satisfies the same differential equation. After all, we made use of the as yet unproven assumption that the two stages of direct form I can be interchanged to create direct form II. To verify direct form II we first express the input signal x and the output signal y in terms of the states z_i, $i = 1, \ldots N$.

The input and output signals are linked through the states at the integrator outputs, but also directly through the uppermost path, and of course we have to consider this path as well. To simplify the notation we introduce another internal signal z_0, and emphasise that it does *not* represent a state.

From the block diagram (Figure 2.3) the relationships

$$ y = \sum_{i=0}^{N} b_i\, z_i \qquad \text{and} \qquad z_0 = \frac{1}{a_0}\left[x - \sum_{i=1}^{N} a_i\, z_i \right] \tag{2.7} $$

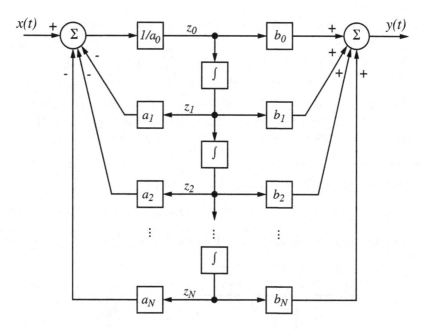

Figure 2.3: An LTI-system in direct form II

are obtained directly. The last relationship can be rewritten as

$$x = \sum_{i=0}^{N} a_i z_i \; .$$

(2.8)

Additionally, every state variable z_i can be obtained by integrating z_0 i times.

$$z_i = \int_{(i)} z_0 \, dt \; .$$

(2.9)

We now insert (2.7) into the left-hand side of the integral equation (2.4) which is equivalent to the differential equation (2.3). By interchanging the order of the summation and integration, and further using (2.9), we obtain

$$\sum_{k=0}^{N} a_k \int_{(k)} y \, dt \;\; = \;\; \sum_{k=0}^{N} a_k \int_{(k)} \sum_{i=0}^{N} b_i z_i \, dt =$$

$$= \;\; \sum_{k=0}^{N} \sum_{i=0}^{N} a_k b_i \int_{(k)} z_i \, dt = \sum_{k=0}^{N} \sum_{i=0}^{N} a_k b_i \int_{(k+i)} z_0 \, dt \; .$$

(2.10)

Integrating z_0 repeatedly $k+i$ times is equivalent to integrating z_k repeatedly

i times. Interchanging the order of integration and summation yields

$$\sum_{k=0}^{N} a_k \int_{(k)} y\, dt = \sum_{k=0}^{N}\sum_{i=0}^{N} a_k b_i \int_{(i)} z_k\, dt = \sum_{i=0}^{N} b_i \int_{(i)} \sum_{k=0}^{N} a_k z_k\, dt\ . \qquad (2.11)$$

The last sum we recognise as x as in (2.8), hence the integral equation (2.4) is satisfied. This shows that the input and output signals of a system in direct form II indeed satisfy the differential equation (2.3), as we had hoped.

── **Example 2.1**

We will construct a block diagram for the equation

$$4\ddot{y} - \dot{y} + 2y = -3\dot{x} + x\ . \qquad (2.12)$$

We can create the direct form II block diagram shown in Figure 2.4 directly from Figure 2.3.

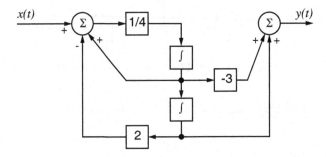

Figure 2.4: Example of a direct form II block diagram [Example 2.1]

─── ■

2.2.3 Direct Form III

The construction of another commonly used structure again originates with the differential equation (2.3). In contrast to direct forms I and II we will make no use of integral equations or rearrangement of block diagrams. Instead we obtain the state variables directly from the differential equation. The transformation takes N steps, each made up of the following elements:

- rearrangement of the differential equation

- introduction of a new state variable

- integration

In the first step the differential equation is rearranged so that all the derivatives are on the left-hand side. The remaining right-hand side becomes the derivative of the new state variable z_N:

$$\sum_{i=1}^{N} \alpha_i \frac{d^i y}{dt^i} - \sum_{k=1}^{N} \beta_k \frac{d^k x}{dt^k} = \beta_0 x - \alpha_0 y = \dot{z}_N \ . \tag{2.13}$$

After integration we obtain

$$\sum_{i=0}^{N-1} \alpha_{i+1} \frac{d^i y}{dt^i} - \sum_{k=0}^{N-1} \beta_{k+1} \frac{d^k x}{dt^k} = z_N \ . \tag{2.14}$$

The second step starts again by collecting all the derivatives on the left-hand side and introducing a new state variable for the right-hand side.

$$\sum_{i=1}^{N-1} \alpha_{i+1} \frac{d^i y}{dt^i} - \sum_{k=1}^{N-1} \beta_{k+1} \frac{d^k x}{dt^k} = z_N + \beta_1 x - \alpha_1 y = \dot{z}_{N-1} \ . \tag{2.15}$$

Integration and changing indices yields

$$\sum_{i=0}^{N-2} \alpha_{i+2} \frac{d^i y}{dt^i} - \sum_{k=0}^{N-2} \beta_{k+2} \frac{d^k x}{dt^k} = z_{N-1} \ . \tag{2.16}$$

With the Nth step only the first derivative with respect to time remains

$$\alpha_N \frac{dy}{dt} - \beta_N \frac{dx}{dt} = z_2 + \beta_{N-1} x - \alpha_{N-1} y = \dot{z}_1 \ . \tag{2.17}$$

The final integration yields

$$\alpha_N y - \beta_N x = z_1 \ . \tag{2.18}$$

Before drawing the block diagram we summarise the important equations and rename the coefficients such that $a_i = \alpha_{N-i}$, $b_k = \beta_{N-k}$

$$
\begin{aligned}
y &= \frac{1}{a_0}[z_1 & + & \quad b_0 x] \\
\dot{z}_1 &= z_2 & + & \quad b_1 x & - & \quad a_1 y \\
\dot{z}_2 &= z_3 & + & \quad b_2 x & - & \quad a_2 y \\
&\ \vdots & \vdots & \quad \ \vdots & & \quad \ \vdots \\
\dot{z}_{N-1} &= z_N & + & \ b_{N-1} x & - & \ a_{N-1} y \\
\dot{z}_N &= & & \ b_N x & - & \quad a_N y \quad .
\end{aligned}
\tag{2.19}
$$

The block diagram representation is shown in Figure 2.5. It can also be obtained graphically from direct form II (Figure 2.3), if

- input x and output y are exchanged,
- all arrows are reversed and
- summation and splitting nodes are exchanged.

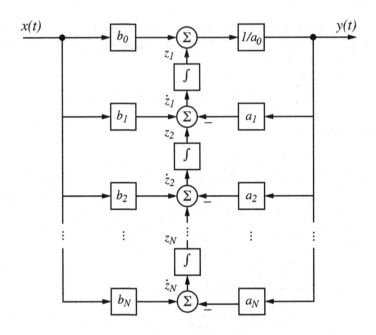

Figure 2.5: LTI-system in direct form III

Example 2.2

The structure of an LTI-system in direct form III can be obtained for differential equation (2.12) from Example 2.1 through the steps described above. As it is a differential equation of the second order, two steps have to be performed transform.

The first step introduces the state variable z_2:

$$4\ddot{y} - \dot{y} + 3\dot{x} \;=\; x - 2y = \dot{z}_2 \qquad\qquad (2.20)$$

$$4\dot{y} - y + 3x \;=\; z_2 \qquad\qquad . \qquad\qquad (2.21)$$

The second step introduces the state variable z_1:

$$4\dot{y} \;=\; z_2 - 3x + y = \dot{z}_1 \qquad\qquad (2.22)$$

$$4y \;=\; z_1 \qquad\qquad . \qquad\qquad (2.23)$$

The block diagram in direct form III can be formed from the equations

$$
\begin{aligned}
y &= \frac{1}{4}z_1 \\
\dot{z}_1 &= z_2 - 3x + y \\
\dot{z}_2 &= \;\; x - 2y
\end{aligned}
\qquad\qquad (2.24)
$$

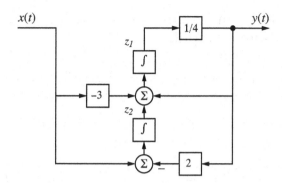

Figure 2.6: A direct form III block diagram from Example 2.2

and is shown in Figure 2.6.

2.2.4 Why not use Differentiators to Build LTI-Systems?

Block diagrams and differential equations are equally valuable forms for modelling LTI-systems. With block diagrams of the direct form I, II and III the correspondence of the block diagram and differential equation is sufficiently strong that the coefficients of both forms agree. The most striking difference is that differential equations are made up of derivatives, whereas block diagrams contain integrators. Why not remove this difference and construct block diagrams using differentiators?

If block diagrams only existed as models of already implemented systems, then using differentiators would make no difference. Block diagrams also have another useful task, however: they serve as a model for the realisation of systems which do not yet exist. The required transfer function can be mathematically formulated as a differential equation and converted into a block diagram, from which the individual components can then be created. In order to decide whether differentiators or integrators are more suitable as a starting point for implementation, the characteristics of the signals involved must be considered.

Every analogue signal is subject to corruption by noise, i.e., it contains unwanted and generally rapidly changing components. Differentiators amplify rapid signal changes and thus increase the level of unwanted noise. Integrators, however, smooth and suppress the undesirable noise. It is for this reason that block diagrams formed with integrators lead to superior, more robust realisations. An example for the physical realisation of an integrator follows in the next section.

2.2.5   Electrical Implementation of an Integrator Using an Operational Amplifier

Operational amplifier (op-amp) circuits are an important example of the electical implementation of integrators. Op-amps are semiconductor amplifiers that usually come in the form of integrated circuits and can be well approximated using a very simple model.

$$U_0 = A(U_+ - U_-)$$

Figure 2.7: Symbolic representation of an op-amp

An ideal op-amp (Figure 2.7) has the following characteristics:

- The input impedance is infinite, i.e., no current flows between the (+) and (-) terminals.

- The output impedance is zero, hence the output is an ideal voltage source.

- The amplification factor A is infinite ($> 10^6$ for a real world op-amp).

- If negative feedback is applied, $U_+ \approx U_-$, i.e., both input terminals are at the same potential.

With these characteristics it is easy to show that an op-amp with capacitor feedback integrates the input voltage $u_1(t)$.

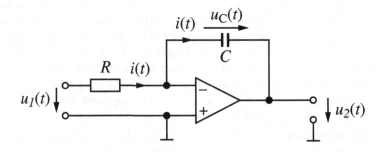

Figure 2.8: Op-amp with feedback circuit

Because of the infinite input impedance the current $i(t)$ through the resistor R is equal to the current through the capacitor C. The infinite amplification forces the two terminals to the same potential, hence the voltage across the resistor is equal to the input voltage $u_1(t)$. The current $i(t)$ is therefore:

$$i(t) = \frac{u_1(t)}{R} = -C\frac{du_2}{dt} \ . \tag{2.25}$$

From that follows the required integral relationship between $u_1(t)$ and $u_2(t)$:

$$u_2(t) = \frac{-1}{RC} \int u_1(t)\, dt \ . \tag{2.26}$$

The circuit given is therefore an integrator of the input voltage. As addition and multiplication are likewise possible using op-amps with other circuit configurations, systems can be implemented as electrical circuits directly from block diagrams.

2.3   State-Space Description of LTI-Systems

In Section 2.2 we had already used the concept of the state of a system loosely. Now we will give a formal definition.

Definition 7:   System state

The system state *is a vector of internal variables with its future values dependent on its current values, which fully captures the effect of the past on the future behaviour of the system.*

We saw with the introduction of block diagrams that the use of internal states leads to a more intuitive system representation than a differential equation alone. The choice of the variables is, however, arbitrary because the differential equation describes the input–output function and does not contain any information about the internal behaviour of the system.

The state-space description offers the possibility of representing the internal form of a system in a standardised form of differential equations. In contrast to the differential equation (2.3), that is a single equation of order N, the corresponding state-space representation is a system of N first-order differential equations. Each differential equation is valid for one of the N so-called state variables (state equation). The output signal is obtained by a linear combination of the states (output equation).

2.3.1   Example of the State Model

The following example shows how the state model is used to represent an electrical circuit. The RLC circuit in Figure 2.9 can be examined using the standard methods of circuit analysis.

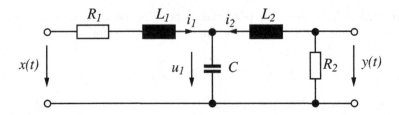

Figure 2.9: RLC circuit

The sum of the voltages in both meshes and the sum of the currents in the node yield

$$L_1 \frac{di_1}{dt} = -R_1 i_1(t) - u_1(t) + x(t) \tag{2.27}$$

$$L_2 \frac{di_2}{dt} = -R_2 i_2(t) - u_1(t) \tag{2.28}$$

$$C \frac{du_1}{dt} = i_1(t) + i_2(t) \qquad . \tag{2.29}$$

The output voltage $y(t)$ is the voltage drop across the resistor R_2:

$$y(t) = -R_2 i_2(t) . \tag{2.30}$$

From these three first-order differential equations we could eliminate i_1, i_2 and u_1 and obtain a third-order differential equation in the form of (2.3). However, this would result in the loss of information about the internal energy stores that determine the system behaviour. Instead, we represent the equations (2.27) to (2.30) in matrix form:

$$\frac{d}{dt} \begin{bmatrix} i_1(t) \\ i_2(t) \\ u_1(t) \end{bmatrix} = \begin{bmatrix} -\dfrac{R_1}{L_1} & 0 & -\dfrac{1}{L_1} \\ 0 & -\dfrac{R_2}{L_2} & -\dfrac{1}{L_2} \\ \dfrac{1}{C} & \dfrac{1}{C} & 0 \end{bmatrix} \begin{bmatrix} i_1(t) \\ i_2(t) \\ u_1(t) \end{bmatrix} + \begin{bmatrix} \dfrac{1}{L_1} \\ 0 \\ 0 \end{bmatrix} x(t) \tag{2.31}$$

$$y(t) = \begin{bmatrix} 0 & -R_2 & 0 \end{bmatrix} \begin{bmatrix} i_1(t) \\ i_2(t) \\ u_1(t) \end{bmatrix} \qquad . \tag{2.32}$$

(2.31) and (2.32) are a system of three connected first-order differential equations and an algebraic equation. The three differential equations can also be viewed as a matrix differential equation for the vector

$$\mathbf{z}(t) = \begin{bmatrix} i_1(t) \\ i_2(t) \\ u_1(t) \end{bmatrix} .$$

It contains the voltage u_1 across the capacitor and the currents i_1 and i_2 through both inductors. They characterise the system's three energy stores and are state-variables.

2.3.2 General Form of the State Model

We can easily generalise the previous example to arrive at the general form of the state-space description. With the abbreviations $\mathbf{A}, \mathbf{B}, \mathbf{C}, \mathbf{D}$ for the vectors and matrices in (2.31) and (2.32) we obtain the following standardised form of the first-order matrix differential equation and of the algebraic equation. It is given here for the general case with M inputs and K outputs

$$\dot{\mathbf{z}} = \mathbf{A}\,\mathbf{z} + \mathbf{B}\,\mathbf{x} \qquad (2.33)$$

$$\mathbf{y} = \mathbf{C}\,\mathbf{z} + \mathbf{D}\,\mathbf{x}\,. \qquad (2.34)$$

Where: $\mathbf{x}$:   column vector M input signals
          $\mathbf{z}$:   column vector N state variables
          $\mathbf{y}$:   column vector K output signals.

The matrix differential equation (2.33) is called the *state equation* and the algebraic equation (2.34) is called the *output equation*. The $N \times N$ matrix $\mathbf{A}$ is called the system matrix. It describes how the change of the state vector $\dot{\mathbf{z}}$ depends on the instantaneous value $\mathbf{z}$. The matrices $\mathbf{B}$ (size $N \times M$) and $\mathbf{C}$ (size $K \times N$) characterise the influence of the input $\mathbf{x}$ on the state $\mathbf{z}$, and the effect of this state on the output $\mathbf{y}$ respectively. The matrix $\mathbf{D}$ (value $K \times M$) describes the direct influence of the input on the output.

In the example in Section 2.3.1 with only one input and one output ($M = K = 1$), $\mathbf{B}$ is a column vector and $\mathbf{C}$ is a row vector, and there is no direct influence of the input on the output ($\mathbf{D} = 0$).

Although in this example we started with a specific realisation of a system in the form of the electrical network in Figure 2.9, the state model does provide a universal representation of a system behaviour. The internal working of the system as given by the details of the network is captured by the states of the three energy stores. Together, these system states contain all relevant information about the past of the system. Starting with an initial state $\mathbf{z}(t_0)$ at time t_0, we can uniquely determine the state of $\mathbf{z}(t)$ at any future time t, if the input signal $x(t)$ is know between t_0 and t.

The state-space description is especially advantageous for systems with many inputs and outputs, for which working with scalar equations with many variables would be very cumbersome.

Of course, there exists a close relationship between the state-space representation, the block diagram and the differential equation of an Nth-order LTI-system. We will examine an example of this in the next section.

2.4 Higher-order Differential Equations, Block Diagrams and the State Model

In order to establish the relationship between the models for LTI-systems previously examined, we consider the direct form II block diagram in Figure 2.3 and look for the corresponding state model. We represent the integrators in Figure 2.3 by the differential equations

$$\dot{z}_i = z_{i-1} \qquad i = 1, \ldots, N \ . \tag{2.35}$$

We have already expressed the input signal of the first integrator through the state variables and the input signal in (2.7). Substituting (2.35) in (2.7) and collecting the scalar variables gives the state equation

$$
\begin{bmatrix} \dot{z}_1 \\ \dot{z}_2 \\ \dot{z}_3 \\ \vdots \\ \dot{z}_N \end{bmatrix}
=
\begin{bmatrix}
-\dfrac{a_1}{a_0} & -\dfrac{a_2}{a_0} & \cdots & -\dfrac{a_{N-1}}{a_0} & -\dfrac{a_N}{a_0} \\
1 & 0 & \cdots & 0 & 0 \\
0 & 1 & \cdots & 0 & 0 \\
\vdots & \vdots & \ddots & \vdots & \vdots \\
0 & 0 & \cdots & 1 & 0
\end{bmatrix}
\begin{bmatrix} z_1 \\ z_2 \\ z_3 \\ \vdots \\ z_N \end{bmatrix}
+
\dfrac{1}{a_0}
\begin{bmatrix} 1 \\ 0 \\ \vdots \\ 0 \\ 0 \end{bmatrix} x \ .
\tag{2.36}
$$

The output equation can be immediately taken from direct form II (Figure 2.3). Note that from every state there are two paths to the output: one through coefficients b_i, and one through a_i, $\dfrac{1}{a_0}$ and b_0. Therefore the output equation becomes

$$
y = \begin{bmatrix} b_1 - \dfrac{b_0}{a_0}a_1 & \cdots & b_N - \dfrac{b_0}{a_0}a_N \end{bmatrix}
\begin{bmatrix} z_1 \\ \vdots \\ z_N \end{bmatrix}
+ \dfrac{b_0}{a_0} x \ .
\tag{2.37}
$$

The equations (2.36) and (2.37) represent the complete state model of the system given as a direct form II block diagram.

As the direct form I realises the input–output behaviour of the integral equation (2.4), or alternatively, differential equation (2.3), a connection is also established between the Nth-order differential equation (2.3) and the matrix differential equation (2.33). We can make this even more explicit by writing the state model with coefficients $\alpha_i = a_{N-i}$ and $\beta_i = b_{N-i}$ of the differential equation (2.3). Correspondingly, we re-index the state variables as $\zeta_i = z_{N-i+1}$.

Then the state model is

$$\dot{\zeta} \;=\; \mathbf{A}\zeta + \mathbf{B}x \tag{2.38}$$

$$y \;=\; \mathbf{C}\zeta + \mathbf{D}x \tag{2.39}$$

with the matrices

$$\mathbf{A} \;=\; \begin{bmatrix} 0 & 1 & \cdots & 0 & 0 \\ \vdots & \vdots & \ddots & \vdots & \vdots \\ 0 & 0 & \cdots & 1 & 0 \\ 0 & 0 & \cdots & 0 & 1 \\ -\dfrac{\alpha_0}{\alpha_N} & -\dfrac{\alpha_1}{\alpha_N} & \cdots & -\dfrac{\alpha_{N-2}}{\alpha_N} & -\dfrac{\alpha_{N-1}}{\alpha_N} \end{bmatrix} \tag{2.40}$$

$$\mathbf{B} \;=\; \frac{1}{\alpha_N}\begin{bmatrix} 0 \\ \vdots \\ 0 \\ 1 \end{bmatrix} \tag{2.41}$$

$$\mathbf{C} \;=\; \begin{bmatrix} \beta_0 - \dfrac{\beta_N}{\alpha_N}\alpha_0 & \cdots & \beta_{N-1} - \dfrac{\beta_N}{\alpha_N}\alpha_{N-1} \end{bmatrix} \tag{2.42}$$

$$\mathbf{D} \;=\; \frac{\beta_N}{\alpha_N}\, . \tag{2.43}$$

The elements of the state matrices $\mathbf{A}$, $\mathbf{B}$, $\mathbf{C}$, $\mathbf{D}$, can be calculated directly from the coefficients of the differential equation (2.3). The special form of the system matrix (2.40) for direct form II is called the *Frobenius matrix*.

2.5   Equivalent State-Space Representations

The state matrices (2.40) to (2.43) describe only one of many possible structures of block diagram, here direct form II. Alternative structures that yield the same input–output behaviour can be obtained by transforming other state variables. This transformation can be formally expressed by the multiplication of a state by a transformation matrix $\mathbf{T}$

$$\mathbf{z} = \mathbf{T}\hat{\mathbf{z}}\, . \tag{2.44}$$

In this equation $\hat{\mathbf{z}}$ is the new state vector. The transformation matrix $\mathbf{T}$ of size $N \times N$ must not be singular, so that inversion is possible, but otherwise there are no restrictions. Substitution into (2.33) and (2.34) and multiplying both values by $\mathbf{T}^{-1}$ gives the new state-space equations

$$\dot{\mathbf{z}} = \hat{\mathbf{A}}\hat{\mathbf{z}} + \hat{\mathbf{B}}x \qquad (2.45)$$

$$y = \hat{\mathbf{C}}\hat{\mathbf{z}} + \hat{\mathbf{D}}x \qquad (2.46)$$

with the matrices

$$\hat{\mathbf{A}} = \mathbf{T}^{-1}\mathbf{A}\mathbf{T} \qquad (2.47)$$

$$\hat{\mathbf{B}} = \mathbf{T}^{-1}\mathbf{B} \qquad (2.48)$$

$$\hat{\mathbf{C}} = \mathbf{C}\mathbf{T} \qquad (2.49)$$

$$\hat{\mathbf{D}} = \mathbf{D} . \qquad (2.50)$$

These are equivalent to (2.33) and (2.34), i.e., they correspond to the same Nth-order differential equation. Through the selection of a transform matrix $\mathbf{T}$ arbitrary state models can be produced for a given differential equation.

Some state models have special properties. For example, if the transformation matrix $\mathbf{T}$ is a modal matrix with regard to matrix $\mathbf{A}$, then the new system matrix $\hat{\mathbf{A}}$ will be diagonal. The corresponding type of structure is called parallel form. An example of this form with one input and one output is shown in Figure 2.10. Here we assume that there is no direct path from input to output, i.e., $\mathbf{D} = \hat{\mathbf{D}} = 0$, and that the $\mathbf{A}$ has only single eigenvalues. The case of multiple eigenvalues is discussed in [19].

As $\hat{\mathbf{A}}$ only contains values in the main diagonal, each state variable is only fed back to itself. $\lambda_1, \lambda_2, \ldots, \lambda_N$ are the eigenvalue matrix of the system that, even with a real matrix $\mathbf{A}$, can appear as complex conjugates. In this case, the corresponding state variables are also complex, but their imaginary parts cancel each other out to yield the real output signal $y(t)$.

The parallel form for a system with M inputs and K outputs is shown in Figure 2.11. Again it is assumed that there is no direct path from input to output, i.e., $\mathbf{D} = \hat{\mathbf{D}} = 0$ and that the system matrix $\mathbf{A}$ only has single eigenvalues. Blocks $\hat{\mathbf{B}}$ und $\hat{\mathbf{C}}$ represent multiplication with the corresponding matrices. The block diagram in Figure 2.10 is included as a special case; the central integrator stage clearly does not depend on how many inputs and outputs the system has.

What use is transforming state-space if all of the resulting structures exhibit the same input–output behaviour? The strict equivalence between the different state models is unfortunately only valid under ideal conditions. Physical implementations are always accompanied by interference that can be thought of as additional signal sources inside the system. For analog electrical systems this interference is

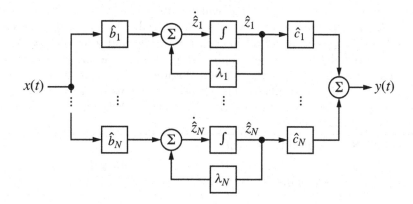

Figure 2.10: The parallel form of a system with one input and one output

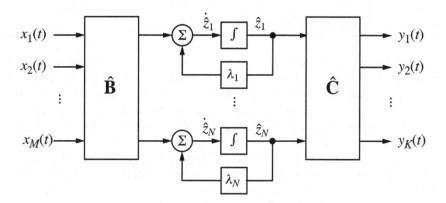

Figure 2.11: The parallel form of a system with multiple inputs and outputs

noise, and for digital circuits and processors in the form of rounding errors. The effects of this interference on the output signal can be minimised through the appropriate choice of internal states. Additionally, some forms of realisation require less hardware or computation than others.

2.6 Controllable and Observable Systems

The parallel form yields a simple way of further classifying the properties of a system with simple eigenvalues. One property is the ability to control a state via the input and another is that a state can be observed by looking at the output. For the structures previously described, it is possible that the input signal can

be cancelled out on its way to a particular state. This state is said to be not
controllable. It can equally occur that a state-variable is cancelled while on its
journey to the output. The output in this case is then said to be not *observable*.

With the parallel form there is only one path from the input to each state
and also from each state to the output (Figure 2.10). Elimination of more
paths cannot occur. Accordingly, this applies to systems with more inputs and
outputs (Figure 2.11). We can thus write the following definitions of the concepts
controllable and *observable*:

Definition 8:   Controllable system

A system is said to be controllable *if after transformation to parallel form, no
elements of the state matrix $\hat{\mathbf{B}}$ are zero.*

Definition 9:   Observable system

A system is said to be observable *if after transformation to the parallel form,
no elements of the state matrix $\hat{\mathbf{C}}$ are zero.*

In contrast, if the element B_{nm} in row n, column m of $\mathbf{B}$ is zero, the state-
variable z_n from input x_m is not controllable. Equally, the state-variable z_n for
output y_k is not observable if the element C_{kn} in row k, column n of $\mathbf{C}$ is zero.

The advantage of this definition of the terms controllable and observable is the
relative ease with which it can be verified from the elements of the state matrices.
The disadvantage is that chosen systems must first be transformed to diagonal
form for these properties to be assessed. There are other possible methods that
can be employed to test for observability and controllability, that work directly
with the given (i.e. not diagonal) state model (e.g. [19, 23]). We now examine the
method given above with the help of an example.

Example 2.3

Figure 2.12 shows a block diagram of a second-order system with two inputs
and one output. At the first glance it seems that all of the states are connected
to both inputs and the output. Whether any eliminations exist is not obvious at
this time.

To determine the state matrices we read from Figure 2.12 an equation for each
of $\dot{z}_1$ and $\dot{z}_2$ at the inputs of the integrator and an equation for y at the output,
and then write these in the form of (2.33), (2.34):

$$\begin{bmatrix} \dot{z}_1 \\ \dot{z}_2 \end{bmatrix} = \begin{bmatrix} -4 & -6 \\ 3 & 5 \end{bmatrix} \begin{bmatrix} z_1 \\ z_2 \end{bmatrix} + \begin{bmatrix} -2 & -3 \\ 2 & 2 \end{bmatrix} \begin{bmatrix} x_1 \\ x_2 \end{bmatrix} \qquad (2.51)$$

$$y = \begin{bmatrix} -2 & -3 \end{bmatrix} \begin{bmatrix} z_1 \\ z_2 \end{bmatrix} . \qquad (2.52)$$

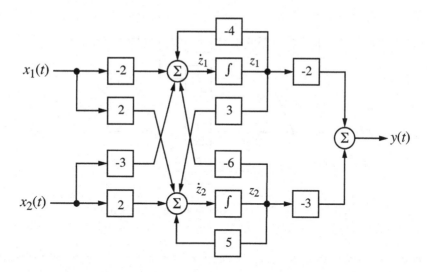

Figure 2.12: Block diagram of the system in Example 2.3

The matrices $\mathbf{A}$, $\mathbf{B}$, $\mathbf{C}$, $\mathbf{D}$ can be immediately written down. The eigenvalues and eigenvectors can be obtained from the equation $(\lambda_i \mathbf{E} - \mathbf{A})\mathbf{t}_i = \mathbf{0}$.

$$\lambda_1 = 2, \quad \lambda_2 = -1, \quad \mathbf{t}_1 = \begin{bmatrix} -1 \\ 1 \end{bmatrix}, \quad \mathbf{t}_2 = \begin{bmatrix} -2 \\ 1 \end{bmatrix}. \tag{2.53}$$

The eigenvectors may, of course, be scaled by a constant factor. They form the modal matrix

$$\mathbf{T} = \begin{bmatrix} -1 & -2 \\ 1 & 1 \end{bmatrix}, \quad \mathbf{T}^{-1} = \begin{bmatrix} 1 & 2 \\ -1 & -1 \end{bmatrix}, \tag{2.54}$$

that is at the same time the transformation matrix in parallel form. From (2.47) - (2.50) follows the state representation also given by the block diagram in Figure 2.13:

$$\begin{bmatrix} \dot{\hat{z}}_1 \\ \dot{\hat{z}}_2 \end{bmatrix} = \begin{bmatrix} 2 & 0 \\ 0 & -1 \end{bmatrix} \begin{bmatrix} \hat{z}_1 \\ \hat{z}_2 \end{bmatrix} + \begin{bmatrix} 2 & 1 \\ 0 & 1 \end{bmatrix} \begin{bmatrix} x_1 \\ x_2 \end{bmatrix} \tag{2.55}$$

$$y = \begin{bmatrix} -1 & 1 \end{bmatrix} \begin{bmatrix} \hat{z}_1 \\ \hat{z}_2 \end{bmatrix}. \tag{2.56}$$

Here the element $\hat{B}_{21} = 0$, and it follows that the second state-variable is not controllable from the first input. No elements of $\hat{\mathbf{C}}$ are equal to zero so it is a completely observable system. This is easily confirmed by the block diagram in Figure 2.13. If we want to control a particular internal state of the system in Figure 2.12 or the equivalent system in Figure 2.13, that is only possible from input signal $x_2(t)$.

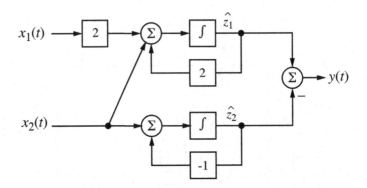

Figure 2.13: Parallel form of the systems in Example 2.3

■

2.7 Summary

At the start of this chapter we aimed to model continuous-time systems in the time domain. We found three types of model: differential equations, block diagrams and state models. The starting point was a system model in which the 'building blocks' were represented by a network of ideal components. Although this is expressed in terms of electrical circuits it can also be applied to other physical structures, providing that the local non-linearities of the building blocks can be disregarded.

We would finally like to combine the results obtained and show the relationships between them. Figure 2.14 shows the fundamental model forms, networks, differential equations, block diagrams and state models.

It is not necessary to examine in more depth here the use of network analysis for the modelling of networks through linear differential equations with constant coefficients. The point we are making here is that from such differential equations we have discovered some extremely useful forms of block diagram. For some of these the coefficients can be taken directly from the differential equation, thus the name direct form is used. We have examined the direct forms I, II and III, of which II and III are canonical forms, i.e. they use the minimum number of integrators (as many as the order of the differential equation). Direct form I requires twice as many integrators, and is therefore not a canonical form. The input–output relationship of a system can be modelled by other forms of block diagram, although the coefficients of these are not immediately interchangeable with those of the differential equation. The parallel form belongs to this group, which is notable as it makes the relationships between the input, intermediate states and the output particularly clear.

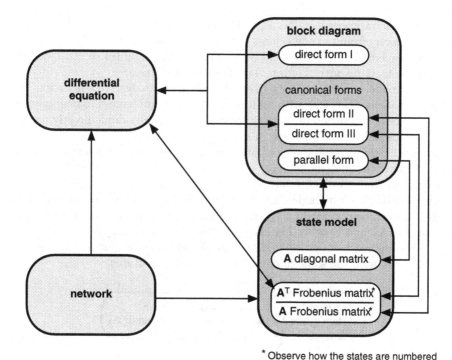

Figure 2.14: Overview of the different model forms of LTI-systems

Closely linked with the block diagrams is the state model. While the original differential equation is an equation of order N, the state model is a system of N first-order differential equations. The numerous coefficients of these equations can be combined in the correct way in a matrix form to give the so-called state matrix. Allocation of internal states is equally as arbitrary as with the block diagrams. We resrict ourselves here to state models with a minimal number of states, as these correspond to block diagrams in canonical form. The elements of the state matrices can, in fact, be used directly to form a corresponding block diagram. Equally, every state model can be easily drawn as an equivalent block diagram. This makes it clear that there must be some kind of relationhip betwen the construction of the block diagrams and the structure of the state matrices. Accordingly the parallel form corresponds to a state model of diagonal matrix form $\mathbf{A}$. The property of the direct form that makes its coefficients interchangeable with those of the differential equation must therefore be present in the state model as well. The matrix $\mathbf{A}$ from the state model for direct form II is a Frobenius matrix (with suitable numbering of states). It contains the coefficients of the differential equation in the bottom row, while all other matrix elements are either 1 or 0. With direct form III, the same is true for the transposed matrix $\mathbf{A}^T$. The elements of the Frobenius matrix can be taken directly from the coefficients of the differential equation, just as with the direct forms. For the non-canonical direct form I there is of course no applicable state-space structure with the minimum number of states.

A state model can, in fact, be derived directly from analysis of a network. This method produces structures with state-variables that correspond to energy stores in the network. Derivation of block diagrams is not usually done directly, instead by way of the differential equation or state-space structure.

Diagonal or Frobenius matrices are indeed useful, but are nevertheless only special forms of the state model. Any number of other state models can be obtained from the same input–output relationship through transformations. As every non-singular matrix $\mathbf{A}$ with single eigenvalues can be transformed to diagonal form, an easy rule can be obtained for the transformation of a non-diagonal state model with matrix $\mathbf{A}_1$ into another likewise non-diagonal matrix $\mathbf{A}_2$ (Figure 2.15). It should be observed in practice that $\mathbf{T}_1$ and $\mathbf{T}_2$ are transformations in the same diagonal form, with the same numeration and scaling of states.

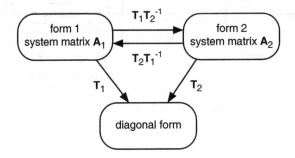

Figure 2.15: Transformation between different state-space representations

The model forms in this chapter are only applicable to continuous-time LTI-systems that can be realised by networks with energy stores, and that can be characterised by ordinary differential equations. It should be mentioned here that LTI-systems exist, which can only be modelled by partial differential equations or difference equations. A special class of system that is modelled by difference equations will be dealt with in Chapter 14.

2.8 Exercises

Exercise 2.1

Show that (2.3) describes an a) time-invariant and b) linear system.

Exercise 2.2

A system is modelled by the linear differential equation

$$0.5\frac{d^3y}{dt^3} - 3\frac{dy}{dt} + y = x + 0.1\frac{dx}{dt}.$$

Of what order is this system? Draw block diagrams for this system in both direct forms I and II. Which of the two forms is canonical? Justify your answer.

Exercise 2.3

The following block diagram is in direct form II. Give a linear differential equation that corresponds to the same system.

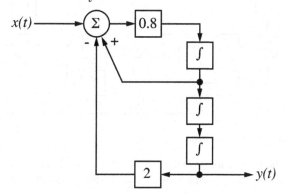

Exercise 2.4

a) Draw a block diagram using differentiators, multipliers and adders that corresponds to the following linear differential equation:

$$a_0\ddot{y} + a_1\dot{y} + a_2 y = b_1\dot{x} + b_2 x$$

Hint: Follow the construction stages for a direct form I block diagram.

b) Redraw the block diagram in canonical form. What condition must be satisfied for this transformation?

Exercise 2.5

For the network in Figure 2.9 the components have the following values: $R_1 = R_2 = 100$, $L_1 = 2$, $L_2 = 5$, $C = 0.01$.

a) Draw a signal flow graph (=block diagram) using integrators, multipliers and adders. Select an integrator output for each state in (2.31) and (2.32).

b) Through elimination of states find a differential equation that models the input–output behaviour of the system.

c) Using the linear differential equation construct a block diagram in direct form II.

d) Are the block diagrams in a) and/or c) canonical?

Exercise 2.6

Transform the following signal flow graph of a system with one input and two outputs into a state-space representation.

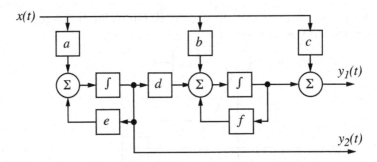

a) Select suitable state-variables.

b) Represent the derivations of the state-variables as dependent on only the non-derived states and the input signal, and give the state-equation in matrix form.

c) Represent both outputs dependent on the non-derived states and the input signal, and give the output equation in matrix form.

d) Which special form does the block diagram have if $d = 0$?

e) Under which conditions is the system controllable for $d = 0$?

f) Is the system completely observable? From which of the outputs is it observable?

Exercise 2.7

A system is modelled by the linear differential equation $\ddot{y} + 4\dot{y} + 5y = 2\ddot{x} + 7x$.

a) From the linear differential equation give a corresponding state-space representation.

b) Transform the state-vector using a suitable matrix $\mathbf{T}$ with $\hat{z} = \mathbf{T}^{-1}z$, so that the system matrix $\hat{\mathbf{A}}$ has a diagonal form. Give the matrix $\mathbf{T}$ and the transformed state equations.

   What are the eigenvalues of the system matrix? Has the transformed system the same input–output behaviour?

c) Is the system controllable, and is it observable?

Exercise 2.8

a) Give the transformation matrix $\mathbf{T}$ and $\mathbf{T}^{-1}$ for the transformation of the states $\mathbf{z}$ in (2.36) and (2.37) to the states ζ in (2.38) and (2.39), so that $\mathbf{z} = \mathbf{T}\zeta$ holds.

b) Verify the equations (2.40) - (2.43) using the relationships $\alpha_i = a_{N-i}$ and $\beta_i = b_{N-i}$ as well as (2.36) and (2.37).

3 Modelling LTI-Systems in the Frequency-Domain

In the last chapter we represented signals exclusively in the time-domain. This has the advantage that differential equations can be used to create system models. LTI-systems of any size can, in fact, be dealt with using these techniques, although calculations become cumbersome and complicated with higher order differential equations or with state models of many variables. It is for this reason that we have not yet derived expressions for the output signal of LTI-systems using differential equations.

Are there any other techniques available for representing signals? As an example of an acoustic signal we will consider a chord; a combination of different notes produced by a musical instrument. The vibrations can be represented as a function of time if a time-domain representation is required, but it is obvious that a chord can also be described by its component notes. The individual notes are characterised by pitch — or expressed technically — by their frequency, as the number of individual oscillations per unit time.

The characterisation of a signal by the the frequencies of the separate oscillations has advantages, not only for musicians, but also for technical reasons. Many systems are known to produce a sinusoidal output signal if a sinusoidal input signal is introduced. The amplitude and phase shift of the output signal may have changed relative to the input, but the frequency remains the same. We will soon see that LTI-systems have exactly this property. In general, the effect on the phase and amplitude are different for each frequency, and the knowledge of this effect for all frequencies presents itself as a further possiblity for modelling systems. This kind of model is called a *frequency-domain* model and is often simpler to handle than a time-domain model.

The response of an LTI-system to an input signal in the frequency-domain can be derived by using the following procedure.

1. Analysis of a sinusoidal input signal for a range of frequencies.

2. Establishing the system response to the individual frequencies by recording the amplitude and phase difference for each one.

3. Combining the individual parts to give a picture of the complete output signal.

These points will each be examined in detail in this and the following chapters, but firstly we will take a general look at the concept of frequency and the representation of signals in the frequency-domain. Mathematical tools for the transition between the time-domain and frequency-domain follow in Chapters 4 and 6 (Laplace Transform) and in Chapter 9 (Fourier Transform).

3.1 Complex Frequencies

3.1.1 What is a Complex Frequency?

The traditional definition of frequency is derived from real values of a sinusoidal signal $x(t)$. Such a signal is characterised by its amplitude $\hat{X}$ and the number of oscillations per unit time f, and the zero crossing relative to time $t = 0$. The frequency f is a real number. Together with the real phase φ it describes exactly the position of all zero crossings:

$$x(t) = \hat{X}\sin(2\pi f t + \varphi)\,. \tag{3.1}$$

The extension of the concept of frequency to a complex value leads to a complex exponential signal of the form

$$x(t) = \hat{X}e^{st}\,. \tag{3.2}$$

In contrast to (3.1) only the time t is real; the amplitude $\hat{X}$ and the frequency $s = \sigma + j\omega$ are complex. Figure 3.1 shows an example of a complex exponential signal with $\hat{X} = 1 + j$ and $s = -0.5 + j5$. It is defined for all points in time, but is only shown here for $t > 0$. For $t = 0$, $x(t)$ takes the value of the complex amplitude $\hat{X}$; for $t > 0$ it is defined by the real (σ) and imaginary (ω) parts of s. The real part is known as the modulus of $x(t)$,

$$|x(t)| = |\hat{X}|e^{\sigma t}, \tag{3.3}$$

while the imaginary part corresponds to the angular frequency $\omega = 2\pi f$, and indicates how fast the complex signal (Figure 3.1) orbits the time axis. The relationship with real oscillations can be seen by separating the real and imaginary parts of $x(t)$, as in Figure 3.2.

3.1.2 Complex Frequency: Example Signals

As examples of the use of complex frequencies we have expressed some real signals by complex exponential function (Table 3.1). For $\hat{X} \in \mathbb{R}$ and for real values of s, $x(t)$ is of course real. In the simplest case $s = 0$ and $\hat{X} = 1$, so $x(t) = 1$, and has become a constant. A real exponential signal is obtained for other real values of s, for example, $x(t) = e^{-3t}$ for $s = -3$. Sinusoidal signals with real values can

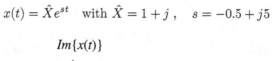

$$x(t) = \hat{X}e^{st} \quad \text{with } \hat{X} = 1 + j, \quad s = -0.5 + j5$$

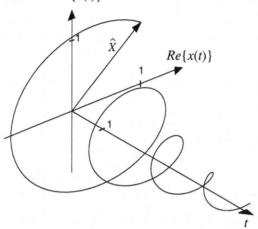

Figure 3.1: Example of a complex exponential signal

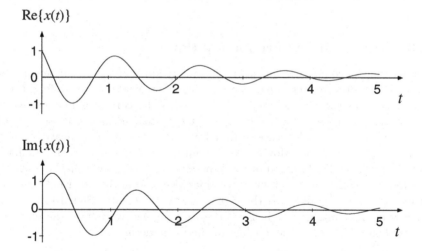

Figure 3.2: Real and imaginary parts of the signal in Figure 3.1

be represented by superimposing two complex exponential functions with purely imaginary frequencies $s = j\omega$ and $s = -j\omega$. Superimposing exponential functions with $\sigma \neq 0$ leads to decaying or growing sinusoidal oscillations.

Table 3.1: Examples of some real signals

	Signal	**Frequency**
Constant	$x(t) = 1$	$s = 0$
Real exponential signal	$x(t) = e^{-3t}$	$s = -3$
Sinusoidal oscillation	$x(t) = \sin 50t$ $= \dfrac{1}{2j}(e^{50jt} - e^{-50jt})$	$s = \pm 50j$
Decaying oscillation	$x(t) = e^{-2t}\sin 50t$ $= \dfrac{1}{2j}\left[e^{(-2+50j)t} - e^{(-2-50j)t}\right]$	$s = -2 \pm 50j$

3.1.3   The Complex Frequency Plane

The advantage of complex frequency is that many kinds of signal can be expressed by a single complex frequency parameter. To give an overview, the different forms of complex exponential function can be assigned the corresponding values in a Gaussian number plane. It is called the *complex frequency plane* or *s-plane*. Figure 3.3 shows the complex exponential functions for different locations in the complex frequency plane. Moving away from the real axis makes the oscillation faster. Looking in the direction of the time axis, in the upper half-plane where ω is positive, the complex oscillations turn clockwise, and in the lower half they turn anticlockwise. On the real axis the signal does not oscillate. Signal forms in the right half of the plane grow – faster if they are further away from the imaginary axis – and signal forms in the left half of the plane decay.

3.2   Eigenfunctions

3.2.1   What are Eigenfunctions?

In general, there is no great similarity between the time behaviour of a system's input and output signals. There are systems, however, that allow certain input

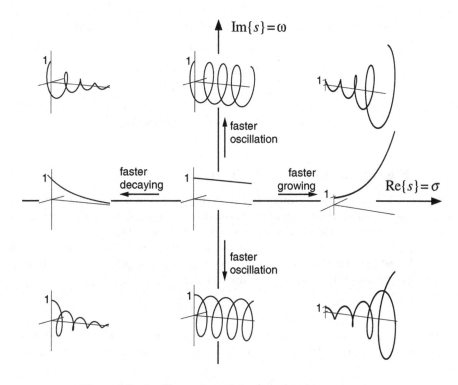

Figure 3.3: An illustration of the complex frequency plane

signals to pass through unchanged, for example, electrical networks made up only of resistances, capacitors and inductors. Their response to a sinusoidal signal is usually (for linear components) another sinusoidal signal, with only the amplitude and phase being different. Sine and cosine functions can be put together from exponential functions (see Figures 3.1 and 3.2), and then the amplitude and phase changes can be expressed by a single factor, the complex amplitude. The output signal can then be obtained from the input signal by multiplication with a complex factor.

A similar phenomenon is known from linear algebra: for certain vectors $\mathbf{x}$, the product of $\mathbf{x}$ with a matrix $\mathbf{A}$ equals a multiple of the vector $\mathbf{x}$: $\mathbf{Ax} = \lambda\mathbf{x}$. Then $\mathbf{x}$ is called eigenvector and λ is called eigenvalue. We will use these notions also for signals which pass through a system without changing their shape.

Definition 10: Eigenfunction

A signal $e(t)$ which when input into a system, produces at the output the response $y(t) = \lambda e(t)$ with the complex constant λ, is called the eigenfunction *of this system.*

Figure 3.4 shows the close relationship between the input and output signal.

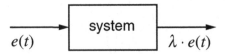

Figure 3.4: Driving a system with eigenfunction $e(t)$

3.2.2 Eigenfunctions of LTI-Systems

As sinusoidal signals pass through linear networks without changing form, and can also be represented by complex exponential functions, we guess that these exponential functions are eigenfunctions of LTI-systems. We have to show that $y(t) = \lambda x(t)$. To prove our theory, we start with an input signal of the form $x(t) = e^{st}$, and look for the corresponding system response $y(t)$, that we can write in a general form as a function of the input signal:

$$y(t) = S\{x(t)\} . \tag{3.4}$$

In the following we use exclusively the stated properties of LTI-systems, time-invariance and linearity. We start with the response to an input signal shifted in time.

$$x(t - \tau) = e^{s(t-\tau)} \tag{3.5}$$

and because of the time-invariance we obtain

$$y(t - \tau) = S\{x(t - \tau)\} = S\{e^{s(t-\tau)}\} = S\{e^{-s\tau}e^{st}\} . \tag{3.6}$$

The factor $e^{-s\tau}$ does not depend on time. Because of the linearity it follows further that

$$y(t - \tau) = S\{e^{-s\tau}e^{st}\} = e^{-s\tau}S\{e^{st}\} = e^{-s\tau}y(t) . \tag{3.7}$$

Now although we do not have $y(t)$ itself, we have a difference equation for $y(t)$ which is

$$y(t - \tau) = e^{-s\tau}y(t) . \tag{3.8}$$

This can be fulfilled by

$$y(t) = \lambda e^{st} \tag{3.9}$$

which can be verified by substituting into (3.8). From Definition 10, $x(t) = e^{st}$ is an eigenfunction of the LTI-system described by S.

The constant λ identifies the behaviour of S. As linearity and time-invariance were the only pre-conditions, we cannot say anything further about λ. In general λ depends on the complex frequency s. We write therefore $\lambda = H(s)$ and call $H(s)$ the *system function* or *transfer function*, as it describes the system and its transfer properties from the input to the output. The connection with the system model in the time-domain will be covered in later sections. Figure 3.5 shows the relationship between eigenfunctions and system functions for LTI-systems.

The scheme given in Figure 3.5 has to be used with care. For many LTI-systems, an analytically determined transfer function $H(s)$ is only valid within a certain region of the complex plane, the so-called region of convergence. Only inside this region, a complex exponential oscillation at the input leads to a finite signal at the output. As an example, consider an integrator with the transfer function $H(s) = \frac{1}{s}$. The response of the integrator to the input signal $x(t) = e^{st}$ is only of finite amplitude, as long as $\mathrm{Re}\{s\} > 0$. The response $y(t) = \int\limits_{-\infty}^{t} e^{s\tau}\, d\tau$ does not converge for $\mathrm{Re}\{s\} \leq 0$. Usually, the indication of the region of convergence for a transfer function is omitted without running into problems. We will discuss this topic further in Chapter 8.4.4.

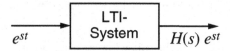

Figure 3.5: Eigenfunction and transfer function of an LTI-system

One more warning: the one-sided exponential function

$$x(t) = \begin{cases} e^{st} & t \geq 0 \\ 0 & \text{otherwise} \end{cases}$$

is in general not an eigenfunction of an LTI-system!

3.2.3   Example: RLC Network

As an example for analysis using complex frequencies we consider the parallel resonant circuit in Figure 3.6

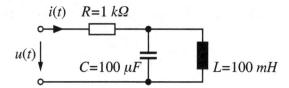

Figure 3.6: Parallel resonant RLC network

The input signal is of the form

$$u(t) = u_0 e^{\sigma_0 t} \cos(\omega_0 t + \varphi_0) . \tag{3.10}$$

To determine the expression for the current $i(t)$, first $u(t)$ is rewritten as two complex exponential functions:

$$u(t) = u_0 e^{\sigma_0 t} \cdot \frac{1}{2} \left[e^{j(\omega_0 t + \varphi_0)} + e^{-j(\omega_0 t + \varphi_0)} \right] = U_1 e^{s_1 t} + U_2 e^{s_2 t} \qquad (3.11)$$

with

$$U_1 = U_2^* = \frac{u_0}{2} e^{j\varphi_0} \; ; \; s_1 = s_2^* = \sigma_0 + j\omega_0 . \qquad (3.12)$$

We can find the expression for current $i(t)$ with the following considerations.

- The input signal is a sum of two exponential functions.

- The electrical network can be modelled by a linear differential equation with constant coefficients, and is therefore an LTI-system. We know that LTI-systems have complex exponential functions as eigenfunctions.

- The output signal may be then be written as the sum of two exponential functions.

For the current we thus make the following starting point:

$$i(t) = I_1 e^{s_1 t} + I_2 e^{s_2 t} . \qquad (3.13)$$

The constants I_1 and I_2 are unknown and must be determined from the model of the network. We put $u(t)$ and $i(t)$ into the differential equation

$$u(t) + LC \frac{d^2 u(t)}{dt^2} = Ri(t) + L\frac{di(t)}{dt} + LCR\frac{d^2 i(t)}{dt^2} \qquad (3.14)$$

that can be immediately written down from the circuit in Figure 3.6. The following equations are then obtained

$$
\begin{aligned}
U_1 e^{s_1 t} + s_1^2 LC U_1 e^{s_1 t} &= RI_1 e^{s_1 t} + Ls_1 I_1 e^{s_1 t} + LCRs_1^2 I_1 e^{s_1 t} \qquad (3.15) \\
U_2 e^{s_2 t} + s_2^2 LC U_2 e^{s_2 t} &= RI_2 e^{s_2 t} + Ls_2 I_2 e^{s_2 t} + LCRs_2^2 I_2 e^{s_2 t} . \qquad (3.16)
\end{aligned}
$$

As U_1 and U_2 are known from (3.12), we can solve (3.15) and (3.16) for I_1 and I_2:

$$I_1 = U_1 \frac{1 + s_1^2 LC}{R + s_1 L + s_1{}^2 LCR} \qquad (3.17)$$

$$I_2 = U_2 \frac{1 + s_2^2 LC}{R + s_2 L + s_2{}^2 LCR} . \qquad (3.18)$$

As with (3.12) $s_1 = s_2^*$, it follows from (3.17) and (3.18) that $I_1 = I_2^*$. The objective is then obtained with (3.13)

$$i(t) = I_1 e^{s_1 t} + I_1^* e^{s_1{}^* t} = 2Re\{I_1 e^{s_1 t}\} . \qquad (3.19)$$

As an example, we input the component values in Figure 3.6

$$u_0 = 20\text{mV}, \quad \omega_0 = 300\text{s}^{-1}, \quad \sigma_0 = -1\text{s}^{-1}, \quad \varphi_0 = \frac{\pi}{4} \approx 0.785$$

into (3.17) and obtain

$$I_1 = 9.75\mu\text{A}\,e^{j0.49}\ .$$

Together with (3.19), the result for $i(t)$ is:

$$i(t) = 19.5\mu\text{A}\,e^{-1\text{s}^{-1}t}\cos(300\text{s}^{-1}t + 0.49)\ .$$

The following observations taken from this simple example are important for the general procedure.

- To obtain the desired output value we only have to determine the factors of the eigenfunctions (here I_1 and I_2).

- These factors can be expressed as algebraic equations (here (3.17) and (3.18)).

- The differential equation that models the network (here (3.14)) does not have to be solved as a whole. Forming the derivative or integral for each individual inductance or capacitance is sufficient. Due to the exponential form of the eigenfunction each derivation or integration results simply in a multiplication or division by the complex frequency s.

- This procedure corresponds to steady-state sinusoidal analysis. Here, however, exponentially decaying or rising oscillations are also included.

The convergence of the system response has not been investigated in this example, but was simply implicitly assumed. We will see in an exercise at the end of Chapter 8 that the system response only converges for a known range of values of σ_0.

3.2.4 Impedance

From the example of the parallel resonant circuit we made the observation that the treatment of the equations for the capacitances and inductances can be broken down to multiplication or division by the complex frequency s. Such networks can be modelled without the use of differential equations or integrals, if every component is given a complex resistance that is dependent on the complex frequency. This complex resistance is called *impedance*. If the voltage and current are in exponential form for the components, Ohm's law can be used to define the relationship between them:

$$U(s) = Z(s) \cdot I(s) \tag{3.20}$$

Table 3.2: Impedance of important components

component		impedance
resistance	R ⊸─▭─⊸	R
capacitance	C ⊸─┤├─⊸	$\dfrac{1}{sC}$
inductivity	L ⊸─▬─⊸	sL

with the impedance $Z(s)$. This relationship is called Ohm's law in impedance form. Table 3.2 shows the impedance functions for the most important components.

Networks that consist of these components can be modelled as an impedance. Taking the equations (3.15) and (3.16) for the parallel resonant circuit we compare with (3.20), and have:

$$U e^{st} = Z(s) \cdot I e^{st} \tag{3.21}$$

where

$$Z(s) = R + \frac{sL \cdot \frac{1}{sC}}{sL + \frac{1}{sC}} = \frac{R + sL + s^2 LCR}{1 + s^2 LC} . \tag{3.22}$$

From section 3.2.2 it follows that the system function of the parallel resonant circuit for the voltage as input signal and current as output signal is:

$$H(s) = \frac{1}{Z(s)} = \frac{1 + s^2 LC}{R + sL + s^2 LCR} . \tag{3.23}$$

The output value of the parallel resonant circuit can also be expressed in a general form:

$$i(t) = H(s_1) U_1 e^{s_1 t} + H(s_2) U_2 e^{s_2 t} . \tag{3.24}$$

The impedance model is only applicable to linear time-invariant circuits (LTI-systems), as only they have exponential functions as eigenfunctions.

3.2.5 Normalisation

A general problem with derivation of the system response and the specification of the system function is the correct consideration of physical units of measurement. There are two fundamental possiblities:

- consistent use of units

- normalisation.

The advantage of consistent use of units is that all values have a physical meaning and that errors in calculation can be detected due to inconsistency of units. Unfortunately this method often leads to unwieldy expressions. We will demonstrate consistency of units with the following simple example.

─── **Example 3.1**

The system function of the network shown in Figure 3.7 will be derived. The voltage $u_1(t)$ is the input signal, $u_2(t)$ the output signal.

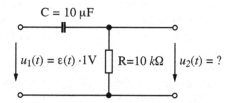

$$C = 10\,\mu F$$

$$u_1(t) = \varepsilon(t)\cdot 1V \quad R = 10\,k\Omega \quad u_2(t) = ?$$

Figure 3.7: RC circuit

Using the complex impedances of the components

$$H(s) = \frac{U_2(s)}{U_1(s)} = \frac{R}{R + \dfrac{1}{sC}} = \frac{s\tau}{s\tau + 1} \quad \text{with } \tau = RC = 0.1\,\text{s} \tag{3.25}$$

is obtained. The complex frequency variable s has dimension 1/unit time, and the time constant $\tau = RC$ the dimension unit time.

─── ∎

The advantage of normalisation is simplicity of expressions, especially when numerical values are given. Admittedly, the physical meaning of the values is no longer immediately apparent, and errors in calculation cannot be found by considering the physical meaning. As dealing with normalised values and their correct interpretation requires some practice, we will examine the relevant concepts here in some detail and finally demostrate them in an example.

Two steps are required to convert a network to normalised values.

- Amplitude normalisation

  With amplitude normalisation, all voltages are expressed as dimensionless multiples of a reference voltage, and all currents as a multiple of a reference current. Amplitude normalisation causes a change in the vertical axis.

- Time normalisation

  With time normalisation, all specified times are expressed as multiples of a reference time. Time normalisation causes a change in the time axis.

Both amplitude normalisation for voltage and current and time normalisation induce a corresponding component normalisation, which expresses all component values as a multiple of the corresponding reference value.

The simplest possibility for the choice of reference voltage, reference current and reference time is the use of the corresponding SI units, 1V, 1A and 1s. From these, reference values for the components are immediately obtained, also in SI units, i.e. 1Ω for resistances, 1F for capacitances and 1H for inductances. Normalisation can thus be carried out so that all values are calculated using SI units. Simply omitting these units gives a normalised representation which can be converted back to a physical representation by adding SI units. However, values for capacities in 1F are very unwieldy, as well as time constants in seconds when dealing with fast microelectronic components.

We therefore discuss in detail amplitude normalisation, time normalisation, and component normalisation for general reference values. From now on, we denote normalised quantities by a tilde (˜). Since we want to represent components by impedances, the following considerations are based on complex frequencies.

We begin with amplitude normalisation and mark the reference voltage with U_0 and the reference current with I_0. Both are dimensionless, time-independent variables. For current and voltage in normalised variables, it follows that

$$\tilde{U} = \frac{U}{U_0}, \qquad \tilde{I} = \frac{I}{I_0}. \tag{3.26}$$

For time normalisation we relate all times t to a reference time t_0 and obtain the dimensionless normalised time $\tilde{t}$

$$\tilde{t} = \frac{t}{t_0}. \tag{3.27}$$

When we are working in the frequency domain, we express the complex frequency likewise with the reference time t_0. We have to make sure that the argument in the exponential function remains dimensionless. We obtain from

$$e^{-st} = e^{-st_0 \cdot t/t_0} = e^{-\tilde{s}\tilde{t}} \tag{3.28}$$

the normalised complex fequency

$$\tilde{s} = st_0. \tag{3.29}$$

The transition from the amplitude and time normalisation to component normalisation is carried out using (3.20). First we express $U(s)$ and $I(s)$ in amplitude and time normalised values with (3.26) and (3.29):

$$\tilde{U}(\tilde{s}) = \frac{U(\tilde{s}/t_0)}{U_0}, \qquad \tilde{I}(\tilde{s}) = \frac{I(\tilde{s}/t_0)}{I_0}. \tag{3.30}$$

If we use the normalised complex frequency variable $\tilde{s}$ in (3.20) and substitute into (3.30), it follows that (3.20) is the corresponding relationship for the normalised value of impedance

$$\tilde{U}(\tilde{s}) = \tilde{Z}(\tilde{s})\tilde{I}(\tilde{s}). \tag{3.31}$$

The normalised impedance $\tilde{Z}(\tilde{s})$ corresponding to (3.30) follows by normalisation of the impedance $Z(s)$ with respect to the reference resistance R_0

$$\tilde{Z}(\tilde{s}) = \frac{Z(\tilde{s}/t_0)}{R_0}, \qquad R_0 = \frac{U_0}{I_0}. \tag{3.32}$$

The reference resistance R_0 is thus clearly set by the amplitude normalised reference voltage U_0 and reference current I_0. The normalised impedances of the frequently occurring components from Table 3.2 are:

$$\tilde{Z}(\tilde{s}) = \begin{cases} \dfrac{R}{R_0} = \tilde{R} & \text{resistance} \\[2ex] \dfrac{t_0}{\tilde{s}R_0 C} = \dfrac{1}{\tilde{s}\tilde{C}} & \text{capacitance} \\[2ex] \tilde{s}\dfrac{L}{R_0 t_0} = \tilde{s}\tilde{L} & \text{inductance} \end{cases} \tag{3.33}$$

The reference resistance R_0 and the reference time $t = 0$ can be combined to give a reference capacitance and a reference inductance

$$C_0 = \frac{t_0}{R_0}, \qquad L_0 = R_0 t_0. \tag{3.34}$$

and the dimensionless normalised component values can then be obtained

$$\tilde{R} = \frac{R}{R_0}, \qquad \tilde{C} = \frac{C}{C_0}, \qquad \tilde{L} = \frac{L}{L_0}. \tag{3.35}$$

They are normalised to the reference values for voltage, current and time chosen at the outset. Conversl, it is also possible to use the three reference values for resistance, capacitance and inductance as a starting point. Then the inverse route has to be followed to interpret the results in the correct dimensions for voltage, current and time. The choice of the reference values yields handy values, when the reference time is in the same order as the typical time constants of the system. Component reference values are chosen such that manageable values result. With some practice, normalisation can be significantly shorter than the detailed representation shown above.

We show the procedure with the same simple network from Example 3.1 that we have already analysed with dimensional values.

Example 3.2

To derive the system function of the network in Figure 3.7, we select reference values for current, voltage and time and calculate from these the reference values of the components. With

$$U_0 = 1\text{V}, \qquad I_0 = 1\text{mA}, \qquad t_0 = 1\text{s}$$

we obtain

$$R_0 = 1\text{k}\Omega, \qquad C_0 = 1\text{mF}$$

for the reference values of the components and from those, the normalised components

$$\tilde{R} = 10, \qquad \tilde{C} = \frac{1}{100}.$$

The system function is then given by

$$\tilde{H}(\tilde{s}) = \frac{\tilde{U}_2(\tilde{s})}{\tilde{U}_1(\tilde{s})} = \frac{\tilde{R}}{\tilde{R} + \dfrac{1}{\tilde{s}\tilde{C}}} = \frac{10}{10 + \dfrac{100}{\tilde{s}}} = \frac{\tilde{s}}{\tilde{s} + 10}. \qquad (3.36)$$

Here all quantities are dimensionless and the result of further calculation can be interpreted respectively as normalisation into volts, milliamperes and seconds.

Usually no separate notation for the normalised quantities (the tilde in our case) is used. The component values are simply given without dimensions. This representation of the dimensionless network is shown in Figure 3.8. From now on, we use both variants for the analysis of systems without denoting normalised and unnormalised quantities differently.

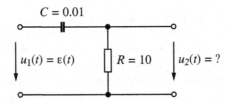

Figure 3.8: Normalised RC circuit

3.3 Exercises

Exercise 3.1

Recognise the frequencies of the following signals in the complex frequency plane.

$$
\begin{aligned}
x_a(t) &= e^{-2t}\left(1 + \sin 5t\right) \\
x_b(t) &= A\cos\omega_0 t + B\cos 2\omega_0 t \\
x_c(t) &= \sin(\omega_0 t + \varphi)
\end{aligned}
$$

Exercise 3.2

Could the following systems be LTI-systems?

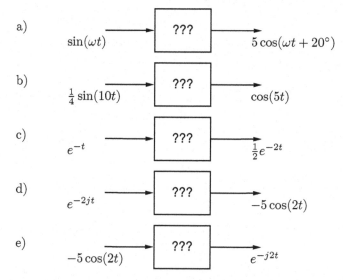

a) $\sin(\omega t)$ → ??? → $5\cos(\omega t + 20°)$

b) $\frac{1}{4}\sin(10t)$ → ??? → $\cos(5t)$

c) e^{-t} → ??? → $\frac{1}{2}e^{-2t}$

d) e^{-2jt} → ??? → $-5\cos(2t)$

e) $-5\cos(2t)$ → ??? → e^{-j2t}

Exercise 3.3

The following simple network will be analysed:

$$L = 1.2\,\text{H} \qquad C = \frac{1}{6}\,\text{F}$$

$u_1(t)$ $R = 1\,\Omega$ $u_2(t)$

$$u_1(t) = e^{-3t\cdot 1/\text{s}}\cos(-4t \cdot 1/\text{s})\cdot 1\,\text{V}$$

a) Normalise the components and the response to 1A, 1V and 1s.

b) Give the transfer function $H(s) = \dfrac{U_2(s)}{U_1(s)}$ by applying Ohm's law in impedance form.

c) Represent $u_1(t)$ as a sum of weighted exponential functions and calculate the system reaction $u_2(t)$ to the exponential excitation.

d) Express $u_2(t)$ using physical dimensions.

Exercise 3.4

Show that

$$x(t) = \begin{cases} e^{st} & t \geq 0 \\ 0 & \text{otherwise} \end{cases}$$

is not an eigenfunction of the LTI-system $y(t) = \displaystyle\int\limits_{-\infty}^{t} x(\tau)d\tau.$

4 Laplace Transform

4.1 The Eigenfunction Formulation

In Chapter 3 we saw that for input signals $x(t) = e^{st}$, that are eigenfunctions of a linear time-invariant system (LTI-system), it is relatively easy to calculate the corresponding output signals. Unfortunately, real world signals (see Chapter 1), are not eigenfunctions of LTI-systems. There is a way around this, however, where any signal $x(t)$ is represented by superimposed eigenfunctions e^{st} with different frequencies. The response of an LTI-system to the individual eigenfunctions can be easily determined and then with the superposition property, the response to the signal $x(t)$ can be put together.

The idea of analysing a signal as its individual components is already familiar from the Fourier series: periodic signals can be represented as a combination of harmonic oscillations. Their frequencies must be integer multiples of the fundamental frequency that represents the periodic signal, and the combination of these frequency components is achieved by summation. We are not restricted to periodic signals, so we must permit analysis in eigenfunctions with any frequency. The superposition then consists of an integral over the possible frequencies. This idea can be put into use in various different ways and it leads to the Laplace and Fourier transforms.

In this chapter we will discuss the Laplace transform, starting with the definitions of the unilateral and bilateral Laplace transforms, and then considering some examples that lead to some general rules for the region of convergence. Finally we will discuss some important laws and properties.

4.2 Definition of the Laplace Transform

In order to be able to represent a signal $x(t)$ as a superposition of individual parts, two mathematical tools are required for:

- the decomposition of $x(t)$ into parts,

- the superposition of the parts of the complete signal $x(t)$.

Mathematical formulation of the decompositon leads to the definition of the Laplace transform

---

Definition 11: Laplace transform

The Laplace transform *of a function* $x(t), t \in \mathbb{R}$ *is*

$$\mathcal{L}\{x(t)\} = X(s) = \int\limits_{-\infty}^{\infty} x(t)e^{-st}\,dt \qquad (4.1)$$

for the the values $s \in \text{ROC} \subset \mathbb{C}$, *for which the improper integral exists.*

The complex function $X(s)$ is called the *Laplace transform* of $x(t)$. Stating the convergence is not necessary as whatever form the function $x(t)$ has, the integral only converges for a bounded region of the complex s-plane. This region is called the *region of convergence* (or ROC). We deal with determining the region of convergence with some examples, when we are more comfortable with the Laplace transform.

In order to reconstruct $x(t)$ from the individual terms, the reverse operation to the Laplace transform must be used. This is the inverse Laplace transform:

$$x(t) = \mathcal{L}^{-1}\{X(s)\} = \frac{1}{2\pi j} \int_{\sigma-j\infty}^{\sigma+j\infty} X(s)e^{st}\,ds \ . \qquad (4.2)$$

The real number σ determines the location of the path of integration within the complex plane. It must lie within the region of convergence, but its exact location and form does not influence the result. An example is given in Figure 4.1. The region of convergence in this case consists of a vertical stripe in the complex plane, and the path of integration (dotted line) runs vertically from top to bottom.

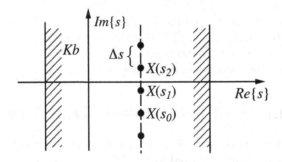

Figure 4.1: Illustration of the Riemann sum

The Laplace transform represents the real or complex function of time $x(t)$ by a complex function $X(s)$ that is defined in the complex frequency plane. Important functions encountered in practice only have one transform, so $X(s)$ fully describes $x(t)$ (see Section 4.6).

The pairing of a function of time $x(t)$ and its Laplace transform $X(s) = \mathcal{L}\{x(t)\}$ can also be identified by the symbol ○—●. The filled circle denotes the frequency-domain quantity:

$$x(t)\circ\!\!-\!\!\bullet X(s) \, . \tag{4.3}$$

The inverse Laplace transform really does represent a function of time as an infinite number of superimposed exponential functions, and this can be made clear by the integral (4.2), sketched in Figure 4.1, and expressed as a Riemann sum:

$$
\begin{aligned}
x(t) &= \frac{1}{2\pi j} \int_{\sigma-j\infty}^{\sigma+j\infty} X(s)e^{st}ds \\
&= \frac{1}{2\pi j} \lim_{\Delta s \to 0} \left\{ \ldots + X(s_0)e^{s_0 t} + X(s_1)e^{s_1 t} + X(s_2)e^{s_2 t} + \ldots \right\} \Delta s \, . \tag{4.4}
\end{aligned}
$$

Every term in (4.4) is a complex exponential function, weighted with a complex factor.

4.3   Unilateral and Bilateral Laplace Transforms

The Laplace transform has more than one version and, as well as Definition 11, the following *unilateral* Laplace transform

$$\boxed{\mathcal{L}_I\{x(t)\} = X(s) = \int_0^\infty x(t)e^{-st}dt \, ,} \tag{4.5}$$

in which the domain of integration only contains the positive values of the time axis, is often used. It assumes that the system is zero for $t < 0$ and is therefore well suited to causal systems and problems with initial conditions. Modelling of a left shift, though, is now problematic.

To distinguish between the two types of transform, the transform in Definition 11 and equation (4.1) is called the *bilateral* Laplace transform. It transforms t for $t \in \mathbb{R}$ and models a left shift of the signal in a simple way. If a causal signal or system is multiplied by the *step function*

$$\varepsilon(t) = \begin{cases} 1 & \text{for } t \geq 0 \\ 0 & \text{otherwise} \end{cases} \tag{4.6}$$

then its value is zero for $t < 0$, and it can be dealt with by the bilateral Laplace transform as with the unilateral Laplace transform. Figure 4.2 shows the step function in the time-domain. Use of the step function will be explained in the following examples.

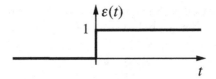

Figure 4.2: Time behaviour of the step function

4.4 Examples of Laplace Transforms

The following examples should clarify the connection between the unilateral and bilateral Laplace transform and the use of the step function. The region of convergence will also be determined.

—————————————————————————————————— **Example 4.1**

We begin with a decaying exponential function that disappears for time $t < 0$. It could, for example, be the impulse response of an RC circuit, i.e. a causal system. The constant a is real:

$$x(t) = e^{-at}\varepsilon(t) \ .$$

Its Laplace transform $X(s) = \mathcal{L}\{x(t)\}$ is required. We obtain through the Laplace integral (4.1):

$$
\begin{aligned}
X(s) &= \int_{-\infty}^{\infty} e^{-at}\varepsilon(t)e^{-st}dt = \int_{0}^{\infty} e^{-(s+a)t}dt \\
&= \left[\frac{-1}{s+a}e^{-(s+a)t}\right]_{t=0}^{t=\infty} = -\frac{1}{s+a}\left[\lim_{t\to\infty} e^{-(s+a)t} - 1\right] = \frac{1}{s+a} \ .
\end{aligned}
$$

We consider the effect of the step function in the bilateral Laplace transform because we evaluate only positive values of time t in the integral, that is, where the step function has the value 1. This has the same meaning as the unilateral Laplace transform of e^{-at}. The limit value of the exponential function for $t \to \infty$ only exists if the real part of the exponential function is negative, i.e. for the values of s where $\text{Re}\{s\} > -a$. For all other values of s the integral does not converge, and it is said that the Laplace transform does not exist. Concisely put:

$$\mathcal{L}\{\varepsilon(t)e^{-at}\} = \frac{1}{s+a} \ ; \quad \text{Re}\{s\} > -a \ . \tag{4.7}$$

The transform clearly has a pole at $s = -a$ (see Section 4.5.2). The region of convergence in the s-plane is represented in Figure 4.3. It contains all values of s for which the real part is greater than $-a$. We can also express this in another way: a vertical line through the pole divides the complex plane into two halves.

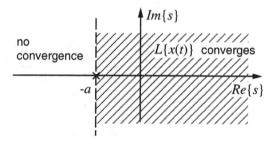

Figure 4.3: Representation of the region of convergence from Example 4.1

The Laplace transform exists for all values of s that are in the right hand side. For $a = 0$ we obtain an important special case: The Laplace transform of the step function $\varepsilon(t)$:

$$\mathcal{L}\{\varepsilon(t)\} = \frac{1}{s} ; \quad \text{Re}\{s\} > 0 . \tag{4.8}$$

Example 4.2

In the second example we consider an exponential function that disappears for $t < 0$:

$$x(t) = -e^{-at}\varepsilon(-t) .$$

To determine the Laplace transform we proceed exactly as in the first example:

$$
\begin{aligned}
X(s) &= -\int_{-\infty}^{\infty} e^{-at}\varepsilon(-t)e^{-st}dt \\
&= -\int_{-\infty}^{0} e^{-(s+a)t}dt = \frac{1}{s+a}\left[1 - \lim_{t\to-\infty} e^{-(s+a)t}\right] = \frac{1}{s+a}
\end{aligned}
$$

and obtain

$$\mathcal{L}\{-\varepsilon(-t)e^{-at}\} = \frac{1}{s+a} ; \quad \text{Re}\{s\} < -a . \tag{4.9}$$

The Laplace transform $X(s)$ has the same form here as in Example 4.1 and so likewise has a pole at $-a$. The difference lies in the region of convergence, which here is the half of the plane on the left side of the vertical line. It is clear that different time functions can have the same Laplace transform, so to invert the transform, the region of convergence must be the deciding factor between the different possibilities.

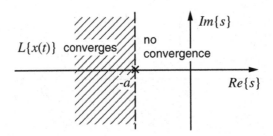

Figure 4.4: Illustration of the region of convergence from Example 4.2

————————————————————————————————————— **Example 4.3**

In this example, we form a right-sided signal from two exponential terms from Example 4.1:

$$x(t) = \varepsilon(t)[e^{-t} + e^{-2t}] \ .$$

To determine the Laplace transform, we split $x(t)$ into two parts and interpret for each part the result of Example 4.1:

$$X(s) = \int_{-\infty}^{\infty} [e^{-t}\varepsilon(t) + e^{-2t}\varepsilon(t)]e^{-st}dt \ = \ \int_{0}^{\infty} e^{-t}e^{-st}dt + \int_{0}^{\infty} e^{-2t}e^{-st}dt$$

$$= \ \frac{1}{s+1} \ + \ \frac{1}{s+2} \ .$$

The Laplace transform $X(s)$ has two poles at $s = -1$ and $s = -2$. As $X(s)$ only exists if both parts converge, the pole with the largest real part determines the region of convergence. Put in other words: the region of convergence lies to the right of a vertical line through the rightmost pole. Through combination of the two terms, it can be seen that $X(s)$ has a zero at $s = -1.5$, but this has no effect on the region of convergence:

$$X(s) = \frac{1}{s+1} + \frac{1}{s+2} = \frac{2s+3}{s^2+3s+2} \ ; \quad \text{Re}\{s\} > -1 \ . \tag{4.10}$$

—————————————————————————————————————— ∎

————————————————————————————————————— **Example 4.4**

Now we consider a signal that is not equal to zero for $-\infty < t < \infty$ and is formed from a right-sided signal for $t > 0$ as in Example 4.1, and a left-sided signal for $t < 0$, as in Example 4.2:

$$x(t) = \varepsilon(t)e^{-2t} - \varepsilon(-t)e^{-t} \ .$$

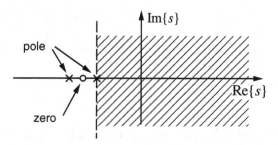

Figure 4.5: Illustration of the region of convergence in Example 4.3

We again express the Laplace transform as its component parts and then have the same form, and so the same poles as in Example 4.3. The region of convergence is made up of the area in the s-plane in which both parts converge, i.e. in the intersection of the two regions of convergence of the exponential parts from Example 4.1 and Example 4.2. The fact that both regions of convergence of the individual terms overlap, means that the the region of convergence lies in the strip between the two poles:

$$X(s) = \frac{1}{s+2} + \frac{1}{s+1} = \frac{2s+3}{s^2+3s+2} \; ; \quad -2 < \text{Re}\{s\} < -1 \; . \qquad (4.11)$$

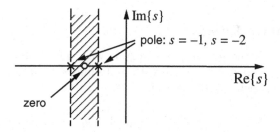

Figure 4.6: Illustration of the region of convergence in Example 4.4

<hr>

Example 4.5

The signal in this example is similar to that in Example 4.4, but the values of the exponents for the left and right sides have been exchanged:

$$x(t) = \varepsilon(t)e^{-t} - \varepsilon(-t)e^{-2t} \; .$$

Each part converges seperately; the positive time part for $\text{Re}\{s\} > -1$, and the negative time part for $\text{Re}\{s\} < -2$. In comparison with Examples 4.1 and 4.2, we see that in this case the regions of convergence do not overlap. This signal therefore has no Laplace transform.

∎

4.5 Region of Convergence of the Laplace Transform

The Examples 4.1-4.5 are concerned with signals of which the Laplace transform is characterised by a few simple poles and zeros. Finding the regions of convergence was relatively simple. The Laplace transform can also characterise, however, more complicated signals and systems, and leads to expressions that can no longer be written as rational functions of s.

That is the case, for example, for all locally distributed systems like electrical circuits (see [19]). We must therefore make the results so far obtained even more general. First of all we will consider the convergence with indefinite integrals. The region of convergence is especially important, because different signals can have the same Laplace transform and are only distinguished by their regions of convergence (see Examples 4.1 and 4.2). Without giving the region of convergence, a unique inverse transform is not possible in general. Further on we will consider general singularities of functions of complex variables.

4.5.1 Indefinite Integrals

First of all we recall the meaning of integrals with infinite limits of integration in (4.1) and (4.5). Unlike an integral with finite limits 0 and $a < \infty$, an integral between 0 and ∞ is defined by the limit

$$\int_0^\infty [\cdot]dt = \lim_{a \to \infty} \int_0^a [\cdot]dt \tag{4.12}$$

It is also called an *indefinite integral*. Correspondingly, the integral of the bilateral Laplace transform is given by two limits

$$\mathcal{L}\{x(t)\} = \int_{-\infty}^\infty x(t)e^{-st}dt = \lim_{A \to -\infty} \int_A^B x(t)e^{-st}dt + \lim_{C \to \infty} \int_B^C x(t)e^{-st}dt. \tag{4.13}$$

Here, each limit must converge individually (in previous examples B was always zero). The region of convergence of the bilateral Laplace transform therefore consists of the intersection between the two regions of convergence for the two limits.

4.5.2 Singularities

The Laplace transforms from the previous examples are functions of the complex variable s. For certain values of s they were singular (e.g. for $s = -a$ in (4.7)). To classify such situations in complex s-plane we must recall some concepts from function theory.

If for a complex function $X(s)$, $s \in \mathbb{C}$ at s_0, the boundary value

$$\lim_{\Delta s \to 0} \frac{X(s_0 + \Delta s) - X(s_0)}{\Delta s} = X'(s_0) \tag{4.14}$$

exists and takes the same value $X'(s_0)$ regardless of the path s_0 is reached by, the function $X(s)$ at the point s_0 is then called an *analytic, regular* or *holomorphic* function. $X(s)$ at the point s_0 is said to be complex differentiable and $X'(s_0)$ is the first *derivative* of $X(s)$. If the first derivative exists, then the higher derivatives also exist at this point. Points for which $X(s)$ is not analytic are called *singularities*.

Functions $X(s)$ that are analytic in the entire complex s-plane apart from some isolated points are often obtained as Laplace transforms. The local behaviour at any isolated point s_0, which has a surrounding $X(s)$ that is analytic, can be classified by the Laurent expansion of $X(s)$ with s_0

$$X(s) = \sum_{n=-\infty}^{\infty} a_n (s - s_0)^n \tag{4.15}$$

The following may occur:

- The Laurent expansion does not contain any elements with negative powers ($a_n = 0$ for $n < 0$, Taylor series). $X(s)$ is then analytic at the point s_0.

- The Laurent expansion contains a finite number M elements with negative powers ($a_n = 0$ for $n < -M < 0$). $X(s)$ is then not analytic at the point s_0. The singularity is then called a *pole of order M*.

- The Laurent expansion contains an infinite number of elements with negative powers. $X(s)$ is then not analytic. The singularity is called an *essential singularity*.

This classification is of little use if the Laurent series of $X(s)$ is not known. The Laurent expansion of functions that occur when LTI-systems are described by the Laplace transform are, however, usually simple to determine. As an example, we will consider the Laplace transform from Example 4.4.

─── **Example 4.6**

To determine how the function $X(s)$ from (4.11) behaves at the point $s = -1$, we re-write the first term and expand it into a geometric series

$$\frac{1}{s+2} = \frac{1}{1 - [-(s+1)]} = \sum_{n=0}^{\infty} [-(s+1)]^n, \quad |s+1| < 1. \tag{4.16}$$

The condition $|s + 1| < 1$ for the convergence of the geometric series is certain to be fulfilled for any small area around $s = -1$. With (4.16) the Laurent expansion of $X(s)$ from (4.11) is yielded:

$$X(s) = \frac{1}{s+1} + \frac{1}{s+2} = \sum_{n=-\infty}^{\infty} a_n(s+1)^n \quad \text{with} \quad a_n = \begin{cases} (-1)^n & n \geq 0 \\ 1 & n = -1 \\ 0 & n < -1 \end{cases}$$

(4.17)

We have found that $a_n = 0$ for $n < -1$, and therefore $X(s)$ has a first-order pole (single pole) at $s = -1$.

In the context of analysing LTI-systems that have a finite number of energy stores, rational fraction Laplace transforms often occur. For a numerator polynomial of order M they can have up to M poles, but no essential singularities.

4.5.3 Properties of the Laplace Transform's Region of Convergence

The properties that we observed in the examples from Section 4.4 can be generalised for more complicated signals. We will list the series of properties that are obtained. They are each concerned with a time signal $x(t)$ and its Laplace transform $X(s)$ (4.1).

1. **The region of convergence is a strip parallel to the imaginary axis in the s-plane.**

   As only the real part of s is responsible for the convergence of the Laplace transform, all points of the s-plane with the same real part have the same convergence properties.

2. **If $x(t)$ is a right-sided signal, then the region of convergence lies to the right of a line through the rightmost singularity.**

   In Example 4.3 we saw that for a right-sided signal with two singularities, the singularity with the greatest real part determines the region of convergence. The same is true for right-sided signals with any number of singularities.

3. **If $x(t)$ is a left-sided signal, the region of convergence lies to the left of a line through the leftmost singularity.**

   Comparing Examples 4.1 and 4.2 shows that for right-sided and left-sided signals, the region of convergence lies to the right or left of the singularities respectively. For multiple singularities, it is again the singularity with the greatest real part that defines the region of convergence.

4. **When $x(t)$ is bilateral – the sum of a right-sided and left-sided signal – the region of convergence is a strip between two singularities, as long as the right-sided and left-sided regions of convergence overlap.**

   The Laplace transform of a bilateral signal can (see (4.13)) be put together from its right-sided and left-sided components. The properties we have just discussed apply to the individual regions of convergence, and the complete region of convergence is their intersection. This intersection is a strip that lies to the left of the rightmost singularity and to the right of the leftmost singularity. The intersection is empty unless all of the singularities in the right-sided component lie to the left of the singularities in the left-sided component.

   In order to fully understand this, we re-examine Examples 4.4 and 4.5. In Example 4.4, the pole of the right-sided component ($s = -2$) lies to the left of the pole of the left-sided component ($s = -1$), and the region of convergence is the strip between these poles. In Example 4.5, the pole of the right-sided component ($s = -1$) lies to the right of the pole of the left-sided component ($s = -2$) and the intersection is empty, so in this case the integral (4.13) does not converge.

5. **The region of convergence does not contain any singularities.**

   Because of the properties we have discussed, it should be clear that the singularities of the Laplace transform either lie to the left of the region of convergence (singularities of the right-sided component) or to the right (singularities of the left-sided component). In the region of convergence itself, there cannot be any singularities.

6. **If $x(t)$ has a finite duration and if $\mathcal{L}\{x(t)\}$ converges for at least one value of s, then the region of convergence is the entire s-plane.**

   If $x(t)$ has finite duration, i.e., when $x(t)$ is only non-zero for $A < t < B$, then we can represent its Laplace transform by only the first integral on the right-hand side, and we do not need to find any limits. All of the considerations concerning the region of convergence are then unnecessary, although $x(t)$ can contain singularities itself, so that an integral over $x(t)$ does not converge and in that case, the Laplace transform for $x(t)$ therefore does not exist.

7. **$\mathcal{L}\{x(t)\}$ can be analysed in the entire region of convergence.**

   Within the region of convergence the derivative of Laplace tranforms can be formed corresponding to the complex frequency (4.14). We will discuss the differentiation theorem in the frequency-domain in Section 4.7.7.

4.6 Existence and Uniqueness of the Laplace Transform

4.6.1 Existence of the Laplace Transform

The previous considerations of convergence of the Laplace transform can be simply summarised: the Laplace transform exists if the region of convergence is not empty. We have already discussed in detail how the region of convergence is defined from the right- and left-sided signal parts. Admittedly, however, finding all singularities is arduous with complicated signals, so we ask the question: is there an alternative to determining the existence of the Laplace transform directly from the time behaviour of $x(t)$?

In order to answer this question we introduce the concept of *exponential order*. If a function is of exponential order, the Laplace integral converges.

Definition 12:   Functions of exponential order

A funtion $x(t)$ is of exponential order for $t \to \infty$, *if, from a defined time point T, it grows at most as fast as an exponential function.*

$$|x(t)| \leq Me^{Ct} \qquad \forall t \geq T . \tag{4.18}$$

A function $x(t)$ is of exponential order for $t \to -\infty$, *if*

$$|x(t)| \leq Me^{Dt} \qquad \forall t \leq -T . \tag{4.19}$$

M, C, D are arbitrary but fixed constants $(M > 0)$.

For functions $x(t)$ that are integrable between any finite limits, we can already make assertions about the existence of their Laplace transforms. If $x(t)$ is replaced by Me^{Ct} and Me^{Dt} in each Laplace transform, we can say from Examples 4.1 and 4.2 in Section 4.4 that:

- if a right-sided function $x(t)$ in accordance with (4.18) is of exponential order for $t \to \infty$, then $\mathcal{L}\{x(t)\}$ exists for $\text{Re}\{s\} > C$,

- if a left-sided function $x(t)$ in accordance with (4.19) is of exponential order for $t \to -\infty$, then $\mathcal{L}\{x(t)\}$ exists for $\text{Re}\{s\} < D$.

These assertions for left-sided and right-sided functions can be put together for bilateral functions:

If for a bilateral function $x(t)$,

1. $|x(t)| \leq Me^{Dt}$ for $t \leq -T$ and

2. $|x(t)| \leq Me^{Ct}$ for $t \geq T$

3. $\int\limits_{-T}^{T} |x(t)|dt < M < \infty$

then $\mathcal{L}\{x(t)\}$ exists for $C < \text{Re}\{s\} < D$.

As the estimates used are sufficient but not necessary conditions for the convergence of integrals, the existence conditions for Laplace transforms derived from them must also be sufficient but not necessary. This means that the region of convergence can be larger than the strip between C and D.

4.6.2 Uniqueness of the Inverse Laplace Transform

With the examples in Section 4.4, we saw that functions of time can be uniquely expressed by their Laplace transforms, and likewise, the Laplace transforms each have a unique function of time, if the region of convergence is correctly taken into consideration. We are interested in knowing whether assigning a function of time to a Laplace transform is unique or does this representation of the function lose information that cannot be recovered by the inverse Laplace transform. The following theorem gives the answer.

If the following conditions for two functions of time are true:

1. $f(t)$ and $g(t)$ are made up of continuous sections

2. $f(t)$ and $g(t)$ are of exponential order $t \rightarrow \infty$ and $t \rightarrow -\infty$

3. the Laplace transforms $F(s) = \mathcal{L}\{f(t)\}$ and $G(s) = \mathcal{L}\{g(t)\}$ exist and have overlapping regions of convergence in the s-plane

4. $F(s) = G(s)$ in the region of convergence

then $f(t) = g(t)$ everywhere where $f(t)$ and $g(t)$ are continuous.

The proof of this theorem can be found [4, Chapter 6]. Instead of examining the details of the proof, we will consider two examples that clarify the conditions of the theorem.

─── **Example 4.7**

The function of time $f(t) = \varepsilon(t)$ and $g(t) = -\varepsilon(-t)$ fulfill both conditions 1 and 2. For the Laplace transform, $F(s) = G(s)$, but with each different region of convergence, so that no overlapping occurs

$$\varepsilon(t) \quad \circ\!\!-\!\!\bullet \quad \frac{1}{s} \quad , \quad \text{Re}\{s\} > 0$$

$$-\varepsilon(-t) \quad \circ\!\!-\!\!\bullet \quad \frac{1}{s} \quad , \quad \text{Re}\{s\} < 0 \quad .$$

This does not fulfill the second part of condition 3, and so we cannot say anything about whether $f(t)$ and $g(t)$ are equal. In fact, $\varepsilon(t) \neq -\varepsilon(-t)$ for all t, even where both functions are continuous. This example emphasises how important the region of convergence is.

■

Example 4.8

Figure 4.7 shows three functions that all fulfill conditions 1 and 2. They are only distinguished by their value at the discontinuity at $t = 1$, which is marked by a black dot. Only the middle function can be expressed by the step function (4.6) $(\varepsilon(t - 1))$. Calculating the Laplace transforms of the three functions shows that conditions 3 and 4 are also fulfilled, and the theorem confirms the observation that all three functions are equal everywhere where they are continuous, that is for $t < 1$ and $t > 1$. The theorem says nothing about the discontinuity at $t = 1$, which would in fact be impossible, as the differences between each two functions for $t < 1$ and $t > 1$ are zero and the integrals over them also have the value zero.

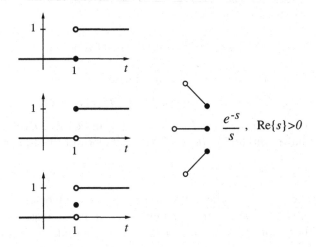

Figure 4.7: Example 2 of the uniqueness of the $\mathcal{L}$ transform

■

The previous example shows that the representation of a function of time by its Laplace transform does actually lose information: the information about function values at discontinuities. Admittedly though, discontinuous functions of time are often only idealised versions of rapidly changing continuous signals. The step function $\varepsilon(t)$ is a good example as it represents an ideal switch operation, for which the assignment of a function value at the time of switching $t = 0$ is arbitrary. A switch at $t = 1$ could be characterised just as well by one of the other signals in Figure 4.7. The loss of information about the amplitude at discontinuities is therefore insignificant for many practical applications. We can also summarise the uniqueness theorem this way:

> **The assignment of a signal to a Laplace transform and vice versa is unique in both directions exclusive of discontinuities.**

For practical problems this property is sufficient.

4.7 Properties of the Laplace Transform

For the simple signals we have considered so far, evaluating the integral (4.1) has been the shortest way to find the Laplace transform. When dealing with complicated signals it is an advantage, however, to be able to use the properties of the Laplace transform, and avoiding having to evaluate the integral directly. Often it is also necessary to perform an operation on a time-domain signal (e.g. differentiation) in the frequency-domain. We will learn the most important properties of the Laplace transform in this section, and formulate theorems for them. The theorems will be needed in later calculations using the Laplace transform.

4.7.1 Linearity of the Laplace Transform

The Laplace transform is linear, so from the Laplace transform of a linear superposition of two functions of time, the Laplace transforms of these functions can be recovered. For any two complex constants a and b:

$$\mathcal{L}\{a \cdot f(t) + b \cdot g(t)\} = a \cdot \mathcal{L}\{f(t)\} + b \cdot \mathcal{L}\{g(t)\} \qquad (4.20)$$

for all values on the complex frequency plane, as long as both $\mathcal{L}\{f(t)\}$ and $\mathcal{L}\{g(t)\}$ exist. The region of convergence for the combined function is the intersection of the regions of convergence for the individual functions. Singularities may be removed by the addition, however, so in general the combined region of convergence is a super set of the intersection between the individual regions of convergence:

$$\text{ROC}\{af + bg\} \supseteq \text{ROC}\{f\} \cap \text{ROC}\{g\} . \qquad (4.21)$$

To show linearity, $a \cdot f(t) + b \cdot g(t)$ is inserted into the definition of the Laplace transform (4.1). From the linearity of the integration, (4.20) is immediately obtained.

Example 4.9

As an example we will evaluate the Laplace transform of $x(t) = \cosh(at) \cdot \varepsilon(t)$. This signal can be split into two exponential functions

$$x(t) = \cosh(at) \cdot \varepsilon(t) = \frac{1}{2}\left(e^{at} + e^{-at}\right) \cdot \varepsilon(t),$$

and now we can use the Laplace transform of each of the exponential functions. With (4.7),

$$X(s) = \frac{1}{2}\left(\frac{1}{s - a} + \frac{1}{s + a}\right) = \frac{s}{s^2 - a^2} \qquad \mathrm{Re}\{s\} > a > 0 \, .$$

The Laplace transform of the cosh function can be traced back without integration to the already known transformation of the exponential function.

4.7.2 Shifting in the Time-Domain and Frequency-Domain

Obtaining the Laplace transform of time signals that have been shifted in time or multiplied by an exponential function is simple. As these two cases are closely related, we will consider them together.

If $X(s) = \mathcal{L}\{x(t)\}$ is the Laplace transform of the time function $x(t)$ and the region of convergence occupies $s \in \mathrm{ROC}\{x\}$, then the *shift theorem*

$$\boxed{\mathcal{L}\{x(t - \tau)\} = e^{-s\tau}X(s)\, , \quad s \in \mathrm{ROC}\{x\}}\tag{4.22}$$

and the *modulation theorem*

$$\boxed{\mathcal{L}\{e^{\alpha t}x(t)\} = X(s - \alpha)\, , \quad s - \mathrm{Re}\{\alpha\} \in \mathrm{ROC}\{x\}\, , \quad \alpha \in \mathbb{C}}\tag{4.23}$$

apply. When multiplied by $e^{\alpha t}$, the region of convergence is displaced by $\mathrm{Re}\{\alpha\}$ to the right. Both theorems can be proven by the substitutions $t = t' - \tau$ and $s = s' - \alpha$ into the definition of the Laplace transform (4.1).

Using the modulation theorem we can also clarify why the region of convergence of the Laplace transform has the form of a vertical strip in the s-plane. A vertical displacement in the s-plane corresponds to a multiplication of the time function by $e^{\alpha t}$ for a purely imaginary value of α. Since these factors have a modulus of one, the muliplication does not change the convergence of the Laplace integral (4.1). The region of convergence must therefore be a vertical strip.

4.7.3   Scaling of the Time Axis or Frequency Plane

An expansion or compression of the time axis occurs, for example, if the unit of measurement of time is changed or if the speed of a magnetic tape changes. From the Laplace transform the original time signal $X(s) = \mathcal{L}\{x(t)\}$, $s \in \text{ROC}\{x\}$ becomes

$$\mathcal{L}\{x(at)\} = \frac{1}{|a|} X\left(\frac{s}{a}\right), \quad a \neq 0 . \tag{4.24}$$

The region of convergence is scaled in the same way, i.e. $s \in \text{ROC}\{x(at)\}$ if $\frac{s}{a} \in \text{ROC}\{x(t)\}$. This relationship is shown through the substitution $t = at'$ into (4.1).

4.7.4   Differentiation and Integration in the Time-Domain

It is necessary, when dealing with differential equations, to know the Laplace transform of the derivation, and the integral of the time function. Construction of this relationship begins with the inverse Laplace transform in (4.2)

$$x(t) = \frac{1}{2\pi j} \int_{\sigma-j\infty}^{\sigma+j\infty} X(s)e^{st}ds .$$

If $x(t)$ can be differentiated, we differentiate by t and obtain

$$\frac{dx(t)}{dt} = \frac{1}{2\pi j} \int_{\sigma-j\infty}^{\sigma+j\infty} X(s)\, s\, e^{st}ds . \tag{4.25}$$

We can take the *differentiation theorem* from this expression:

$$\boxed{\mathcal{L}\left\{\frac{dx(t)}{dt}\right\} = s\,X(s) ; \quad s \in \text{ROC} \supseteq \text{ROC}\{x\} .} \tag{4.26}$$

Through integration of $x(t)$ the *integration theorem* is obtained:

$$\boxed{\mathcal{L}\left\{\int_{-\infty}^{t} x(\tau)d\tau\right\} = \frac{1}{s} X(s), s \in \text{ROC} \supseteq \text{ROC}\{x\} \cap \{s : \text{Re}\{s\} > 0\} .} \tag{4.27}$$

The integration produces a pole at $s = 0$ that has an effect on the region of convergence.

Although the differentiation theorem (4.26) is very simple and easy to remember, in this form it is only valid for functions $x(t)$ that can be differentiated for all t. Many important signal forms, like square wave or saw-tooth signals are excluded by this condition. For this reason we will discuss some extensions of the differentiation rule that can also handle instantaneous signal changes.

Next we consider right-sided signals, that are only non-zero for $t \geq 0$, and that may have a step at $t = 0$. This leads to the differentiation rule of the unilateral Laplace transform, that is important for the solution of the initial conditions problem (Chapter 7). Then we investigate signals that may have any number of steps.

4.7.5 Differentiation Theorem and Integration Theorem for the Unilateral Laplace Transform

Systems that are switched on at a specific time point (e.g. $t = 0$) have an output signal $x(t)$ typically like that shown in Figure 4.8.

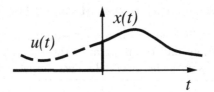

Figure 4.8: Signal with and without step at $t = 0$

Before the turn-on point $(t < 0)$, $x(t)$ is zero and afterwards $(t > 0)$, $x(t)$ is any function that can be differentiated. For $t = 0$, the function can jump instantaneously and so $x(t)$ as a whole cannot be differentiated, and the conditions for the use of the differentiation theorem in the form of (4.26) are not satisfied. In order to give a useable equation, however, we represent the right-sided signal $x(t)$ by a bilateral signal $u(t)$ which can be differentiated, and the step function $\varepsilon(t)$ (see Figure 4.8):

$$x(t) = u(t)\varepsilon(t) . \tag{4.28}$$

In the following construction we use the signal $u(t)$ in place of the signal $x(t)$. This replacement is necessary for unique formulation of the differentiation theorem.

The Laplace transform of $x(t)$ (4.1) is

$$X(s) = \int_{-\infty}^{\infty} x(t)e^{-st}\, dt = \int_{0}^{\infty} u(t)e^{-st}\, dt . \tag{4.29}$$

Partial integration of the second integral expression yields

$$X(s) = -\frac{1}{s}\, u(t)e^{-st}\Big|_{0}^{\infty} + \frac{1}{s}\int_{0}^{\infty} \dot{u}(t)e^{-st}\, dt . \tag{4.30}$$

If $x(t)$ and therefore also $u(t)$ has exponential order for $t \to \infty$, then $s \in \text{ROC}\{x\}$

$$sX(s) = u(0) + \int_0^\infty \dot{u}(t)e^{-st}\, dt \,. \tag{4.31}$$

Unfortunately we cannot simply replace $\dot{u}(t)$ with $\dot{x}(t)$, as the derivative of $x(t)$ at $t = 0$ does not exist. To avoid this difficulty without using complicated mathematics, we introduce the function

$$x°(t) = \begin{cases} \dot{u}(t) & t > 0 \\ 0 & t < 0 \end{cases} = \begin{cases} \dot{x}(t) & t > 0 \\ 0 & t < 0 \end{cases} = \dot{x}(t) \quad \forall\, t \neq 0. \tag{4.32}$$

It is the same as the derivative of $x(t)$ where $x(t)$ can be differentiated ($t \neq 0$) and it is not defined at $t = 0$. This hole in the defined region is not a problem, as we have already seen in Section 4.6.2 that deviances at isolated points do not influence the value of the Laplace transform.

It should also be noted that at $t = 0$ the value $u(0)$ is equal to the right-sided limit of $x(t)$ at $t \to 0$. This value is also known as $x(0+)$:

$$u(0) = \lim_{\delta \to 0} x(0 + \delta) = x(0+), \qquad \delta > 0\,. \tag{4.33}$$

We have now obtained the differentiation theorem of the (bilateral) Laplace transform for right-sided signals:

$$\mathcal{L}\{x°(t)\} = sX(s) - x(0+)\,. \tag{4.34}$$

It expresses the function $x°(t)$ – which like (4.32) represents the derivative of $x(t)$ – by the Laplace transform of $x(t)$ and the $x(0+)$.

For signals that are initially only defined for $t > 0$, the *unilateral* Laplace transform $\mathcal{L}_I$, which integrates only for positive time $0 < t < \infty$, is often used. For unilateral signals it is not distinguishable from the bilateral Laplace transform, and therefore has the same differentiation theorem (4.34). It is often given in the form:

$$\boxed{\mathcal{L}_I\{\dot{x}(t)\} = sX(s) - x(0)} \tag{4.35}$$

where it should be noted that the function $x(t)$ and its derivative is only of interest for $t > 0$ and the value of $x(0)$ should be understood to be the right-hand side limit (4.33).

The integration theorem of the unilateral Laplace transform matches the integration theorem for the unilateral Laplace transform (4.27) as $x(t)$ does not have to be differentiable:

$$\boxed{\mathcal{L}_I\left\{ \int_0^t x(\tau)d\tau \right\} = \frac{1}{s}X(s), \quad s \in \text{ROC} \supseteq \text{ROC}\{x\} \cap \{s : \text{Re}\{s\} > 0\}\,.} \tag{4.36}$$

The unilateral Laplace transform is often used when dealing with initial value problems as its differentiation theorem explicitly takes into consideration the function value at $t = 0$. We will continue using the bilateral Laplace transform, however, and take into account the special properties of unilateral functions where appropriate (see (4.34)).

4.7.6 Differentiation Theorem for Piecewise Continuous Signals

The case of the unilateral function just considered is a special case of a piecewise continuous signal. The term 'piecewise continuous' means that the signal may have steps but inbetween it is differentiable. Figure 4.9 shows an example of such a signal with steps at times t_1, t_2, t_3. The heights of the steps $S(t_i)$ are positive for upward steps and negative for downward steps. We will restrict ourselves here to signals with a finite number of steps.

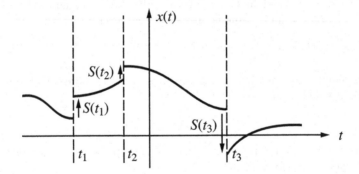

Figure 4.9: Example of a piecewise continuous signal

As in (4.28) we assemble the piecewise continuous signal $x(t)$ from the differentiable functions $u_i(t)$, that each are switched on at the time of the respective step. The heights $S(t_i)$ are equal to the values of the new functions $u_i(t_i)$:

$$x(t) = u_0(t) + \sum_{i=1}^{n} u_i(t)\varepsilon(t - t_i) \quad \text{and} \quad S(t_i) = u_i(t_i) \,. \tag{4.37}$$

Using the Laplace transform and partial integration (4.29, 4.30) for all of the summed terms in (4.37) yields

$$X(s) = \frac{1}{s} \int_{-\infty}^{\infty} \left[\dot{u}_0(t) + \sum_{i=1}^{n} \dot{u}_i(t)\varepsilon(t - t_i) \right] e^{-st}\, dt - \frac{1}{s} \sum_{i=1}^{n} u_i(t)e^{-st} \Big|_{t_i}^{\infty} \,. \tag{4.38}$$

The expression in the square brackets of the first integral corresponds to the derivative of $x(t)$ where the signal is continuous (see (4.32)), that is, excluding the steps:

$$x^\circ(t) = \dot{u}_0(t) + \sum_{i=1}^{n} \dot{u}_i(t)\varepsilon(t - t_i) = \dot{x}(t) \quad t \neq t_i, \quad i = 1, \ldots, n \,. \tag{4.39}$$

With the step heights $S(t_i)$ from (4.37), the Laplace transform of $X(s)$ is finally,

$$X(s) = \frac{1}{s}\mathcal{L}\{x^\circ(t)\} + \frac{1}{s}\sum_{i=1}^{n} S(t_i)e^{-st_i} \,. \tag{4.40}$$

Rearranging yields the differentiation theorem for piecewise continuous signals:

$$\boxed{\mathcal{L}\{x^\circ(t)\} = s\,X(s) - \sum_{i=1}^{n} e^{-st_i}\,S(t_i) \,.} \tag{4.41}$$

It contains the differentiation theorem (4.34) for unilateral functions with a step at $t = 0$ as a special case.

——————————————————————————————— **Example 4.10**

We will now use the differentiation theorem for piecewise continuous signals (4.41) to calculate the Laplace transform of a triangular impulse $x(t)$ shown in Figure 4.10.

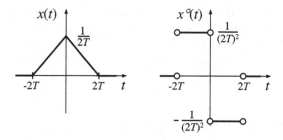

Figure 4.10: A triangular impulse and its derivative

From (4.41) we obtain

$$\mathcal{L}\{x^\circ(t)\} = s\,X(s) \,, \tag{4.42}$$

as the triangle impulse is not differentiable, but without steps. $X(s)$ is the unknown Laplace transform of the triangle signal. The function $x^\circ(t)$, which contains the derivative of the differentiable region of $x(t)$ is here the easiest to determine and is shown in Figure 4.10. Instead of evaluating $\mathcal{L}\{x^\circ(t)\}$ using the Laplace integral (4.1) we use the differentiation theorem (4.41) a second time, and obtain

$$\mathcal{L}\{x^{\circ\circ}(t)\} = s\,\mathcal{L}\{x^\circ(t)\} - \frac{1}{(2T)^2}e^{2sT} + \frac{2}{(2T)^2} - \frac{1}{(2T)^2}e^{-2sT} . \tag{4.43}$$

As $x^{\circ\circ}(t)$ is zero at all points where it is defined, with (4.42) and some rearrangement,

$$0 = s^2\,X(s) - \frac{1}{(2T)^2}\left[e^{sT} - e^{-sT}\right]^2 = s^2\,X(s) - \frac{1}{T^2}\left[\sinh(sT)\right]^2 . \tag{4.44}$$

The Laplace transform of the triangle signal is therefore

$$X(s) = \left[\frac{\sinh(sT)}{sT}\right]^2 . \tag{4.45}$$

Using the differentiation theorem for piecewise continuous signals (4.41), we avoided having to do any integration.

■

4.7.7   Differentiation in the Frequency-Domain

The relationships for the Laplace transforms of differentiated functions of time that we have so far dealt with in detail, will be used in Chapter 6 to analyse LTI-systems in the frequency-domain. In the same way, relationships for the derivatives of Laplace transforms can be found. They are mainly useful for deriving transform pairs for frequently occurring functions.

As a Laplace transform is analytic in its region of convergence, it can be differentiated any number of times. As in (4.25), we obtain with complex differentiation of $X(s)$ from (4.1),

$$\frac{dX(s)}{ds} = \int\limits_{-\infty}^{\infty} x(t)\,(-t)\,e^{-st}dt \tag{4.46}$$

and we can then obtain a theorem for the Laplace transform of functions of time $x(t)$ that have been multiplied by the time variable t:

$$\boxed{\mathcal{L}\{t\,x(t)\} = -\frac{dX(s)}{ds} \; ; \quad s \in \mathrm{ROC}\{x\} .} \tag{4.47}$$

——————————————————————————————————— Example 4.11

As an example we will calculate the Laplace transform of $x(t) = te^{-at} \cdot \varepsilon(t)$. We will be using the transform pair

$$\mathcal{L}\left\{e^{-at}\varepsilon(t)\right\} = \frac{1}{s+a}, \quad \mathrm{Re}\{s\} > -a .$$

derived in Example 4.1. With (4.47) we obtain

$$\mathcal{L}\left\{te^{-at}\varepsilon(t)\right\} = -\frac{d}{ds}\left[\frac{1}{s+a}\right] = \frac{1}{(s+a)^2} . \qquad (4.48)$$

Repeated use yields

$$\mathcal{L}\left\{t^n e^{-at}\varepsilon(t)\right\} = \frac{(n)!}{(s+a)^{n+1}} . \qquad (4.49)$$

■

4.7.8 Table of the Most Important Laplace Transforms

When dealing with simple LTI-systems it is sufficient just to know some of the more frequently occurring transform pairs. The most important are summarised in Table 4.1. All of the pairs listed (and many more) can be derived from the relationships introduced in the previous sections. More comprehensive tables can be found in Appendix Appendix B.1, the relevant textbooks or in special collections of Laplace transforms [14]. When using such tables always take note of whether the unilateral or bilateral Laplace transform was specified. Most tables in textbooks give the unilateral Laplace transforms, but they can be used to find bilateral Laplace transforms if the function of time is completed by multiplying it with the step function $\varepsilon(t)$ (see Section 4.3).

4.8 Exercises

Exercise 4.1

Calculate the Laplace transforms of the following signals using (4.1), as long as they exist. Determine the region of convergence for each, when you determine a condition for s for which the improper integral converges.

a) $x(t) = \sin(t)\,\varepsilon(t)$

b) $x(t) = \sin(t)$

Table 4.1: Some important transform pairs for the Laplace transform

$x(t)$	$X(s)$	ROC
$\varepsilon(t)$	$\dfrac{1}{s}$	$\mathrm{Re}\{s\} > 0$
$e^{-at}\varepsilon(t)$	$\dfrac{1}{s+a}$	$\mathrm{Re}\{s\} > \mathrm{Re}\{-a\}$
$-e^{-at}\varepsilon(-t)$	$\dfrac{1}{s+a}$	$\mathrm{Re}\{s\} < \mathrm{Re}\{-a\}$
$te^{-at}\varepsilon(t)$	$\dfrac{1}{(s+a)^2}$	$\mathrm{Re}\{s\} > \mathrm{Re}\{-a\}$
$t^n e^{-at}\varepsilon(t)$	$\dfrac{n!}{(s+a)^{n+1}}$	$\mathrm{Re}\{s\} > \mathrm{Re}\{-a\}$
$(\sin \omega_0 t)\varepsilon(t)$	$\dfrac{\omega_0}{s^2 + \omega_0{}^2}$	$\mathrm{Re}\{s\} > 0$
$(\cos \omega_0 t)\varepsilon(t)$	$\dfrac{s}{s^2 + \omega_0{}^2}$	$\mathrm{Re}\{s\} > 0$

c) $x(t) = e^{2t}\,\varepsilon(t - T)$

d) $x(t) = t\,e^{2t}\,\varepsilon(t)$

e) $x(t) = \sinh(2t)\,\varepsilon(-t)$

Check the regions of convergence with the properties given in Section 4.5.3

Exercise 4.2

Which of the following functions are of exponential order for $t \to \infty$?
     a) t^2 b) $t^5 + 20t^2 + 7$ c) e^{5t} d) $t^2\,e^{5t}$ e) e^{t^2} f) $\sin(t)$.

Exercise 4.3

For which of the following functions can the bilateral Laplace transform exist?
Check whether the functions are of exponential order for $t \to \infty$ and $t \to -\infty$.
     a) $\sin(t)$ b) $\sin(t)\,\varepsilon(t)$ c) t^2 d) $5\,e^{2t}$

Exercise 4.4

Prove the linearity of the Laplace transform (4.20) using (4.1).

Exercise 4.5

$F(s) = \dfrac{2s + 3}{s^2 + 3s + 2}$ and $G(s) = \dfrac{3s + 1}{s^2 + 4s + 3}$ are the Laplace transforms of two right-sided signals.

a) Find the poles of $F(s)$ and give the region of convergence.

b) Find the poles of $G(s)$ and give the region of convergence.

c) Find the poles and zeros of $F(s) + G(s)$ and give the region of convergence.

Exercise 4.6

Prove

a) the shift theorem (4.22)

b) the modulation theorem (4.23),

by substituting (4.1) $t = t' - \tau$ and $s = s' - \alpha$.

Exercise 4.7

Prove equation (4.24) (time and frequency scaling) by substituting $t = at'$ into (4.1).

Exercise 4.8

Derive the transform pairs in Table 4.1 starting with the transform pair $\varepsilon(t) \circ\!\!-\!\!\bullet\, X(s) = \dfrac{1}{s}$, $\text{Re}\{s\} > 0$ using the theorems from Section 4.7.

5 Complex Analysis and the Inverse Laplace Transform

In Chapter 4 we started with time functions and through evaluation of the integral definition in (4.1), verified the Laplace transform. For the reverse, the so-called *inverse Laplace transform* we gave the equation (4.2) without giving a justification. In this chapter we will look at the inverse transform in depth with the help of complex function theory. We restrict ourselves to Laplace transforms that only have poles, and have no significant singularities. It will be shown that for these cases, carrying out the inverse Laplace transform leads to simple calculations that can be performed without having to continually refer back to function theory. The first step is to commit the important results from complex analysis to memory.

5.1   Path Integrals in the Complex Plane

We consider the complex function

$$Q(s) = Q_r(s) + jQ_i(s) \tag{5.1}$$

of the complex variable s. The real and imaginary parts of $Q(s)$ are denoted by $Q_r(s)$ and $Q_i(s)$.

In order to describe the integration of such a function it is not sufficient to give two limits of integration. As s may take any value in the complex plane, all of the values on the path between the start and end points must be given. Such an integral is written

$$I = \int_W Q(s)\, ds\,, \tag{5.2}$$

where W is the path of integration in the complex plane. To define this path precisely, the values of s that lie on the path are given as a function of the real parameters ν. This is called a *parametric curve*. Figure 5.1 shows such a parametric path of integration in the s-plane. For $\nu = \nu_A$, $s(\nu)$ takes the complex value of the start point A, likewise for $\nu = \nu_B$ $s(\nu)$ takes the value of the endpoint B. For $\nu_A < \nu < \nu_B$, $s(\nu)$ follows the desired path of integration W. The integral (5.2) means that the value of $Q(s)ds$ is accumulated from all of the infinitesimal elements ds on the length of the path W.

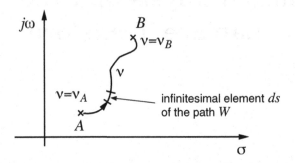

$$W = \{s : s(\nu) = \sigma(\nu) + j\omega(\nu) \wedge \nu_A \le \nu \le \nu_B\}$$

Figure 5.1: Path of integration in the s-plane

For evaluation of the path integral not only is $Q(s)$ represented by real and imaginary parts as in (5.1), but also $s(\nu)$.

$$s(\nu) = \sigma(\nu) + j\omega(\nu) \, . \tag{5.3}$$

Inserting (5.3) into equation (5.2), and using the substitution $s = s(\nu)$ gives (after multiplying out and collecting terms) two real integrals with respect to the real parameter ν:

$$\int_W Q(s)\, ds \;=\; \int_{\nu_A}^{\nu_B} Q(s)\frac{ds}{d\nu}\,d\nu \tag{5.4}$$

$$= \int_{\nu_A}^{\nu_B}\left[Q_r(s)\frac{d\sigma}{d\nu} - Q_i(s)\frac{d\omega}{d\nu}\right]d\nu + j\int_{\nu_A}^{\nu_B}\left[Q_i(s)\frac{d\sigma}{d\nu} + Q_r(s)\frac{d\omega}{d\nu}\right]d\nu\,.$$

The path integral in the complex plane (5.2) is actually only a compact form of the lengthy expression in (5.4). As (5.2) can be viewed as a sum of many real value integrals, the known rule for real integrals is also valid for the path integral in the complex plane.

5.2   The Main Principle of Complex Analysis

If $Q(s)$ is analytic (i.e. free of singularities) in a region G then the integral

$$\int_A^B Q(s)ds = \int_{W_1} Q(s)ds = \int_{W_2} Q(s)ds \tag{5.5}$$

is independent from the path, as long as both paths W_1 and W_2 run completely within G and there are no singularities between them. Figure 5.2 shows two paths of integration that produce the same value when integrated. An important result is

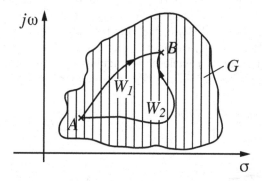

Figure 5.2: Independent path integrals W_1 and W_2

obtained if the integration stretches from A to B along the path W_1 and then along the path W_2 against the direction of arrow back to A. Such an integral around a closed path is called a circular integral and is denoted by an integral sign with a circle. If the integration along W_1 and W_2 produces the same value, the circular integral along W_1 in the direction of the arrow and W_2 against the direction of the arrow must give the value zero. This statement is valid for any paths W_1 and W_2 as long as they are inside G. The main principle of complex analysis follows directly from this: every circular integral inside the region G disappears if the path of integration does not include any singularities:

$$\oint Q(s)\, ds = 0 \,. \tag{5.6}$$

5.3 Circular Integrals that Enclose Singularities

The previous statement refers to paths of integration that do not enclose singularities. For the characterisation of complex functions, however, the singularities are important. We therefore also need relationships for circular integrals that contain singularities.

Next we clarify whether the value of a circular integral containing a singularity depends on the route of the path of integration. We consider Figure 5.3 which shows two different closed paths of integration W_1 and W_2 within the analytic region G. Both enclose a singularity that lies within the white region in the middle which does not belong to G. For the calculation of the difference between the two circular integrals along W_1 and W_2 we can refer back to the calculation of two circular integrals without a singularity. In place of the calculation of the integrals along W_1 and W_2 we go from W_1 to W_2, and then back again along W_1. The new circular integral along W_α does not contain the singularity. The same procedure is repeated for the untraversed parts of W_1 and W_2 and the path of integration W_β is obtained. If the transition from W_1 to W_2 and vice versa, is chosen so that the contributions to W_α and W_β cancel due to the opposite orientation, then the integration along W_α and W_β equals the difference of the integrals along W_1 and W_2. As neither W_α nor W_β encloses a singularity, both integrals are zero and therefore the integrals along W_1 and W_2 have equal value.

$$\oint_{W_1} Q(s)\,ds - \oint_{W_2} Q(s)\,ds = \oint_{W_\alpha} Q(s)\,ds + \oint_{W_\beta} Q(s)\,ds = 0. \qquad (5.7)$$

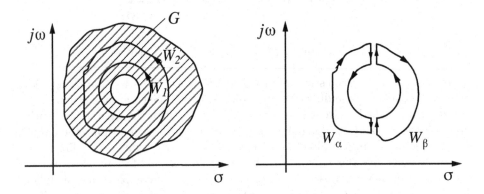

Figure 5.3: Equal value integrals around a singularity

This argument can be used for any two ring integrals around a singularity in G. The value of the integral does not depend on the form of the path containing the singular inner region.

If a closed path of integration W contains multiple singularities (see Figure 5.4) we can show in the same way that the circular integral along W is the same as the sum of the circular integrals around the individual singularities (W_1, W_2 and W_3 in Figure 5.4).

In a multiset connected analytic G, the value of a circular integral only depends on which singularities are enclosed by the path of integration

$$\oint_W Q(s)\,ds = \sum_{\mu=1}^{N} \oint_{W_\mu} Q(s)\,ds. \qquad (5.8)$$

The value of the circular integral around an individual singularity, independent from the path of integration W_μ, can also be expressed by the corresponding *residuum:*

$$R_\mu = \frac{1}{2\pi j} \oint_{W_\mu} Q(s)\, ds\,. \tag{5.9}$$

The equation

$$\oint_W Q(s)\, ds = 2\pi j \sum_{\mu=1}^{N} R_\mu \tag{5.10}$$

is called the *residue theorem*. We will show how to calculate residues in the next section.

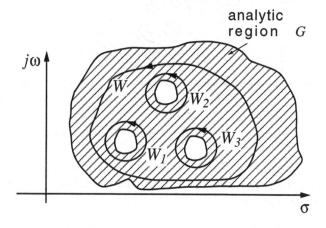

Figure 5.4: Regular area with multiple singularities

5.4 Cauchy Integrals

Now that integration around multiple singularities has been reduced to calculating circular integrals around individual singularities, it still remains to be determined what value an integral around an individual singularity has. That is equivalent to calculating the residues in (5.9). To this end we consider a singularly connected analytic region G of a complex function $F(s)$ according to section 5.2

$$\oint_W F(s)\, ds = 0 \quad \forall\, W \subset G\,. \tag{5.11}$$

The function

$$Q(s) = \frac{F(s)}{s - s_0}\,, \quad s_0 \in G \tag{5.12}$$

is then formed, which has a simple pole at $s = s_0$. For the circular integral of $Q(s)$ around the pole s_0 (see Fig. 5.5), the *Cauchy integral* is

$$\oint_W Q(s)\,ds = \oint_W \frac{F(s)}{s - s_0}\,ds = 2\pi j\, F(s_0). \tag{5.13}$$

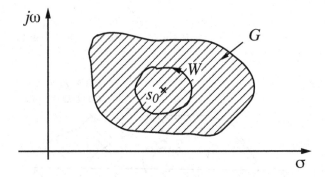

Figure 5.5: Singularity s_0 in the regular area G

The Cauchy integral can easily be proven by integration of an infinitesimal circle around s_0, that is parametricised, for example, by $s(\nu) = \delta e^{j2\pi\nu} + s_0$. For any small radius δ, along this path: $F(s) = F(s_0)$.

5.4.1   Residue Calculation

With the Cauchy integral we can easily give the values of the residues (5.9). For a simple pole of $Q(s)$ at s_μ,

$$R_\mu = \frac{1}{2\pi j} \oint_{W_\mu} Q(s)\,ds = F(s_\mu). \tag{5.14}$$

Although it is not necessary, the residues can be found using complex integration. If only $Q(s)$ is known, the residuum R_μ at s_μ (5.12) can also be found using the limit:

$$R_\mu = \lim_{s \to s_\mu} [Q(s)(s - s_\mu)]. \tag{5.15}$$

The calculation of a circular integral for a complex function with many simple poles can be done using the residue theorem (5.10) and (5.15). This technique is particularly appropriate when $Q(s)$ is a rational fraction function.

Calculation of the residues has so far only been shown for simple poles. To calculate multiple pole residues, we need some further results from the Cauchy integral. We now write it in the form

$$F(s) = \frac{1}{2\pi j} \oint_W \frac{F(w)}{w - s} \, dw \, . \tag{5.16}$$

It can be shown that regularity (i.e. being analytic) of the complex derivatives of $F(s)$ entails regularity of $F(s)$ itself. Then

$$F'(s) \;=\; \frac{dF(s)}{ds} = \frac{d}{ds} \frac{1}{2\pi j} \oint_W \frac{F(w)}{w - s} \, dw = \frac{1}{2\pi j} \oint_W \frac{d}{ds} \frac{F(w)}{w - s} \, dw =$$

$$= \frac{1}{2\pi j} \oint_W \frac{F(w)}{(w - s)^2} \, dw \, . \tag{5.17}$$

Interchanging integration and derivation with respect to s is permitted as the integral converges uniformly. Deriving $(n-1)$-times with respect to s gives

$$F^{(n-1)}(s) = \frac{d^{(n-1)}}{ds^{(n-1)}} F(s) = \frac{(n-1)!}{2\pi j} \oint_W \frac{F(w)}{(w - s)^n} \, ds \, . \tag{5.18}$$

With the function

$$Q_n(s) = \frac{F(s)}{(s - s_0)^n} \tag{5.19}$$

and the correspondingly altered notation in (5.18), we obtain a formula for multiple poles that is equivalent to (5.13)

$$\oint_W Q_n(s) \, ds = \oint_W \frac{F(s)}{(s - s_0)^n} \, ds = 2\pi j \frac{1}{(n-1)!} F^{(n-1)}(s_0) \, . \tag{5.20}$$

The residuum at the location of the n-order pole s_μ of $Q_n(s)$

$$R_\mu = \frac{1}{2\pi j} \oint_{W_\mu} Q_n(s) \, ds = \frac{1}{(n-1)!} F^{(n-1)}(s_\mu) \tag{5.21}$$

is again obtained without complex integration, from (5.19)

$$R_\mu = \frac{1}{(n-1)!} \lim_{s \to s_\mu} \left[\frac{d^{n-1}}{ds^{n-1}} [Q_n(s)(s - s_\mu)^n] \right] \, . \tag{5.22}$$

5.4.2 Integration Parallel to the Imaginary Axis

The last section showed that the Cauchy integral significantly simplified the evaluation of circular integrals around single and multiple poles, using the residue theorem. In this section we will discuss another simplification that leads to a form of the inverse Laplace transform. It can be used not only for poles, but also for essential singularities, but first we have to make some restrictions. We consider an analytic function $F(s)$ within a region, that decays by at least $1/|s|$ for sufficiently large values of $|s|$

$$|F(s)| < \frac{M}{|s|} \,. \tag{5.23}$$

M is any positive real number. Then

$$\left| \frac{F(s)}{s - s_0} \right| < \frac{\tilde{M}}{|s|^2} \tag{5.24}$$

with the positive real number $\tilde{M}$. To use this decrease for large values of $|s|$ we choose the circular integral around s_0 as in Fig. 5.6. The analytic region contains all values of s where $\mathrm{Re}\{s\} > \sigma_{\min}$. The path of integration consists of the path W_p, which runs parallel to the imaginary axis at a distance $\sigma_p > \sigma_{\min}$, and the path W_r, which is an arc with radius R around the origin

$$2\pi j F(s_0) = \oint_{W_p + W_r} \frac{F(s)}{s - s_0} \, ds = \int_{W_p} \frac{F(s)}{s - s_0} \, ds + \int_{W_r} \frac{F(s)}{s - s_0} \, ds \,. \tag{5.25}$$

We estimate the integral over W_r with (5.24)

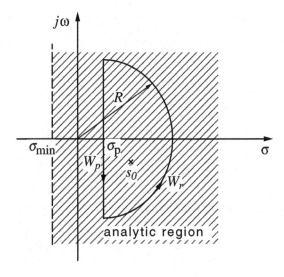

Figure 5.6: Path of integration with curve segment W_p parallel to the imaginary axis

$$\left| \int_{W_r} \frac{F(s)}{s - s_0} \, ds \right| \leq \int_{W_r} \left| \frac{F(s)}{s - s_0} \right| |ds| \leq \tilde{M} \int_{W_r} \frac{1}{|s|^2} |ds| \,. \tag{5.26}$$

The last integral can be parametricised with $s(\nu) = \mathrm{Re}^{j\nu}$, and we then obtain

$$\tilde{M} \int\limits_{W_r} \frac{1}{|s|^2} |ds| = \tilde{M} \int\limits_{-\pi/2}^{\pi/2} \frac{1}{|s(\nu)|^2} \left| \frac{ds(\nu)}{d\nu} \right| d\nu = \tilde{M} \int\limits_{-\pi/2}^{\pi/2} \frac{1}{R^2} \left| j\mathrm{Re}^{j\nu} \right| d\nu = \frac{\pi\tilde{M}}{R} .$$
(5.27)

For $R \to \infty$ the value of this integral approaches zero, so that in (5.25), only the integral via the path W_p remains:

$$F(s_0) = \frac{1}{2\pi j} \int\limits_{\sigma_p+j\infty}^{\sigma_p-j\infty} \frac{F(s)}{s - s_0} ds = \frac{1}{2\pi j} \int\limits_{\sigma_p-j\infty}^{\sigma_p+j\infty} \frac{F(s)}{s_0 - s} ds \quad \text{with} \quad \mathrm{Re}\{s_0\} > \sigma_p .$$
(5.28)

Thus under the condition (5.23) the circular integral is calculated by integration parallel to the imaginary axis.

5.4.3   Importance of the Cauchy Integral

If an analytic function $F(s)$ is known along a closed path W, and $F(s)$ is analytic everywhere in the enclosed region (Fig. 5.5), the Cauchy integral (5.16) can be used to calculate $F(s)$ everywhere in the enclosed region. The pole in (5.16) is moved to a location of interest. It is therefore completely sufficient to know $F(s)$ on the border of an analytic region. It is likewise sufficient to know $F(s)$ along a line parallel to the imaginary axis (5.28), to be able to calculate every value to the right of the line. Because it is differentiable, an analytic function has a strong inner structure. In fact, $F(s)$ can even be analytically continued outside of a closed path W, and also when $F(s)$ is only known along a section of the path.

Now that we have found the Cauchy integral, the review of complex analysis is concluded. We will now use the result to derive the formula for the inverse Laplace transform and give simple ways to perform it. In this respect, the Cauchy integral is important for two reasons:

- The simplification of the path of integration considered in section 5.4.2 leads directly to the formula for the inverse Laplace transform.

- The residue theorem (section 5.4.1) allows a simplified calculation of the inverse transform (without complex integration) for systems with single or multiple poles.

5.5   Inverse Laplace Transform

To derive the inverse Laplace transform we start with the Cauchy integral in the form of (5.28). The condition (5.23) is always fulfilled by a rational fraction

Laplace transform, if the numerator is of a higher degree than the denominator. Otherwise the inverse transform leads to distributions in the time-domain that will be introduced in Chapter 8. We will begin with the derivation for the unilateral Laplace transform and then extend the result to the bilateral Laplace transform.

5.5.1   Inverse Unilateral Laplace Transform

To derive the inverse unilateral Laplace transform we use the Cauchy integral for a path of integration parallel to the imaginary axis (5.28), where we rename s as s' and s_0 as s

$$F(s) = \frac{1}{2\pi j} \int_{\sigma-j\infty}^{\sigma+j\infty} \frac{F(s')}{s-s'}\, ds' \quad \text{with } \mathrm{Re}\{s\} > \sigma. \tag{5.29}$$

We continue using the Laplace transform of a unilateral exponential function (compare Example 4.1)

$$\mathcal{L}\{e^{s't}\,\varepsilon(t)\} = \mathcal{L}_I\{e^{s't}\} = \int_0^\infty e^{s't}e^{-st}\, dt = \frac{1}{s-s'}, \quad \mathrm{Re}\{s\} > \mathrm{Re}\{s'\}. \tag{5.30}$$

Putting (5.30) into (5.29) yields

$$\begin{aligned}
F(s) &= \frac{1}{2\pi j} \int_{\sigma-j\infty}^{\sigma+j\infty} F(s') \left[\int_0^\infty e^{s't}e^{-st}dt \right] ds' \\
&= \int_0^\infty \underbrace{\left[\frac{1}{2\pi j} \int_{\sigma-j\infty}^{\sigma+j\infty} F(s')e^{s't}\, ds' \right]}_{f(t)} e^{-st}\, dt
\end{aligned} \tag{5.31}$$

by swapping the integrals under the condition that the convergence is of the same type. Comparison with the unilateral Laplace transform in (4.5)

$$F(s) = \mathcal{L}_I\{f(t)\} = \int_0^\infty [f(t)]e^{-st}dt$$

shows that $F(s)$ is the Laplace transform of the time function inside the square brackets in (5.31). Consequently, the bracketed expression represents the inverse unilateral Laplace transform:

$$f(t) = \frac{1}{2\pi j} \int_{\sigma-j\infty}^{\sigma+j\infty} F(s)e^{st}ds. \tag{5.32}$$

5.5.2   Inverse Bilateral Laplace Transform

The inverse bilateral Laplace transform can be formed by combining the inverse transform already obtained for right-sided signals and the corresponding inverse transform for left-sided signals (see Example 4.2). The derivation for left-sided signals is carried out as in (5.29) to (5.32), so that we can avoid repeating steps. The two steps are distinguished from each other by the direction of integration of the Cauchy integral, and the sign of the left-sided function.

As the region of convergence for left-sided time functions lies left of a vertical line in the s-plane (see Fig. 4.4), the circular integral must also be contained within the left half of the s-plane. Since, by definition, the positive orientation of complex contour integrals is counter-clockwise, the parallel to the imaginary axis has to be traversed in the opposite direction to (5.29). This leads to a change of the sign compared to (5.29) for the Laplace transform of left-sided exponential functions

$$\mathcal{L}\{-\varepsilon(-t)\, e^{s't}\} = -\int\limits_{-\infty}^{0} e^{s't} e^{-st}\, dt = \frac{1}{s - s'}\,. \tag{5.33}$$

This is the same in Examples 4.1 and 4.2, where a left-sided and a right-sided function of time have the same form of Laplace transform. Note that here there is a minus sign in front of the left-sided exponential function.

Putting (5.33) into the Cauchy integral for left-sided signals compensates for both changes of sign and we likewise obtain for the inverse transform equation (5.32). This shows that the inverse of right-sided and bilateral Laplace transforms have the same form. The region of convergence should be considered, however, for the bilateral Laplace transform .

A bilateral signal can be split into left-sided and right-sided components, so the inverse bilateral Laplace transform can be expressed as

$$f(t) = \mathcal{L}^{-1}\{F(s)\} = \frac{1}{2\pi j}\int\limits_{\sigma-j\infty}^{\sigma+j\infty} F(s) e^{st}\, ds\,. \tag{5.34}$$

5.5.3   Path of Integration for the Inverse Laplace Transform

The choice of the path of integration W_p in Fig. 5.6 is appropriate because it can be easily parametricised, but alternative choices would also be possible. Also, every other path of integration within the region of convergence of $F(s)$ gives the same result, since $F(s)e^{st}$ is analytic. Fig. 5.7 shows several permitted paths of integration. The multiple possiblities are only useful when the complex integration is actually carried out by parametrisation. In most cases it is simpler to use the residue theorem.

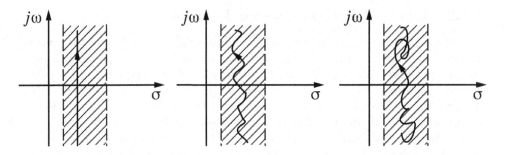

Figure 5.7: Various paths of integration for an inverse Laplace transform

5.5.4   Calculating the Inverse Laplace Transform with the Residue Theorem

Calculating the inverse transform is simplest with the residue theorem. It reduces the calculation effort to partial fraction expansion, which will be discussed later.

On first glance, it seems impossible to carry out the inverse Laplace transform with the residue theorem. To use the residue theorem, a path of integration must be chosen (see Section 5.3) for which the function is analytic and the singularities are enclosed. On the other hand, we know from Chapter 4.5.3, that the Laplace transform is only defined within its region of convergence, which is a strip on the complex plane that contains no singularities. A path of integration that encloses all singularities must therefore run outside the region of convergence, where the Laplace integral diverges.

To solve this conflict, we use the method of *analytic continuation*, which is related to the discussion in Section 5.4.3. We will explain this method with Example 4.1, where we will calculate the Laplace transform of a unilateral exponential function $f(t) = e^{-at}\varepsilon(t)$. The result is (4.7)

$$F(s) = \mathcal{L}\{f(t)\} = \mathcal{L}\{\varepsilon(t)e^{-at}\} = \frac{1}{s+a} \; ; \quad \mathrm{Re}\{s\} > -a \; .$$

Outside of the region of convergence (for $\mathrm{Re}\{s\} \leq -a$) the Laplace transform does not exist. The function

$$A(s) = \frac{1}{s+a} \tag{5.35}$$

is analytic in the entire complex plane, except for the points $s = -a$. Within the region of convergence of $F(s)$,

$$F(s) = A(s) \; ; \quad \mathrm{Re}\{s\} > -a \; . \tag{5.36}$$

In contrast to $F(s)$, it is possible to choose a closed path of integration with a field of regularity that encloses the pole $s = -a$ (compare Fig. 5.4). The value of this integral can be calculated using the residue theorem. $A(s)$ is known as the *analytic continuation* of $F(s)$. In the same way, every other rational fraction Laplace transform can also be assigned an analytic continuation. The difficulties encountered when using the residue theorem have now been removed, as we can now calculate the residues using analytic continuation. In terms of the actual calculation, the switch to analytic continuation makes no difference as it has the same form as the Laplace transform.

We will limit ourselves to right-sided functions at first, for the sake of simplicity. These functions have, of course, a Laplace transform with a right-sided region of convergence, so all the singularities lie left of the path of integration for the inverse Laplace transform (5.34). By combining (5.34) with the residue theorem (5.10), we obtain

$$f(t) = \mathcal{L}^{-1}\{F(s)\} = \frac{1}{2\pi j} \int\limits_{\sigma-j\infty}^{\sigma+j\infty} F(s)e^{st}\,ds = \sum_{\mu=1}^{N} R_\mu \,. \tag{5.37}$$

The path of integration is completed from $s = \sigma - j\infty$ to $s = \sigma + j\infty$ by a very large (approaching infinitely large) arc through the left half of the s-plane. If $F(s)e^{st}$ decays more quickly than $\frac{1}{R}$ with growing radius R, the contribution of the arc to the integral disappears, so its value is not changed by closing the path of integration. For a closed path of integration, however, the residue theorem (5.10) can be used, and (5.37) can be immediately obtained. For a rational fraction $F(s)$, for which the denominator is greater than the numerator, (5.37) is valid for $t > 0$. If $F(s)$ contains exponential terms, for example, the disappearance of the arc's contribution to the integral dependent on t must be investigated separately.

For bilateral functions of time we have to complete the path of integration with a second arc through the right half of the s-plane, which then encloses the singularities to the right of the region of convergence. The sum of the residues of $F(s)e^{st}$ in accordance with (5.37) then yields for a rational fraction $F(s)$, the time function $f(t)$ for $t < 0$, because the contribution of the arc to the integral for $\mathrm{Re}\{s\} > 0$ only disappears when $t < 0$. Using the observations in Section 5.4.1, the residues can be obtained by integrating the analytic continuation of $F(s)e^{st}$ on a path around all poles. For simple poles the residues R_μ of $F(s)e^{st}$ (5.15) are

$$R_\mu = \lim_{s \to s_\mu} [F(s)e^{st}(s - s_\mu)] = P_\mu e^{s_\mu t} \,. \tag{5.38}$$

These can also be expressed by the residues P_μ of $F(s)$

$$P_\mu = \lim_{s \to s_\mu} [F(s)(s - s_\mu)] \tag{5.39}$$

as e^{st} is analytic in the entire complex plane. Calculation of the inverse Laplace transform for a rational fraction $F(s)$ with simple poles is then reduced to

$$f(t) = \mathcal{L}^{-1}\{F(s)\} = \sum_{\mu=1}^{N^+} P_\mu^+ e^{s_\mu t}\varepsilon(t) + \sum_{\nu=1}^{N^-} P_\nu^- e^{s_\nu t}\varepsilon(-t) . \tag{5.40}$$

Here, P_μ^+ are the residues (5.39) of $F(s)$ for poles left of the region of convergence and P_ν^- are the residues for poles right of the region of convergence. The summation covers all N^+ to N^- poles.

The same result is obtained for a rational fraction function $F(s)$, when we write $F(s)$ as a partial fraction expansion

$$F(s) = \sum_{\mu=1}^{N} \frac{P_\mu}{s - s_\mu} . \tag{5.41}$$

Piece-by-piece inverse transforming, while bearing in mind the location of the poles relative to the region of convergence leads exactly to (5.40). Of course, this is not a coincidence because (5.39) is the formula for calculating partial fraction coefficients for simple poles. There are also other methods available for calculating partial fraction coefficients, for example, equating coefficients.

For multiple poles, the residues R_μ of $F(s)e^{st}$ are obtained from (5.22)

$$R_\mu = \lim_{s \to s_\mu} \frac{1}{(n-1)!} \frac{d^{(n-1)}}{ds^{(n-1)}} \left[F(s)e^{st}(s - s_\mu)^n \right] . \tag{5.42}$$

The use of this formula is demonstrated in an example.

── **Example 5.1**

The Laplace transform

$$F(s) = \frac{1}{(s+1)(s+2)^2} , \qquad \mathrm{Re}\{s\} > -1 \tag{5.43}$$

has a simple pole at $s_1 = -1$ and a double pole at $s_2 = -2$. The inverse transform requires that the residues are calculated at both these points:

$$f(t) = \mathcal{L}^{-1}\{F(s)\} = R_1 + R_2 \tag{5.44}$$

with

$$R_1 = \left. F(s)e^{st}(s+1) \right|_{s=-1} = \left. \frac{e^{st}}{(s+2)^2} \right|_{s=-1} = e^{-t} \tag{5.45}$$

$$R_2 = \left. \frac{1}{(2-1)!} \frac{d}{ds} \left[F(s)e^{st}(s+2)^2 \right] \right|_{s=-2} = \left. \frac{d}{ds} \left[\frac{e^{st}}{(s+1)} \right] \right|_{s=-2}$$

$$= \left. \left[-\frac{e^{st}}{(s+1)^2} + \frac{te^{st}}{(s+1)} \right] \right|_{s=-2} = -e^{-2t} - te^{-2t} . \tag{5.46}$$

From the two residues the function of time can now be put together. As $F(s)$ is in rational fraction form and converges in the right half-plane, the inverse transform leads to a right-sided function of time:

$$f(t) = \mathcal{L}^{-1}\{F(s)\} = \left[e^{-t} - e^{-2t} - te^{-2t} \right] \varepsilon(t) . \tag{5.47}$$

∎

A relationship to the partial fraction expansion can be constructed here too for multiple poles. An example of this is given in the next section.

5.5.5 Practical Calculation of the Inverse Laplace Transform

Using the residue theorem avoids evaluation of complex integrals, but for complex poles, it requires calculation of complex residues. For rational fraction Laplace transforms with real numerator and denominator coefficients, the corresponding function of time is likewise real, even when complex poles appear (here in complex conjugate form). It would therefore be convenient if the inverse transform could be carried out purely with real calculations. This can be done using an appropriate combination of partial fraction expansions and the modulation theorem. We will show the procedure in an example.

── **Example 5.2**

We want to find the inverse Laplace transform of

$$F(s) = \frac{222}{(s+3)(s^2+4s+40)} , \qquad \mathrm{Re}\{s\} > 0 . \tag{5.48}$$

We can see from the region of convergence that we are dealing with a right-sided function of time. Starting with the partial fraction expansion

$$F(s) = \frac{222}{(s+3)(s^2+4s+40)} = \frac{A}{s+3} + \frac{Bs+C}{s^2+4s+40} \tag{5.49}$$

and with (5.15), we obtain

$$A = \lim_{s \to -3} [F(s)(s+3)] = \frac{222}{s^2+4s+40}\bigg|_{s=-3} = \frac{222}{9-12+40} = \frac{222}{37} = 6 . \tag{5.50}$$

As the other poles of $F(s)$ are complex, we determine B and C by equating coefficients, to avoid having to calculate complex residues. From

$$6(s^2+4s+40) + (Bs+C)(s+3) = (6+B)s^2 + (24+3B+C)s + (240+3C) = 222 \tag{5.51}$$

we obtain $B = C = -6$.

We can expect that for a second-order denominator with real coefficients and complex conjugate poles, that the corresponding function of time is composed of sin and cos terms. We therefore want to bring the second term of the partial fraction expansion (5.49) into a form that corresponds to sine and cosine functions of time. This can be done by completing the square

$$\frac{Bs + C}{s^2 + 4s + 40} = \frac{-6s - 6}{s^2 + 4s + 40} = \frac{-6(s + 2) + 6}{(s + 2)^2 + 36}. \tag{5.52}$$

Putting this into (5.49) gives the representation

$$F(s) = \frac{6}{s + 3} - 6\frac{s + 2}{(s + 2)^2 + 6^2} + \frac{6}{(s + 2)^2 + 6^2}. \tag{5.53}$$

With Table 4.1 and the modulation theorem (4.23) we obtain the function of time

$$f(t) = 6e^{-3t} \cdot \varepsilon(t) - 6e^{-2t} \cos 6t \cdot \varepsilon(t) + e^{-2t} \sin 6t \cdot \varepsilon(t). \tag{5.54}$$

∎

Partial fraction expansions are also advantageous for multiple poles. They avoid multiple derivations with respect to s when calculating the residues (5.42), but the partial fraction coefficients have to be calculated instead. To compare the two possibilities we will consider the function from Example 5.1 once more.

—————————————————————————————————— **Example 5.3**

The partial fraction expansion of the function $F(s)$ from Example 5.1 needs three coefficients:

$$F(s) = \frac{1}{(s + 1)(s + 2)^2} = \frac{A}{s + 1} + \frac{B_1}{s + 2} + \frac{B_2}{(s + 2)^2}. \tag{5.55}$$

For simple poles the coefficients can be calculated (5.39):

$$A = \lim_{s \to -1}[F(s)(s + 1)] = 1. \tag{5.56}$$

To understand how B_1 and B_2 are determined, we consider the term

$$F(s)(s + 1) = A + B_1\frac{s + 1}{s + 2} + B_2\frac{s + 1}{(s + 2)^2} \tag{5.57}$$

again. Substituting $s = -1$ gives the exact value of A because the other two terms become zero. This procedure also works for B_2, so

$$F(s)(s + 2)^2 = A\frac{(s + 2)^2}{s + 1} + B_1(s + 2) + B_2 \tag{5.58}$$

and setting $s = -2$ yields

$$B_2 = \lim_{s \to -2}[F(s)\,(s+2)^2] = -1\,. \tag{5.59}$$

We obtain B_1 in the same way if we form

$$\frac{d}{ds}[F(s)\,(s+2)^2] = \frac{d}{ds}\left[A\frac{(s+2)^2}{s+1} + B_1(s+2) + B_2\right] = A\frac{d}{ds}\left[\frac{(s+2)^2}{s+1}\right] + 1\cdot B_1 \tag{5.60}$$

and put in $s = -2$:

$$B_1 = \lim_{s \to -2}\frac{d}{ds}[F(s)\,(s+2)^2] = -1\,. \tag{5.61}$$

The partial fraction expansion of $F(s)$ is then

$$F(s) = \frac{1}{(s+1)(s+2)^2} = \frac{1}{s+1} - \frac{1}{s+2} - \frac{1}{(s+2)^2}\,. \tag{5.62}$$

Inverse transforming term-by-term, for example, using Table 4.1 gives the same result that we obtained in Example 5.1 with the residue calculation, as long as we bear in mind the region of convergence:

$$f(t) = \mathcal{L}^{-1}\{F(s)\} = \left[e^{-t} - e^{-2t} - te^{-2t}\right]\varepsilon(t)\,. \tag{5.63}$$

The principle used in the previous Example 5.3 can be extended to cover order m poles. Partial fraction coefficients $B_1 \dots B_m$ are then found with

$$B_\mu = \frac{1}{(m-\mu)!}\lim_{s \to s_p}\frac{d^{m-\mu}}{ds^{m-\mu}}[F(s)\,(s-s_p)^m] \qquad \mu = 1\dots m\,. \tag{5.64}$$

5.6 Exercises

Exercise 5.1

Find the Cauchy integral (5.13) by integrating an infinitesimal circle around s_0, which should be parametricised by $s(\nu) = s_0 + \delta\,e^{j2\pi\nu}$, $0 \le \nu < 1$, where δ can be any small real number.

Exercise 5.2

Find the inverse Laplace transform $f(t) = \mathcal{L}^{-1}\{F(s)\}$ of
$$F(s) = \frac{2-2s}{(s+1)(s+2)(s+5)}, \qquad \mathrm{Re}\{s\} > -1.$$

Exercise 5.3

Find the inverse Laplace transform $f(t) = \mathcal{L}^{-1}\{F(s)\}$ of

$$F(s) = \frac{2s - 1}{(s+1)^3(s+4)}, \quad \text{Re}\{s\} > -1.$$ Compare different methods of partial fraction expansion.

a) Calculate all partial fraction coefficients with (5.64).

b) Calculate the partial fraction coefficients of $(s+4)$ and $(s+1)^3$ with (5.64) or (5.39) and the others by equating coefficients.

c) Calculate all four partial fraction coefficients by equating coefficients.

Exercise 5.4

$$F(s) = \frac{s+3}{s^2 + 2s + 5}$$ has complex conjugate poles and a right-sided region of convergence. Find its inverse Laplace transform $f(t) = \mathcal{L}^{-1}\{F(s)\}$ and represent the result as sine and cosine terms.

a) Use the complex partial fraction expansion $F(s) = \dfrac{A}{s - s_p} + \dfrac{A^*}{s - s_p^*}$.

b) Complete the square, then use the shift theorem and Table 4.1.

Exercise 5.5

Find the inverse Laplace transform $f(t) = \mathcal{L}^{-1}\{F(s)\}$ of

$$F(s) = \frac{s}{(s+2)(s^2 + w_0^2)}, \quad \text{Re}\{s\} > 0$$ using partial fraction expansion with real values.

Exercise 5.6

Find the inverse Laplace transform of

$$F(s) = \frac{s^2 - s - 2}{(s^3 + 4s^2 + s - 6) \cdot s^2}, \quad \text{Re}\{s\} > 1.$$

Exercise 5.7

Let $F(s) = \dfrac{-3s^3 - 12s^2 - 16s - 5}{s^4 + 7s^3 + 17s^2 + 17s + 6}, \quad \text{Re}\{s\} > -1.$

Determine the inverse Laplace transform $f(t) = \mathcal{L}^{-1}\{F(s)\}$. (Note that there is a double pole at -1.)

6 Analysis of Continuous-Time LTI-Systems with the Laplace Transform

The Laplace transform introduced in Chapter 4 does not only serve to characterise signals; above all it provides an elegant description of the properties of LTI-systems. In system and network theory, it is the standard method for deriving the system response. In particular, given initial values of the output signal and initial states of systems can be considered using it.

We are now concerned with the derivation of the system response to bilateral signals and with this we can determine the transfer (or system) function. At the end of the chapter we will extend the results to cover combinations of LTI-systems. The solution of initial value problems will then be dealt with in Chapter 7.

6.1 System Response to Bilateral Input Signals

In this section, we consider the response of LTI-systems to input signals, on which we make no restrictions, except that their Laplace transforms must exist. In particular, bilateral signals are also permitted.

Next we recall the definition of the Laplace transform from Chapter 4.2, where we had interpreted it as the analysis of a function in exponential terms, and its inverse as the recombining of these terms. Now we put a further operation in between these two steps, that we have already encountered in Chapter 3.2.2: determining the response of an LTI-system from its eigenfunctions. With this, the three stage system analysis that we outlined in Chapter 3 is complete:

1. Analysis of the input signal in exponential terms can now be accomplished with the Laplace transform.

2. Determining the system response from the individual terms, with help from the system function.

3. Combining the individual components at the system output to form the complete output signal is achieved with the inverse Laplace transform.

To mathematically formulate the system analysis we look back to the inverse Laplace transform, expressed as a Riemann sum (4.4). For output signal $y(t)$,

$$y(t) = \frac{1}{2\pi j} \lim_{\Delta s \to 0} \left\{ ... + Y(s_0)e^{s_0 t} + Y(s_1)e^{s_1 t} + Y(s_2)e^{s_2 t} + ... \right\} \Delta s. \qquad (6.1)$$

The individual terms of the summation are single complex exponential functions, and so also eigenfunctions of the LTI-system. As in Chapter 3.2, we can express this by multiplication of the exponential terms $X(s_i)e^{s_i t}$ of the input signal $x(t)$ (see (4.4)) with the system function $H(s)$ (see Figure 6.1)

$$Y(s_i)e^{s_i t} = H(s_i)X(s_i)e^{s_i t} \qquad (6.2)$$

and thus obtain the output signal

$$y(t) = \frac{1}{2\pi j} \lim_{\Delta s \to 0} \left\{ ... + H(s_0)X(s_0)e^{s_0 t} \quad + \quad H(s_1)X(s_1)e^{s_1 t} \right. \qquad (6.3)$$
$$\left. + \quad H(s_2)X(s_2)e^{s_2 t} + ... \right\} \Delta s.$$

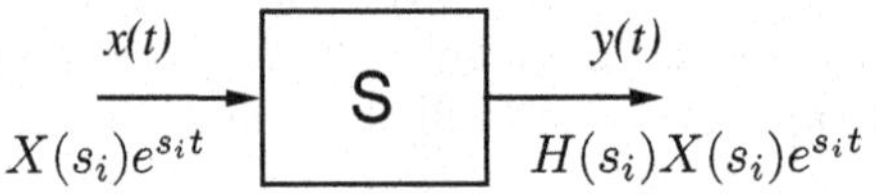

Figure 6.1: Eigenfunctions of an LTI-system

Instead of the Riemann sum we now write the corresponding complex integral

$$\frac{1}{2\pi j} \int_{\sigma - j\infty}^{\sigma + j\infty} H(s)X(s)e^{st} ds = \frac{1}{2\pi j} \int_{\sigma - j\infty}^{\sigma + j\infty} Y(s)e^{st} ds. \qquad (6.4)$$

From this follows the fundamental relationship between the input and output signals of LTI-systems (see Figure 6.2)

$$\boxed{Y(s) = H(s)X(s).} \qquad (6.5)$$

$H(s)$ is the *system function* or *transfer function*, that we already know from Chapter 3.2. In contrast to Chapter 3.2, however, it is not now only defined as the ratio of exponential oscillations, but also more generally as the ratio of the Laplace transforms of the input and output signals. The system function is the key to finding the system response, because it represents a complete description of an LTI-system. There are two fundamental ways of finding the system function:

1. Forming the quotient of $Y(s)$ and $X(s)$, if the input and output signals are known.

2. Analysing the system, if its internal structure is known.

The first possiblity is a case of system *identification*, the second is system *analysis*. Using some simple networks, we will show the procedure for system analysis.

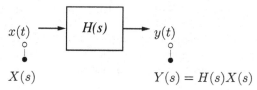

$$Y(s) = H(s)X(s)$$

Figure 6.2: Determining the system response with the system (or transfer) function

6.2 Finding the System Function

We will now demonstrate how to find a system function, using two electrical networks as examples. In Chapter 3.2.5 we learnt two ways of avoiding physical dimensions. We will perform both of these methods here — in Examples 6.1 to 6.5 we will work with normalised equations, and in Examples 7.1 to 7.5 in the next chapter, we will demonstrate the procedure with physical units.

———————————————————————————————— **Example 6.1**

The first network we will consider is the RC-circuit depicted in Figure 6.3. The components have already been normalised so that they have helpful values.

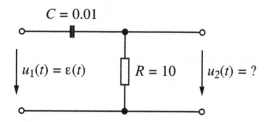

Figure 6.3: RC-circuit

Analysis of this network can be carried out directly in the frequency-domain, if the components are given their complex impedances (from Table 3.2). The system function is immediately obtained if the Laplace transform of the output signal is divided by the Laplace transform of the input signal:

$$H(s) = \frac{U_2(s)}{U_1(s)} = \frac{R}{R + \dfrac{1}{sC}} = \frac{s}{s + 10} . \tag{6.6}$$

With a known input signal, the Laplace transform of the output signal follows immediately with (6.5) as $U_2(s) = H(s)U_1(s)$. For a step function $\varepsilon(t)$ at the input we obtain (compare Table 4.1)

$$U_2(s) = H(s)\frac{1}{s} = \frac{1}{s+10}, \quad \text{Re}\{s\} > -10 \qquad \bullet\!\!-\!\!\circ \qquad u_2(t) = e^{-10t}\varepsilon(t). \quad (6.7)$$

The input–output relationships in the time-domain and the frequency-domain are shown in Figure 6.4. The response to the step function $\varepsilon(t)$ is called the *step response*. Figure 6.5 shows the step response for the RC-circuit.

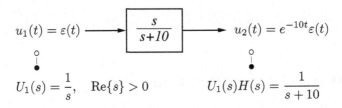

$$u_1(t) = \varepsilon(t) \longrightarrow \boxed{\dfrac{s}{s+10}} \longrightarrow u_2(t) = e^{-10t}\varepsilon(t)$$

$$U_1(s) = \frac{1}{s}, \quad \text{Re}\{s\} > 0 \qquad\qquad U_1(s)H(s) = \frac{1}{s+10}$$

Figure 6.4: System function and system response for the RC-circuit

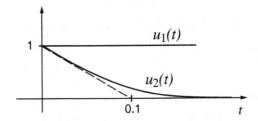

Figure 6.5: Step function and step response for the RC-circuit

As the input signal is explicitly left as a bilateral signal, the output signal it defines contains the entire past history of the system since $t = -\infty$, for every point in time. Additional knowledge of initial conditions or states of energy stores at certain times is not required.

Example 6.2

We consider for the second network, two independent RC-circuits as shown in Figure 6.6. As the voltage $u_2(t)$ is coupled to the right part of the circuit by the voltage source, the system function is

$$H(s) = \frac{U_3(s)}{U_1(s)} = \frac{U_3(s)}{U_2(s)} \cdot \frac{U_2(s)}{U_1(s)} = \frac{s^2}{(s+10)^2}. \qquad (6.8)$$

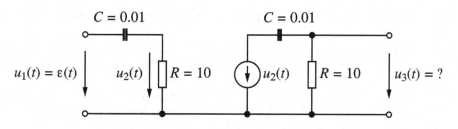

Figure 6.6: Independent RC-circuits

$$u_1(t) = \varepsilon(t) \longrightarrow \boxed{\dfrac{s^2}{(s+10)^2}} \longrightarrow u_3(t) = (1 - 10t)e^{-10t}\varepsilon(t)$$

$$U_1(s) = \frac{1}{s} \qquad\qquad U_1(s)H(s) = \frac{s}{(s+10)^2}$$

Figure 6.7: System function and system response of the independent RC-circuits from Figure 6.6

Driving the circuit with a step function gives a transform of the output signal similar to (6.7). A simple partial fraction expansion and inverse transformation of both terms with Table 4.1 yields

$$U_3(s) = \frac{s}{(s+10)^2} = \frac{1}{s+10} - \frac{10}{(s+10)^2} \quad \bullet\!-\!\circ \quad u_3(t) = (1-10t)e^{-10t}\varepsilon(t). \quad (6.9)$$

Figure 6.7 shows the input–output relationship in the time-domain and the frequency-domain. The behaviour over time of the step function at the input and of the voltages $u_2(t)$ and $u_3(t)$ is depicted in Figure 6.8.

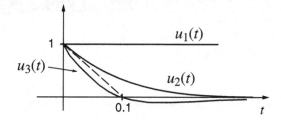

Figure 6.8: Step function and step response of the independent RC-circuits from Fig 6.6

6.3   Poles and Zeros of the System Function

The system function is a completely different kind of system description of the networks in Figures 6.3 and 6.6. Even so, if the circuits are described by fewer components, then the corresponding system function will also be simpler. The system functions here can be defined by giving all of their poles and zeros together with a constant factor. Only rational fraction system functions will occur when we are dealing with LTI-systems with a finite number of energy stores.

As an example we consider the system function of the RC-circuit in (6.6). Figure 6.9 shows its magnitude on the complex frequency plane. The zero at $s = 0$ and the pole at $s = -10$ are easily recogniseable. Their location defines the value of $H(s)$ (both magnitude and phase) at all other points on the s-plane. It is therefore sufficient to give these points to define $H(s)$ to a constant factor. A plot of these points in the s-plane is known as a pole-zero diagram.

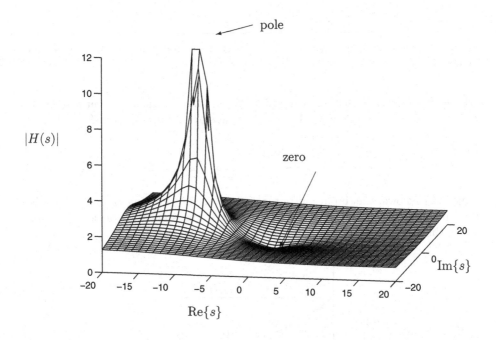

Figure 6.9: The magnitude of a system function

Two examples of pole-zero diagrams are shown in Figure 6.10. The upper diagram represents the system function from Figure 6.9, and the lower diagram represents a system function with double poles and zeros at the same locations.

Poles and zeros as roots of the numerator and denominator polynomials of a

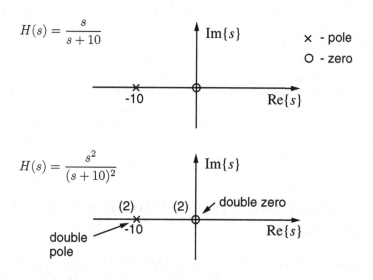

$$H(s) = \frac{s}{s+10}$$

$$H(s) = \frac{s^2}{(s+10)^2}$$

Figure 6.10: Examples of pole-zero diagrams

rational fraction system function can also take complex values. For polynomials with real coefficients, however, they always occur in complex conjugate pairs. Figure 6.11 shows a simple example.

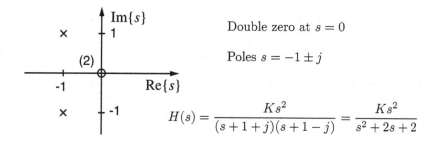

Double zero at $s = 0$

Poles $s = -1 \pm j$

$$H(s) = \frac{Ks^2}{(s+1+j)(s+1-j)} = \frac{Ks^2}{s^2 + 2s + 2}$$

Figure 6.11: Pole-zero diagram with conjugated complex poles

The order of a system is given by the number of poles it has, which is equal to the number of independent energy stores. The number of zeros has no influence on the order of a system.

We have already briefly remarked in Chapter 3.2.2 that the system function also has a region of convergence. Often this relates to causal systems, for which the response at the output cannot occur earlier than the cause at the input. For causal systems the region of convergence lies to the right of a vertical line through the rightmost singularity, and with a rational fraction system function, to the right of the rightmost pole. The location of zeros has no influence on the region of convergence. In order that the Laplace transform of the output signal $Y(s)$ in Figure 6.2 exists, the region of convergence of the input signal $X(s)$ and the system function $H(s)$ must overlap. As the regions of convergence for right-sided input signals and causal systems always overlap, if $\mathrm{Re}\{s\}$ is chosen to be large enough, the region of convergence of the system function is not normally of interest. The region of convergence for system functions in general, not just causal systems, will emerge later from the convolution theorem (Chapter 8.4.2).

6.4   Determining the System Function from Differential Equations

The system function was found relatively easily from the network in Section 6.2 through the use of complex impedances. It is even easier to determine the transfer function if an LTI-system is given as a differential equation (with constant coefficients). Use of the differentiation theorem (4.26) replaces every derivative of a function of time by a product of a power of s and the Laplace transform of the function of time. The differential equation then becomes an algebraic equation, which immediately yields the transfer function. We will show how this is done with two examples.

Example 6.3

From the differential equation

$$2\ddot{y} - 3\dot{y} + 5y = 10\dot{x} - 7x\,, \tag{6.10}$$

use of the differentiation theorem (4.26) yields the algebraic equation

$$2s^2Y(s) - 3sY(s) + 5Y(s) = 10sX(s) - 7X(s)\,, \tag{6.11}$$

from which we can obtain the transfer function

$$H(s) = \frac{Y(s)}{X(s)} = \frac{10s - 7}{2s^2 - 3s + 5}\,. \tag{6.12}$$

The coefficients of the differential equation come directly from the coefficients of the transfer function.

■

Example 6.4

The reverse is also possible; obtaining a differential equation from a transfer function. From the transfer function

$$H(s) = \frac{Y(s)}{X(s)} = \frac{s^2}{(s+10)^2} = \frac{s^2}{s^2 + 20s + 100} \tag{6.13}$$

we obtain the algebraic equation

$$(s^2 + 20s + 100)\, Y(s) = s^2\, X(s), \tag{6.14}$$

which, using the differentiation theorem (4.26), corresponds to the differential equation

$$\ddot{y} + 20\dot{y} + 100y = \ddot{x}. \tag{6.15}$$

■

If the input signal cannot be differentiated for all values, the differentiation theorem for signals with continuous sections (4.41) must be used. This will be covered in detail in Chapter 7.

The region of convergence of the system function cannot be found from the differential equation alone. To determine it, we require further information about the causality or stability of the system. Initial condition problems, which will be dealt with in Chapter 7, imply that the system we are dealing with is causal.

6.5 Summarising Example

In the previous examples only the step response was determined, that is, the response to a signal that takes the value zero for $t < 0$. It represents a special case of bilateral signal, that does not excite a system for $t < 0$. Now we will show how to determine the system response to a signal that is always non-zero.

Example 6.5

We consider the system response of the RC-circuit from Example 6.1 to the input signal

$$x(t) = 3e^{-5t}\varepsilon(t) + 3e^{5t}\varepsilon(-t) = 3e^{-5|t|} \tag{6.16}$$

$$\circ\!\!-\!\!\bullet$$

$$X(s) = \frac{3}{s+5} - \frac{3}{s-5}, \qquad -5 < \text{Re}\{s\} < 5. \tag{6.17}$$

With partial fraction expansion, the Laplace transform of the output signal is given by

$$Y(s) = \frac{s}{s+10}X(s) = \frac{4}{s+10} - \frac{3}{s+5} - \frac{1}{s-5} \tag{6.18}$$

The region of convergence (ROC) of $X(s)$ lies completely in the region of convergence of the system function for the causal RC-circuit $\text{Re}\{s\} > -10$. The inverse transform can take place as integration within the ROC $-5 < \text{Re}\{s\} < 5$, so that the first two terms lie left of the ROC and lead to right-sided functions of time because of their poles. The third term leads to a left-sided function of time.

$$y(t) = 4e^{-10t}\varepsilon(t) - 3e^{-5t}\varepsilon(t) + e^{5t}\varepsilon(-t). \tag{6.19}$$

The signals $x(t)$ and $y(t)$ are illustrated in Figure 6.12. To test this, we begin

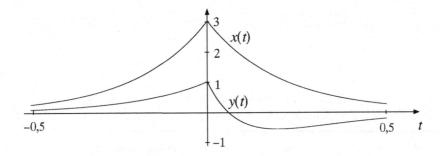

Figure 6.12: Input and output signals in the time-domain

with the differential equation of the RC-circuit that can be either taken directly from the network in Figure 6.3, or alternatively from the transfer function as in Example 6.4:

$$\dot{y}(t) + 10y(t) = \dot{x}(t). \tag{6.20}$$

To show that $y(t)$ (6.19) fulfills the differential equation, we separate the cases $t < 0$ and $t > 0$. For $t < 0$,

$$x(t) = 3e^{5t}, \qquad y(t) = e^{5t}, \qquad t < 0 \tag{6.21}$$

and therefore

$$\dot{y}(t) + 10y(t) = 15e^{5t} = \dot{x}(t), \qquad t < 0. \tag{6.22}$$

For $t > 0$,

$$x(t) = 3e^{-5t}, \qquad y(t) = 4e^{-10t} - 3e^{-5t}, \qquad t > 0 \tag{6.23}$$

and therefore

$$\dot{y}(t) + 10y(t) = -15e^{-5t} = \dot{x}(t), \qquad t > 0. \tag{6.24}$$

The output signal (6.19) satisfies the differential equation of the RC-circuit for all time t where the input signal can be differentiated. The differential equation is, however, not sufficient to confirm the validity of the solution. For example,

$$y(t) = -4e^{-10t}\varepsilon(-t) - 3e^{-5t}\varepsilon(t) + e^{5t}\varepsilon(-t)$$

also fulfills equation (6.20), but does not correspond to the causal RC-circuit. The two solutions can be distinguished by the region of convergence of their Laplace transforms.

■

6.6 Combining Simple LTI-Systems

Until now, we were only concerned with individual systems and their system functions, each determined by a given system description or differential equation. In Section 1.2.2, we had formulated the goal of system theory as being abstract from the details of the system implementation. When describing a system with a system function, we therefore do not always want to first describe all subsystems, or recreate the complete differential equation or state-space description. Instead it would be more appropriate to obtain the system function directly from the known system functions of the subsystems. To do this we only need to know the simple relationships between systems connected in common forms. These forms are series coupled, parallel coupled and feedback coupled systems.

6.6.1 Series Coupling

Figure 6.13 shows two systems with the system functions $H_1(s)$ and $H_2(s)$, that are coupled in series so that the output signal of $H_1(s)$ is the input signal of $H_2(s)$. From the system functions

$$H_1(s) = \frac{Y_1(s)}{X(s)}, \qquad H_2(s) = \frac{Y(s)}{Y_1(s)} \tag{6.25}$$

we can immediately find the system function of the whole system

$$H(s) = \frac{Y(s)}{X(s)} = \frac{Y(s)}{Y_1(s)} \cdot \frac{Y_1(s)}{X(s)} = H_2(s)H_1(s) = H_1(s)H_2(s). \tag{6.26}$$

The output signal of a system with system function $H(s) = H_1(s)H_2(s)$ is therefore identical to the output signal of the two systems $H_1(s)$ and $H_2(s)$ in series. Both system components can be interchanged without altering the output signal.

Example 6.6

The direct form I shown in Figure 2.1 represents the series coupling of two systems. The system functions of the component systems are

$$H_1(s) \;=\; \frac{b_0 s^N + \ldots + b_{N-1}s + b_N}{s^N},$$

$$\tag{6.27}$$

$$H_2(s) \;=\; \frac{s^N}{a_0 s^N + \ldots + a_{N-1}s + a_N}.$$

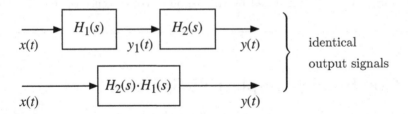

Figure 6.13: Systems in series

Their product represents the desired system function for the differential equation (2.3):

$$H(s) = H_1(s)H_2(s) = \frac{b_0 s^N + \; \ldots \; + b_{N-1}s + b_N}{a_0 s^N + \; \ldots \; + a_{N-1}s + a_N}. \qquad (6.28)$$

As both component systems are interchangeable, we could carry out the transformation from direct form I to direct form II shown in Figure 2.2.

■

── Example 6.7

The two independent RC-circuits from Example 6.2 are an example of series coupling of systems. Their transfer function is the product of the transfer function of the two individual RC-circuits from Example 6.1. The decoupling of the two circuits with the controlled voltage source (Figure 6.6) is necessary to prevent the second system from affecting the first system.

■

6.6.2 Parallel Coupling

The parallel circuit in Figure 6.14 has two component systems with the same input. Because the system is linear,

$$Y(s) = H_1(s)X(s) + H_2(s)X(s) = [H_1(s) + H_2(s)]X(s) = H(s)X(s). \qquad (6.29)$$

The output signal of the parallel cicuit is therefore identical to the output signal of a system with system function $H(s) = H_1(s) + H_2(s)$. We have already made use of this relationship with partial fraction expansion. Expanding a system function into partial fractions is nothing more than the dividing up of a system into simpler parts. The parallel form in Figure 2.10 is likewise a division of an order N system into N first-order systems.

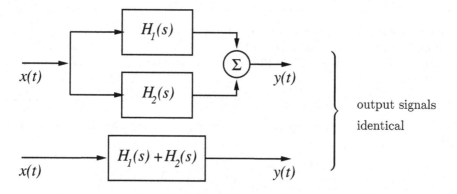

Figure 6.14: A parallel system

6.6.3   Feedback

Systems with feedback like Figure 6.15 are very important in control systems and have many useful applications. At the output of feedback systems we find the expression

$$Y(s) = F(s)[X(s) + G(s)Y(s)]. \tag{6.30}$$

This yields the system function

$$\boxed{H(s) = \frac{Y(s)}{X(s)} = \frac{F(s)}{1 - F(s)G(s)}} \cdot \tag{6.31}$$

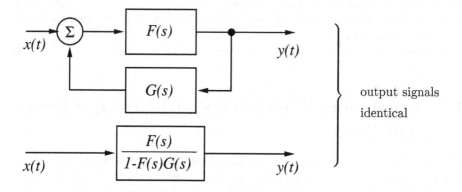

Figure 6.15: A feedback system

Example 6.8

Every branch of the parallel form in Figure 2.10 contains a feedback system of the kind in Figure 6.15. The transfer function of the integrators comes from the integration theorem (4.27), while the transfer function in the feedback path is a constant:

$$F(s) = \frac{1}{s} \qquad G(s) = \lambda_i, \quad i = 1, \dots, N. \tag{6.32}$$

Together with the relationship for series coupling, the transfer function of each path is

$$H_i(s) = b_i \frac{\frac{1}{s}}{1 - \frac{1}{s}\lambda_i} c_i = \frac{b_i c_i}{s - \lambda_i}. \tag{6.33}$$

The entire transfer function can be obtained with the parallel coupling relationship, as the sum of all partial transfer functions $H_i(s)$, $i = 1, \dots, N$.

Example 6.9

If a system with transfer function $F(s)$ is an ideal amplifier with amplification factor V, then $F(s) = v$. The transfer function of the feedback system is then

$$H(s) = \frac{V}{1 - VG(s)} = \frac{1}{\frac{1}{V} - G(s)}. \tag{6.34}$$

If the amplification is very high, then the approximation

$$H(s) = -\frac{1}{G(s)} \tag{6.35}$$

can be used. This means that the inverse of a transfer function $G(s)$ can be implemented by a feedback system. The high amplification required can be achieved using operational amplifiers (see Chapter 2.2.5).

6.7 Combining LTI-Systems with Multiple Inputs and Outputs

The rules discussed in the last section for series, parallel and feedback coupling of systems can also be combined and used to analyse complicated block circuit diagrams. We are still confined, however, to systems with only one input and one output. We now would like to extend these rules to cover systems with multiple inputs and outputs.

We have already met systems with multiple inputs and outputs in Chapters 2.3.2, 2.5 and 2.6. The transfer function of a system with M input signals and K output signals is a $K \times M$ matrix $\mathbf{H}(s)$ which combines the vectors $\mathbf{X}(s)$ of the transformed input signals and the vectors $\mathbf{Y}(s)$ of the transformed output signals (see (6.5)).

$$\boxed{\mathbf{Y}(s) = \mathbf{H}(s)\mathbf{X}(s) \, .} \tag{6.36}$$

The individual elements of the matrix $\mathbf{H}(s)$ are the scalar transfer functions between the individual components of the input and output vectors $\mathbf{X}(s)$ and $\mathbf{Y}(s)$. The element in row κ, column μ is the transfer function between the input numbered μ and the output numbered κ

$$Y_\kappa(s) = H_{\kappa\mu}(s)X_\mu(s) \, . \tag{6.37}$$

The rules for combining systems with multiple inputs and outputs are obtained with the laws of matrix calculation in the same way as in the last section for only one input and output.

6.7.1 Series Circuits

The series circuit for two systems with multiple inputs and outputs is shown in Figure 6.16. Of course, the number of inputs of the second system must be the same as the number of outputs of the first system – then the matrices of the transfer functions $\mathbf{H}_1(s)$ and $\mathbf{H}_2(s)$ are compatible, and we can express the transfer function of the complete system with $\mathbf{H}_1(s)$ and $\mathbf{H}_2(s)$. With

$$\mathbf{Y}_1(s) = \mathbf{H}_1(s)\mathbf{X}(s) \, , \qquad \mathbf{Y}(s) = \mathbf{H}_2(s)\mathbf{Y}_1(s) \tag{6.38}$$

we find for $\mathbf{H}(s)$

$$\mathbf{Y}(s) = \mathbf{H}_2(s)\mathbf{Y}_1(s) = \underbrace{\mathbf{H}_2(s)\mathbf{H}_1(s)}_{\mathbf{H}(s)} \mathbf{X}(s) = \mathbf{H}(s)\mathbf{X}(s) \tag{6.39}$$

and therefore

$$\boxed{\mathbf{H}(s) = \mathbf{H}_2(s)\mathbf{H}_1(s) \, .} \tag{6.40}$$

In contrast to systems with only one input and output, the matrices $\mathbf{H}_1(s)$ and $\mathbf{H}_2(s)$ cannot be interchanged.

6.7.2 Parallel Circuits

If two systems are to be connected in parallel (Figure 6.17), the number of both inputs and outputs must be the same for both systems. Under these conditions, the matrices $\mathbf{H}_1(s)$ and $\mathbf{H}_2(s)$ each have the same number of rows and the same

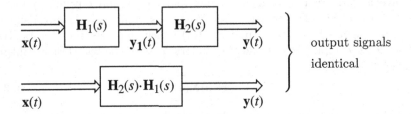

Figure 6.16: Series coupling of systems

number of columns. We can then express the addition of the output signals by an addition of the system functions

$$\mathbf{Y}(s) = \mathbf{H}_1(s)\mathbf{X}(s) + \mathbf{H}_2(s)\mathbf{X}(s) = \underbrace{[\mathbf{H}_1(s) + \mathbf{H}_2(s)]}_{\mathbf{H}(s)}\mathbf{X}(s) = \mathbf{H}(s)\mathbf{X}(s). \quad (6.41)$$

This yields

$$\boxed{\mathbf{H}(s) = \mathbf{H}_1(s) + \mathbf{H}_2(s).} \quad (6.42)$$

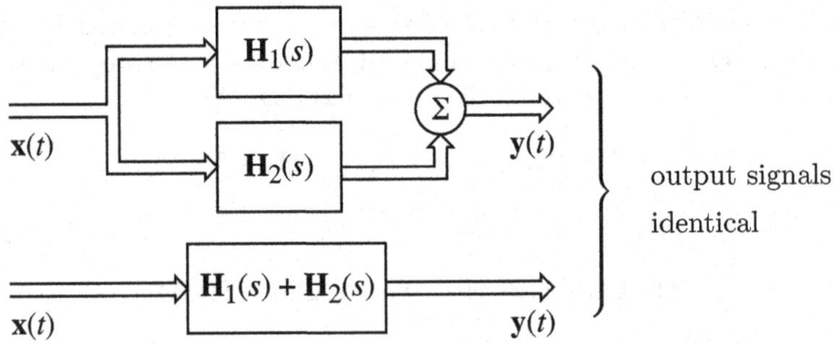

Figure 6.17: Systems coupled in parallel

6.7.3   Feedback

Figure 6.18 shows a feedback system with multiple inputs and outputs. Here we must take special care with the dimensions of the matrices. We begin with addition at the input. All vectors here must have the same number of elements. If input vector $\mathbf{X}(s)$ has M elements, then matrix $\mathbf{F}(s)$ must have M columns, and matrix $\mathbf{G}(s)$ must have M rows. At the output, the signal $\mathbf{Y}(s)$ from the output of $\mathbf{F}(s)$ is combined with the input of $\mathbf{G}(s)$, so if $\mathbf{Y}(s)$ has K elements, $\mathbf{F}(s)$ must have K

columns. Therefore $\mathbf{F}(s)$ must be a $K \times M$ matrix, and $\mathbf{G}(s)$ must be a $M \times K$ matrix.

To determine the transfer function, we consider the output (as in (6.30)). For $\mathbf{Y}(s)$ we can write

$$\mathbf{Y}(s) = \mathbf{F}(s)[\mathbf{X}(s) + \mathbf{G}(s)\mathbf{Y}(s)]. \tag{6.43}$$

We bring the terms containing $\mathbf{Y}(s)$ to the right-hand side and factorise $\mathbf{Y}(s)$

$$[\mathbf{I} - \mathbf{F}(s)\mathbf{G}(s)]\mathbf{Y}(s) = \mathbf{F}(s)\mathbf{X}(s). \tag{6.44}$$

Here, $\mathbf{I}$ is a $K \times K$ unity matrix, so the product $\mathbf{F}(s)\mathbf{G}(s)$ likewise has dimensions $K \times K$. Again special care must be taken over the correct sequence of the matrices. As the matrix in the square brackets is quadratic $(K \times K)$, it can be inverted, as long as none of its eigenvalues are zero. With this condition,

$$\mathbf{Y}(s) = \underbrace{[\mathbf{I} - \mathbf{F}(s)\mathbf{G}(s)]^{-1}\mathbf{F}(s)}_{\mathbf{H}(s)}\mathbf{X}(s) = \mathbf{H}(s)\mathbf{X}(s). \tag{6.45}$$

and

$$\boxed{\mathbf{H}(s) = [\mathbf{I} - \mathbf{F}(s)\mathbf{G}(s)]^{-1}\mathbf{F}(s).} \tag{6.46}$$

The transfer function matrix $\mathbf{H}(s)$ is given by the transfer functions $\mathbf{F}(s)$ in the forward branch and $\mathbf{G}(s)$ in the reverse branch, as in the scalar case if we replace the division in (6.31) by a matrix inversion and note that the order of the matrix multiplication $\mathbf{F}(s)\mathbf{G}(s)$ cannot be changed.

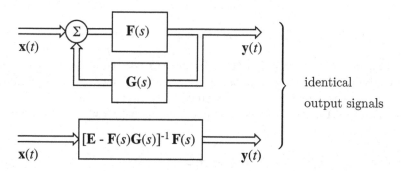

Figure 6.18: A feedback system

6.8   Analysis of State-Space Descriptions

The rules we have considered for the parallel, series and feedback circuits for systems with multiple inputs and outputs are entirely sufficient for analysis of many LTI-systems. As an example for their systematic use, we will calculate the transfer function of a general state-space representation for a system with multiple inputs and outputs as shown in the block diagram in Figure 6.19. It represents the state-equations (2.33,2.34) described in Chapter 2.3.2. As before, we start with

   a vector $\mathbf{x}(t)$ with M input signals,

   a vector $\mathbf{y}(t)$ with K output signals and

   a vector $\mathbf{z}(t)$ with N state variables

The matrices of the state-space model then have dimensions

$$
\begin{aligned}
\mathbf{A} \quad & N \times N\,, \\
\mathbf{B} \quad & N \times M\,, \\
\mathbf{C} \quad & K \times N\,, \\
\mathbf{D} \quad & K \times M\,.
\end{aligned}
$$

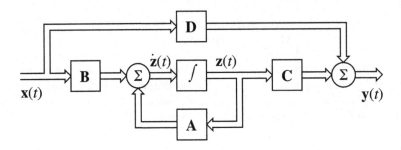

Figure 6.19: Block diagram of a state-space representation

The complete transfer function $\mathbf{H}(s)$ for this state-space representation is obtained by suitable analysis of component transfer functions and by applying the rules for combining LTI-systems step-by-step. First of all, we recognise a parallel circuit in Figure 6.19, with two component systems with the transfer functions $\mathbf{H}_1(s)$ and $\mathbf{H}_2(s)$.

$$\mathbf{H}_1(s) = \mathbf{D}\,, \tag{6.47}$$

which means that the transfer functions in the direct path between input and output are all constant. From (6.42) we know that for the transfer function $\mathbf{H}(s)$,

$$\mathbf{H}(s) = \mathbf{H}_1(s) + \mathbf{H}_2(s)\,. \tag{6.48}$$

The component transfer function $\mathbf{H}_2(s)$ must be determined from the lower path. It consists of three systems in series (6.40)

$$\mathbf{H}_2(s) = \mathbf{H}_{23}(s)\mathbf{H}_{22}(s)\mathbf{H}_{21}(s) \tag{6.49}$$

with

$$\mathbf{H}_{21}(s) = \mathbf{B}, \qquad \mathbf{H}_{23}(s) = \mathbf{C}. \tag{6.50}$$

The component transfer function $\mathbf{H}_{22}(s)$ is obtained by analysing the feedback system that consists of the loops $\dot{\mathbf{z}}(t)$ and $\mathbf{z}(t)$. with the notation

$$\mathbf{F}(s) = \frac{1}{s}\mathbf{I}, \qquad \mathbf{G}(s) = \mathbf{A} \tag{6.51}$$

we obtain with (6.46),

$$\mathbf{H}_{22}(s) = \left[\mathbf{I} - \frac{1}{s}\mathbf{I}\mathbf{A}\right]^{-1}\frac{1}{s}\mathbf{I} = [s\mathbf{I} - \mathbf{A}]^{-1}. \tag{6.52}$$

$\mathbf{I}$ is a unity matrix, size $N \times N$. Now all component transfer functions have been found, we obtain from (6.48) and (6.49), the complete transfer function is

$$\mathbf{H}(s) = \mathbf{H}_{23}(s)\mathbf{H}_{22}(s)\mathbf{H}_{21}(s) + \mathbf{H}_1(s) \tag{6.53}$$

or with (6.47), (6.50) and (6.52)

$$\boxed{\mathbf{H}(s) = \mathbf{C}\,[s\mathbf{E} - \mathbf{A}]^{-1}\mathbf{B} + \mathbf{D}.} \tag{6.54}$$

By applying the rules for combining LTI-systems, we have found the fundamental connection between the matrices of the state-space model, and the transfer function of a system with multiple inputs and outputs.

6.9   Exercises

Exercise 6.1

Determine the transfer function $H(s)$ of the following series resonant circuit using complex impedances.

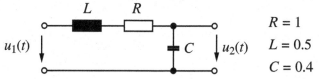

$R = 1$

$L = 0.5$

$C = 0.4$

Calculate the step response $u_2(t)$.

Exercise 6.2

Draw the function $u(t) = \varepsilon(t)\,e^{-\frac{t}{T}}$ and the tangent at the point $u(0)$. The tangent can be used to help construct the graph of exponential functions, if its point of intersection with the x-axis is known. Determine the point of intersection.

Exercise 6.3

Determine the transfer function that has the following pole-zero diagram.

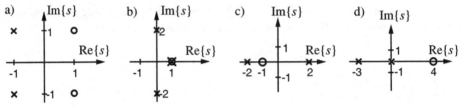

a)   Im{s}   b)   Im{s}   c)   Im{s}   d)   Im{s}

Exercise 6.4

Two systems are given by differential equations:

a) $\ddot{y} + 2\dot{y} + y = \ddot{x} + 4\dot{x} + 4x$

b) $\dfrac{d^3y}{dt^3} + 3\dfrac{d^2y}{dt^2} + 25\dfrac{dy}{dt} + 75y = \dfrac{d^2x}{dt^2} + 4\dfrac{dx}{dt} - 21x$

Determine the transfer functions $H(s)$ and draw the pole-zero diagrams of the systems.

Exercise 6.5

Determine the transfer function and the differential equation of the system with the following pole-zero diagram and $H(0) = 1$

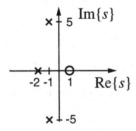

7 Solving Initial Condition Problems with the Laplace Transform

We assumed when we considered the response of systems to bilateral signals, that the input signal of a system for time $-\infty < t < \infty$ is known. The system response depended exclusively on the input signal, and the case where the input signal is zero for $t < 0$ and the system is at rest for $t < 0$, was included.

In many cases, observation of a system is started at a particular point in time. The input signal before the point is unknown, but all that is needed to represent the past history of a system is knowledge of the system state at the time when observation begins. Evaluation of the system response must therefore rely on the state found before the start, and the behaviour of the input signal since then. In the terminology of differential calculus this is an *initial condition problem*.

The way the system state can be given depends on the available description of the system. For a description of a physical structure, for example, an electrical network, the state of the energy stores is available. For a description that uses block diagrams or a state-space structure, the states of the integrators or state-values can be given. If only the differential equation of a system is known, such internal values cannot be observed, but instead, the past history can be represented by the value of the output and its derivatives, at the time that the observation starts.

In order to discuss the solution of initial condition problems in depth, we begin with first-order systems because they make the general principle clear. Second-order and higher-order systems follow on from there.

7.1 First-Order Systems

First-order systems will be described by differential equations like (2.3) for $N = M = 1$:

$$\alpha_1 \dot{y}(t) + \alpha_0 y(t) = \beta_1 \dot{x}(t) + \beta_0 x(t). \tag{7.1}$$

The input variable $x(t)$ and the output variable $y(t)$ are only defined for $t > 0$, and the values $x(0)$ and $y(0)$ at the start of observation are known and can have any value. The inital value of the output signal $y(0)$ is the result of the unkown past history of the system. The input variable $x(t)$ is given, and we want to find the system response $y(t)$ for $t > 0$.

We start with the classical procedure for solving initial condition problems, and later we will use the Laplace transform.

7.1.1 Classical Solution of Initial Condition Problems

The starting point for the classical solution to the problem described above consists of calculating the solution of the homogenous problem and of a particular solution of the non-homogenous problem. A general solution is formed from the sum of the two, in which the previously open parameters are defined, so that the solution fulfills the initial conditions. The following example shows this classical solution for a first-order system.

─── **Example 7.1**

We simplify the problem by specialising the coefficients in (7.1) and examining the differential equation

$$\begin{align}
\tau\dot{y}(t) + 2y(t) &= \tau\dot{x}(t) + x(t), \quad t > 0 \tag{7.2}\\
x(t) &= x_0 \cos\omega_0 t, \quad t > 0, x_0 \in \mathbb{R} \tag{7.3}\\
y(0) &= y_0. \quad y_0 \in \mathbb{R} \tag{7.4}
\end{align}$$

The homogenous solution $y_h(t)$ must fulfill the differential equation $\tau\dot{y}_h(t) + 2y_h(t) = 0$ and in this case it is

$$y_h(t) = C e^{-2t/\tau}. \tag{7.5}$$

It can be easily verified by inserting it into the differential equation. C is any constant, and will be given a value later, by the inital condition (7.4).

To calculate a specific solution $y_s(t)$ of the non-homogenous problem (7.2), (7.3), we use the harmonic character of $x(t)$ (see (7.3)), and write $x(t)$ and $y_s(t)$ for $t > 0$ as

$$\begin{align}
x(t) &= \text{Re}\left\{x_0 e^{j\omega_0 t}\right\} \tag{7.6}\\
y_s(t) &= \text{Re}\left\{Y e^{j\omega_0 t}\right\}, \tag{7.7}
\end{align}$$

with the complex amplitude Y not yet determined. Inserting into (7.2) yields

$$(2 + j\omega_0\tau)Y e^{j\omega_0 t} = (1 + j\omega_0\tau)x_0 e^{j\omega_0 t}. \tag{7.8}$$

The complex amplitude is

$$Y = \frac{1 + j\omega_0\tau}{2 + j\omega_0\tau}x_0 = P(\omega_0)x_0\, e^{j\Theta(\omega_0)}, \tag{7.9}$$

with

$$P(\omega_0) = \sqrt{\frac{1 + (\omega_0\tau)^2}{4 + (\omega_0\tau)^2}}, \qquad \Theta(\omega_0) = \arctan(\omega_0\tau) - \arctan(\omega_0\tau/2). \qquad (7.10)$$

Substituting in (7.7) gives a particular solution

$$y_s(t) = P(\omega_0)x_0 \cos(\omega_0 t + \Theta(\omega_0)). \qquad (7.11)$$

The general solution is obtained by adding the homogenous solution with the yet undetermined constant C and the particular solution of the non-homogenous problem:

$$y(t) = y_h(t) + y_s(t) = C\,e^{-2t/\tau} + P(\omega_0)x_0 \cos(\omega_0 t + \Theta(\omega_0)). \qquad (7.12)$$

It fulfills the inital conditions (7.4) for

$$C = y_0 - P(\omega_0)x_0 \cos(\Theta(\omega_0)). \qquad (7.13)$$

Finally, we have found the output signal

$$y(t) = y_0\,e^{-2t/\tau} + P(\omega_0)x_0 \left[\cos(\omega_0 t + \Theta(\omega_0)) - e^{-2t/\tau}\cos(\Theta(\omega_0))\right]. \qquad (7.14)$$

∎

To assess this kind of solution, the following points must be noted.

- The homogenous solution can only be found this easily for first-order systems. In the general case we must first determine the complete set of characteristic frequencies of the system.

- The harmonic form of the input signal $x(t)$ was used to determine the particular solution. For other input signals this can be much more difficult.

If prior knowledge of the characteristic frequencies of the system is necessary and an input signal of harmonic form is advantageous, a method like the Laplace transform, that represents signals and LTI-systems by exponential frequencies sounds ideal. We will show this in the next section, by comparing how a first-order system is analysed with the Laplace transform.

7.1.2 External and Internal Parts of the Solution

To analyse a first-order system with the Laplace transform we start with the differential equation (7.1). For the input signal $x(t)$ we do not take any particular function, and instead we set the following requirements

- $x(t)$ is a right-sided signal, with $x(t) = 0$ for $t < 0$,

- $x(t)$ can be differentiated for $t > 0$,

- $x(t)$ is of exponential order for $t \to \infty$.

We can now be sure that the Laplace transform of $x(t)$ exists, although we must consider a possible step at $t = 0$, so that the differentiation theorem (4.34) can be used. If we only consider the differential equation for $t > 0$, then $\dot{x}(t) = x^\circ(t)$. We can then use the differentiation theorem in the form (4.35) for the initial condition problem.

$$
\begin{aligned}
\alpha_1 \dot{y}(t) + \alpha_0 y(t) &= \beta_1 \dot{x}(t) + \beta_0 x(t), \quad t > 0 \\
y(0) &= y_0 \qquad\qquad\qquad t = 0
\end{aligned}
\tag{7.15}
$$

This leads to the algebraic equation

$$
\alpha_1 [sY(s) - y(0)] + \alpha_0 Y(s) = \beta_1 [sX(s) - x(0)] + \beta_0 X(s),
\tag{7.16}
$$

which can be solved for the Laplace transform of the output signal:

$$
\begin{aligned}
Y(s) &= \frac{\beta_1 s + \beta_0}{\alpha_1 s + \alpha_0} X(s) + \frac{1}{\alpha_1 s + \alpha_0} [\alpha_1 y(0) - \beta_1 x(0)] \\
&= H(s) X(s) + \frac{1}{\alpha_1 s + \alpha_0} [\alpha_1 y(0) - \beta_1 x(0)].
\end{aligned}
\tag{7.17}
$$

We can already see in the frequency-domain that the output signal consists of two parts. The first part contains the input signal for $t > 0$, weighted with the transfer function

$$
H(s) = \frac{\beta_1 s + \beta_0}{\alpha_1 s + \alpha_0}
\tag{7.18}
$$

and is known as the *external part*. The second part is given by the values of the input and output signals at time $t = 0$ and is divided by the denominator polynomial of the transfer function. The numerator of the second part does not depend on s. We will call this the *internal part*, because as we will soon see, it corresponds to the internal initial state $z(0)$ of the system. The output signal itself is the inverse Laplace transform of both parts.

The result of combining the two parts is clarified in Figure 7.1. The external part is calculated in the frequency-domain with $Y_{ext}(s) = H(s)X(s)$, and the internal part depends on a further transfer function $G(s)$ of the system state at $t = 0$. At the input of $G(s)$, the initial state $z(0)$ appears, which does not depend on the complex frequency, but instead on a constant in the Laplace-domain[1]. In the following section we will meet a procedure which allows us to determine $G(s)$ in a similarly simple way as $H(s)$, without having to worry about the differentiation theorem for signals with discontinuities. The combination of internal and external parts as shown in Figure 7.1 also works for higher-order systems as we will show in Section 7.3. The initial state $z(0)$ is in that case, of course, a vector, and $G(s)$ is an LTI-system with multiple inputs.

---

[1] We will see in Chapter 9 that this corresponds to a delta impulse in the time-domain.

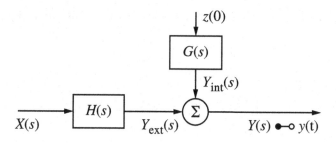

Figure 7.1: Combination of the internal and external part

7.1.3   Initial Values and Initial States

It is important to have a good grasp of the difference between the terms *initial value* and *initial state*.

- The initial value is the value of the *output signal* at time $t = 0$. For signals with a step at $t = 0$, this is the value of the right-hand limit $y(0+)$, regardless of whether or not the plus sign is written.

- The initial state is the value of the *internal states* at time $t = 0$. It can be interpreted as the contents of the energy stores and is usually constant for physical reasons, especially at $t = 0$.

While the initial value is the same for all different realisations of a system, different initial states correspond to different state representations of the same differential equation.

If a system is initially at rest, then $z(0) = 0$ and correspondingly $Y_{int}(s) = 0$, and $y_{int}(t) = 0 \, \forall t$. The solution of the initial condition problem then only consists of $y(t) = y_{ext}(t)$ and the initial value is $y(0+) = y_{ext}(0+)$. The initial value will only arise when the input signal $x(t)$ is turned on. On the other hand, if we have an initial condition problem where $y(0+) = y_{ext}(0+)$, this implies that the system is initially at rest. We therefore say that these are *natural* initial conditions.

_____ **Example 7.2**

Figure 7.2 shows a first-order system with the differential equation (7.1) in a direct form II structure (compare with Figure 2.3). In contrast to having just a differential equation, we now also know the internal structure of the system. We must not forget, however, that there are many different structures that represent the same differential equation (see Chapter 2.5).

As we want to know the situation when the input signal is applied (at time $t = 0$), we express the output signal for this time point by the input signal $x(0)$

and the initial state by $z(0)$. The relationship can be taken from Figure 7.2:

$$y(0) = [x(0) - \alpha_0 z(0)]\frac{1}{\alpha_1}\beta_1 + \beta_0 z(0).\qquad(7.19)$$

By rearranging we can obtain

$$\alpha_1 y(0) - \beta_1 x(0) = z(0)[\alpha_1 \beta_0 - \alpha_0 \beta_1].\qquad(7.20)$$

Putting this into (7.17) yields

$$Y(s) = H(s)X(s) + G(s)z(0) \qquad \text{with} \qquad G(s) = \frac{\alpha_1 \beta_0 - \alpha_0 \beta_1}{\alpha_1 s + \alpha_0}.\qquad(7.21)$$

This shows that the internal part is determined by the initial state of the integrator in Figure 7.2.

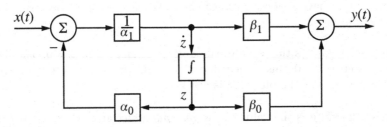

Figure 7.2: First-order system in direct form II

<hr />

Example 7.3

In order to further clarify the meaning of the initial state, we consider the RC-circuit from Example 6.1. From the node equation

$$(\dot{u}_1 - \dot{u}_2)C = \frac{1}{R}u_2\qquad(7.22)$$

and the time constant $\tau = RC$, the differential equation

$$\dot{u}_2\tau + u_2 = \dot{u}_1\tau\qquad(7.23)$$

follows immediately, and from that, using the differentiation theorem, we obtain

$$U_2(s) = \frac{s\tau}{s\tau + 1}U_1(s) + \frac{\tau}{s\tau + 1}[u_2(0) - u_1(0)].\qquad(7.24)$$

Note that the content of the square brackets is the value of the voltage across the capacitor

$$u_C(t) = u_2(t) - u_1(t)\qquad(7.25)$$

at $t = 0$, so we can also express the Laplace transform of $u_2(t)$ by $u_C(0)$:

$$U_2(s) = \frac{s\tau}{s\tau + 1} U_1(s) + \frac{\tau}{s\tau + 1} u_C(0). \tag{7.26}$$

Comparison with (7.21) shows that in this case the initial state corresponds to the voltage across the capacitor at $t = 0$.

∎

If this initial state is equal to zero, the system description has the form (6.5), that we know from bilateral signals. To explain this, we consider the system from the standpoint of a bilateral signal: when a right-sided signal is made bilateral, and it is zero for $t < 0$, the energy stores at $t = 0$ can have no non-zero initial state, and we say that the system is initially at rest. Whether considered as a right-sided or a bilateral signal, for a system initially at rest, we get the same result.

It can be seen in Example 7.3 that the initial state and initial value should not be confused. If the capacitor is not loaded initially, $(u_C(0) = 0)$, the initial state is clearly zero, but the initial value of the output is $u_2(0) = u_1(0)$.

Under what conditions is the initial value of the output signal zero, if the system is energy-less at $t = 0$? To answer this, we integrate the differential equation (7.1) from $-\infty$ to 0, and obtain (compare (2.4))

$$\alpha_1 y(0) + \alpha_0 \int_{-\infty}^{0} y(t)\,dt = \beta_1 x(0) + \beta_0 \int_{-\infty}^{0} x(t)\,dt. \tag{7.27}$$

If the input signal for $t = 0$ is zero, and the system is causal, this is also true for the output signal. Both integrals then disappear, but the relationship for the natural initial conditions of a first-order initial value problem remains

$$\alpha_1 y(0) = \beta_1 x(0). \tag{7.28}$$

We have already obtained this result for the special case of the direct form II realisation in Example 7.2, when the initial state disappeared. From (7.28) it can be read that $y(0) = 0$ only when $\beta_1 = 0$, regardless of the input signal. Put differently: a first-order system with no direct path from input to output($\beta_1 = 0$, Figure 7.2) and without excitation before time $t = 0$, has an output signal $y(0)$ equal to zero. For $\beta_1 \neq 0$, the initial value $y(0)$ depends on the value $x(0)$ of the input signal (7.28).

Example 7.4

What significance does the initial value have for systems which are not intially $(t = 0)$ free of energy? To answer this question, we consider the system from Example 7.1 and apply a bilateral signal

$$\tilde{x}(t) = x_0 \cos \omega_0 t, \qquad -\infty < t < \infty \tag{7.29}$$

and a unilateral signal $x(t) = \tilde{x}(t)$, $t > 0$ to it (7.3). The response $\tilde{y}(t)$ to $\tilde{x}(t)$ can be obtained, for example, using complex amplitudes. Its form can be taken from Example 7.1, where we ensured that the general solution (7.12) was valid for input signals that can be differentiated, even for $t \leq 0$. As the input signal consists of two eigenfunctions of a LTI-system, the output signal $\tilde{y}(t)$ must be made up of the same eigenfunctions, and therefore yields from (7.12) with $C = 0$

$$\tilde{y}(t) = P(\omega_0)x_0 \cos(\omega_0 t + \Theta(\omega_0)) \,. \tag{7.30}$$

The response $y(t)$ to the unilateral signal $x(t)$ has already been determined (7.14). We express it here somewhat differently:

$$y(t) = P(\omega_0)x_0 \cos(\omega_0 t + \Theta(\omega_0)) + [y_0 - P(\omega_0)x_0 \cos\Theta(\omega_0)]\, e^{-2t/\tau}, \quad t > 0 \,. \tag{7.31}$$

If we choose the initial value y_0 in (7.31) to be the same as the value $\tilde{y}(0)$ in (7.30), the second part then disappears from (7.31) and for $t > 0$, $\tilde{y}(t) = y(t)$.

This means that we can also determine the response to a bilateral signal $\tilde{x}(t)$ for $t > 0$ by applying a unilateral signal $x(t) = \tilde{x}(t)\varepsilon(t)$, if the initial value $y(0)$ is set equal to the response to $\tilde{x}(t)$ at time $t = 0$. In other words, we choose the initial value y_0 so that it contains the complete past history of the system, and thus avoid transient behaviour for $t > 0$.

■

7.1.4 Example: Initial Condition Problem with a Sinusoidal Signal

We can now use the Laplace transform to solve initial condition problems. In electrical engineering, initial condition problems with sinusoidal signals are particularly important.

── Example 7.5

As an example we consider the exercise from Example 7.1, which used the classical method of solution. We take the Laplace transform $X(s)$●─○$x(t)$ of the input signal from Table 4.1 and then obtain with (7.17),

$$
\begin{aligned}
Y(s) &= Y_{\text{ext}}(s) \quad + \quad Y_{\text{int}}(s) \\[2mm]
&= \frac{\tau s + 1}{\tau s + 2} \cdot \frac{x_0 s}{s^2 + \omega_0^2} + \frac{\tau}{\tau s + 2}[y_0 - x_0] \\[2mm]
&= \frac{s - s_0}{s - s_\infty} \cdot \frac{x_0 s}{s^2 + \omega_0^2} + \frac{1}{s - s_\infty}[y_0 - x_0] \tag{7.32} \\[2mm]
s_0 &= -\frac{1}{\tau} \qquad s_\infty = -\frac{2}{\tau} \,. \tag{7.33}
\end{aligned}
$$

The output signal is given by the inverse Laplace transform, for which we treat both terms separately. For the first term $Y_{\text{ext}}(s) = H(s)X(s)$, the inverse transform is easiest when we represent it as a partial fraction, and the calculation of the partial fraction coefficients can be simplified by first of all representing $H(s)$ and $X(s)$ with partial fractions:

$$H(s) = \frac{s - s_0}{s - s_\infty} = 1 + \frac{A}{s - s_\infty} \qquad\qquad A = s_\infty - s_0 = -\frac{1}{\tau}$$

$$X(s) = \frac{x_0\, s}{s^2 + \omega_0^2} = \left[\frac{B}{s - j\omega_0} + \frac{B^*}{s + j\omega_0}\right] x_0 \qquad B = \frac{1}{2} \tag{7.34}$$

From here the partial fraction coefficients C and D of $H(s)X(s)$ follow immediately

$$H(s)X(s) \;=\; \frac{C}{s - s_\infty} + \frac{D}{s - j\omega_0} + \frac{D^*}{s + j\omega_0} \tag{7.35}$$

$$C \;=\; (s - s_\infty)H(s)X(s)\big|_{s=s_\infty} = A\,X(s_\infty) = \frac{2x_0}{4 + (\omega_0\tau)^2} \tag{7.36}$$

$$D \;=\; (s - j\omega_0)H(s)X(s)\big|_{s=j\omega_0} = H(j\omega_0)Bx_0 \tag{7.37}$$

$$\;=\; \frac{x_0}{2}\frac{1 + j\omega_0\tau}{2 + j\omega_0\tau}.$$

The inverse transformation is now easy to perform using Table 4.1:

$$y_{\text{ext}}(t) = \mathcal{L}^{-1}\{H(s)X(s)\} = C\,e^{-2t/\tau} + D\,e^{j\omega_0 t} + D^*\,e^{-j\omega_0 t}. \tag{7.38}$$

In order to bring the complex conjugates together, it is useful to express the coefficients D by their phase and magnitude. Then it will occur to us that $H(j\omega_0)x_0$ is identical with the complex amplitude Y (7.9). That is no coincidence, as the complex amplitude of the output signal for a sinusoidal input signal corresponds to the value of the transfer function at this location. The magnitude and phase of $H(j\omega_0)x_0$ are therefore the same for $P(\omega_0)$ and $\Theta(\omega_0)$ (see (7.10)).

$$|H(j\omega_0)| \;=\; P(\omega_0) = \sqrt{\frac{1 + (\omega_0\tau)^2}{4 + (\omega_0\tau)^2}}, \tag{7.39}$$

$$\arg\{H(j\omega_0)\} \;=\; \Theta(\omega_0) = \arctan(\omega_0\tau) - \arctan(\omega_0\tau/2). \tag{7.40}$$

With

$$D = |H(j\omega_0)|e^{j\Theta(\omega_0)} \cdot \frac{1}{2}x_0 \tag{7.41}$$

we obtain

$$y_{\text{ext}}(t) = C\,e^{-2t/\tau} + |H(j\omega_0)|x_0\cos(\omega_0 t + \Theta(\omega_0)). \tag{7.42}$$

The internal term is obtained by inverse transforming the second term in (7.32):

$$y_{\text{int}}(t) = \mathcal{L}^{-1}\left\{\frac{1}{s - s_\infty}[y_0 - x_0]\right\} = [y_0 - x_0]e^{-2t/\tau}. \tag{7.43}$$

With (7.36), (7.43) it now follows that the output signal for $t > 0$ is

$$y(t) = \underbrace{\frac{2x_0}{4 + (\omega_0 \tau)^2} e^{-2t/\tau} + |H(j\omega_0)|x_0 \cos(\omega_0 t + \Theta(\omega_0))}_{\text{external part}} + \underbrace{[y_0 - x_0]e^{-2t/\tau}}_{\text{internal part}}.$$

$$(7.44)$$

To conclude, we express this result in the same form as the result of the classical solution (7.44). The terms with $x_0 e^{-2t/\tau}$ are collected together, giving

$$\left[1 - \frac{2}{4 + (\omega_0 \tau)^2}\right] = \left[\frac{2 + (\omega_0 \tau)^2}{4 + (\omega_0 \tau)^2}\right] = \text{Re}\{H(j\omega_0)\} = P(\omega_0) \cos \Theta(\omega_0) \qquad (7.45)$$

and the result is then put into (7.44). The final result corresponds to (7.14). ∎

For a sinusoidal input signal, as considered in the previous example, we can further divide up the external term of the system response so that (7.44) consists of three parts

$$y(t) = \underbrace{\frac{2x_0}{4 + (\omega_0 \tau)^2} e^{-2t/\tau}}_{\text{transient part}} + \underbrace{|H(j\omega_0)|x_0 \cos(\omega_0 t + \Theta(\omega_0))}_{\text{excitation part}} + \underbrace{[y_0 - x_0]e^{-2t/\tau}}_{\text{decay part}}.$$

$$(7.46)$$

These terms can be interpreted as follows:

- The *transient part* determines how the system responds to an input starting at $t = 0$. It is formed from the characteristic frequency of the system, the complex frequency s_∞ ($A e^{s_\infty t} \circ\!\!-\!\!\bullet A/(s - s_\infty)$) and the value of the Laplace transform of the input signal $X(s_\infty)$ at the system's characteristic frequency. For stable systems (Re$\{s_\infty\} < 0$), this part decays over time.

- The *excitation part* is the part of the input signal that appears at the output of the system. It determines the steady-state response of the system and neglects the single-sided character of the input signal. It is identical to the result obtained by using complex amplitudes. The excitation part is defined in terms of magnitude and phase of the frequency response. They determine the amplitude and phase of the output signal.

- The *decay part* is identical to the internal part and represents the response to the initial state. Similar to the transient part, it decays with the complex frequency of the system.

The transient part and the excitation part comprise the first term in (7.17, 7.44) ($\mathcal{L}\{H(s)X(s)\}$), i.e. the external part. The separation of the external part into transient and excitation parts is performed by partial fraction expansion. The pole of the system $(s = s_\infty)$ determines the transient part and the poles of the excitation $(s = \pm j\omega_0)$ determine the excitation part. As already mentioned above, the decay part is identical to the second term in (7.17, 7.44), i.e. the internal part.

7.1.5 Summary

The results of our intensive investigation of first-order initial condition problems can be summarised by the following points:

- The Laplace transform allows the response of an LTI-system to unilateral signals to be determined just as elegantly as with bilateral signals.

- Use of the differentiation theorem for right-sided signals treats a step in the input signal at $t = 0$ as a pre-determined initial value of the output signal.

- The response to a right-sided signal can be split into an internal and an external term.

- If the input signal is sinusoidal, the external term can be further split into a transient part and an excitation part.

- If both the differential equation *and* the internal structure of an LTI-system are known, the internal term can be interpreted as the response of the system to the initial state.

- The initial state of a system completely summarises the effects of its past history.

These results have been derived from first-order systems but they also apply to higher-order systems. In the following sections, we will show this for second-order systems and general systems of any order in the state-space description.

7.2 Second-Order Systems

The differential equation(2.3) is used for second-order systems

$$\alpha_2 \ddot{y} + \alpha_1 \dot{y} + \alpha_0 y = \beta_2 \ddot{x} + \beta_1 \dot{x} + \beta_0 x. \qquad (7.47)$$

Analysis with the Laplace transform uses the differentiation theorem for right-sided signals (4.34), for the first and second derivatives. For $t > 0$, $x^\circ(t) = \dot{x}(t)$, and we obtain

$$\mathcal{L}\{\dot{x}\} = s\mathcal{L}\{x\} - x(0) = sX(s) - x(0) \qquad (7.48)$$
$$\mathcal{L}\{\ddot{x}\} = s\mathcal{L}\{\dot{x}\} - \dot{x}(0) = s^2X(s) - [sx(0) + \dot{x}(0)]. \qquad (7.49)$$

The bounds $x(0)$ and $\dot{x}(0)$ are again right-sided bounds. Using (7.48, 7.49) on (7.47) yields after rearrangement:

$$[\alpha_2 s^2 + \alpha_1 s + \alpha_0]Y(s) - s\alpha_2 y(0) - [\alpha_1 y(0) + \alpha_2 \dot{y}(0)] =$$
$$[\beta_2 s^2 + \beta_1 s + \beta_0]X(s) - s\beta_2 x(0) - [\beta_1 x(0) + \beta_2 \dot{x}(0)] . \tag{7.50}$$

For a unique solution we require two initial conditions, the value of the output signal $y(0)$ and its first derivative $\dot{y}(0)$. The corresponding values $x(0)$ and $\dot{x}(0)$ of the input signals are known. We can then solve (7.50) for the Laplace transform of the unknown solution $Y(s)$ and obtain with the system function $H(s)$

$$H(s) = \frac{\beta_2 s^2 + \beta_1 s + \beta_0}{\alpha_2 s^2 + \alpha_1 s + \alpha_0} \tag{7.51}$$

the solution

$$Y(s) = H(s)X(s)+$$
$$\frac{s[\alpha_2 y(0) - \beta_2 x(0)] + [\alpha_1 y(0) + \alpha_2 \dot{y}(0) - \beta_1 x(0) - \beta_2 \dot{x}(0)]}{\alpha_2 s^2 + \alpha_1 s + \alpha_0} . \tag{7.52}$$

The output signal $y(t)$ in the time-domain is found from (7.52) using the inverse Laplace transform. The first term gives the external part of the system response and the second term gives the internal part.

If the internal structure of the system is known, we can represent the internal part by the initial states. Depending on the structure of the system, different expressions are obtained. The relationships are especially simple with direct form III, which is shown in Figure 7.3 for the second order differential equation (7.47). In contrast to Fig 2.5 for order N systems, the original notation (α_i, β_k) for the coefficients is kept. At the output of Figure 7.3, we can read that (compare (2.18) for $N = 2$)

$$y(t) = \frac{1}{\alpha_2} z_1(t) + \frac{\beta_2}{\alpha_2} x(t) . \tag{7.53}$$

At the middle summing node (compare (2.15) for $N = 2$)

$$\dot{z}_1(t) = z_2(t) + \beta_1 x(t) - \alpha_1 y(t) . \tag{7.54}$$

Differentiating (7.53) and substituting in (7.54) yields

$$\dot{y}(t) = \frac{\beta_1}{\alpha_2} x(t) + \frac{\beta_2}{\alpha_2} \dot{x}(t) + \frac{1}{\alpha_2} z_2(t) - \frac{\alpha_1}{\alpha_2} y(t) . \tag{7.55}$$

From (7.53) and (7.55) we solve for the states when $t = 0$

$$z_1(0) = \alpha_2 y(0) - \beta_2 x(0) \tag{7.56}$$
$$z_2(0) = \alpha_1 y(0) + \alpha_2 \dot{y}(0) - \beta_1 x(0) - \beta_2 \dot{x}(0) . \tag{7.57}$$

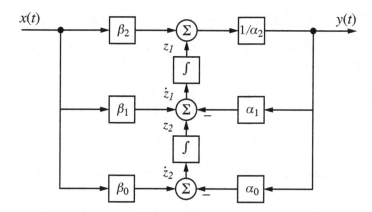

Figure 7.3: Second-order system in direct form III

We can now express the cumbersome terms in (7.52) much more simply using both of the initial states:

$$Y(s) = H(s)X(s) + \frac{sz_1(0) + z_2(0)}{\alpha_2 s^2 + \alpha_1 s + \alpha_0}. \tag{7.58}$$

Replacing initial values by initial states is also possible with other structures, although the resulting expressions are not always as simple as in (7.58).

7.3 Higher-Order Systems

The Laplace transform has so far been shown to be an effective tool for solving initial condition problems for first- and second-order systems. It can be used in the same way for higher-order systems. For this purpose, the differentiation theorem as in (7.48) and (7.49) must be extended to higher derivatives. From the differential equation of order N in (2.3), we can obtain an algebraic equation that leads to a system function $H(s)$ with a polynomial denominator of order N. The internal part contains N initial values $y(0), \dot{y}(0), \ddot{y}(0), \ldots, y^{(N-1)}(0)$, up to the $(N-1)$th derivative and also the corresponding values of the input signal. As the expressions that arise rapidly become involved, however, it is helpful to switch over to the state-space description (see (2.40) to (2.43)). This can be readily done, when the internal structure of the system is known. Otherwise, the matrices $\mathbf{A}$, $\mathbf{B}$, $\mathbf{C}$, $\mathbf{D}$ can be written down from the coefficients of the differential equation for direct forms I, II and III. The arbitrary states thus obtained have no physical meaning, but the advantage of the simplified notation still remains.

7.3.1   Solution of the State-Space Differential Equation

We start with the order N system with one input and one output that is given by
the state-space description (2.33), (2.34). It could have any internal structure, and
so also any coefficients for matrices $\mathbf{A}$, $\mathbf{B}$, $\mathbf{C}$, $\mathbf{D}$. To use the Laplace transform on
this system of N first-order differential equations we only need the differentiation
theorem for the first derivative. Used on the state vector $\mathbf{z}(t)$ it is

$$\mathcal{L}\{\dot{\mathbf{z}}(t)\} = s\mathbf{Z}(s) - \mathbf{z}(0)\,. \qquad (7.59)$$

The Laplace transform here refers to the individual components of the state vectors

$$\mathcal{L}\{\mathbf{z}(t)\} = \mathbf{Z}(s) = \begin{bmatrix} Z_1(s) \\ \vdots \\ Z_N(s) \end{bmatrix} \quad \text{with} \quad Z_i(s) = \mathcal{L}\{z_i(t)\}\,. \qquad (7.60)$$

The initial states are collected together as vector $\mathbf{z}(0)^2$. The state-space description
in the time-domain (see (2.33, 2.34)) becomes

$$\begin{aligned} s\mathbf{Z}(s) - \mathbf{z}(0) &= \mathbf{A}\mathbf{Z}(s) + \mathbf{B}X(s) & (7.61) \\ Y(s) &= \mathbf{C}\mathbf{Z}(s) + \mathbf{D}X(s) & (7.62) \end{aligned}$$

in the frequency-domain. The differential equation then becomes the algebraic
equation system (7.61), which we can solve for the state vector transform $\mathbf{Z}(s)$.
Collecting the terms with $\mathbf{Z}(s)$ together, we notice that $s\mathbf{Z}(s) - \mathbf{A}\mathbf{Z}(s) = (s\mathbf{I} - \mathbf{A})\mathbf{Z}(s)$, where $\mathbf{I}$ is the unity matrix. We obtain

$$\mathbf{Z}(s) = (s\mathbf{I} - \mathbf{A})^{-1}\mathbf{B}X(s) + (s\mathbf{I} - \mathbf{A})^{-1}\mathbf{z}(0)\,. \qquad (7.63)$$

The inverse of the matrix $(s\mathbf{I} - \mathbf{A})$ must stay on the left because the matrix product
is not commutative.

By inverse Laplace transformation (7.63), the time behaviour of the state variables can be determined if it is of interest. Here we are looking for the output
signal $y(t)$, however, and so we substitute (7.63) into (7.62). The result is

$$\boxed{Y(s) = H(s)X(s) + \mathbf{G}(s)\mathbf{z}(0)} \qquad (7.64)$$

---

^{2}At first glance, it is surprising that for the state $\mathbf{z}(t)$ we use the differentiation theorem for
functions discontinuous at $t = 0$ (7.59), while $\mathbf{z}(t)$ is usually continuous for physical reasons.
However, we do not know $\mathbf{z}(t)$ for $t < 0$, so we arbitrarily set $\mathbf{z}(t) = 0 \; \forall \; t < 0$ and let the state
at time $t = 0$ jump to $\mathbf{z}(0)$.

with system function $H(s)$ and column vector $\mathbf{G}(s)$

$$
\begin{aligned}
H(s) &= \mathbf{C}(s\mathbf{I} - \mathbf{A})^{-1}\mathbf{B} + \mathbf{D} &\qquad (7.65)\\
\mathbf{G}(s) &= \mathbf{C}(s\mathbf{I} - \mathbf{A})^{-1}. &\qquad (7.66)
\end{aligned}
$$

The equation (7.64) has the same structure as (7.21) and (7.58); it is just the initial states combined as a vector. Equation (7.65) describes in a general form how to calculate the transfer function from the input to the output of a state-space structure, from the matrices $\mathbf{A}$, $\mathbf{B}$, $\mathbf{C}$, $\mathbf{D}$. This problem has already been dealt with in Chapter 6.8. $\mathbf{G}(s)$ can in any case be interpreted as a vector of transfer functions. The individual components describe the transfer behaviour from the input of the individual integrators to the output of the whole system. The expression $s\,\mathbf{G}(s)$ is the vector of the transfer functions from the state variables $\mathbf{Z}(s)$ to the output of the whole system, that arise at the integrator outputs.

The important equation (7.64) is illustrated in Figure 7.4. The response of the Nth order LTI-system with initial conditions to an input signal $x(t)\circ\!\!-\!\!\bullet X(s)$ is obtained by computing the response $X(s) \cdot H(s)\bullet\!\!-\!\!\circ y_{\text{ext}}(t)$ of a system initially at rest and adding $\mathbf{G}(s)\mathbf{z}(0)\bullet\!\!-\!\!\circ y_{\text{int}}(t)$. The N initial states $\mathbf{z}(0)$ must be chosen such that the complete solution $y(t) = y_{\text{ext}}(t) + y_{\text{int}}(t)$ fulfills the initial conditions.

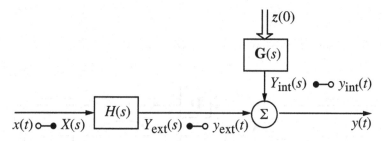

Figure 7.4: Solution of initial condition problems in the Laplace-domain with (7.64)

Calculating the system function $H(s)$ with (7.65) seems to be contradictory: while the system function is clearly fixed by the given differential equation (see Examples 6.3 and 6.4), in (7.65) matrices occur that depend on the chosen state-space model. Does the transfer function therefore depend on the internal structure? To show that the transfer function really is the same for all equivalent state-space structures, we insert the state matrices of an equivalent structure (2.47)–(2.50) into (7.65), and obtain

$$
\begin{aligned}
H(s) &= \mathbf{\hat{C}}(s\mathbf{I} - \mathbf{\hat{A}})^{-1}\mathbf{\hat{B}} + \mathbf{\hat{D}} = \mathbf{CT}(s\mathbf{T}^{-1}\mathbf{T} - \mathbf{T}^{-1}\mathbf{AT})^{-1}\mathbf{T}^{-1}\mathbf{B} + \mathbf{D} =\\
&= \mathbf{C}(s\mathbf{I} - \mathbf{A})^{-1}\mathbf{B} + \mathbf{D}. &\qquad (7.67)
\end{aligned}
$$

The transformation matrix $\mathbf{T}$ disappears when the system function is calculated and so all equivalent state-space structures have the same system function. The matrix expression on the right-hand side (7.65) is said to be invariant for similarity transformations. However, the vector of transfer functions $\mathbf{G}(s)$ from the states to the output depends on the structure of the system. A similarity transformation changes it to

$$
\begin{aligned}
\hat{\mathbf{G}}(s) &= \hat{\mathbf{C}}(s\mathbf{I} - \hat{\mathbf{A}})^{-1} = \mathbf{CT}(s\mathbf{T}^{-1}\mathbf{T} - \mathbf{T}^{-1}\mathbf{AT})^{-1} = \\
&= \mathbf{C}(s\mathbf{I} - \mathbf{A})^{-1}\mathbf{T} = \mathbf{G}(s)\mathbf{T}.
\end{aligned} \tag{7.68}
$$

(7.67) can be made much clearer by manipulating the block diagrams in Figures 7.5

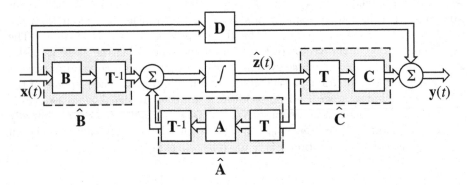

Figure 7.5: Block diagram with transformed system matrices

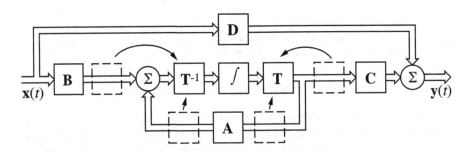

Figure 7.6: Moving the transformation matrices

and 7.6. In Figure 7.5 the matrix multiplications are first represented by cascading multiplications by $\hat{\mathbf{A}}$, $\hat{\mathbf{B}}$ and $\hat{\mathbf{C}}$ with $\mathbf{T}$ and $\mathbf{T}^{-1}$ in accordance with (2.47) - (2.50). The linearity of $\mathbf{T}$ and $\mathbf{T}^{-1}$ allows these blocks to be shifted as shown in Figure 7.6, without changing the transfer behaviour of the system. In the central path of Figure 7.6 , we see the transfer function $\mathbf{T}s\mathbf{IT}^{-1} = s\mathbf{ITT}^{-1} = s\mathbf{I}$, which does not depend on $\mathbf{T}$. We have now shown that $H(s)$ does not depend on the similarity transformation $\mathbf{T}$. This is also true, as shown in Figures 7.5 and 7.6, for systems with any number of inputs and outputs. $\hat{\mathbf{G}}(s)$ can be interpreted in Figure 7.6 as a vector of transfer functions from the input of the integrator to the output and (7.68) can be directly written down from this.

We will clarify the connection between determining the system function from a state-space structure as in (7.65) and determining it from a differential equation, with the next few examples.

Example 7.6

A first order system is represented in direct form II in Example 7.2. The state-space description can be derived directly from Figure 7.2:

$$\dot{z} = -\frac{\alpha_0}{\alpha_1} z + \frac{1}{\alpha_1} x \tag{7.69}$$

$$y = \left(\beta_0 - \alpha_0 \frac{\beta_1}{\alpha_1} \right) z + \frac{\beta_1}{\alpha_1} x . \tag{7.70}$$

The matrices of the state-space description in this case have dimension 1×1:

$$\mathbf{A} = -\frac{\alpha_0}{\alpha_1}, \quad \mathbf{B} = \frac{1}{\alpha_1}, \quad \mathbf{C} = \frac{\alpha_1 \beta_0 - \alpha_0 \beta_1}{\alpha_1}, \quad \mathbf{D} = \frac{\beta_1}{\alpha_1} . \tag{7.71}$$

From here we obtain the same result with (7.65) and (7.66), that we obtained in (7.21) directly from the differential equation:

$$
\begin{aligned}
H(s) &= \frac{\alpha_1 \beta_0 - \alpha_0 \beta_1}{\alpha_1} \left(s + \frac{\alpha_0}{\alpha_1} \right)^{-1} \frac{1}{\alpha_1} + \frac{\beta_1}{\alpha_1} = \frac{1}{\alpha_1} \frac{\alpha_1 \beta_0 - \alpha_0 \beta_1}{\alpha_1 s + \alpha_0} + \frac{\beta_1}{\alpha_1} \\
&= \frac{\beta_1 s + \beta_0}{\alpha_1 s + \alpha_0}
\end{aligned}
$$

$$
G(s) = \frac{\alpha_1 \beta_0 - \alpha_0 \beta_1}{\alpha_1} \left(s + \frac{\alpha_0}{\alpha_1} \right)^{-1} = \frac{\alpha_1 \beta_0 - \alpha_0 \beta_1}{\alpha_1 s + \alpha_0} .
$$

$$\tag{7.72}$$

◼

Example 7.7

The state-space description of a second-order system in direct form III (see Figure 7.3) is:

$$\dot{z}_1 = -\frac{\alpha_1}{\alpha_2} z_1 + z_2 + \frac{\alpha_2 \beta_1 - \alpha_1 \beta_2}{\alpha_2} x \tag{7.73}$$

$$\dot{z}_2 = -\frac{\alpha_0}{\alpha_2} z_1 \quad + \frac{\alpha_2 \beta_0 - \alpha_0 \beta_2}{\alpha_2} x \tag{7.74}$$

$$y = \frac{1}{\alpha_2} z_1 \quad + \frac{\beta_2}{\alpha_2} x . \tag{7.75}$$

We write down the system matrices:

$$\mathbf{A} = \frac{1}{\alpha_2} \begin{bmatrix} -\alpha_1 & \alpha_2 \\ -\alpha_0 & 0 \end{bmatrix} \qquad \mathbf{B} = \frac{1}{\alpha_2} \begin{bmatrix} \alpha_2\beta_1 - \alpha_1\beta_2 \\ \alpha_2\beta_0 - \alpha_0\beta_2 \end{bmatrix} \tag{7.76}$$

$$\mathbf{C} = \frac{1}{\alpha_2}[1 \quad 0] \qquad \mathbf{D} = \frac{\beta_2}{\alpha_2}.$$

With

$$(s\mathbf{I} - \mathbf{A}) = \frac{1}{\alpha_2} \begin{bmatrix} \alpha_2 s + \alpha_1 & -\alpha_2 \\ \alpha_0 & \alpha_2 s \end{bmatrix} \tag{7.77}$$

$$(s\mathbf{I} - \mathbf{A})^{-1} = \frac{1}{\alpha_2 s^2 + \alpha_1 s + \alpha_0} \begin{bmatrix} \alpha_2 s & \alpha_2 \\ -\alpha_0 & \alpha_2 s + \alpha_1 \end{bmatrix} \tag{7.78}$$

and (7.65) the system function follows after some rearrangements. $\mathbf{G}(s)$ is then obtained as

$$\mathbf{G}(s) = \frac{1}{\alpha_2 s^2 + \alpha_1 s + \alpha_0}[s \quad 1] \tag{7.79}$$

and with (7.64) follows the result in (7.58). ∎

For systems of any degree, the determinant of $(s\mathbf{I} - \mathbf{A})$ becomes the denominator polynomial of $H(s)$ and $\mathbf{G}(s)$.

7.3.2   Determining the Initial State from the Initial Values

The equations (7.64) to (7.66) generate a concise formulation of the input–output relationship for systems of any degree in a general state-space structure. If a direct form II or III state-space desription is chosen, however, only to benefit from the advantages of matrix notation, the initial state $\mathbf{z}(0)$ is not available. It is necessary therefore, to find the connection between the initial state and initial values for systems of all degrees.

In order to do so, we also use the advantages of the state-space description. From the output equation(2.34) in the time-domain, we find through differentiation and use of the state equations (2.33) at $t = 0$:

$$y(0) \quad = \quad \mathbf{Cz}(0) \quad + \quad \mathbf{D}x(0)$$

$$\dot{y}(0) \quad = \quad \mathbf{CAz}(0) \quad + \quad \mathbf{CB}x(0) \quad + \quad \mathbf{D}\dot{x}(0)$$

$$\ddot{y}(0) \quad = \quad \mathbf{CA}^2\mathbf{z}(0) \quad + \quad \mathbf{CAB}x(0) \quad + \quad \mathbf{CB}\dot{x}(0) \quad +\mathbf{D}\ddot{x}(0)$$

$$\vdots$$

After the $N-1$ derivatives of $y(t)$ at $t=0$ have been calculated, the results are summarised in matrix form

$$\mathbf{y}(0) = \mathbf{W}\mathbf{z}(0) + \mathbf{V}\mathbf{x}(0) \tag{7.80}$$

with the vectors

$$\mathbf{x}(0) = \begin{bmatrix} x(0) \\ \dot{x}(0) \\ \ddot{x}(0) \\ \vdots \\ x^{(N-1)}(0) \end{bmatrix}, \quad \mathbf{y}(0) = \begin{bmatrix} y(0) \\ \dot{y}(0) \\ \ddot{y}(0) \\ \vdots \\ y^{(N-1)}(0) \end{bmatrix} \tag{7.81}$$

and the matrices

$$\mathbf{W} = \begin{bmatrix} \mathbf{C} \\ \mathbf{CA} \\ \mathbf{CA}^2 \\ \vdots \\ \mathbf{CA}^{N-1} \end{bmatrix}, \quad \mathbf{V} = \begin{bmatrix} \mathbf{D} & \mathbf{0} & \mathbf{0} & \cdots & \mathbf{0} \\ \mathbf{CB} & \mathbf{D} & \mathbf{0} & \cdots & \\ \mathbf{CAB} & \mathbf{CB} & \mathbf{D} & \cdots & \mathbf{0} \\ \vdots & \vdots & \vdots & \ddots & \\ \mathbf{CA}^{N-2}\mathbf{B} & \mathbf{CA}^{N-3}\mathbf{B} & \mathbf{CA}^{N-4}\mathbf{B} & & \mathbf{D} \end{bmatrix}. \tag{7.82}$$

Now the initial state can be expressed by the vector of the initial values $\mathbf{y}(0)$ and $\mathbf{x}(0)$:

$$\boxed{\mathbf{z}(0) = \mathbf{W}^{-1}[\mathbf{y}(0) - \mathbf{V}\mathbf{x}(0)].} \tag{7.83}$$

With (7.64) follows for the output of an LTI-system described by an Nth-order differential equation like (2.3), with given initial conditions $y(0)$ to $y^{(N-1)}(0)$, and the input signal $x(t)$:

$$\boxed{Y(s) = H(s)X(s) + G(s)\mathbf{W}^{-1}[\mathbf{y}(0) - \mathbf{V}\mathbf{x}(0)].} \tag{7.84}$$

Example 7.8

As an illustration, we continue Example 7.7. The matrices $\mathbf{W}$ and $\mathbf{V}$ are in this case

$$\mathbf{W} = \frac{1}{\alpha_2}\begin{bmatrix} 1 & 0 \\ -\dfrac{\alpha_1}{\alpha_2} & 1 \end{bmatrix}, \quad \mathbf{W}^{-1} = \begin{bmatrix} \alpha_2 & 0 \\ \alpha_1 & \alpha_2 \end{bmatrix} \tag{7.85}$$

$$\mathbf{V} = \begin{bmatrix} \mathbf{D} & \mathbf{0} \\ \mathbf{CB} & \mathbf{D} \end{bmatrix} = \frac{1}{\alpha_2}\begin{bmatrix} \beta_2 & 0 \\ \dfrac{\alpha_2\beta_1 - \alpha_1\beta_2}{\alpha_2} & \beta_2 \end{bmatrix}. \tag{7.86}$$

By substitution, we obtain the result (7.52), which has been shown before only for a second order system. Here, it follows as a special case of an Nth-order system.

7.3.3 Determining the Internal Part in the Time-Domain

So far we have learnt two methods to determine the output signal of an LTI-system with a known right-sided input signal and known initial values: the classical method with a general homogenous and particular non-homogenous solution (Section 7.1.1), and system analysis using the Laplace transform (Section 7.1.2). We will now examine a third possiblity, that falls somewhere between the other two methods. It can be seen as a special case of the solution using the Laplace transform and is particularly simple to use, if a system has been given as a differential equation and initial conditions at $t = 0$:

$$\sum_{i=0}^{N} \alpha_i \frac{d^i y}{dt^i} = \sum_{k=0}^{N} \beta_k \frac{d^k x}{dt^k} \tag{7.87}$$

$$y^{(i)}(0) = y_i \qquad i = 0, 1, \ldots, N-1 .$$

If the numerical values of the coefficients of the differential equation are given, then this method is generally the quickest solution. It starts by separating the solution into the external and internal parts

$$Y(s) = \underbrace{H(s)X(s)}_{Y_{\text{ext}}(s)} + \underbrace{\mathbf{G}(s)\mathbf{z}(0)}_{Y_{\text{int}}(s)} . \tag{7.88}$$

As before, the external part is the inverse Laplace transform of $H(s)X(s)$

$$Y_{\text{ext}}(s) = H(s)X(s) . \tag{7.89}$$

To determine the internal part

$$Y_{\text{int}}(s) = \mathbf{G}(s)\mathbf{z}(0) = \mathbf{G}(s)\mathbf{W}^{-1}[\mathbf{y}(0) - \mathbf{V}x(0)] \tag{7.90}$$

we have so far had to start with a defined system structure and

- either choose the initial states $\mathbf{z}(0)$ so that they are compatible with initial values $\mathbf{y}(0)$,

- or additionally to $\mathbf{G}(s)$, determine matrices $\mathbf{W}^{-1}$ and $\mathbf{V}$.

Both possiblities are cumbersome for higher-order systems, if the system model only consists of a differential equation (7.87) and initial values (see (7.52) for a second-order system). In order to avoid this, we combine the advantages of the Laplace transform with the simplicity of the classical homogeneous solution (7.5) in Section 7.1.1.

At first, we note that the order of the numerator of the transfer function $\mathbf{G}(s)$ exceeds the order of the denominator by at least one. That can be illustrated in two ways.

- The inverse matrix $(s\mathbf{I} - \mathbf{A})^{-1}$ can also be expressed as the adjunct matrix $\mathrm{adj}(s\mathbf{I} - \mathbf{A})$ and the determinant $\det(s\mathbf{I} - \mathbf{A})$

$$(s\mathbf{I} - \mathbf{A})^{-1} = \frac{\mathrm{adj}(s\mathbf{I} - \mathbf{A})}{\det(s\mathbf{I} - \mathbf{A})} . \qquad (7.91)$$

  The adjunct matrix contains the determinants of matrices that are formed from $(s\mathbf{I} - \mathbf{A})$ by deleting one row and one column. For a system of order N, the highest degree of numerator polynomial in s that can arise is $N - 1$.

- $\mathbf{G}(s)$ describes the internal feedback of the state in the system through the integrators and matrix $\mathbf{A}$. This feedback cannot go along any path that does not contain an integrator with the transfer function $\frac{1}{s}$. The order of the numerator must therefore be less than the order of the denominator.

From the representation of $(s\mathbf{I} - \mathbf{A})^{-1}$ it is evident from (7.91) that $\mathbf{G}(s)$ and $H(s)$ both have the same denominator polynomial, $\det(s\mathbf{I} - \mathbf{A})$, in fact. We can therefore represent the internal part by partial fractions:

$$Y_{\mathrm{int}}(s) = \mathbf{G}(s)\mathbf{z}(0) = \sum_{i=1}^{N} \frac{A_i}{s - s_i} . \qquad (7.92)$$

The poles s_i are equal to the poles of the system function

$$H(s) = \frac{A(s)}{B(s)} \qquad (7.93)$$

with the denominator polynomial $B(s) = \det(s\mathbf{I} - \mathbf{A})$. The partial fraction expansion in (7.92) only accounts for single poles for simplicity. The expansion for multiple poles has to be calculated according to (5.64).

In contrast to before, we will *not* determine the partial fraction coefficients A_i and with them the numerator polynomial of $\mathbf{G}(s)\mathbf{x}(0)$ from the initial conditions in a more or less complicated way in the frequency domain. Instead, we determine the internal part in the time domain and obtain immediately from (7.92),

$$y_{\mathrm{int}}(t) = \sum_{i=1}^{N} A_i e^{s_i t}, \quad t > 0 . \qquad (7.94)$$

This defines the internal part in a general form, but we still do not have the partial fraction coefficients A_i. So we take the first $N-1$ derivatives of the desired solution

$$
\begin{array}{rcl}
y(t) & = & y_{\mathrm{ext}}(t) \quad + \quad y_{\mathrm{int}}(t) \\
\dot{y}(t) & = & \dot{y}_{\mathrm{ext}}(t) \quad + \quad \dot{y}_{\mathrm{int}}(t) \\
\ddot{y}(t) & = & \ddot{y}_{\mathrm{ext}}(t) \quad + \quad \ddot{y}_{\mathrm{int}}(t) \\
& \vdots & \\
y^{(N-1)}(t) & = & y_{\mathrm{ext}}^{(N-1)}(t) \quad + \quad y_{\mathrm{int}}^{(N-1)}(t) .
\end{array}
\qquad (7.95)
$$

Here the external part $y_{\text{ext}}(t) = \mathcal{L}^{-1}\{H(s)X(s)\}$ is already known and so are its derivatives. The internal part is given by (7.94), or another corresponding form in the case of multiple poles. For simple poles, the derivatives of the internal part are

$$y_{\text{int}}^{(n)}(t) = \sum_{i=1}^{N} A_i s_i^n e^{s_i t}, \quad t > 0. \tag{7.96}$$

In the solution $y(t)$ and its derivatives we only know the initial values y_i, $i = 1, \ldots, N$ at $t = 0$. That is sufficient, however, to set up a system of linear equations for the partial fraction coefficients A_i from (7.95)

$$
\begin{bmatrix}
1 & 1 & \cdots & 1 \\
s_1 & s_2 & \cdots & s_N \\
s_1^2 & s_2^2 & \cdots & s_N^2 \\
\vdots & \vdots & & \vdots \\
s_1^{N-1} & s_2^{N-1} & \cdots & s_N^{N-1}
\end{bmatrix}
\begin{bmatrix}
A_1 \\
A_2 \\
\vdots \\
A_N
\end{bmatrix}
=
\begin{bmatrix}
y_0 - y_{\text{ext}}(0) \\
y_1 - \dot{y}_{\text{ext}}(0) \\
y_2 - \ddot{y}_{\text{ext}}(0) \\
\vdots \\
y_{N-1} - y_{\text{ext}}^{(N-1)}(0)
\end{bmatrix}. \tag{7.97}
$$

After solving these equations by standard methods, we obtain the solution for the output signal in the form

$$\boxed{y(t) = y_{\text{ext}}(t) + y_{\text{int}}(t) = \mathcal{L}^{-1}\{H(s)X(s)\} + \sum_{i=1}^{N} A_i e^{s_i t}, \quad t > 0.} \tag{7.98}$$

The advantage of determining the internal part in the time-domain is that we only need the information that is provided directly by the problem, in the form of (7.87).

- The system function $H(s)$ can be obtained directly from the differential equation (see Chapter 6.4).

- The external part is given by the inverse Laplace transform of the system function and the Laplace transform of the input signal $y_{\text{ext}}(t) = \mathcal{L}^{-1}\{H(s)X(s)\}$. The inverse transform can be carried out using the partial fraction form.

- A general form of the internal part with unknown factors can be obtained from the denominator polynomial by partial fraction analysis and inverse Laplace tranformation.

- The unknown factors are determined so that the solution fulfills the initial conditions set in (7.87).

Most of the effort required to carry out this procedure is in the formation of the derivatives of the external part $y_{\text{ext}}(t)$ and in the solution of the equation system. Both are relatively easy to carry out, if numerical values are given for the coefficients of the differential equation.

--- **Example 7.9**

We are looking for the output signal $y(t)$ of a first-order system for $t > 0$, for a given input signal $x(t)$:

$$\dot{y} + 0.1\,y = x, \qquad y(0) = 8, \qquad x(t) = 10.1\sin t, \quad t > 0. \qquad (7.99)$$

The system function is

$$H(s) = \frac{1}{s + 0.1} \qquad \text{Re}\{s\} > -0.1. \qquad (7.100)$$

The external part follows with

$$x(t) = 10.1\sin t \cdot \varepsilon(t) \qquad \circ\!\!-\!\!\bullet \qquad X(s) = \frac{10.1}{s^2 + 1} \qquad \text{Re}\{s\} > 0 \qquad (7.101)$$

by splitting $X(s)H(s)$ into partial fractions:

$$Y_{\text{ext}}(s) = X(s)H(s) = \frac{10.1}{(s^2 + 1)(s + 0.1)} = \frac{A}{s + 0.1} + \frac{Bs + C}{s^2 + 1}. \qquad (7.102)$$

The coefficients are obtained either as in Example 7.5, or by equating coefficients and solution of a linear 3×3 equation system. The result is $A = 10$, $B = -10$, $C = 1$. Inverse transformation yields the external part:

$$Y_{\text{ext}}(s) \quad = \quad \frac{10}{s + 0.1} \quad - \quad \frac{10s}{s^2 + 1} \quad + \quad \frac{1}{s^2 + 1}$$

$$\qquad\qquad\qquad\qquad\quad \bullet \qquad\qquad \bullet \qquad\qquad \bullet \qquad\qquad\qquad (7.103)$$
$$\qquad\qquad\qquad\qquad\quad \big\downarrow \qquad\qquad \big\downarrow \qquad\qquad \big\downarrow$$
$$\qquad\qquad\qquad\qquad\quad \circ \qquad\qquad \circ \qquad\qquad \circ$$

$$y_{\text{ext}}(t) \quad = \quad 10e^{-0.1t} \quad - \quad 10\cos t \quad + \quad \sin t \quad , \quad t > 0.$$

The general form of the internal part is obtained from the denominator of the system function:

$$Y_{\text{int}}(s) = \frac{a}{s + 0.1} \qquad \bullet\!\!-\!\!\circ \qquad y_{\text{int}}(t) = ae^{-0.1t}, \quad t > 0. \qquad (7.104)$$

The complete solution is the sum of both parts:

$$y(t) = y_{\text{ext}}(t) + y_{\text{int}}(t) = (10 + a)e^{-0.1t} - 10\cos t + \sin t, \quad t > 0. \qquad (7.105)$$

If fulfills the differential equation and the initial conditions for $a = 8$. The Figures 7.7, 7.8 and 7.9 show the internal part, the external part and the complete solution.

--- ∎

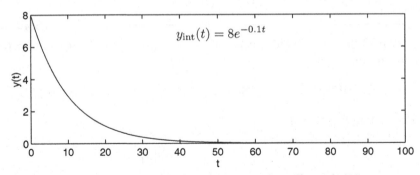

Figure 7.7: Internal part of the system response from Example 7.9

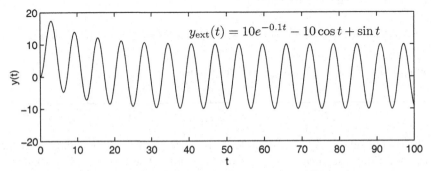

Figure 7.8: External part of the system response from Example 7.9

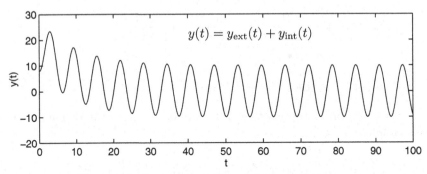

Figure 7.9: Complete solution of Example 7.9

7.4 Assessment of the Procedures for Solving Initial Condition Problems

We now know three procedures for dealing with LTI-systems with right-sided input signals and known initial conditions:

1. the classical solution of initial condition problems in the time-domain (Sec-

tion 7.1.1),

2. system analysis with the Laplace transform completely in the frequency-domain (Sections 7.1.4, 7.2, 7.3.1, 7.3.2),

3. system analysis with the Laplace transform in the frequency-domain and calculation of the internal part in the time-domain (Section 7.3.3).

The classical solution is certainly the most complicated method in terms of calculation effort, as it works exclusively in the time-domain. It is difficult to guess a particular solution for higher-order systems. If the Laplace transform is used to find the particular solution, the classical solution turns into the third procedure.

System analysis with the Laplace transform completely in the frequency-domain is the most suitable procedure if the internal structure of the system is known, for example, an electrical or state-space representation. Then the internal part can be determined either from the initial state or the initial values. The transfer function $\mathbf{G}(s)$ and matrices $\mathbf{V}$ and $\mathbf{W}$ are obtained directly from the state-space representation.

Calculation of the internal part in the time-domain is suitable if the internal structure is unknown, and only the differential equation and the initial conditions have been given. The necessary calculation steps are simple to carry out, if the numerical values of the differential equation coefficients have been provided.

7.5 Exercises

Exercise 7.1

A causal system with transfer function $H(s) = \dfrac{1}{s+1}$ has the input signal $x(t) = -\sin(\omega_0 t)\,\varepsilon(-t) + t\,e^{-2t}\,\varepsilon(t)$. At $t \to -\infty$ the energy stores in the system have been empty.

a) Explain why only the external part and can determined in this case, and not the internal part.

b) Give the response $y(t)$ of the system.

Exercise 7.2

Solve the initial condition problem

$$\dot{y}(t) + 3y(t) = x(t), \quad t > 0$$
$$x(t) = 10\cos(4t), \quad t > 0$$
$$y(0+) = y_0$$

with the 'classical method'. Determine

a) the homogenous solution $y_h(t)$

b) a particular solution $y_s(t)$

c) the complete solution $y(t)$.

Indicate the internal and external parts of the solution.

Exercise 7.3

Solve the initial condition problem

$$\dot{y}(t) + 3y(t) = x(t), \quad t > 0$$
$$x(t) = 10 \cos(4t), \quad t > 0$$
$$y(0+) = y_0$$

with the Laplace transform. Consider the initial conditions using the differentiation theorem for the unilateral Laplace transform (4.34), as shown in Section 7.1.2.

Determine

a) $H(s)$

b) $Y(s)$

c) $y(t)$ for $t > 0$

Exercise 7.4

Derive equation (7.16) from (7.15) using the differentiation theorem of the Laplace trransform (4.34).

Exercise 7.5

The following first-order system is given as a block diagram:

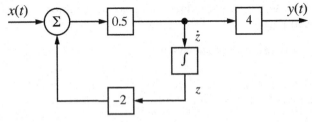

$$x(t) = \varepsilon(t) \cdot (1 - t) + \varepsilon(t - 1)(t - 1)$$
$$z(0) = z_0$$

Determine

a) the initial values $y(0-)$ and $y(0+)$.

b) the output signal $y(t)$.

Exercise 7.6

The following second-order system is given in direct form III:

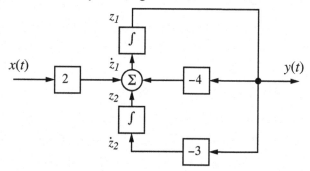

$$x(t) = \varepsilon(t) - \varepsilon(t-2)$$
$$z_1(0) = 2a$$
$$z_2(0) = 2b$$

a) Determine $y(t)$ with (7.58).

b) Can (7.58) also be used, if the block diagram is also given in direct form II?

8 Convolution and Impulse Response

8.1 Motivation

In Chapter 6 we gave the fundamental relationship

$$\boxed{Y(s) = H(s)X(s)} \tag{8.1}$$

in accordance with (6.5) between the input and output signals of an LTI-system. The system function $H(s)$ and the transform of the input signals $X(s)$ stand adjacent as Laplace transforms. They are even interchangeable. With our current knowledge, however, these seem to be two completely different kinds of functions. $X(s)$ is the Laplace transform of a function of time and can be determined from the input signal (see (4.1)): $X(s) = \mathcal{L}\{x(t)\}$, but the system function $H(s)$ is obtained, in contrast, from the given system model in the form of a differential equation, an electrical network or another form of notation. In spite of this obvious difference, however, $H(s)$ and $X(s)$ can take the same form, as we will show in two short examples.

Example 8.1

The function of time

$$x(t) = ae^{-at}\varepsilon(t), \quad a \in \mathbb{R}, \quad a > 0$$

has (see Example 4.1 or Table 4.1) the Laplace transform

$$X(s) = \mathcal{L}\{x(t)\} = \frac{a}{s+a}. \tag{8.2}$$

∎

Example 8.2

The RC-network from Figure 8.1 has transfer function

$$H(s) = \frac{Y(s)}{X(s)} = \frac{\frac{1}{sC}}{R + \frac{1}{sC}} = \frac{\frac{1}{T}}{s + \frac{1}{T}} = \frac{a}{s+a}, \qquad a = \frac{1}{T} = \frac{1}{RC} \qquad (8.3)$$

Clearly the Laplace transform of the signal $x(t) = ae^{-at}\varepsilon(t)$ and the system function of the RC-circuit have the same form. This leads us to assume that relationships exist here that we have not yet discovered. In order to shed some light on these relationships, we will first ask some questions.

1. Can the system function $H(s)$ be assigned a function of time $h(t)$, such that $H(s) = \mathcal{L}\{h(t)\}$?

2. Is it possible to excite a system with a particular input signal that produces the time function $h(t)$ as the output signal? What would this special input signal look like?

3. Can a system with system function $H(s)$ also be uniquely identified by the corresponding function of time $h(t)$ if it exists? What does the function $h(t)$ say about a system?

4. Is there an equivalent relationship to $Y(s) = H(s)X(s)$ in the time-domain, between $y(t)$, $x(t)$ and $h(t)$?

We will answer these questions in the following sections.

8.2   Time Behaviour of an RC-Circuit

8.2.1   System Function

To approach the answers to the above questions, we consider in this section the RC-circuit from Figure 8.1, with the system function $H(s)$ given in (8.3). First of all we use the inverse Laplace transform on the system function $H(s)$ and obtain the inverse of (8.2)

$$h(t) = \mathcal{L}^{-1}\{H(s)\} = \mathcal{L}^{-1}\left\{\frac{a}{s+a}\right\} = ae^{-at}\varepsilon(t) \,. \qquad (8.4)$$

Figure 8.2 shows the time behaviour.

By inverse transforming the system function $H(s)$, we have assigned it a function of time $h(t)$, although we do not actually know what it means. Taking a right-sided function for $h(t)$ is at this time arbitrary as we have not yet specified

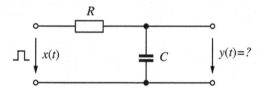

Figure 8.1: RC-circuit with rectangular input signal

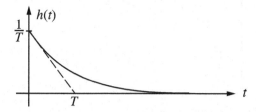

Figure 8.2: Time behaviour of the function $h(t)$ from equation (8.4) $(a = \frac{1}{T})$

a region of convergence, but we will give a justification for this choice later, for physical reasons. All the same, we can now answer question 1 in Section 8.1, with yes.

8.2.2 Response to a Rectangular Impulse

To explain the meaning of the function $h(t)$, we evaluate the response of the RC-circuit to a rectangular impulse $x(t)$ shown in Figure 8.3. The function of time and its Laplace transform are

$$x(t) = \begin{cases} \dfrac{1}{T_0} & \text{for } 0 \le t < T_0 \\ 0 & \text{otherwise} \end{cases} \qquad \circ\!\!-\!\!\bullet \qquad X(s) = \frac{1}{sT_0}\left(1 - e^{-sT_0}\right) . \qquad (8.5)$$

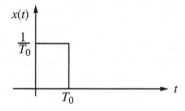

Figure 8.3: Rectangular impulse

From the methods in Chapter 6 we obtain the Laplace transform of the output signal $Y(s)$

$$Y(s) = H(s)X(s) = \frac{a}{s+a}\frac{1}{sT_0}\left(1 - e^{-sT_0}\right) = \frac{1}{T_0}\left[\frac{1}{s} - \frac{1}{s+a}\right]\left(1 - e^{-sT_0}\right) \quad (8.6)$$

and from the inverse Laplace transform, we obtain the output signal $y(t)$ itself

$$
\begin{aligned}
y(t) &= \frac{1}{T_0}\left[1 - e^{-at}\right]\varepsilon(t) - \frac{1}{T_0}\left[1 - e^{-a(t-T_0)}\right]\varepsilon(t - T_0) = \\
&= \begin{cases} \dfrac{1}{T_0}\left[1 - e^{-t/T}\right] & 0 \le t < T_0 \\[2mm] \dfrac{1}{T_0}\left[e^{T_0/T} - 1\right]e^{-t/T} & T_0 \le t \\[2mm] 0 & \text{otherwise} \end{cases}
\end{aligned} \quad (8.7)
$$

Figure 8.4 shows the behaviour of $y(t)$ for a certain value T_0.

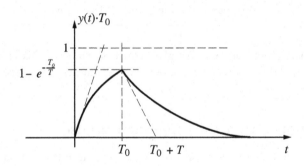

Figure 8.4: Response of the RC-circuit to a rectangular impulse

8.2.3   Response to a Very Short Rectangular Impulse

We now vary the width of the rectangular impulse at the input and investigate the effects on the output signal. Figure 8.5 shows various rectangle impulses $x_i(t)$, $i = 1, 2, 3, 4$ from (8.5) for the values $T_0 = 1$; 0.5; 0.2; 0.05. As the height of the rectangle is the reciprocal of the width, all of the impulses have unit area.

The corresponding output signals $y_i(t)$, $i = 1, 2, 3, 4$ are shown in Figure 8.6. It seems as though for smaller and smaller values of T_0, $y(t)$ becomes more and more similar to the form of $h(t)$ shown in Figure 8.2. In fact, the first term ($0 \le t < T_0$) becomes shorter, and the form of $y(t)$ becomes mainly defined by the second term ($T_0 \le t$). With the limit $T_0 \to 0$,

$$y(t) = \lim_{T_0 \to 0}\frac{e^{aT_0} - 1}{T_0}e^{-at}\varepsilon(t) = ae^{-at}\varepsilon(t) = h(t) . \quad (8.8)$$

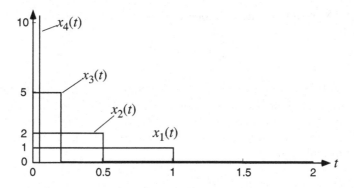

Figure 8.5: Various rectangular impulses with unit area

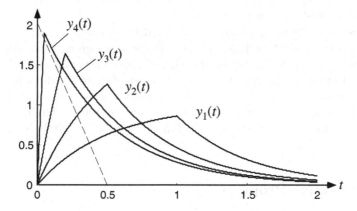

Figure 8.6: Response of the RC-circuit to different rectangle impulses

Before we answer the second question in Section 8.1 with yes, we need some insight into the nature of the input signal $x(t)$, that produces the response $y(t) = h(t)$.

8.3 The Delta Impulse

8.3.1 Introduction

In order to determine the input signal $x(t)$ that leads to the output signal $y(t) = h(t)$, we will try to use the limit $T_0 \to 0$ directly on the rectangular impulse $x(t)$

from (8.5). The result is called the delta impulse $\delta(t)$

$$\lim_{T_0 \to 0} x(t) = \delta(t) \, . \tag{8.9}$$

It is also known as the *Dirac delta function*, *Dirac impulse* and *unit impulse*.

The delta impulse $\delta(t)$ is not a function in the usual sense. Attempting to define $\delta(t)$ by a rule that assigns a value $\delta(t)$ to every time point t leads to a function that takes the value zero for $t \neq 0$ and grows beyond all limits at $t = 0$. The casual formulation

$$\delta(t) = \begin{cases} 0 & \text{for} \quad t \neq 0 \\ \infty & \text{for} \quad t = 0 \end{cases} \tag{8.10}$$

although not wrong, is of little use as it does not explain how $\delta(t)$ can be mathematically combined with other functions.

A mathematically precise definition of the delta impulse as a reliable function for which every value of an independent variable is assigned a function value is not possible. It is necessary to use the idea of *distributions* instead [17, 19]. We will refrain from a mathematically thorough formulation of the delta impulse and related distributions, and instead we will illustrate some important properties that will be useful when dealing with signals and systems. The delta impulse will be dealt with using the analogy of infinitely thin and infinitely high rectangle impulses.

8.3.2   Selective Property

The principle for use of the delta impulse and other distributions is that they are not described by their undetermineable properties, rather by their effects on other functions. Instead of describing the delta impulse by an assignment (for instance (8.10)), we see whether the value of the integral

$$\int_{-\infty}^{\infty} f(t)\delta(t)\, dt \, ,$$

describes the effect of the delta impulse on the function $f(t)$. First we abandon the delta impulse in favour of the rectangle impulse $x(t)$ from (8.5), evaluate the integral and carry out the limit $T_0 \to 0$. With (8.5),

$$\int_{-\infty}^{\infty} f(t)x(t)\, dt = \frac{1}{T_0} \int_{0}^{T_0} f(t)\, dt = \frac{F(T_0) - F(0)}{T_0} \, , \tag{8.11}$$

where $F(t)$ is an antiderivative of $f(t)$ $(F'(t) = f(t))$. The limit $T_0 \to 0$ leads to the differential quotient of $F(t)$ at $t = 0$ and yields

$$\lim_{T_0 \to 0} \int_{-\infty}^{\infty} f(t)x(t)\, dt = \lim_{T_0 \to 0} \frac{F(T_0) - F(0)}{T_0} = F'(t)\Big|_{t=0} = f(0)\,. \tag{8.12}$$

With (8.9) we obtain

$$\int_{-\infty}^{\infty} f(t)\delta(t)\, dt = f(0)\,. \tag{8.13}$$

This relationship describes the *selective property* of the delta impulse. It means that the integral of the product of a function and the delta impulse removes all function values $f(t)$ for $t \neq 0$, and selects the value $f(0)$. Note that $f(t)$ must be continuous at $t = 0$.

Figure 8.7 describes this situation. The delta impulse is represented by an arrow pointing upwards at $t = 0$. It means that $\delta(t)$ disappears for $t \neq 0$ and grows beyond all limits at $t = 0$ (compare with (8.10)).

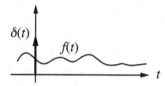

Figure 8.7: Selection of the function value $f(0)$

If $f(t) = 1$ $\forall t$, we know immediately from the selective property (8.13) that

$$\int_{-\infty}^{\infty} \delta(t)\, dt = 1\,. \tag{8.14}$$

This result is clear if the delta impulse is imagined to be the limit value of a unit area rectangle. We also say that the delta impulse has an *area* or better, a *weight* of one.

The delta impulse is not restricted to selecting a value at $t = 0$, in fact, it can be used to select any point. In the same way as in equations (8.11) to (8.13) follows

$$\int_{-\infty}^{\infty} f(t)\delta(t - t_0)\, dt = f(t_0)\,. \tag{8.15}$$

This general form of the selective property is shown in Figure 8.8.

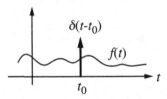

Figure 8.8: Selection of the function value $f(t_0)$

The selection property is the most important part of the definition of the delta impulse in distribution theory. With its help we can show all of the important properties of the delta impulse.

8.3.3 Impulse Response

Before considering further properties of the delta impulse, we will summarise the results obtained so far. They allow us to answer the second question in Section 8.1: it *is* possible to cause the RC-circuit depicted in Figure 8.1 to produce the function $h(t)$ at the output. The input signal that does this is the delta impulse $\delta(t)$. $h(t)$ is therefore called the *impulse response* to the RC-circuit.

The function $h(t) = \mathcal{L}^{-1}\{H(s)\}$ that was previously anonymous, is slowly revealing itself to be a powerful tool in identifying systems, but to further investigate its properties we need more information about the delta impulse.

8.3.4 Calculations with the Delta Impulse

The following derivations of calulation rules for the delta impulse depend on two principles.

- The delta impulse can only be dealt with when it is within an integral. Its properties can be expressed in terms of the effects of the selective property on other functions of time.

- Calculations with the delta impulse must be consistent with the calulation rules for ordinary functions.

We will investigate the properties of delta impulses using these two principles. Another possiblity would be using rectangle functions of width T_0 and the limit $T_0 \to 0$, as we did when deriving the selective property(see (8.11) to (8.13)). Using this limit is cumbersome, however, so we will stick with the elegance of the selective property.

8.3.4.1   Linear Combination of Delta Impulses

The most elementary property of delta impulses is their behaviour when added and multiplied with factors, i.e., in expressions of the form $a\delta(t) + b\delta(t)$. The properties of these linear combinations can be investigated using the selective property

$$\int_{-\infty}^{\infty} [a\delta(t) + b\delta(t)]f(t)dt = a\int_{-\infty}^{\infty} \delta(t)f(t)dt + b\int_{-\infty}^{\infty} \delta(t)f(t)dt$$

$$= af(0) + bf(0) = (a+b)f(0) . \tag{8.16}$$

The linear combination $a\delta(t) + b\delta(t)$ has the same effect on a function $f(t)$ as an individual delta impulse $(a+b)\delta(t)$ with weight $(a+b)$. Therefore

$$\boxed{a\delta(t) + b\delta(t) = (a+b)\delta(t) .} \tag{8.17}$$

8.3.4.2   Scaling the Time-Axis

Scaling the time-axis occurs when the unit of time is changed or with normalisation. It may seem that the delta impulse is not affected by such operations as it is zero except at $t = 0$. Attempting to derive the properties of the delta impulse from its degenerate time behaviour is deceptive, however, as we will show in the following investigation using the selective property.

To explain the properties of a delta impulse with the argument at, we start with (8.13) and with the substitution $\tau = at$, $a \in \mathbb{R}$, we obtain

$$\int_{-\infty}^{\infty} \delta(at)f(t)dt = \frac{1}{|a|}\int_{-\infty}^{\infty} \delta(\tau)f\left(\frac{\tau}{a}\right)d\tau = \frac{1}{|a|}f(0) . \tag{8.18}$$

Taking the magnitude of a is necessary, which is clear if the substitution $\tau = at$ is carried out with different signs of a.

The delta impulse $\delta(at)$ has the same effect on the function $f(t)$ as a delta impulse $\frac{1}{|a|}\delta(t)$ with weight $\frac{1}{|a|}$. Therefore,

$$\boxed{\delta(at) = \frac{1}{|a|}\delta(t) .} \tag{8.19}$$

Delta impulses with weighting factors are depicted as a vertical arrow showing the corresponding weight. The scaled delta impulse from (8.19) has the representation shown in Figure 8.9.

From (8.19) it is immediately yielded when $a = -1$, that the delta impulse is even, so

$$\boxed{\delta(t) = \delta(-t) .} \tag{8.20}$$

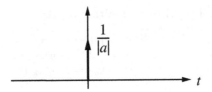

Figure 8.9: Scaled delta impulse

8.3.4.3 Multiplication with a Continuous Function

From the selective property,

$$\int\limits_{-\infty}^{\infty} f(t)\delta(t)dt = \int\limits_{-\infty}^{\infty} f(0)\delta(t)dt = f(0) \,. \tag{8.21}$$

As all values of $f(t)$ at $t \neq 0$ are removed, multiplying a function by a delta impulse corresponds to weighting the delta impulse with the function value at $t = 0$.

$$\boxed{f(t)\delta(t) = f(0)\delta(t)} \tag{8.22}$$

This statement is only true – just like selective property (8.13) – for functions $f(t)$ that are continuous at $t = 0$. In particular, the product of two delta impulses $\delta(t) \cdot \delta(t)$ is not permitted.

8.3.4.4 Derivation

The investigation of the delta impulse with the selective property can be expanded to other properties, for example, forming a derivative. Differentiation of $\delta(t)$ by forming the differential quotient is of course not possible, as the delta impulse cannot be differentiated. Despite this, a comparable operation can, in fact, be used on distributions. To distinguish this from differentiating ordinary functions, however, the term *derivation* is used.

The derivative of the delta impulse has the same notation as the symbol for a derivative with respect to time, written $\dot{\delta}(t)$. To explain what should be understood by this, we will again use the selective property. Under the integral, the derivative of $\delta(t)$ can be transformed using partial integration of the corresponding continuous function, and the selective property of the undefined function $\dot{\delta}(t)$ can be represented by the effect of a delta impulse $\delta(t)$.

$$\int\limits_{-\infty}^{\infty} \dot{\delta}(t)f(t)dt = \delta(t)f(t)\Big|_{-\infty}^{\infty} - \int\limits_{-\infty}^{\infty} \delta(t)\dot{f}(t)dt = -\dot{f}(0) \,. \tag{8.23}$$

As the term $\delta(t)f(t)$ disappears for $t \to \pm\infty$, only the negative derivative of $f(t)$ at $t = 0$ remains. We can interpret the derivative $\dot{\delta}(t)$ as a distribution that forms the value of the derivative $-\dot{f}(0)$ from a function $f(t)$, using the selective property. The derivation of $\delta(t)$ is revealed by its effect on the continuous function $f(t)$.

Unfortunately we cannot use rectangle functions with a limit to illustrate $\dot{\delta}(t)$, as they are not differentiable, but we could, however, have introduced the delta impulse as another impulse function with a limit. We only chose the rectangle function because it was particularly simple to integrate (8.11).

The corresponding limit for a bell-shaped impulse function $d(t)$ with characteristic width T is illustrated in Figure 8.10 (top). In contrast to the rectangle function, the derivative $\dot{d}(t)$ of the bell-shaped function can be formed. The limit for $T \to 0$ then leads to the derivative $\dot{\delta}(t)$ of $\delta(t)$ shown in Figure 8.10 (below). Because of the double impulse upwards and downwards it is also called a *doublette* [19]. It should also be emphasised here that the graphical representaion with arrows on the right-hand side of Fig 8.10 is for illustrative purposes only. An exact description of distributions is only possible using their effects on ordinary functions (see (8.13) and (8.23)).

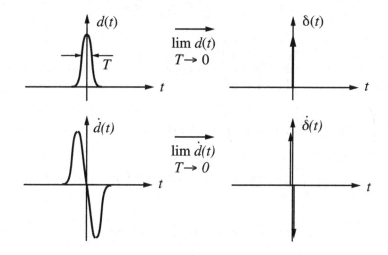

Figure 8.10: Graphical illustration of $\dot{\delta}(t)$

Higher derivatives of $\delta^{(n)}(t)$ can also be introduced in the same way as $\dot{\delta}(t)$. They form the nth derivative of a function $f(t)$ at the point $t = 0$ (with a change of signs where appropriate)

$$\int_{-\infty}^{\infty} \delta^{(n)}(t)f(t)\,dt = (-1)^n f^{(n)}(0)\,. \tag{8.24}$$

By choosing certain functions $f(t)$, different relationships between the delta impulse $\delta(t)$ and its derivatives can be obtained. For example, from (8.23) we let $f(t) = -t$, and equating the integrands we obtain the interesting relation

$$-t\dot{\delta}(t) = \delta(t) \ . \tag{8.25}$$

As $\delta(t)$ is an even distribution and $-t$ is an odd function, $\dot{\delta}(t)$ must be odd:

$$\dot{\delta}(-t) = -\dot{\delta}(t) \ . \tag{8.26}$$

Of course, we could have guessed that this was the case from Figure 8.10.

8.3.4.5 Integration

We will consider integration of the delta impulse as the inverse of derivation

$$\int\limits_{-\infty}^{t} \delta(\tau)d\tau = \int\limits_{-\infty}^{\infty} \delta(\tau)\varepsilon(t - \tau)d\tau \ . \tag{8.27}$$

The inclusion of the step function $\varepsilon(t)$ makes it possible to form the integral without a variable upper limit. The integrals appearing within the selective property can be interchanged, and then,

$$\int\limits_{-\infty}^{\infty} \left(\int\limits_{-\infty}^{t} \delta(\tau)\,d\tau \right) f(t)\,dt = \int\limits_{-\infty}^{\infty}\int\limits_{-\infty}^{\infty} \delta(\tau)\varepsilon(t - \tau)d\tau\,f(t)\,dt = \tag{8.28}$$

$$= \int\limits_{-\infty}^{\infty} \delta(\tau) \int\limits_{-\infty}^{\infty} \varepsilon(t - \tau)f(t)\,dt d\tau = \int\limits_{-\infty}^{\infty} \delta(\tau) \int\limits_{\tau}^{\infty} f(t)\,dt d\tau = \int\limits_{0}^{\infty} f(t)\,dt \ .$$

The same effect on a function $f(t)$ can be achieved with the step function $\varepsilon(t)$, however,

$$\int\limits_{-\infty}^{\infty} \varepsilon(t)f(t)dt = \int\limits_{0}^{\infty} f(t)dt \ . \tag{8.29}$$

We can therefore identify the integral over the delta impulse $\delta(t)$ as the step function $\varepsilon(t)$

$$\int\limits_{-\infty}^{t} \delta(\tau)d\tau = \varepsilon(t) \ . \tag{8.30}$$

Reversed, the delta impulse is the derivative of the step function

$$\boxed{\dot{\varepsilon}(t) = \delta(t) \; .}$$ (8.31)

Figure 8.11 shows the relationship between the delta impulse and the step function.

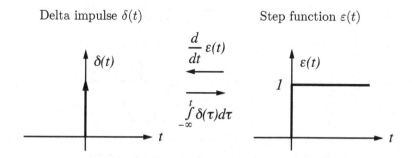

Figure 8.11: Delta impulse and step function

8.3.5   Using Delta Impulses

The calculation rules we have just learnt show that the delta impulse can be dealt with almost like an ordinary function. With this extended repertoire of functions, many problems can be solved more elegantly, where before, it was cumbersome using exclusively ordinary functions. We will examine two such examples.

-- **Example 8.3**

Figure 8.12 shows a signal $x(t)$ (top left), that cannot be differentiated because it is discontinuous at one point. Forming a derivative in the normal way is not possible. If the right-sided and left-sided limits of the differential quotient (not the function itself!) are equal at the point of discontinuity, the signal $x(t)$ can, however, be represented as the sum of a function that can be differentiated and the step function (Figure 8.12 top right). The step height a corresponds to the difference at the discontinuity. With differentiation of the continuous part and derivation of the step function, the derivative of the continuous part and a delta impulse with weight a is obtained (Figure 8.12 bottom right). Both terms can be put together to give the derivation of the discontinuous signal, which has no derivative in the normal sense. The discontinuity with step height a appears in the derivation as a delta impulse with weight a (Figure 8.12 bottom left).

-- ∎

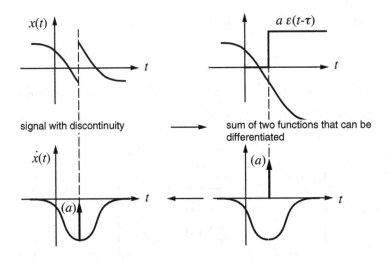

Figure 8.12: Analysing signals with discontinuities

Example 8.4

The equation for the motion of a puck that slides on a frictionless surface comes directly from the formula Force = Mass × Acceleration:

$$\ddot{y}(t) = \frac{1}{m}x(t) , \qquad (8.32)$$

where $x(t)$ is the force on the puck, m is its mass and $y(t)$ is its displacement (see Figure 8.13). Both force and displacement are independent quantities.

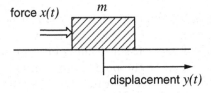

Figure 8.13: A puck sliding on ice

If the puck is uniformly set in motion at $t = 0$ by an ice hockey stick, the displacement increases proportionally with time (Figure 8.14 top). The force could be found by differentiating the displacement twice, if $y(t)$ could be differentiated. Because of the step at $t = 0$, however, this is not the case, and again we need the help of distributions. Derivation of the displacement yields the velocity in the form of a step function. Repeating the derivation then yields force in the form of a delta impulse. The delta impulse here is the idealized strike of the puck with the hockey stick.

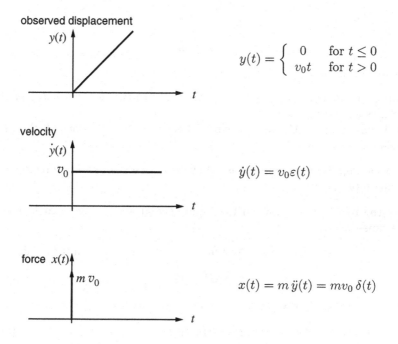

$$y(t) = \begin{cases} 0 & \text{for } t \leq 0 \\ v_0 t & \text{for } t > 0 \end{cases}$$

$$\dot{y}(t) = v_0 \varepsilon(t)$$

$$x(t) = m\,\ddot{y}(t) = mv_0\,\delta(t)$$

Figure 8.14: Displacement, velocity and applied force of a struck puck

8.4 Convolution

After we have got to know the delta impulse in detail, we will return to answer the remaining questions in Section 8.1. The impulse response briefly introduced in Section 8.3 will play an important part in this.

8.4.1 Describing Systems with the Impulse Response

In Chapter 3 we were interested for the first time in describing the transfer properties of LTI-systems. We could show only using the definitions of linearity and time-invariance that functions of the form e^{st} are eigenfunctions of LTI-systems. This leads directly to the system function as a system model of LTI-systems in the frequency-domain.

After we identified the function of time $h(t) = \mathcal{L}^{-1}\{H(s)\}$ as another system description and interpreted it as the system response to a delta impulse, we now derive an equivalent relationship to (8.1) in the time-domain. To do this we only need to use

- the impulse response $h(t)$ of an LTI-system,

- the selective property of the delta impulse,

- the definitions of linearity and time-invariance.

The result is then general and does not only apply to the network considered in Section 8.2 as an introduction.

We will start with an LTI-system with a known impulse response $h(t)$ and we want to show

- that the impulse response $h(t)$ is sufficient to describe the system's reaction to any input signal,

- how the relationship between the inputs signal and output signal looks in the time-domain.

First of all we describe the LTI-system with its reaction to an input signal

$$y(t) = \mathcal{S}\{x(t)\} .\tag{8.33}$$

In particular, the reaction to a delta impulse is

$$h(t) = \mathcal{S}\{\delta(t)\} .\tag{8.34}$$

Then we use the selective property (8.15), combined with (8.20) to express the input signal as a function of itself

$$x(t) = \int_{-\infty}^{\infty} x(\tau)\delta(t - \tau)d\tau .\tag{8.35}$$

With (8.33) we find the output signal

$$y(t) = \mathcal{S}\left\{ \int_{-\infty}^{\infty} x(\tau)\delta(t - \tau)d\tau \right\} .\tag{8.36}$$

In the integral only $\delta(t - \tau)$ depends on t; the values $x(\tau)$ are in terms of t only weighting factors. Because of the linearity of the system

$$y(t) = \int_{-\infty}^{\infty} x(\tau)\mathcal{S}\{\delta(t - \tau)\}d\tau .\tag{8.37}$$

Finally, from the time-invariance and (8.34),

$$y(t) = \int_{-\infty}^{\infty} x(\tau)h(t-\tau)d\tau \,. \tag{8.38}$$

This is our goal, as (8.38) shows that only knowledge of the impulse response of an LTI-system is needed to find its reaction to any input signal $x(t)$.

The combination of two functions of time $f(t)$ and $g(t)$ (8.38) is called *convolution* and is denoted by *.

$$\int_{-\infty}^{\infty} f(\tau)g(t-\tau)d\tau = f(t) * g(t)$$

With the substitution $\tau' = t - \tau$ it is easy to show that convolution is commutative

$$f(t) * g(t) = g(t) * f(t) \,.$$

8.4.2 Impulse Response and System Function

The answer to question 4 in Section 8.1 is now clear: just as the transform of the output signal $Y(s)$ can be obtained in the frequency-domain by multiplying the transform of the input signal $X(s)$ with the system function $H(s)$, the output signal in the time-domain is given by the convolution of the input signal with the impulse response.

$$X(s) \longrightarrow \boxed{H(s)} \longrightarrow Y(s) = H(s) \cdot X(s)$$

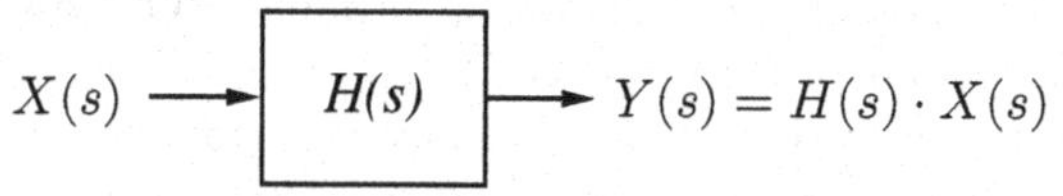

Figure 8.15: A system in the frequency-domain

Figures 8.15 and 8.16 show how multiplication with the system function in the frequency-domain and convolution with the impulse response in the time-domain are equivalent. The connection can also be formally expressed. The relation we already know in the frequency domain follows from the convolution relation

$$y(t) = h(t) * x(t) = \int_{-\infty}^{\infty} h(t-\tau)x(\tau)\,d\tau \tag{8.39}$$

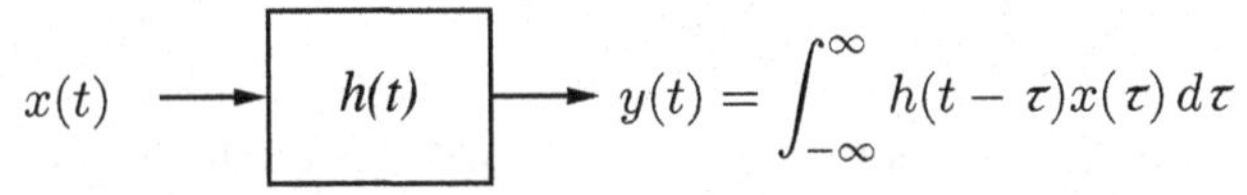

$$x(t) \longrightarrow \boxed{h(t)} \longrightarrow y(t) = \int_{-\infty}^{\infty} h(t-\tau)x(\tau)\,d\tau$$

Figure 8.16: A system in the time-domain

using the Laplace transform, swapping the integrals and collecting together the terms.

$$
\begin{aligned}
\mathcal{L}\{y(t)\} &= \int_{-\infty}^{\infty}\left[\int_{-\infty}^{\infty} h(t-\tau)x(\tau)\,d\tau\right]e^{-st}\,dt \\
&= \int_{-\infty}^{\infty}\int_{-\infty}^{\infty} h(t-\tau)e^{-st}\,dt\, x(\tau)\,d\tau \\
&= \int_{-\infty}^{\infty} H(s)e^{-s\tau}x(\tau)\,d\tau = H(s)\int_{-\infty}^{\infty} e^{-st}x(t)\,dt = H(s)\cdot X(s)
\end{aligned}
$$

$$(8.40)$$

We have now shown the *convolution theorem* of the Laplace transform:

$$\boxed{h(t)*x(t) \quad \circ\!\!-\!\!\bullet \quad H(s)X(s)\,, \quad s\in \mathrm{ROC}\{h*x\} \supseteq \mathrm{ROC}\{x\}\cap \mathrm{ROC}\{h\}\,.}$$

$$(8.41)$$

It is not only true when $h(t)$ is an impulse response and $x(t)$ is an input signal, but also generally for all pairs of functions of time whose Laplace transforms exist.

The region of convergence of the convolution product $\mathrm{ROC}\{h*x\}$ is the intersection between the regions of convergence $\mathrm{ROC}\{x\}$ and $\mathrm{ROC}\{h\}$. These two regions of convergence must overlap for the Laplace transform of the convolution product to exist. In (8.40) we had set out the condition that the complex frequency variable s can only take values for which both $X(s)$ and $H(s)$ exist. The region of convergence $\mathrm{ROC}\{h*x\}$ can, however, actually be larger than the intersection of $\mathrm{ROC}\{x\}$ and $\mathrm{ROC}\{h\}$ if, for example, poles are cancelled out by zeros. Often when $\mathcal{L}\{h(t)*x(t)\}$ does not exist, it means that the convolution product (8.38) does not exist itself. Expression (8.38) is indeed an improper integral that only converges under certain conditions.

The $\mathrm{ROC}\{h\}$ of the system function $H(s)$ that we have so far not paid much attention to, can be derived from the connection between the system function and impulse response

$$H(s) = \mathcal{L}\{h(t)\}\,. \tag{8.42}$$

The normal rules governing the region of the Laplace transform apply in this case, and as they are all given in Chapter 4.5.3 we will not repeat them here.

An interesting consequence can be obtained from the convolution theorem (8.41), if a delta impulse is input as $x(t)$.

$$h(t) * \delta(t) \circ\!\!-\!\!\bullet H(s)\mathcal{L}\{\delta(t)\} \tag{8.43}$$

As $h(t) * \delta(t) = h(t)$, however,

$$\boxed{\mathcal{L}\{\delta(t)\} = 1 \,.} \tag{8.44}$$

This can be immediately confirmed with the selective property

$$\mathcal{L}\{\delta(t)\} = \int_{-\infty}^{\infty} \delta(t)\, e^{-st}\, dt = 1, \qquad \text{ROC} = \mathbb{C} \,. \tag{8.45}$$

The region of convergence encloses the entire complex plane, as the delta impulse is a signal of finite duration.

Just as the number 1 is the unity element of multiplication, the delta impulse is the unity element of convolution. Figure 8.17 illustrates this.

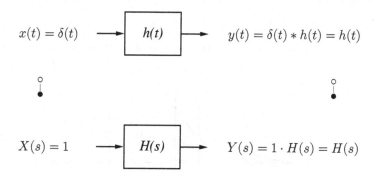

$$x(t) = \delta(t) \quad\longrightarrow\quad \boxed{h(t)} \quad\longrightarrow\quad y(t) = \delta(t) * h(t) = h(t)$$

$$X(s) = 1 \quad\longrightarrow\quad \boxed{H(s)} \quad\longrightarrow\quad Y(s) = 1 \cdot H(s) = H(s)$$

Figure 8.17: LTI-system with a delta impulse as input function

With the convolution theorem we can also easily confirm that e^{st} is the eigenfunction of LTI-systems. In Chapter 3 we derived this property from the concepts of linearity and time-invariance. Convolution for $x(t) = e^{st}$ gives

$$y(t) = \int_{-\infty}^{\infty} h(\tau)\, e^{s(t-\tau)}\, d\tau = e^{st} \int_{-\infty}^{\infty} h(\tau)\, e^{-s\tau}\, d\tau = e^{st} H(s) \,, \quad s \in \text{ROC}\{h\} \,. \tag{8.46}$$

The response to an exponential function is likewise an exponential function, multiplied by the system function $H(s)$. The convolution integral clearly converges when the complex frequency s of the the input signal $x(t)$ lies in the region of convergence of the system function.

8.4.3 Calculating the Convolution Integral

Calculation of the convolution integral (8.39) requires some practice. It is especially important to make the distinction between the time variable t and the integration variable τ in (8.39). We will deal with this subtle but important difference in detail, in a simple example where we will calculate the response of an RC-circuit to a rectangle impulse. In Section 8.2.2 this problem has already been solved in the frequency-domain and the result is shown in Figure 8.4. This time we will stay in the time-domain and use the convolution theorem.

First we consider the function of time $h(t)$ from Figure 8.2, which we already know represents the impulse response. In Figure 8.18 above, $h(\tau)$ is plotted against the variable of integration τ. To change it to the form $h(t - \tau)$ from (8.39), we first have to turn the right-sided function to the left $(h(-\tau))$ and shift it by t $(h(t - \tau))$. Figure 8.18 (middle) shows the result for various values of t. Note that t takes a fixed value while being integrated over τ. The result $y(t)$ is dependent on t because the convolution integral is calculated for many values of t.

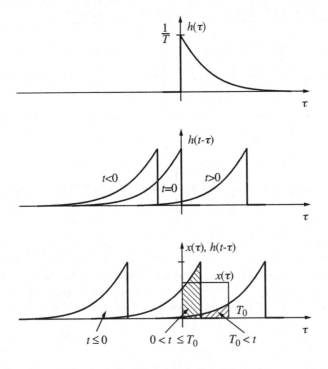

Figure 8.18: The convolution integral

The integrand $h(t - \tau)x(\tau)$ is obtained by multiplication with $x(\tau)$. The rectangular form of $x(\tau)$ means that the integral can only have non-zero values for

$0 \leq \tau \leq T_0$. Here there are three cases to be distinguished (Figure 8.18 bottom).

- For $t \leq 0$ there is no overlap of $h(t - \tau)$ and $x(\tau)$. The product $h(t - \tau)x(\tau)$ and likewise the integral have the value zero.

- For $0 < t < T_0$ $h(t - \tau)$ and $x(\tau)$ partially overlap so the upper limit of the integral depends on t. The result is

$$
\begin{aligned}
y(t) &= \int_0^t h(t - \tau) \frac{1}{T_0} \, d\tau = \frac{1}{T_0} \int_0^t \frac{1}{T} e^{-(t-\tau)/T} \, d\tau = \\
&= \frac{1}{T_0} \frac{1}{T} e^{-t/T} \int_0^t e^{\tau/T} \, d\tau = \\
&= \frac{1}{T_0} \frac{1}{T} e^{-t/T} T e^{\tau/T} \Big|_0^t = \frac{1}{T_0} \left[1 - e^{-t/T} \right] .
\end{aligned}
\tag{8.47}
$$

- For $T_0 \leq t$, $h(t - \tau)$ and $x(\tau)$ overlap completely and the integration must be carried out between 0 and T_0. We obtain:

$$
\begin{aligned}
y(t) &= \int_0^{T_0} h(t - \tau) \frac{1}{T_0} \, d\tau = \frac{1}{T_0} \int_0^{T_0} \frac{1}{T} e^{-(t-\tau)/T} \, d\tau = \\
&= \frac{1}{T_0} \frac{1}{T} e^{-t/T} \int_0^{T_0} e^{\tau/T} \, d\tau = \\
&= \frac{1}{T_0} \frac{1}{T} e^{-t/T} T e^{\tau/T} \Big|_0^{T_0} = \frac{1}{T_0} \left[e^{T_0/T} - 1 \right] e^{-t/T} .
\end{aligned}
\tag{8.48}
$$

Summarised, the result of convolution is

$$
\begin{aligned}
y(t) &= \frac{1}{T_0} \left[1 - e^{-t/T} \right] \varepsilon(t) - \frac{1}{T_0} \left[1 - e^{-(t-T_0)/T} \right] \varepsilon(t - T_0) = \\
&= \begin{cases} \frac{1}{T_0} \left[1 - e^{-t/T} \right] & 0 \leq t < T_0 \\ \frac{1}{T_0} \left[e^{T_0/T} - 1 \right] e^{-t/T} & T_0 \leq t \\ 0 & \text{otherwise}, \end{cases}
\end{aligned}
\tag{8.49}
$$

as we obtained in the frequency-domain calculation in (8.7). Because of the three cases, evaluating the convolution integral in the time-domain is significantly more complicated even for the simple case considered here.

8.4.4   Impulse Response of Special Systems

Now that we have investigated the relationship between the system function $H(s)$ in the frequency-domain and the impulse response $h(t)$ in the time-domain, we will consider the impulse responses of some special systems.

8.4.4.1   Integrators

First of all we would like to know about the properties of a system that has the step function $\varepsilon(t)$ as its impulse response. From the convolution of $\varepsilon(t)$ and $x(t)$, we find that with (8.39) the output signal is

$$y(t) = \varepsilon(t) * x(t) = \int_{-\infty}^{\infty} \varepsilon(t - \tau)x(\tau)\,d\tau\,. \tag{8.50}$$

Since

$$\varepsilon(\tau) = \begin{cases} 1 & \tau \geq 0 \\ 0 & \tau < 0 \end{cases} \quad \text{and} \quad \varepsilon(t - \tau) = \begin{cases} 1 & \tau \leq t \\ 0 & t < \tau \end{cases} \tag{8.51}$$

we can leave out the step function under the integral in (8.50) and instead only integrate over the values of τ for which $\varepsilon(t-\tau) = 1$, that is, between $-\infty < \tau \leq t$. It should again be noted that for integration over τ, the time t should be seen as a fixed parameter. We obtain the system description

$$y(t) = \varepsilon(t) * x(t) = \int_{-\infty}^{t} x(\tau)\,d\tau\,. \tag{8.52}$$

A system with the step response $\varepsilon(t)$ as impulse response therefore leads to an integration of the input signal, or in short: it is an integrator.

This result was also found in Section 8.3.4.5, where we said that integrating a delta impulse yielded a step function. Figure 8.19 shows a block diagram of an integrator and also its response to a delta impulse at the input. The convolution

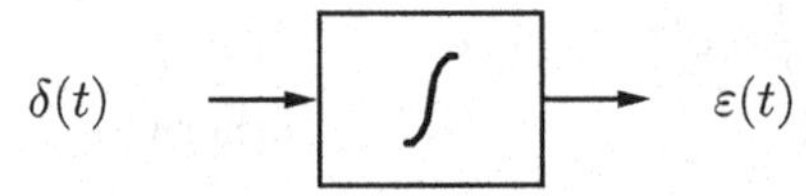

$$\delta(t) \quad \longrightarrow \quad \boxed{\int} \quad \longrightarrow \quad \varepsilon(t)$$

Figure 8.19: An integrator responds to a delta impulse with a step function.

theorem (8.41) for $h(t) = \varepsilon(t)$ is

$$\varepsilon(t) * x(t) \quad \circ\!\!-\!\!\bullet \quad \frac{1}{s}X(s), \quad s \in \mathrm{ROC} \supseteq \mathrm{ROC}\{x\} \cap \{s : \mathrm{Re}\{b\} > 0\}\,.$$

(8.53)

This corresponds to the integration theorem for the Laplace transform (4.27). The transfer function of the integrator is

$$H(s) = \mathcal{L}\{\varepsilon(t)\} = \frac{1}{s}\,, \quad \mathrm{Re}\{s\} > 0\,. \tag{8.54}$$

8.4.4.2 Differentiator

In Section 8.3.4.4 we got to know $\dot{\delta}(t)$ as the derivative of the delta impulse with respect to time. We would therefore guess that $\dot{\delta}(t)$ would be the impulse response for a differentiator. We can confirm that this is the case if we find the convolution of an input signal $x(t)$ with the suggested impulse response:

$$y(t) \;\; = \dot{\delta}(t) * x(t) = \int\limits_{-\infty}^{\infty} \dot{\delta}(t-\tau)x(\tau)d\tau$$

(8.55)

$$= -\int\limits_{-\infty}^{\infty} \dot{\delta}(\tau-t)x(\tau)d\tau = \dot{x}(t)$$

In (8.55) we used the selective property of $\dot{\delta}(t)$ (8.23) as well as the fact that we know $\dot{\delta}(t)$ is odd (8.26). Convolution with the impulse response $\dot{\delta}(t)$ yields the derivative of a function with respect to time.

The system function for a differentiator is

$$H(s) = \mathcal{L}\{\dot{\delta}(t)\} = \int\limits_{-\infty}^{\infty} \dot{\delta}(t)e^{-st}dt = s\,, \quad s \in \mathbf{C}\,. \tag{8.56}$$

We can now re-write the differentiation theorem of the Laplace transform (4.26) directly as a special case of the convolution theorem (8.41):

$$\dot{x}(t) = \dot{\delta}(t) * x(t) \quad \circ\!\!-\!\!\bullet \quad sX(s), \quad s \in \mathrm{ROC} \supseteq \mathrm{ROC}\{x\}\,. \tag{8.57}$$

Interestingly enough we can also use (8.57) on signals with discontinuities or on distributions, if we interpret $\dot{x}(t)$ as the derivation of $x(t)$, which may contain steps, delta impulses and derivatives of delta impulses. We can then avoid having to deal with the initial value $x(0+)$ separately in the differentiation theorem for right-sided signals (4.34) or the step in the differentiation theorem (4.41). This is illustrated in the following example.

Example 8.5

Here we will calculate the Laplace transform of the triangular impulse (Example 4.10) once more. First we form the second derivation of $x(t)$.

$$\dot{x}(t) \quad = \quad x(t) * \dot{\delta}(t) \quad = \quad \frac{1}{(2T)^2}\varepsilon(t - 2T) - \frac{2}{(2T)^2}\varepsilon(t) + \frac{1}{(2T)^2}\varepsilon(t + 2T)$$

$$\ddot{x}(t) \quad = \quad \dot{x}(t) * \dot{\delta}(t) \quad = \quad \frac{1}{(2T)^2}\delta(t - 2T) - \frac{2}{(2T)^2}\delta(t) + \frac{1}{(2T)^2}\delta(t + 2T)$$

The Laplace transform yields:

$$s^2 X(s) = \frac{1}{(2T)^2}\left(e^{2sT} - 2 + e^{-2sT}\right) = \frac{1}{(2T)^2}\left[e^{sT} - e^{-sT}\right]^2 .$$

The result

$$X(s) = \left[\frac{\sinh(sT)}{sT}\right]^2$$

is then immediately obtained, and it corresponds to (4.45).

8.4.4.3   Delay Circuits

A delay circuit is a system that reproduces the input signal at the output after a fixed time delay t_0. Apart from the delay, the signal remains unchanged. Therefore the system response of a delay circuit to a delta impulse $\delta(t)$ must be a delta impulse delayed by t_0 (see Figure 8.20)

$$h(t) = \delta(t - t_0) . \tag{8.58}$$

Figure 8.20: Impulse response of a delay circuit

That an LTI-system with this impulse response also delays all other input signals by t_0, follows from the selective property and convolution

$$y(t) = \delta(t - t_0) * x(t) = \int_{-\infty}^{\infty} \delta(t - \tau - t_0)x(\tau)\,d\tau = x(t - t_0) . \tag{8.59}$$

This relationship is depicted in Figure 8.21.

The system function

$$H(s) = \mathcal{L}\{\delta(t - t_0)\} = \int\limits_{-\infty}^{\infty} \delta(t - t_0)e^{-st}dt = e^{-st_0}, \quad s \in \mathbb{C} \tag{8.60}$$

represents a delay circuit. It can be used to show how the shift theorem of the Laplace transform (4.22) can be interpreted as a special case of the convolution theorem (8.41), just like the integration theorem and differentiation theorem. Note that the system function (8.60) is not a rational fraction function, as an ideal delay circuit cannot be implemented with a finite number of initial states or energy stores.

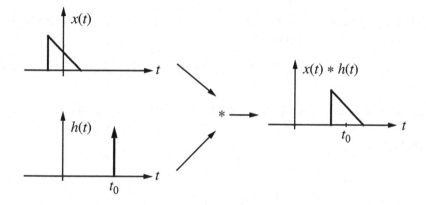

Figure 8.21: Convolution with a shifted delta impulse

A system can also consist of multiple delay circuits in parallel. It is then possible, for example, to describe propagation of sound waves or electromagnetic waves with multiple echoes in an ideal form. Each echo is represented by its own delay circuit, for which the delay time corresponds to the propagation delay of the signal. Different levels of damping can be represented by weighting factors for the delta impulses.

Figure 8.22 shows the impulse response of such a system and the result of convolution with a triangular input signal $x(t)$. The impulse with weight 2 is a negative delay. This is of course impossible for real-world propagation of waves. This example is only intended to show that system theory has no problem describing idealised systems that are impossible to implement.

In Chapter 11 we will be investigating the sampling of continuous signals. Convolution with an impulse train will play an important role there as it can be viewed as a superposition of delayed impulses. The delay times are each a multiple

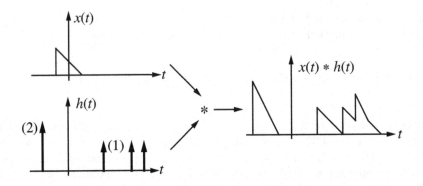

Figure 8.22: Convolution with multiple shifted delta impulses

of a fundamental delay time T, which is the sampling interval. The impulse train
can be written as a distribution

$$\sum_{i=-\infty}^{\infty} \delta(t - iT) \,. \tag{8.61}$$

An impulse train is shown in Figure 8.23 (bottom left). Convolution of a signal
$x(t)$ with an impulse train leads to a periodic repetition of $x(t)$ with period T (see
Figure 8.23).

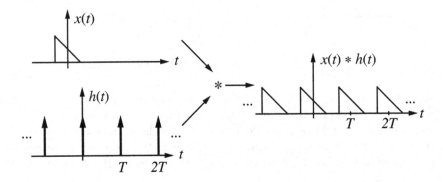

Figure 8.23: Convolution!with an impulse train

8.4.4.4 Control Engineering

The principle of signals and systems, to describe systems using their input–output
behaviour (system function, impulse response, step response) and not their implementation has already been used for a long time in control engineering. Control

methods were at first (starting with steam engines) developed separately for each
of the different areas (mechanical, electrical, chemical engineering, etc.) but it
was soon recognised that the control procedures for very different problems all
followed the same principles. Ways of describing systems independent of their
implementation were then investigated, the goal being to represent only the significant relationships between cause and effect.

It was customary in control engineering to use the step response to characterise
LTI-systems instead of the impulse response. This is partly because the theory of
distributions was developed later, but also because the step response is easier to
measure than the impulse response. That is understandable as not every sensitive
technical or biological system can deal with a powerful spike (as an approximation
for a delta impulse) as easily as the puck in Figure 8.13.

The most important elements of control engineering are depicted in Figure 8.24.
The graphical symbols each show the fundamental behaviour of the impulse response. The formulae for the impulse responses are shown next to them.

The I-circuit is an integrator with a factor in front of it. It responds to a step
with a steadily climbing ramp. Its impulse response is a scaled step function. The
I-circuit defines systems that can store energy, for example, capacitors.

The P-circuit is a multiplier that is defined by its factor. Its impulse response
is a scaled impulse.

A PI-circuit consists of a P-circuit and an I-circuit connected in parallel. The
step and impulse responses are each the sum of the corresponding functions for
the P-circuit and I-circuit.

A PT_1-circuit represents a differential equation of the form

$$\dot{y}(t) + \frac{1}{T}y(t) = x(t) \ . \tag{8.62}$$

An example is the RC-network from Figure 8.1 (to a constant factor). The step
response exponentially approaches a value that depends on a time constant T.
The impulse response is a decaying exponential function.

A DT_1-circuit represents the differential equation

$$\dot{y}(t) + \frac{1}{T}y(t) = \dot{x}(t) \ . \tag{8.63}$$

An example is the RC-network from Figure 6.3 with the step response shown in
Figure 6.5.

8.4.5   Combinations of Simple LTI-Systems

In Section 6.6 we investigated the combination of simple LTI-systems that were
connected in parallel or series. We saw that the combinations of LTI-systems led
to more complex LTI-systems whose system functions could be found easily by
combining the component's individual system functions. For impulse responses of

Name	Block diagram	Impulse Response

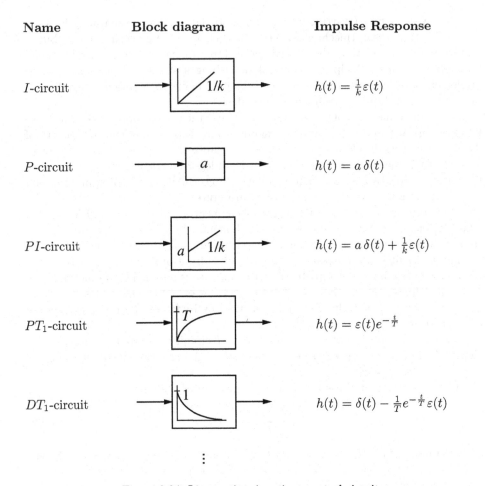

I-circuit $\qquad$ $1/k$ $\qquad$ $h(t) = \frac{1}{k}\varepsilon(t)$

P-circuit $\qquad$ a $\qquad$ $h(t) = a\,\delta(t)$

PI-circuit $\qquad$ a $\quad$ $1/k$ $\qquad$ $h(t) = a\,\delta(t) + \frac{1}{k}\varepsilon(t)$

PT_1-circuit $\qquad$ T $\qquad$ $h(t) = \varepsilon(t)e^{-\frac{t}{T}}$

DT_1-circuit $\qquad$ 1 $\qquad$ $h(t) = \delta(t) - \frac{1}{T}e^{-\frac{t}{T}}\varepsilon(t)$

$\vdots$

Figure 8.24: Linear, time-invariant control circuits

systems connected in parallel or series, as shown in Figures 8.25 and 8.26, there are also similarly simple relationships.

For the series coupling, with (6.26) and the convolution theorem,

$$H(s) = H_1(s) \cdot H_2(s) \quad \bullet\!\!-\!\!\circ \quad h(t) = h_1(t) * h_2(t)\,. \qquad (8.64)$$

For the parallel coupling, with (6.29),

$$H(s) = H_1(s) + H_2(s) \quad \bullet\!\!-\!\!\circ \quad h(t) = h_1(t) + h_2(t) \qquad (8.65)$$

because of the linearity of the Laplace transform. A similarly simple relationship for the impulse response of a feedback system, that would correspond to the

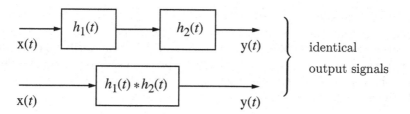

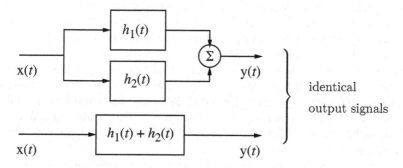

Figure 8.25: Systems in series

Figure 8.26: Systems in parallel

frequency-domain description (6.31) unfortunately does not exist.

Combining the step responses, for example, to characterise LTI-systems in control engineering (compare Section 8.4.4), is somewhat more involved than (8.64) and (8.65). If $s_1(t)$ and $s_2(t)$ are the step responses of systems with the impulse responses $h_1(t)$ and $h_2(t)$, then

$$s_1(t) = \varepsilon(t) * h_1(t)$$

$$s_2(t) = \varepsilon(t) * h_2(t),$$

(8.66)

and the complete system step response for the parallel coupling is

$$s(t) = \varepsilon(t) * (h_1(t) + h_2(t)) = s_1(t) + s_2(t).$$

For the series coupling, however, the step functions must be combined in accordance with

$$s(t) = \varepsilon(t) * h_1(t) * h_2(t) \quad = \dot{\delta}(t) * [\varepsilon(t) * h_1(t)] * [\varepsilon(t) * h_2(t)]$$

$$= \dot{\delta}(t) * s_1(t) * s_2(t) = \dot{s}_1(t) * s_2(t) = s_1(t) * \dot{s}_2(t)$$

(8.67)

One of the two step responses must also be differentiated before convolution takes place. Alternatively, the convolution of the step response can be carried out, and then the result can be differentiated. Equation (8.67) can easily be expanded to cover N cascaded systems. The complete step response is obtained by differentiating the component step responses or the complete product of convolution $(N-1)$ times.

8.4.6 Convolution by Inspection

In Section 8.4.3 we briefly dealt with the calculation of the convolution integral. This method always works, but it is cumbersome for functions that are defined by multiple cases, as the convolution integral takes a different form for each case, and each must be calculated individually.

In this section we will describe a simple method that is well suited to signals with constant sections, as it is not necessary to evaluate the convolution integral for these sections. With some practice it is possible to find the convolution product with this method, called *convolution by inspection*. The reader should try to acquire this skill, because it brings an intuitive understanding of the convolution operation which is essential for practical work.

To demonstrate the process, we will consider the two rectangle signals from Figure 8.27 and first evaluate the convolution integral as in Section 8.4.3.

_____   **Example 8.6**

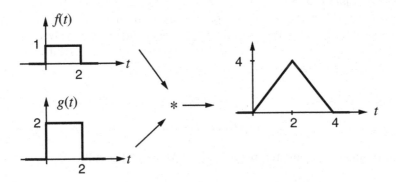

Figure 8.27: Example 8.6 of convolution by inspection

The two signals from Figure 8.27 are defined by the functions

$$f(t) = \begin{cases} 1 & 0 \le t \le 2 \\ 0 & \text{otherwise} \end{cases} \quad \text{and} \quad g(t) = \begin{cases} 2 & 0 \le t \le 2 \\ 0 & \text{otherwise} \end{cases} . \tag{8.68}$$

To calculate the convolution integral, however, both signals must be functions of

τ and additionally, one must be inverted and shifted in time:

$$f(\tau) = \begin{cases} 1 & 0 \le \tau \le 2 \\ 0 & \text{otherwise} \end{cases} \quad \text{and} \quad g(t - \tau) = \begin{cases} 2 & 0 \le t - \tau \le 2 \quad \text{or} \\ & (t-2) \le \tau \le t \\ 0 & \text{otherwise} \end{cases} \qquad (8.69)$$

The product $f(\tau)\,g(t - \tau)$ is zero for certain sections of t:

$$f(\tau)\,g(t - \tau) = \begin{cases} 0 & t < 0 \\ \begin{cases} 2 & 0 \le \tau \le t \\ 0 & \text{otherwise} \end{cases} & 0 \le t < 2 \\ \begin{cases} 2 & (t-2) \le \tau \le 2 \\ 0 & \text{otherwise} \end{cases} & 2 \le t < 4 \\ 0 & 4 \le t \end{cases} \qquad (8.70)$$

The convolution integral takes the values

$$y(t) = \int_{-\infty}^{\infty} f(\tau)\,g(t - \tau)\,d\tau \; = \; \begin{cases} \int_0^t 2\,d\tau & 0 \le t < 2 \\ \int_{t-2}^2 2\,d\tau & 2 \le t < 4 \\ 0 & \text{otherwise} \end{cases}$$

$$= \begin{cases} 2t & 0 \le t < 2 \\ 4 - 2t & 2 \le t < 4 \\ 0 & \text{otherwise} \end{cases} \qquad (8.71)$$

This defines the triangular signal shown in Figure 8.27.

■

The method of calculation we have shown here is correct, but because we had to consider various different cases it is unnecessarily cumbersome. Looking at the simplicity of the result, it seems likely that there is a simpler way of performing this convolution.

The first property that is noticeable, is that because the signals $f(t)$ and $g(t)$ are piecewise constant, the product $f(\tau)\,g(t - \tau)$ can also only have a constant value for some sections. Consequently, the convolution integral gives either the value zero (when $f(\tau)\,g(t - \tau) = 0$) or a linearly growing or decaying function. It is therefore sufficient to only calculate the value of the convolution integral at certain points, where the value of $f(\tau)\,g(t-\tau)$ changes. The intermediate values of the convolution integral can be obtained by linking the points by straight lines. To explain we will repeat the last example with the graphical method just described.

Example 8.7

Determining the convolution of $f(t)$ and $g(t)$ from Figure 8.27 requires that we first take g as a function of τ, and invert $g(\tau)$ to give $g(-\tau)$. Shifting backwards and forwards is performed with $g(t-\tau)$ and various values of t. For $t < 0$, $f(\tau)$ and $g(t-\tau)$ do not overlap, but at $t = 0$ both rectangle functions start at the same point, $\tau = 0$. For $0 \le t < 2$ the two rectangles overlap in the region $0 \le \tau \le t$. At $t = 2$, $f(\tau)$ and $g(2-\tau)$ cover each other comletely. For $2 \le t < 4$, the overlap is only in the region $t - 2 < \tau < 2$, and at $t = 4$ the rectangle functions $f(\tau)$ and $g(4-\tau)$ only border each other, at $\tau = 2$. When $t > 4$ their is no longer any overlapping. The critical points are therefore:

- $t = 0$ no overlap until this point,
- $t = 2$ complete overlap,
- $t = 4$ no longer any overlap.

At $t = 0$ and $t = 4$ the value of the convolution integral is zero $(f(\tau) g(-\tau) = 0$ and $f(\tau) g(4-\tau) = 0)$. At $t = 2$ the product $f(\tau) g(2-\tau)$ has the value $1 \times 2 = 2$; the value of the convolution integral is equal to the area of a rectangle with height 1×2 and width 2, i.e., 4. The result of the convolution of $f(t)$ and $g(t)$ can be obtained in a graphical way by linking the individual points $(0,0)$, $(2,4)$ and $(4,0)$ with straight lines. For $t = 0$ and $4 < t$ the value of the convolution integral is zero. This forms the triangle in Figure 8.27 without any integration being necessary. ∎

The procedure we have just described can be summarised as a general method for graphically determining the convolution of signals with constant sections.

1. Turn one of the signals around and imagine shifting it. As convolution is commutative, this can be done with the simpler of the two signals.

2. Recognise and note the the displacements where changes in the overlap between the two signals occur.

3. Determine the convolution integral at these points by forming the product of both signals and calculating the areas from height × width.

4. Mark these points on the time axis (t) and link them with straight lines. Sections on the time axis where no overlapping occurs are given the value zero.

We will now give two examples of how this procedure should be followed.

Example 8.8

We want to perform the convolution of the two signals from Figure 8.28.

1. We choose signal $g(t)$ to be turned around and shifted because it starts at $t = 0$.

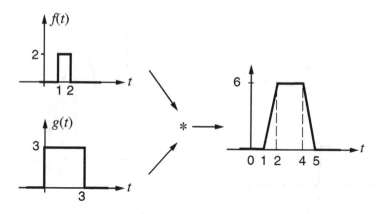

Figure 8.28: Convolution by inspection (Example 8.8)

2. The recogniseable points on the time axis are:

   • at $t = 1$ $f(\tau)$ and $g(1 - \tau)$ start to overlap,
   • at $t = 2$ $g(2 - \tau)$ completely covers the rectangle $f(\tau)$,
   • it remains completely covered while $2 \le t \le 4$ and stops at $t = 4$,
   • at $t = 4$, $g(t - \tau)$ only partially overlaps $f(\tau)$,
   • at $t = 5$ the functions no longer overlap.

3. While $f(\tau)$ is completely covered($2 \le t \le 4$), the value of the product $f(\tau)g(t - \tau)$ is $2 \times 3 = 6$. The convolution integral then has the value of a rectangle with height 6 and width 1, which is $1 \times 6 = 6$.

4. Marking in the points (1,0), (2,6), (4,6) and (5,0) and linking them forms the trapezium shown Figure 8.28.

■

Example 8.9

Figure 8.29 shows the convolution of a rectangle signal with a non-rectangular signal, which is, however, constant in sections. Using our prescribed procedure leads to the following results.

1. The signal $g(t)$ is the simplest and so it will be reversed and shifted.

2. The characteristic points are:

   • $t = 1$ overlapping starts,

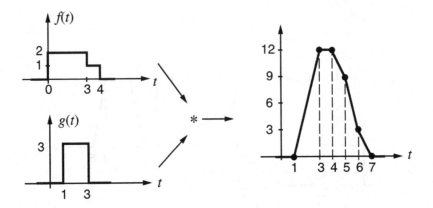

Figure 8.29: Convolution by inspection (Example 8.9)

- $t = 3$ the rectangle is completely superimposed over the higher part of $f(\tau)$,

- $t = 4$ the rectangle stops being completely superimposed over the higher part of $f(\tau)$ and starts being partially superimposed over the lower part of $f(\tau)$,

- $t = 5$ the lower part of $f(\tau)$ is completely covered by the rectangle,

- $t = 6$ the rectangle no longer covers any part of the higher part, and only partially overlaps the lower part,

- $t = 7$ all overlapping with $f(\tau)$ stops.

3. By multiplying the signal with constant sections at the characteristic points, constant products are obtained. Determining the areas gives the following values for the convolution integral:

- $t = 1$ $\int_{-\infty}^{\infty} f(-\tau)\, g(\tau)\, d\tau = 0$

- $t = 3$ $\int_{-\infty}^{\infty} f(2-\tau)\, g(\tau)\, d\tau = 12$

- $t = 4$ $\int_{-\infty}^{\infty} f(3-\tau)\, g(\tau)\, d\tau = 12$

- $t = 5$ $\int_{-\infty}^{\infty} f(4-\tau)\, g(\tau)\, d\tau = 6+3 = 9$

- $t = 6$ $\int_{-\infty}^{\infty} f(5-\tau)\, g(\tau)\, d\tau = 3$

- $t = 7$ $\int\limits_{-\infty}^{\infty} f(6-\tau)\,g(\tau)\,d\tau = 0$

4. Marking the points (1,0), (3,12), (4,12), (5,9), (6,3), (7,0) and linking them with straight lines forms the result depicted in Figure 8.29.   ■

Comparing Figures 8.27, 8.28 and 8.29 it becomes clear that the length of the convolution product is equal to the sum of the lenghts of the two signals. The points at the start and end of the convolution product are also the sums of the individual start and end points. It can easily be shown that this is generally true for the convolution of all time bounded functions (not just those with constant sections), by using point 2 of our procedure on the start and end points of overlap for general signals, as in Figure 8.30.

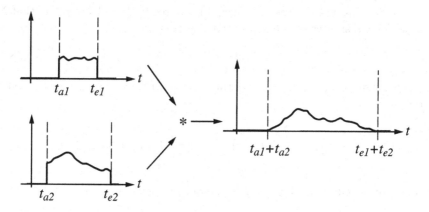

Figure 8.30: Length of the convolution product of two finite functions

The prescribed procedure for signals with constant sections can be used on all signals that can be described in sections by polynomials. The convolution of these signals with degrees K and L leads to a convolution product with order $K + L + 1$. That means that convolution of a triangular signal $(K = 1)$ with a rectangular signal $(L = 0)$ would have a convolution product formed from second-order parabolic sections $(K+L+1 = 2)$. The convolution of two triangular signals $(K = L = 1)$ gives a third-order convolution product $(K + L + 1 = 3)$.

8.5   Applications

Many interesting uses of signal processing are based on convolution relationships. Here we will examine two examples, matched filters and de-convolution.

8.5.1 Matched Filter

In many practical tasks, the aim is to recognise known signal forms, which means identifying what point in time they occur and distinguishing them from other signal forms. Examples of these tasks are speech recognition, object recognition in images, etc.

Principle: In the simplest case we want to recognise when a signal component $m(t)$ occurs. That means we are dealing with a signal of the form

$$x(t) = m(t - t_0). \tag{8.72}$$

where the function $m(t)$ is known but the time t_0 by which $m(t)$ is shifted is not known. For example, we could transmit an acoustic signal $m(t)$, and with the echo $x(t)$, determine the unknown time taken for it to return.

To achieve this, the received signal $x(t)$ is filtered with an LTI-system that has an impulse response derived from the signal form $m(t)$:

$$h(t) = m(-t). \tag{8.73}$$

The minus sign in $m(-t)$ cancels out the reversal of the signal in convolution, so the output signal is

$$y(t) = h(t) * x(t) = \int_{-\infty}^{\infty} h(t - \tau)\, x(\tau)\, d\tau = \int_{-\infty}^{\infty} m(\tau - t)\, x(\tau)\, d\tau. \tag{8.74}$$

For an input signal of the form (8.72), the output signal is

$$y(t) = \int_{-\infty}^{\infty} m(\tau - t)\, m(\tau - t_0)\, d\tau. \tag{8.75}$$

The output signal will be maximal at the unknown time t_0:

$$y(t_0) = \int_{-\infty}^{\infty} m^2(\tau - t_0)\, d\tau \geq \int_{-\infty}^{\infty} m(\tau - t)\, m(\tau - t_0)\, d\tau. \tag{8.76}$$

The integral over the always positive value $m^2(\tau - t_0)$ must be at least greater than $y(t)$ except at $t = t_0$. Suitable choice of the signal $m(t)$ can make the value of $y(t_0)$ much greater than all other values of $y(t)$, and then the unknown time t_0 can then be recognised as the peak in the signal $y(t)$. Figure 8.31 shows the corresponding arrangement of a filter with impulse response (8.73) and a peak detector. A filter of this kind is called a *matched filter*.

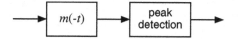

Figure 8.31: Finding a signal pattern $m(t)$ using convolution with $m(-t)$

Example 8.10

As an example of a possible application for a matched filter, we will consider the arrangement shown in Figure 8.32, intended to count and roughly measure the speed of vehicles. The only source of information available is the signal from the distance sensor, which gives the distance between itself and an object below it for all t. The problem is to decide what kind of vehicle is passing, work out its speed and count the number of each kind of vehicle. Typical forms of signal for cars and lorries moving quickly or slowly are shown in Figure 8.33.

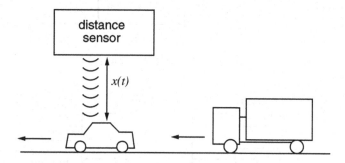

Figure 8.32: Counting and measuring the speed of vehicles

A system that performs the counting function with the signal $x(t)$ from the distance sensor is depicted in Figure 8.34. The sensor signal runs through a bank of four filters that each responds to a different class of vehicle using the templates in Figure 8.33. For each filter, the maximum detector recognises if and when its signal form has occurred, and these events are automatically counted to give the desired result.

■

8.5.2   Deconvolution

Sometimes the problem does not require that a signal is changed by a filter, but instead that an already filtered signal is returned to its originial form. Problems of this kind often occur.

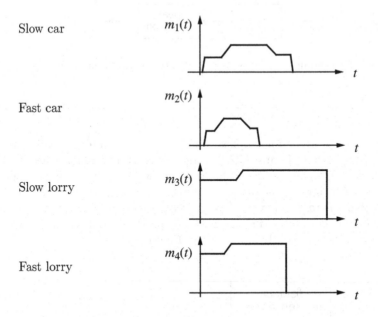

Slow car

Fast car

Slow lorry

Fast lorry

Figure 8.33: Typical signal forms

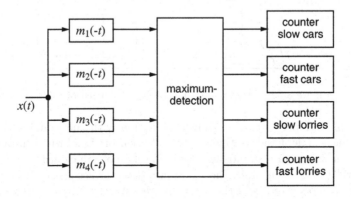

Figure 8.34: Block diagram of the detector

- A microphone recording made in a room is the convolution of the sound signal produced by the instrument or voice with the impulse response of the surrounding space. The impulse response of a room can by heard as the response to a clap or the crack of a whip. This convolution must be done in reverse to remove the interference.

- A blurred image from a camera can be interpreted as convolution with a smothing system. The impulse response is the blurred spot that would be a sharp point on the original image.

- A radio channel with multi-path propagation can be described by an impulse response that consists of multiple time-shifted echos (see Chapter 8.4.4.3).

- The inertia of measuring instruments has a smoothing effect on the measured signals and smooths sharp steps into slowly rising slopes. The impulse response of such systems can often be modelled by an RC-circuit with a large time-constant.

If the impulse response $h(t)$ is known for such a system, then it is possible to attempt to invert the influence of $h(t)$ by convolution with a second LTI-system $g(t)$. In the ideal case the result is again a delta impulse:

$$g(t) * h(t) = \delta(t) . \tag{8.77}$$

The system function $G(s) = \mathcal{L}\{g(t)\}$ of the second system can be expressed by the system function $H(s) = \mathcal{L}\{h(t)\}$, using the convolution theorem:

$$G(s) = \frac{1}{H(s)} . \tag{8.78}$$

The effect of a system with system function $G(s)$ is called *deconvolution*. It is rarely possible to carry out in the ideal form of (8.78), because the measured signals usually have additional noise interference. In addition, signal components that were suppressed completely by the first system (zeros of the system function), cannot be recovered from the second system. The filter that represents the best compromise between deconvolution and noise reduction will be introduced in Chapter 18 as the Wiener filter. A system with system function $G(s)$ (8.78) could also be unstable. This problem will be dealt with in Section 16.3.

8.6   Exercises

Exercise 8.1

Show that the response of an RC-circuit to a short rectangle impulse (8.7) turns into the impulse response (8.4) for $T_0 \to 0$.

Exercise 8.2

Using the calculation rules for the delta impulse, determine:

a) $f_a = \int\limits_{-\infty}^{\infty} e^{-t}\delta(t) \, dt$

b) $f_b = \int\limits_{-\infty}^{\infty} e^{-t}\delta(t-\tau)\,dt$

c) $f_c = \int\limits_{-\infty}^{\infty} (t^2 - 2)\,\delta(3t)\,dt$

d) $f_d = \int\limits_{-\infty}^{\infty} t\,e^{-t}\,\delta(4 - 2t)\,dt$

Exercise 8.3

The following signals are each turned on at $t = 0$:

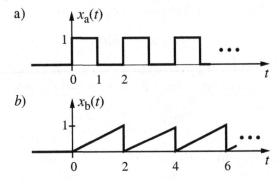

Give $x_a(t)$ and $x_b(t)$ using the step function $\varepsilon(t)$ for both, then form the derivations and sketch them.

Exercise 8.4

Form the derivation of $f(t) = \varepsilon(-t)$.

Exercise 8.5

Form the derivation of $f(t) = \varepsilon(at)$.

Exercise 8.6

You are given the functions $f(t) = t\,\varepsilon(t)$ and $g(t) = \varepsilon(t) - \varepsilon(t - 4)$. Calculate $y(t) = f(t) * g(t)$ using the convolution integral.

Exercise 8.7

Examine the following signals:

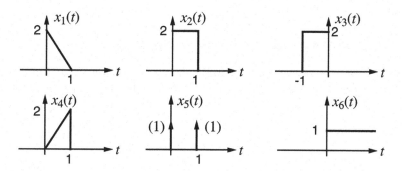

The following convolution products are also given:

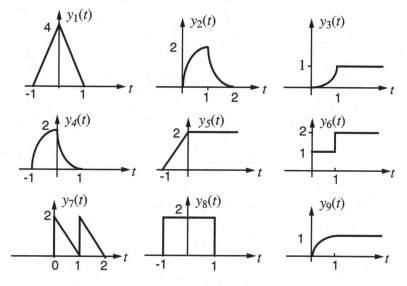

For each of the convolution products, determine the pair of signals x_i and x_j that created this result.

Exercise 8.8

Find the response of the RC-network shown in Figure 8.1 to the rectangle impulse shown in Figure 8.3 by solving the convolution integral $\int\limits_{-\infty}^{\infty} h(\tau)\, x(t-\tau)\, d\tau$.

Note: see Section 8.4.3.

Exercise 8.9

For the example of an RLC-network (Chapter 3.2.3), give

a) the system function (note that $i(t)$ is the output variable)

b) the impulse response

c) the region of convergence of the system function

d) Which values can the parameters σ_0 and ω_0 of the input signal $u(t)$ take, so that the system response $i(t)$ converges? Assume that $u(t)$ is right sided.

9 The Fourier Transform

Like the Laplace transform, the Fourier transform plays an important role in system analysis and the modelling of LTI-systems. Its defintion, properties and usage will be examined in this chapter.

After we have already proven that the Laplace transform is very useful, one may ask why there is a need for another transform. We will therefore start the chapter with a critical review of the Laplace transform.

9.1 Review of the Laplace Transform

In the previous chapters we got to know the Laplace transform as a model of LTI-systems that were described by differential equations. It allowed us to evaluate the system response in an elegant way, in particular if there were initial conditions to consider.

The obvious advantages of the Laplace transform only come with LTI-systems that are characterised by a clear number of poles and zeros, or equivalently, are described by a differential equation with constant coefficients. Other important LTI-systems also exist, however, that cannot be characterised by a few poles and zeros, for example the delay circuit in Chapter 8.4.4.3. We will learn further important examples in this chapter. Attempting to describe such LTI-systems using the Laplace transform leads to system functions with an infinite number of poles and zeros, or with essential singularities. It is actually possible to work around these difficulties, but the mathematically correct methods require significantly more advanced knowledge of function theory than we have discussed in Chapter 5. Even worse is the fact that the elegance of the method – modelling systems as a few complex natural resonances in the time-domain and frequency-domain – is lost. The Laplace transform is particularly suitable if we are concerned with signals of finite duration, or unilateral signals. We have represented these signals with the Laplace transform as overlapping eigenfunctions of LTI-systems which have the form e^{st}, although $\mathcal{L}\{e^{st}\}$ does not exist. The same also goes for $\mathcal{L}\{\sin \omega t\}$, $\mathcal{L}\{\cos \omega t\}$ and $\mathcal{L}\{1\}$. In this chapter we will see how elegantly the Fourier transform overcomes this problem if generalised functions in the frequency-domain are permitted.

There are more aspects of our previous dealings with the Laplace transform that we have not yet made fully clear: putting the system model in the frequency-domain has the primary goal of making calculations simpler to carry out than they would be in the time-domain. The first example of this is the evaluation of the system response, which can be represented in the frequency-domain as a multiplication by the system function, but in the time-domain requires the use of the convolution integral. As a second example, we give the analysis in natural resonances which is so simple in the frequency-domain using partial fractions, that we have never done it in the time-domain. In all cases, however, we have returned to the time-domain after using the advantages of the frequency-domain representation.

We have rarely attempted to characterise or design systems directly through their properties in the frequency-domain. The only time being the description of a system function by the poles and zeros in the complex s-plane (see Chapter 6.3). The interpretation of such pole-zero plots is not so simple, however, because the system function is a complex function of complex variables. It fails completely for systems that cannot be described by a differential equation with constant coefficients, for example, a delay circuit. Furthermore, because the system function $H(s)$ has to be analytic in the region of convergence, designing a system in the s-plane is difficult, as this property may be violated.

The following point summarise the disadvantages of the Laplace transform.

- It only provides a simple and elegant system model for LTI-systems that can be described by ordinary differential equations with constant coefficients.

- The Laplace transform does not really exist for signals e^{st}, that are eigen-functions of LTI-systems.

- From the system function, the properties of a system in the frequency domain cannot easily be seen.

As an alternative to the Laplace transform, we now consider the Fourier transform. We will see that it overcomes the stated disadvantages of the Laplace transform, while also keeping many of the merits.

9.2 Definition of the Fourier Transform

9.2.1 Forward Transformation

At first, the definition of the Fourier transform looks similar to the Laplace transform. A signal is also projected with complex exponential oscillations by an integral transformation. The complex frequency of these exponential oscillations is however, purely imaginary, so no decaying or growing oscillations occur.

$$\boxed{X(j\omega) = \mathcal{F}\{x(t)\} = \int_{-\infty}^{\infty} x(t)e^{-j\omega t}dt} \qquad (9.1)$$

The result of the transformation $X(j\omega)$ is called the *Fourier Transform*, *Fourier spectrum* or *complex amplitude spectrum*. The complex exponential oscillation $e^{-j\omega t}$ is known as a *complex function of time*. The real variable ω is the *frequency parameter* or the *frequency*. The Fourier transform therefore depends on the real frequency parameter, but can itself take a complex value

$$X(j\omega) = |X(j\omega)| \cdot e^{j\varphi(j\omega)} . \tag{9.2}$$

Its magnitude $|X(j\omega)|$ is called the *magnitude spectrum*. The *phase spectrum* $e^{j\varphi(j\omega)}$ is usually expressed by the *phase* $\varphi(j\omega)$. At the beginning it is confusing that one writes $X(j\omega)$ and $\varphi(j\omega)$, and not $X(\omega)$ and $\varphi(\omega)$. The argument $j\omega$ is just a convention, which means that a one-dimensional complex function in the Gaussian number plane on the imaginary axis is being defined – it could also be defined as $X(\omega)$ on the real axis, and there are actually several books that do this – the dependency on ω is unchanged. The sense of the $j\omega$ convention becomes immediately clear, if a relationship between Fourier and Laplace transforms is made (Section 9.3). For the Fourier transform, the abbreviated correspondency form

$$X(j\omega) \bullet\!\!-\!\!\circ x(t) \tag{9.3}$$

will be used. We use the same symbol $\bullet\!\!-\!\!\circ$ for different transforms, for example, the Laplace, Fourier and later, the z-transform. This symbol is not a strict mathematical assignment, but typographical short-hand for 'corresponding to each other in the original and transform domain'. Some textbooks attempt to use different symbols for different transforms, in order to give a strict mathematical meaning.

9.2.2   Existence of the Fourier Transform

Like the Laplace transform, the Fourier transform only exists for a certain class of signals. If a function of time $x(t)$ is *absolutely integrable*, that is sufficient for the convergence of the Fourier integral. If

$$\int_{-\infty}^{\infty} |x(t)| \, dt < \infty , \tag{9.4}$$

then

$$|X(j\omega)| \leq \lim_{\substack{A \to -\infty \\ B \to +\infty}} \int_A^B \left| x(t)e^{-j\omega t} \right| \, dt = \lim_{\substack{A \to -\infty \\ B \to +\infty}} \int_A^B |x(t)| \, dt . \tag{9.5}$$

This condition is sufficient but not necessary, and there are in fact time functions that although not integrable, have a Fourier transform. Examples of such time functions and their Fourier transforms are

$$x(t) = \frac{\sin t}{t} \qquad \circ\!\!-\!\!\bullet \qquad X(j\omega) = \begin{cases} \pi & \text{for } |\omega| < 1 \\ 0 & \text{for } |\omega| > 1 \end{cases} \tag{9.6}$$

and the step function $\varepsilon(t)$

$$x(t) = \varepsilon(t) \qquad \circ\!\!-\!\!\bullet \qquad X(j\omega) = \pi\delta(\omega) + \frac{1}{j\omega}. \qquad (9.7)$$

We will deal with the evaluation of these Fourier tranforms in the course of this chapter.

The Fourier transform of the step function contains a delta impulse $\delta(\omega)$. Up to now, we have encountered it in the time-domain, but not in the frequency-domain. As Laplace transforms in the region of convergence are analytic functions, there is no room there for generalised functions. However, with Fourier transforms that are functions of a real parameter ω (just as with functions of time), the extension to cover generalised functions does make sense. The popularity of the Fourier transform comes from the possiblity of assigning a spectrum with the help of the delta impulse to many important practical functions for which there is no Laplace transform. Some elementary examples are constants (e.g. $x(t) = 1$) and the trigonometric functions $\sin\omega_0 t$ and $\cos\omega_0 t$.

9.3 Similarities and Differences between Fourier and Laplace Transforms

The differences between Fourier and Laplace transforms were previously discussed as the Fourier transform was introduced as an independent transformation. There are many cases, however, where the Fourier and Laplace transform formulae agree. Comparing the Fourier transform and the bilateral Laplace transform,

$$\mathcal{F}\{x(t)\} = \int_{-\infty}^{\infty} x(t)e^{-j\omega t}dt \qquad \mathcal{L}\{x(t)\} = \int_{-\infty}^{\infty} x(t)e^{-st}dt \qquad (9.8)$$

it can be seen immediately that the Fourier transform of a function of time is the same as its Laplace transform on the imaginary axis $s = j\omega$ of the complex plane. That is of course only true if the Laplace transform exists as well. The connection between Laplace and Fourier transforms is: if the region of convergence of $\mathcal{L}\{x(t)\}$ contains the imaginary axis $s = j\omega$, then

$$\boxed{\mathcal{F}\{x(t)\} = \mathcal{L}\{x(t)\}\Big|_{s=j\omega}} \qquad (9.9)$$

With (9.9) the convention should now make sense, that the Fourier transform is defined on the imaginary axis and not on the real axis.

Example 9.1

The function

$$x(t) = \varepsilon(t)e^{-at}, \qquad a > 0 \tag{9.10}$$

has the Laplace transform

$$X(s) = \frac{1}{s+a}, \qquad \text{Re}\{s\} > -a. \tag{9.11}$$

The region of convergence lies to the right of the negative number $-a$, and thus contains the imaginary axis of the s-plane with $\text{Re}\{s\} = 0$.

The Fourier transform is obtained by substituting (9.10) into the defintion of the Fourier transform (9.1) and evaluating the integral

$$X(j\omega) = \frac{1}{j\omega + a}. \tag{9.12}$$

This clearly shows the connection (9.9) between the Fourier and Laplace transforms.

If, however, $a < 0$ in (9.10), the Fourier transform does not exist, because the Fourier integral does not converge. The Laplace transform, on the other hand is unchanged (9.11), although the region of convergence now lies to the right of the positive number $-a$ and no longer contains the imaginary axis.

■

If the region of convergence of the Laplace transform contains the imaginary axis, the Fourier transform (9.9) can then be differentiated any number of times. The Laplace transform can therefore easily be obtained from the Fourier transform using analytic continuation (Chapter 5.4.3). In practice, this means that $j\omega$ is replaced by s and the region of convergence defines the Laplace transform.

Example 9.2

What is the Laplace transform of the signal with Fourier transform

$$X(j\omega) = \frac{1}{1 + \omega^2} ? \tag{9.13}$$

Expression (9.13) has no singularities for real values of ω and can be differentiated at all points. We can perform analytic continuation on $X(j\omega)$, where we write

$$X(j\omega) = \frac{1}{1 - (j\omega)^2} \tag{9.14}$$

$$X(s) = \frac{1}{1 - s^2} = \frac{-1}{(s-1)(s+1)}. \tag{9.15}$$

$X(s)$ has poles at $s = -1$ and $s = +1$, therefore

$$-1 < Re\{s\} < 1,$$

as the ROC must contain the imaginary axis.

■

In Example 9.1 we saw that signals exist which have a Laplace transform, but no Fourier transform. We will also see that signals like $x(t) = e^{j\omega t}$ and others derived from it (constants, trigonometric functions, etc.) have a Fourier transform, but no Laplace transform. The Fourier transform is therefore a necessary and meaningful companion of the Laplace transform. This is particularly apparent for selective systems (filters) that are characterised by their frequency behaviour, because then the system properties can be read immediately from the Fourier transform of the system response.

The Fourier transform also has great importance for digital signal processing, as its discrete counterpart, the discrete Fourier transform (DFT), can be implemented directly as a computer program or as a circuit. With it, discrete-time signals can be worked with directly in the frequency-domain.

9.4   Examples of the Fourier Transform

In this section we will find the Fourier transform of some important signals, which we will refer to later.

9.4.1   The Fourier Transform of the Delta Impulse

The Fourier transform of the delta impulse is obtained by inserting it into the Fourier integral (9.1), and using its selective property.

$$X(j\omega) = \int_{-\infty}^{\infty} \delta(t)e^{-j\omega t}dt = 1. \tag{9.16}$$

The result corresponds to the Laplace transform, as in this case the conditions for (9.9) have been fulfilled. Figure 9.1 shows the corresponding transform pair

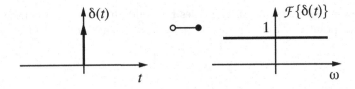

Figure 9.1: The delta impulse and its Fourier transform

We obtain the Fourier transform of a shifted delta impulse the same way:

$$X(j\omega) = \int_{-\infty}^{\infty} \delta(t - \tau)e^{-j\omega t}dt = e^{-j\omega\tau}. \qquad (9.17)$$

In contrast to the Laplace transform, for which the function's behaviour as a complex function is difficult to represent, we can more easily represent the Fourier transform graphically, as a complex function of a real variable. This is most simply done by splitting it into real and imaginary parts, or into magnitude and phase of the complex transform (Figure 9.2). The shift of the delta impulse does not change the magnitude of the Fourier transform, but instead changes the phase $\varphi(j\omega) = \arg\{\mathcal{F}\{\delta(t - \tau)\}\}$ so that it decays linearly. The gradient of the phase corresponds to the amount of displacement τ. The phase of $\mathcal{F}\{\delta(t)\}$ has the value zero, and is not shown in Figure 9.1.

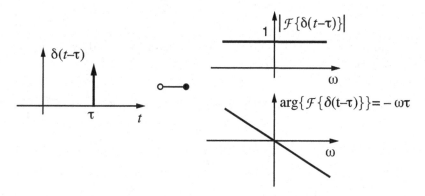

Figure 9.2: Shifted delta impulse and its Fourier transform

9.4.2 Fourier Transform of the Rectangle Function

To find the Fourier transform of the rectangle function, we first introduce an abbreviation, of which we will make widespread use. A rectangle function is easy to draw (see Figure 9.3), but dealing with the piecewise definition in (9.18) is not suitable because of the necessary case distinction. We therefore use the notation $\text{rect}(t)$ to indicate the rectangle function, and deal with the symbol rect as the function defined in (9.18).

$$x(t) = \text{rect}(t) = \begin{cases} 1 & \text{for } |t| \leq \dfrac{1}{2} \\ 0 & \text{otherwise} \end{cases} \qquad (9.18)$$

The limits at $t = \pm\frac{1}{2}$ are chosen so that the rectangle $\text{rect}(t)$ has unit height, width and area. Rectangle impulses with other widths will be described by scaling

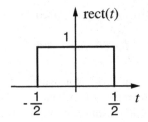

Figure 9.3: The rectangle function

with a factor $a > 0$:

$$\text{rect}(at) = \begin{cases} 1 & \text{for } |t| \leq \dfrac{1}{2a} \\ 0 & \text{otherwise} \end{cases} \tag{9.19}$$

The Fourier transform of the rectangle function is obtained by evaluating the Fourier integral

$$\mathcal{F}\{\text{rect}(t)\} = \int_{-\infty}^{\infty} \text{rect}(t)e^{-j\omega t}dt = \int_{-\frac{1}{2}}^{+\frac{1}{2}} e^{-j\omega t}dt = \left. \frac{1}{-j\omega}e^{-j\omega t}\right|_{-\frac{1}{2}}^{+\frac{1}{2}}$$

$$= -\frac{1}{j\omega}\left(e^{-\frac{j\omega}{2}} - e^{\frac{j\omega}{2}}\right) = \frac{\sin\dfrac{\omega}{2}}{\dfrac{\omega}{2}} \ . \tag{9.20}$$

The plot of the transform $\mathcal{F}\{\text{rect}(t)\}$ is shown in Figure 9.4.

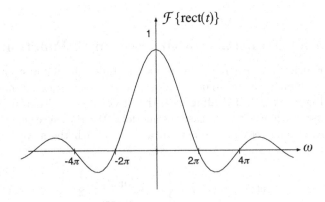

Figure 9.4: Fourier transform of the rectangle function

We will also introduce an abbreviation for this characteristic function. We call it the si-function and define it as

$$\text{si}(\nu) = \begin{cases} \dfrac{\sin \nu}{\nu} & \text{for } \nu \neq 0 \\[2mm] 1 & \text{for } \nu = 0. \end{cases} \tag{9.21}$$

In spite of the value ν in the denominator, $\text{si}(\nu)$ is continuous, so $\nu = 0$ corresponds to the limit that we obtained at this point from the l'Hopital rule.

Figure 9.5 shows the course of $\text{si}(\nu)$, dependent on the dimensionless variable ν.

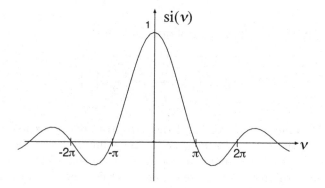

Figure 9.5: The si-function

For the entire area under the integral

$$\int_{-\infty}^{\infty} \text{si}(\nu) d\nu = \pi. \tag{9.22}$$

This area corresponds exactly to the area of a triangle between the main peak and the first two zero points to the right and left of $\nu = 0$ (see Figure 9.6). This rule is very practical as it can be also used for any other si-function, whether it is scaled in the horizontal or the vertical axis.

With the si-function, we can very elegantly formulate the correspondence between the rectangle function and its Fourier transform. By comparing (9.20) and (9.21), we find that

$$\boxed{\text{rect}(t) \quad \circ\!\!-\!\!\bullet \quad \text{si}\left(\frac{\omega}{2}\right).} \tag{9.23}$$

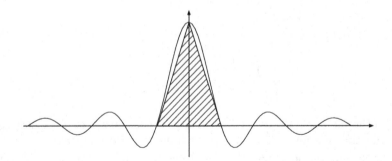

Figure 9.6: The area of the triangle corresponds to the integral of the si-function

For rectangles with any width a, we obtain from (9.19)

$$\text{rect}(at) \quad \circ\!\!-\!\!\bullet \quad \frac{1}{|a|}\,\text{si}\left(\frac{\omega}{2a}\right). \tag{9.24}$$

In Figure 9.7, rectangle functions and their spectra are represented for different values of a. As the scaling factor a occurs as a multiplier in the time-domain and divisor in the frequency-domain, the spectra become wider as the rectangle function becomes thinner. We will see later that this effect is a general principle, and occurs with many other functions of time.

9.4.3   The Fourier Transform of a Complex Exponential Function

The previous Fourier transforms could be solved by evaluating the Fourier integral (9.1). This process is unsuccessful for the complex exponential function

$$x(t) = e^{j\omega_0 t} \tag{9.25}$$

as the integral that arises

$$\int_{-\infty}^{\infty} e^{j\omega_0 t} e^{-j\omega t}\, dt = \int_{-\infty}^{\infty} e^{j(\omega_0-\omega)t}\, dt$$

clearly does not converge for $\omega = \omega_0$. Nevertheless, the Fourier transform $e^{j\omega_0 t}$ can be given in the form of a distribution. To derive this, we use the si-function that we have just introduced.

In order to overcome the mentioned convergence problem for integration of an oscillation of infinite duration, we consider first of all a section of the oscillation

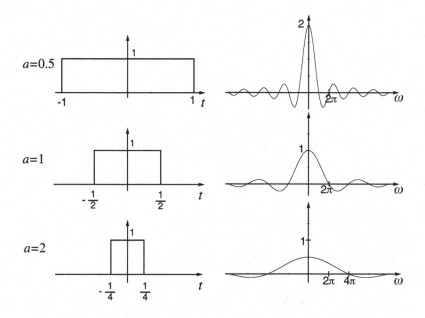

Figure 9.7: Various rectangle functions rect(at) and their Fourier transforms

with finite duration:

$$x_T(t) = \begin{cases} e^{j\omega_0 t} & \text{for } |t| \leq T \\ \\ 0 & \text{otherwise} \end{cases} \tag{9.26}$$

The Fourier transform of this can be found easily, and the steps required are the same as for the Fourier transform of the rectangle function, carried out in (9.20). They both lead to a result that can be expressed by the si-function.

$$
\begin{aligned}
X_T(j\omega) &= \int_{-T}^{T} e^{j\omega_0 t} e^{-j\omega t} dt = \int_{-T}^{T} e^{j(\omega_0 - \omega)t} = \\
&= \frac{1}{j(\omega_0 - \omega)} e^{j(\omega_0 - \omega)t} \Big|_{-T}^{T} = \frac{1}{j(\omega_0 - \omega)} \left(e^{j(\omega_0 - \omega)T} - e^{-j(\omega_0 - \omega)T} \right) \\
&= \frac{2\sin(\omega - \omega_0)T}{\omega - \omega_0} = 2T \, \text{si}\big((\omega - \omega_0)T\big) \quad .
\end{aligned}
\tag{9.27}
$$

Figure 9.8 shows a plot of this spectrum. It is the same as the spectrum shown in Figure 9.7, but is shifted by ω_0 on the frequency axis.

Since the complex exponential function in (9.25) consists of the signal of finite duration $x_T(t)$, with the limit $T \to \infty$ the spectrum of $x(t)$ is also obtained from

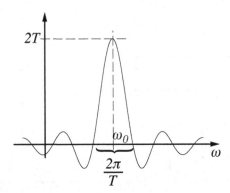

Figure 9.8: Spectrum of a complex exponential function of finite duration

$X_T(j\omega)$ with this limit. If we write $X_T(j\omega)$ with $T = \frac{1}{2a}$ as

$$X_T(j\omega) = \frac{1}{|a|} \, \mathrm{si}\left(\frac{\omega - \omega_0}{2a}\right), \tag{9.28}$$

we recognise from Figure 9.7, that $X_T(j\omega)$ becomes a spike with height approaching infinity and width approaching zero, for $T \to \infty$ (so $a \to 0$). $X_T(j\omega)$ can clearly not be represented by an ordinary function, so we ask whether it can be expressed by a delta impulse. To do this we must test $X_T(j\omega)$ for the selective property of the shifted delta impulse (8.15) with the limit $T \to \infty$. We form the product of $X_T(j\omega)$ and another function $F(j\omega)$, that is constant at $\omega = \omega_0$, but can otherwise be any function. For large values of T, the function $X_T(j\omega)$ becomes ever smaller (apart from at ω_0), (see Figures 9.7 and 9.8), and so when the product $X_T(j\omega)F(j\omega)$ for $T \to \infty$ is integrated, only the value $F(j\omega_0)$ is contributed:

$$\int_{-\infty}^{\infty} \lim_{T \to \infty} X_T(j\omega)F(j\omega)\,d\omega = F(j\omega_0) \int_{-\infty}^{\infty} X_T(j\omega)\,d\omega. \tag{9.29}$$

The area under the si-function is, however, the same as the area under an isosceles triangle with its base between the two zero points next to the maximum of the si-function (see Figure 9.6). That means here from Figure 9.8 that,

$$\int_{-\infty}^{\infty} X_T(j\omega)\,d\omega = \int_{-\infty}^{\infty} 2T\,\mathrm{si}\big((\omega - \omega_0)T)\big)\,d\omega = \frac{1}{2}\,2T \cdot \frac{2\pi}{T} = 2\pi. \tag{9.30}$$

Equation (9.29) thus becomes

$$\int_{-\infty}^{\infty} \frac{1}{2\pi} \lim_{T \to \infty} X_T(j\omega)F(j\omega)\,d\omega = F(j\omega_0). \tag{9.31}$$

The function $\frac{1}{2\pi} \lim_{T \to \infty} X_T(j\omega)$ clearly removes all values of $F(j\omega)$ except at $\omega = \omega_0$. That is exactly the property that identifies the delta impulse. We can therefore write

$$\lim_{T \to \infty} \frac{1}{2\pi} X_T(j\omega) = \delta(\omega - \omega_0). \tag{9.32}$$

Combining these results yields

$$\mathcal{F}\{e^{j\omega_0 t}\} = \mathcal{F}\{\lim_{T \to \infty} x_T(t)\} = \lim_{T \to \infty} X_T(j\omega) = 2\pi\, \delta(\omega - \omega_0). \tag{9.33}$$

We have now obtained the transform pair for the complex exponential function:

$$\boxed{e^{j\omega_0 t} \circ\!\!-\!\!\bullet\; 2\pi\, \delta(\omega - \omega_0).} \tag{9.34}$$

It is similar to the transform pair for an impulse shifted in the time-domain (9.17):

$$\delta(t - \tau) \circ\!\!-\!\!\bullet\; e^{-j\omega\tau}. \tag{9.35}$$

9.4.4 Fourier Transform of $\frac{1}{t}$

A further example for a time function whose spectrum cannot be found by simple evaluation of the Fourier integral is the function

$$x(t) = \frac{1}{t}, \tag{9.36}$$

shown in graphical form in Figure 9.9. As the time function at $t = 0$ grows

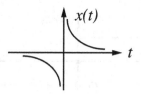

Figure 9.9: Temporal behaviour of the function $\frac{1}{t}$

beyond limit, it is not easy to integrate over this point. It helps to split the integral into two parts for $t < 0$ and $t > 0$. The upper integral is solved first for the finite limits ε and T, with $0 < \varepsilon < T < \infty$, and the lower integral is then solved correspondingly for $-T$ and $-\varepsilon$. From the results of these two integrals, the solution can be obtained by the two limits $T \to \infty$ and $\varepsilon \to 0$, as long as these two limits can be evaluated. This procedure is also known as calculating the

Cauchy's principle value. For $x(t)$ according to (9.36) the necessary steps are

$$\mathcal{F}\{x(t)\} = X(j\omega) = \int_{-\infty}^{\infty} \frac{1}{t} e^{-j\omega t} dt = \lim_{\substack{\varepsilon\to 0 \\ T\to\infty}} \left[\int_{-T}^{-\varepsilon} \frac{1}{t} e^{-j\omega t} dt + \int_{\varepsilon}^{T} \frac{1}{t} e^{-j\omega t} dt \right] =$$

$$= \lim_{\substack{\varepsilon\to 0 \\ T\to\infty}} \int_{\varepsilon}^{T} \frac{1}{t}(e^{-j\omega t} - e^{j\omega t}) dt = \lim_{\substack{\varepsilon\to 0 \\ T\to\infty}} -2j \int_{\varepsilon}^{T} \frac{\sin\omega t}{t} dt . \qquad (9.37)$$

To evaluate this improper integral, we can use (9.6), or even better the triangle rule (Figure 9.6). Please note that a change of sign of ω also leads to a change of sign of the amplitude of the si-function. For $\omega = 0$ the calculation is trivial. We obtain

$$\mathcal{F}\{x(t)\} = \lim_{\substack{\varepsilon\to 0 \\ T\to\infty}} -2j \int_{\varepsilon}^{T} \frac{\sin\omega t}{t} dt = \begin{cases} -j\pi & \text{for } \omega > 0 \\ 0 & \text{for } \omega = 0 \\ j\pi & \text{for } \omega < 0 \end{cases} . \qquad (9.38)$$

The three different cases for the sign of ω can also be expressed by the signum-function $\text{sign}(\omega)$. We can now give the transform pair in the simple form

$$\frac{1}{t} \circ\!\!-\!\!\bullet -j\pi\,\text{sign}(\omega) . \qquad (9.39)$$

A purely imaginary spectrum is obtained that has a constant value for both positive and negative values of frequency ω (see Figure 9.10). Because of the discontinuity, $X(j\omega)$ cannot be analytically continued. The function $x(t) = \frac{1}{t}$ cannot be absolutely integrated (9.4), although its Fourier transform can be determined. This illustrates once again, that if a function can be integrated, that is a sufficient but not necessary condition for the existence of the Fourier transform.

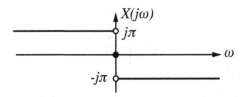

Figure 9.10: Spectrum of the function $\frac{1}{t}$

9.5 Symmetries of the Fourier Transform

From the examples of Fourier transforms of simple signals that we have looked at so far, we can see that it is possible to determine the spectra of more complicated signals without having to do more integration, if the properties of the Fourier transform are known. That is why symmetry relationships are so useful. If certain properties of a signal are known, the appearance of the spectrum can be determined and vice versa. To describe the symmetries of a signal, we will first of all introduce the concepts of *even* and *odd* functions, and investigate how they relate to the corresponding spectra.

9.5.1 Even and Odd Functions

Even and odd functions are characterised by their behaviour when the signs of their arguments are changed.

Definition 13: Even and odd functions

Two real functions $x_e(t)$ and $x_o(t)$ are called even *or* odd *functions when:*

$$x_e(t) \;=\; x_e(-t) \tag{9.40}$$
$$x_o(t) \;=\; -x_o(-t) \;. \tag{9.41}$$

Every function can be split into an even and an odd part

$$x(t) \;=\; x_e(t) + x_o(t), \tag{9.42}$$

where the terms are given by

$$x_e(t) \;=\; \frac{1}{2}\big(x(t) + x(-t)\big) \tag{9.43}$$

$$x_o(t) \;=\; \frac{1}{2}\big(x(t) - x(-t)\big). \tag{9.44}$$

With (9.40), (9.41) it is easy to confirm that $x_e(t)$ and $x_o(t)$ have the assumed symmetries.

Example 9.3

For the right-sided function $x(t)$ from Figure 9.11, the even and odd terms are both bilateral functions, the sum of which disappears for $t < 0$.

■

The even and odd symmetry that we introduced here with functions of time can, of course, also be carried over to functions of frequency ω. Even and odd spectra are identified by

$$X_e(j\omega) \;=\; X_e(-j\omega) \tag{9.45}$$
$$X_o(j\omega) \;=\; -X_o(-j\omega). \tag{9.46}$$

The definitions for even and odd functions are equally valid for both real and complex signals.

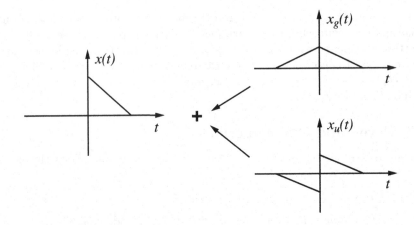

Figure 9.11: Combining the even and odd parts of a function

9.5.2 Conjugate Symmetry

For the real and imaginary parts of complex signals, there are often different symmetries that can be summarised by the concept of conjugate symmetry.

Definition 14:   Conjugate Symmetry

A complex function $x(t)$ has conjugate symmetry *when:*

$$x(t) = x^*(-t) . \qquad (9.47)$$

Here $x^*(t)$ means the complex conjugate of the function $x(t)$.

By splitting $x(t) = \text{Re}\{x(t)\} + j\text{Im}\{x(t)\}$ into real and imaginary parts, it is immediately clear that a function with conjugate symmetry has an even function as the real part and an odd function as the imaginary part.

Of course, this definition also applies to functions of frequency. Correspondingly, for a spectrum with conjugate symmetry,

$$X(j\omega) = X^*(-j\omega) . \qquad (9.48)$$

9.5.3 Symmetry Relationships for Signals with Real Values

We will now use the symmetry relationships just introduced to describe the connection between time signals and spectra, starting with real-value time signals and investigate how these properties manifest themselves in the frequency-domain. We will apply both operations associated with conjugate symmetry, one after the other to the Fourier integral (9.1) for a real function of time $x(t)$:

$$X(j\omega) = \int_{-\infty}^{\infty} x(t)e^{-j\omega t}\, dt$$

$$X(-j\omega) = \int_{-\infty}^{\infty} x(t)e^{j\omega t}\, dt$$

$$X^*(-j\omega) = \int_{-\infty}^{\infty} x^*(t)e^{-j\omega t}\, dt = \int_{-\infty}^{\infty} x(t)e^{-j\omega t}\, dt\ .$$

As the real function $x(t)$ does not change when transformed to the complex conjugate function $x^*(t)$, the operations cancel each other out. Real signals thus have conjugate symmetrical spectra

$$\boxed{x(t)\ \text{real} \ \longleftrightarrow\ X(j\omega) = X^*(-j\omega)\ .}\tag{9.49}$$

As spectra with complex values can be represented as real values of imaginary and real parts, or alternatively as magnitude and phase, conjugate symmetry of $X(j\omega)$ can be expressed by even or odd symmetry of these components

$$
\begin{aligned}
X(j\omega) &= X^*(-j\omega) & &\text{(9.50)}\\
\mathrm{Re}\{X(j\omega)\} &= \mathrm{Re}\{X(-j\omega)\} & \text{real part even,} & \text{(9.51)}\\
\mathrm{Im}\{X(j\omega)\} &= -\mathrm{Im}\{X(-j\omega)\} & \text{imaginary part odd,} & \text{(9.52)}\\
|X(j\omega)| &= |X(-j\omega)| & \text{magnitude even,} & \text{(9.53)}\\
\arg\{X(j\omega)\} &= -\arg\{X(-j\omega)\} & \text{phase odd.} & \text{(9.54)}
\end{aligned}
$$

We would further like to know which parts of the time signal correspond to the real and imaginary parts of the spectrum. Starting with the real part of the spectrum, with the Fourier integral (9.1) we substitute $\tau = -t$ for the time variable, bearing in mind that $\mathrm{Re}\{e^{-j\omega t}\} = \cos(\omega t)$ is an even function:

$$\mathrm{Re}\{X(j\omega)\} = \int_{-\infty}^{\infty} x(t)\cos(\omega t)\, dt$$

$$\mathrm{Re}\{X(j\omega)\} = \int_{-\infty}^{\infty} x(-\tau)\cos(-\omega\tau)\, d\tau$$

$$\mathrm{Re}\{X(j\omega)\} = \int_{-\infty}^{\infty} x(-t)\cos(\omega t)\, dt\ \ .$$

In the last row we have again set $\tau = t$, as the value of the integral does not depend on the notation used for the integration variables. All right sides are identical, so for a real function of time with a real spectrum it must be true that $x(t) = x(-t)$. According to Definition 13, $x(t)$ is an even function.

The same deductions can be made for the imaginary part of the spectrum, which brings us to the conclusion that a real function of time with an imaginary spectrum must be an odd function. We have now shown the following symmetry relationships for real signals:

$$
\begin{array}{lll}
x_e(t) \text{ real, even} & \longleftrightarrow & X(j\omega) \text{ real, even} \\
x_o(t) \text{ real, odd} & \longleftrightarrow & X(j\omega) \text{ imaginary,odd.}
\end{array}
$$

(9.55)
(9.56)

These general principles are revealed by our observations of the transforms of certain signals.

———————————————————————— **Example 9.4**

The rectangle function is real and even. The same is true for its spectrum.

$$
\text{rect}(t) \circ\!\!\!-\!\!\!\bullet \ \text{si}\left(\frac{\omega}{2}\right)
$$

(9.57)

———————————————————————— ∎

———————————————————————— **Example 9.5**

The real and odd function of time $\frac{1}{t}$ has an imaginary and odd spectrum.

$$
\frac{1}{t} \circ\!\!\!-\!\!\!\bullet \ -j\pi\text{sign}(\omega)
$$

(9.58)

———————————————————————— ∎

9.5.4 Symmetry of Imaginary Signals

The same deductions for time signals with real values can also be made for imaginary time signals. The equivalent relationships to (9.55), (9.56) are

$$
\begin{array}{lll}
x_e(t) \text{ imaginary, even} & \longleftrightarrow & X(j\omega) \text{ imaginary, even} \\
x_o(t) \text{ imaginary, odd} & \longleftrightarrow & X(j\omega) \text{ real, odd.}
\end{array}
$$

(9.59)
(9.60)

The imaginary function $x(t)$ and its real imaginary part $\text{Im}\{x(t)\}$ should not be confused. For purely imaginary functions $x(t) = j\text{Im}\{x(t)\}$.

9.5.5 Symmetry of Complex Signals

The previous results for real and for purely imaginary signals can now be united to give the symmetry for general complex signals. Every signal can be split into its even and odd parts, and its real and imaginary parts, and this gives four terms in total for the time signal, and also the spectrum. The symmetry relationships (9.55), (9.56) for real signals can be used for the real part, and (9.59), (9.60) for imaginary signals can be used for the imaginary part. The scheme (9.61) can be formed from these, for the symmetry between real and imaginary parts of the even and odd parts of a time signal and spectrum [19]. Although the symmetries seem to get more and more complicated, in fact, the general complex case is surprisingly simple, logical, and easy to remember.

$$
\begin{aligned}
x(t) &= \mathrm{Re}\{x_e(t)\} + \mathrm{Re}\{x_o(t)\} + j\mathrm{Im}\{x_e(t)\} + j\mathrm{Im}\{x_o(t)\}
\end{aligned}
$$

$$
X(j\omega) = \mathrm{Re}\{X_e(j\omega)\} + \mathrm{Re}\{X_o(j\omega)\} + j\mathrm{Im}\{X_e(j\omega)\} + j\mathrm{Im}\{X_o(j\omega)\}
$$

(9.61)

Example 9.6

A complex signal $x(t)$ has the Fourier transform $X(j\omega)$. What is $\mathcal{F}\{x^*(t)\}$?

It can be taken from (9.61) that changing the sign of the imaginary part in the time-domain has the following effect in the frequency-domain:

$$
\mathcal{F}\{x^*(t)\} = \mathrm{Re}\{X_e(j\omega)\} - \mathrm{Re}\{X_o(j\omega)\} - j\mathrm{Im}\{X_e(j\omega)\} + j\mathrm{Im}\{X_o(j\omega)\}.
$$

Using the symmetry of the even and odd parts this can be written more concisely:

$$
\begin{aligned}
\mathcal{F}\{x^*(t)\} &= \mathrm{Re}\{X_e(-j\omega)\} + \mathrm{Re}\{X_o(-j\omega)\} - j\mathrm{Im}\{X_e(-j\omega)\} - j\mathrm{Im}\{X_o(-j\omega)\} \\
&= X^*(-j\omega).
\end{aligned}
$$

Equation (9.61) yields the important transform pair

$$
x^*(t) \circ\!\!-\!\!\bullet\, X^*(-j\omega)
$$

■

9.6 Inverse Fourier Transform

The inverse of the Fourier transform is an integral expression that has a lot of similarity with the Fourier integral (9.1):

$$
x(t) = \mathcal{F}^{-1}\{X(j\omega)\} = \frac{1}{2\pi} \int_{-\infty}^{\infty} X(j\omega)e^{j\omega t}d\omega .
$$

(9.62)

The cause of this similarity is that both the time and frequency parameters are real values. The significant differences between the transform and its inverse are the integration variables (time / frequency), the sign in front of the exponential function $(-/+)$ and the factor $1/2\pi$ before the integral. Like the inverse Laplace transform, the inverse Fourier transform can represent a superimposition of eigen-functions $e^{j\omega t}$ of an LTI-system, but in this case only undamped oscillations are permitted.

In fact, (9.62) can be interpreted as an inverse Laplace transform (5.34) for which the imaginary axis $s = j\omega$ has been chosen as the path of integration. This interpretation is valid if the conditions for $X(s) = X(j\omega)|_{s=j\omega}$ (9.9) are fulfilled; the region of convergence for $X(s)$ encloses the imaginary axis and $X(j\omega)$ can be differentiated any number of times.

To verify that (9.62) actually yields the corresponding function of time for a possibly discontinuous spectrum $X(j\omega)$, we put the definition of $X(j\omega)$ into (9.62) in accordance with (9.1), and exchange the sequence of integrations. There is an expression (in square brackets) within the outer integral that we can interpret (see Section 9.4.3) as a delta impulse shifted in the time-domain. In contrast to Section 9.4.3, the roles of ω and t have been swapped. With the selective property, we end up with $x(t)$. Thus we have seen that the integral in (9.62) leads to a function of time from which (with (9.1)) $X(j\omega)$ can be calculated.

$$
\begin{aligned}
\frac{1}{2\pi} \int_{-\infty}^{\infty} X(j\omega)e^{j\omega t}\, d\omega &= \frac{1}{2\pi} \int_{-\infty}^{\infty} \int_{-\infty}^{\infty} x(\tau)e^{-j\omega\tau}\, d\tau\, e^{j\omega t}\, d\omega = \\
&= \frac{1}{2\pi} \int_{-\infty}^{\infty} \int_{-\infty}^{\infty} x(\tau)e^{j\omega(t-\tau)}\, d\tau\, d\omega = \\
&= \int_{-\infty}^{\infty} \underbrace{\left[\frac{1}{2\pi} \int_{-\infty}^{\infty} e^{j\omega(t-\tau)}\, d\omega \right]}_{\delta(t-\tau)} x(\tau)\, d\tau = \\
&= x(t) \quad\quad\quad\quad\quad\quad\quad\quad\quad\quad\quad\quad\quad\quad (9.63)
\end{aligned}
$$

This derivation requires that both improper integrals can be swapped, which is the case for normal convergence of integrals.

Just as for the Laplace transform, the assignment of a time signal to its Fourier transform is unambiguous, if discontinuities are foreseen and dealt with (Chapter 4.6.2). Deviations at individual points of discontinuities do not change the value of the integrals in (9.63), and when solving practical problems, this degree of unambiguousness is sufficient.

9.7   Properties of the Fourier Transform

As well as the symmetry properties that we have already dealt with, the Fourier transform has many properties that need to be known to take advantage of the frequency-domain representation, without having to solve integrals over and over again.

9.7.1 Linearity of the Fourier Transform

It follows directly from the linearity of the integration that the principle of superposition applies to the Fourier transform and its inverse:

$$
\begin{aligned}
\mathcal{F}\{a\,f(t)+b\,g(t)\} &= a\,\mathcal{F}\{f(t)\}+b\,\mathcal{F}\{g(t)\} \\
\mathcal{F}^{-1}\{c\,F(j\omega)+d\,G(j\omega)\} &= c\,\mathcal{F}^{-1}\{F(j\omega)\}+d\,\mathcal{F}^{-1}\{G(j\omega)\}.
\end{aligned}
\tag{9.64}
$$

a, b, c and d can be any real or complex constants.

———————————————————————————————— **Example 9.7**

The Fourier transform of a delta impulse pair as shown in Figure 9.12 can be obtained as the sum of the transforms of two individual impulses that are each shifted by $\pm\tau$:

$$
\delta(t+\tau)+\delta(t-\tau) \quad\circ\!\!-\!\!\bullet\quad e^{-j\omega\tau}+e^{j\omega\tau}=2\cos\omega\tau.
\tag{9.65}
$$

∎

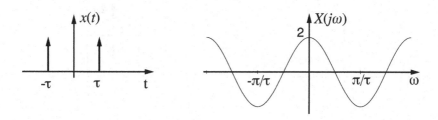

Figure 9.12: The impulse pair and its Fourier transform

9.7.2 Duality

Because of the similarity between the formulae for the Fourier transform and its inverse, a correspondence between the frequency-domain and time-domain can be derived by carefully choosing the arguments of a given transform pair. With appropriate substitutions into the integrals for the transform and its inverse, it can be shown that for two functions f_1 and f_2 that

$$
\begin{aligned}
f_1(t) &\quad\circ\!\!-\!\!\bullet\quad f_2(\omega) \\
f_2(t) &\quad\circ\!\!-\!\!\bullet\quad 2\pi f_1(-\omega).
\end{aligned}
\tag{9.66}
$$

This relationship is called *duality*.

Example 9.8

Using the property of duality we can obtain from the relationship (9.23)

$$\mathrm{rect}(t) \quad \circ\!\!-\!\!\bullet \quad \mathrm{si}\left(\frac{\omega}{2}\right)$$

that we already know as the transform pair for a rectangle-shaped spectrum

$$\Rightarrow \quad \mathrm{si}\left(\frac{t}{2}\right) \quad \circ\!\!-\!\!\bullet \quad 2\pi\,\mathrm{rect}(-\omega) \quad = \quad 2\pi\,\mathrm{rect}(\omega)\,. \tag{9.67}$$

Figure 9.13 shows the duality between the si-function and the rectangle function

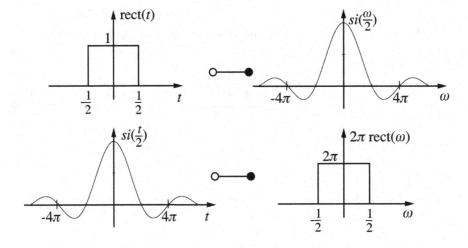

Figure 9.13: Duality of the si-function and rectangle function

Example 9.9

From (9.65) we obtain using the duality property of the Fourier transform of a sinusoidal function of time:

$$\cos\omega_0 t \quad \circ\!\!-\!\!\bullet \quad \pi\big[\delta(\omega + \omega_0) + \delta(\omega - \omega_0)\big]\,. \tag{9.68}$$

The Fourier transform consists of a pair of delta impulses. For such functions of time there is a Fourier transform, but no Laplace transform. This is because there is no analytical continuation into the complex s-plane for the delta impulses on the imaginary axis.

9.7.3 Similarity Theorem

We saw when dealing with rectangle functions that thin rectangles have wide spectra, and vice versa (Figure 9.7). This is also true for all functions of time and their corresponding spectra.

The substitution $t' = at$ and $\omega' = \omega/a$ in the integrals of the transform and inverse transform show that compressing the time axis corresponds to expanding the frequency axis, and vice versa. The factor a may be negative, but not complex:

$$x(at) \quad \circ\!\!-\!\!\bullet \quad \frac{1}{|a|} X\left(\frac{j\omega}{a}\right) \qquad a \in \mathbb{R}\backslash\{0\}\,. \tag{9.69}$$

This theorem has an important impact on the so-called 'time-bandwidth product', which will be discussed in Section 9.10.

9.7.4 Convolution Theorem of the Fourier Transform

As with the Laplace transform, convolution corresponds to multiplication of the Fourier transforms:

$$
\begin{aligned}
y(t) &= x(t) * h(t) = \int\limits_{-\infty}^{\infty} x(\tau)h(t-\tau)d\tau \\
&\quad\quad\quad \circ\!\!\!\!\!\! \atop \bullet \\
Y(j\omega) &= X(j\omega)\, H(j\omega) \quad .
\end{aligned}
\tag{9.70}
$$

The proof is carried out in much the same way as for the convolution property for the Laplace transform (compare Chapter 8.4.2).

The most important use is the calculation of the output signals of LTI-systems. The Fourier transforms of the impulse response is called the *frequency response*. Figure 9.14 shows the relationship between convolution with the impulse response and multiplication with the frequency response.

─── **Example 9.10**

As an example for the use of the convolution property, we find the Fourier transform of the triangular impulse shown in Figure 9.15, which is defined as a function of time by

$$x(t) = \begin{cases} t+1 & \text{for } -1 \le t < 0 \\ -t+1 & \text{for } 0 \le t < 1 \\ 0 & \text{otherwise} \end{cases} \tag{9.71}$$

Calculating the Fourier transform $X(j\omega) = \mathcal{F}\{x(t)\}$ with the Fourier integral is indeed not so difficult, but it is somewhat tedious. It is much simpler to notice that

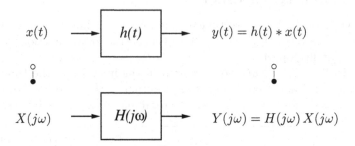

Figure 9.14: The relationship between convolution and multiplication

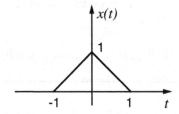

Figure 9.15: Triangular impulse

a triangular impulse is the result of the convolution of two rectangular impulses. With Figure 9.16 this can be easily confirmed by inspection. This connection can

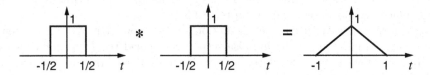

Figure 9.16: Triangular impulse as yielded by the convolution of two rectangular impulses

also be expressed by (9.72).

$$\text{rect}(t) \; * \; \text{rect}(t) \;\; = \;\; x(t) \tag{9.72}$$

$$\text{si}\left(\frac{\omega}{2}\right) \cdot \text{si}\left(\frac{\omega}{2}\right) \;\; = \;\; X(j\omega). \tag{9.73}$$

With the convolution property, (9.73) follows immediately. The Fourier transform we were looking for is already there:

$$X(j\omega) = \text{si}^2\left(\frac{\omega}{2}\right). \tag{9.74}$$

Figure 9.17 shows its plot. As the triangular impulse is real and even, its spectrum has the same properties because of the symmetry relationships.

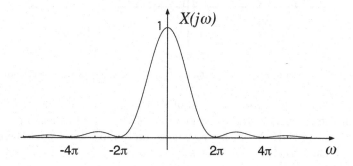

Figure 9.17: Fourier transform of the triangular impulse

9.7.5 Multiplication Property

By using the duality of $\mathcal{F}$ und $\mathcal{F}^{-1}$ (9.66), we obtain from the convolution property, the relationship for the Fourier transform of a signal $y(t)$, that can be written as the product of two signals $f(t)$ and $g(t)$,

$$\begin{array}{rcl} y(t) & = & f(t) \cdot g(t) \\ & \circ\!\!-\!\!\bullet & \\ Y(j\omega) & = & \dfrac{1}{2\pi} F(j\omega) * G(j\omega) \end{array} \tag{9.75}$$

as convolution of the spectra $F(j\omega)$ and $G(j\omega)$.

─── **Example 9.11**

As a more detailed example, we consider a typical spectral analysis problem. The spectrum of two superimposed sinusoidal signals with almost the same frequency is to be represented in such a way that both signals can be easily distinguished in the frequency-domain. The signal can only be observed for a limited duration, and we wish to determine for how long it needs to be observed for us to able to distinguish the two signals from each other. Figure 9.18 shows the two signals whose superimposition we will be observing from time $-T/2$ to $T/2$.

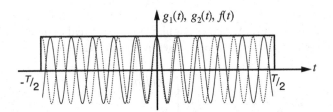

Figure 9.18: Sinusoidal signals $g_1(t)$, $g_2(t)$ and the window of observation $f(t)$

We already know that the spectrum of a sinusoidal signal consists of two delta impulses on the frequency axis. The location of the impulses corresponds to the frequency of the signal. As both frequencies are almost the same, the delta impulses will lie close to each other:

$$g(t) = g_1(t) + g_2(t) = \cos \omega_1 t + \cos \omega_2 t$$

$$\circ\!\!\!-\!\!\!\bullet$$

$$(9.76)$$

$$G(j\omega) = \pi \left(\delta(\omega - \omega_1) + \delta(\omega + \omega_1) + \delta(\omega - \omega_2) + \delta(\omega + \omega_2) \right).$$

The finite observation time can be expressed mathematically as multiplication by a rectangle function in the time-domain. The function describes the finite window of observation through which we see the sinusoidal function.

$$f(t) = \text{rect}\left(\frac{t}{T} \right) \quad \circ\!\!\!-\!\!\!\bullet \quad F(j\omega) = T \, \text{si}\left(\frac{\omega T}{2} \right) \qquad (9.77)$$

We already know the spectrum of the window: a si-function. The observed signal is thus

$$y(t) = f(t) \cdot g(t) = f(t) \cdot [g_1(t) + g_2(t)], \qquad (9.78)$$

as shown in Figure 9.19. The spectrum actually observed can be represented with the multiplication rule, as the convolution of the spectrum $G(j\omega)$ which contains the ideal delta impulse with the spectrum of the window of observation. Because of the convolution with four shifted delta impulses, the spectrum of the measured signal contains four terms that each have the same form as the spectrum of the window of observation and are located at the signal frequency.

$$Y(j\omega) = \frac{1}{2\pi} F(j\omega) * G(j\omega) \qquad (9.79)$$

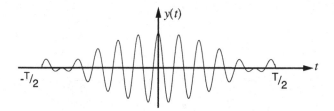

Figure 9.19: Signal $y(t)$ weighted with the observation window

$$= \frac{T}{2}\left[\text{si}\left((\omega - \omega_1)\frac{T}{2}\right) + \text{si}\left((\omega + \omega_1)\frac{T}{2}\right)\right.$$
$$\left. + \text{si}\left((\omega - \omega_2)\frac{T}{2}\right) + \text{si}\left((\omega + \omega_2)\frac{T}{2}\right)\right]$$

Figure 9.20 shows the spectrum of the reference signal for three different widths of the window of observation. From the similarity property we know that the wider the window, the thinner the spectrum of the window, that means the longer the duration of measurement. A window of observation with length $T = 1.2$ has a spectrum wide enough so that the components of both frequencies can no longer be distinguished from each other, and appear as a single signal. The two components first start to become distinguishable at a window width of $T = 1.5$, but are still not recogniseable as two separate impulses. Only by significantly increasing the duration of measurement, for example, to $T = 10$, can the two signals be clearly identified. At least several periods of the frequency difference between the two signals must be observed for them to be spectrally separated.   ∎

9.7.6   Shift and Modulation

If we set one convolution partner equal to $\delta(t - \tau)$ in the convolution theorem (9.70), we obtain the *shift theorem* with $x(t) * \delta(t - \tau) = x(t - \tau)$:

$$\boxed{x(t - \tau) \circ\!\!-\!\!\bullet\, e^{-j\omega\tau}X(j\omega)\,.} \qquad (9.80)$$

─── **Example 9.12**

In Figure 9.21 (top), the equation $x(t) = \text{si}(\pi t) \circ\!\!-\!\!\bullet X(j\omega) = \text{rect}(\frac{\omega}{2\pi})$ is shown. The Fourier transform of $x(t - 1)$ and $x(t - 5)$ can be given directly by the shift theorem:

$$\text{si}(\pi(t - 1)) \circ\!\!-\!\!\bullet\, e^{-j\omega} \cdot \text{rect}(\frac{\omega}{2\pi}) \qquad (9.81)$$

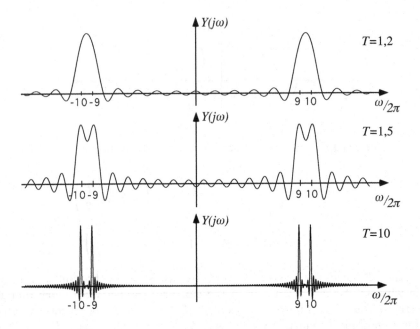

Figure 9.20: Spectrum $Y(j\omega)$ for various observation windows

and

$$\operatorname{si}(\pi(t-5)) \circ\!\!-\!\!\bullet\; e^{-j5\omega} \cdot \operatorname{rect}(\frac{\omega}{2\pi}). \qquad (9.82)$$

Both pairs are likewise illustrated in Figure 9.21 (centre and bottom).

■

The above example shows an interesting property of the shift theorem: shifting the signal does not change the magnitude spectrum, as the spectrum is only multiplied by a complex exponential function.

A shift in the frequency-domain is described by the *modulation theorem*

$$\boxed{e^{j\omega_0 t}x(t) \circ\!\!-\!\!\bullet\; X(j(\omega - \omega_0))} \qquad (9.83)$$

where the multiplication of a time signal with a complex exponential function is known as *modulation*. The modulation theorem is a special case of the multiplication theorem (9.75) for multiplication with

$$e^{j\omega_0 t} \circ\!\!-\!\!\bullet\; 2\pi\delta(\omega - \omega_0). \qquad (9.84)$$

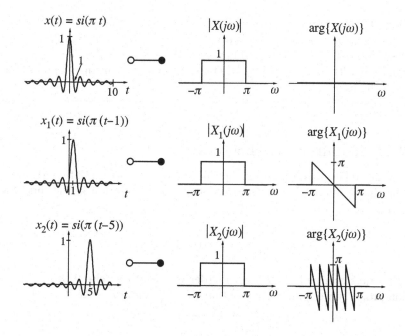

Figure 9.21: Example of the shift theorem

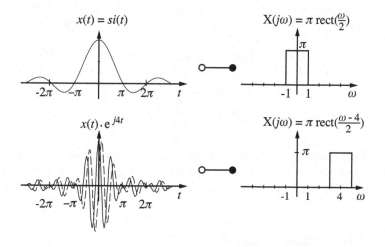

Figure 9.22: An example of the modulation theorem

Example 9.13

Figure 9.22 illustrates the effect of modulating the signal $x(t) = \mathrm{si}(t)$ with e^{j4t}. The time signal is complex after the modulation, so its real part is represented by the solid line (–), and its imaginary part by the dotted line (– –). Note that a modulation with frequency $j\omega_0 = j4$ corresponds to shifting the spectrum by $\omega_0 = 4$ to the right.

9.7.7 Differentiation Theorem

As with the bilateral Laplace transform, there is a differentiation theorem for both the time-domain and frequency-domain. Differentiation in the time-domain corresponds to convolution with $\dot{\delta}(t)$:

$$\frac{dx(t)}{dt} = x(t) * \dot{\delta}(t) \tag{9.85}$$

With the transform pair $\mathcal{F}\{\dot{\delta}(t)\} = j\omega$ (derivation in Exercise 9.13) and the convolution property we immediately obtain the *differentiation theorem*

$$\boxed{\frac{dx(t)}{dt} \circ\!\!-\!\!\bullet\ j\omega X(j\omega).} \tag{9.86}$$

As might be expected, it can be derived from the differentiation theorem of the bilateral Laplace transform (compare Chapter 4.7.4) by substituting s with $j\omega$. This requires $x(t)$ to be differentiable.

The differentiation of the Fourier spectrum corresponds as with the Laplace transform to a multiplication of the time signal with $-t$ (compare Chapter 4.7.7):

$$\boxed{-tx(t) \circ\!\!-\!\!\bullet\ \frac{dX(j\omega)}{d(j\omega)}.} \tag{9.87}$$

This theorem is therefore also called *'multiplication with t'*. It can easily be demonstrated by differentiating the defining equation of the Fourier transform (9.1) with respect to $j\omega$ (see Exercise 9.14). Note that it must be possible to differentiate $X(j\omega)$.

9.7.8 Integration Theorem

Here we are only interested in integrating in the time-domain. This corresponds to a convolution with the unit step function

$$\int_{-\infty}^{t} x(\tau)d\tau = x(t) * \varepsilon(t). \tag{9.88}$$

To derive the theorem we need the transform $\mathcal{F}\{\varepsilon(t)\}$, which is easy to calculate if we split $\varepsilon(t)$ into its even and odd parts:

$$\varepsilon(t) = \frac{1}{2} + \frac{1}{2}\text{sign}(t). \tag{9.89}$$

Using the principle of duality and the transform pair $\delta(t) \circ\!\!-\!\!\bullet\, 1$ (9.16) and $\frac{1}{t} \circ\!\!-\!\!\bullet -j\pi\text{sign}(\omega)$ (9.39), the even term is obtained as

$$\mathcal{F}\left\{\frac{1}{2}\right\} = \pi\delta(\omega) \tag{9.90}$$

and the odd term as

$$\mathcal{F}\left\{\frac{1}{2}\text{sign}(t)\right\} = \frac{1}{j\omega}. \tag{9.91}$$

Thus

$$\varepsilon(t) \circ\!\!-\!\!\bullet\, \pi\delta(\omega) + \frac{1}{j\omega}, \tag{9.92}$$

and using the convolution theorem, we finally reach the *integration theorem*

$$\boxed{\int_{-\infty}^{t} x(\tau)\,d\tau \quad\circ\!\!-\!\!\bullet\quad X(j\omega)\left[\pi\delta(\omega) + \frac{1}{j\omega}\right] = \frac{1}{j\omega}X(j\omega) + \pi X(0)\,\delta(\omega)\,.} \tag{9.93}$$

The integration theorem of the Fourier transform cannot be obtained simply by inserting $s = j\omega$ into the integration theorem of the Laplace transform (4.27). The form thus obtained is incomplete and is only valid for signals with zero mean $X(0) = 0$. In fact, the imaginary axis according to (4.27) is not part of the region of convergence of the Laplace transform, as integration generates a pole at $s = 0$. (9.93) can therefore not be analytically continued since it contains a delta impulse at $\omega = 0$ for signals with a non-zero mean.

9.8 Parseval's Theorem

A further property of the Fourier transform expresses *Parseval's theorem*. It says that the integral of the product of two functions of time can also be expressed as the integral of the product of their spectra. To derive this theorem we start with the multiplication theorem (9.75) and write out the Fourier integral for the functions of time and the convolution in detail

$$\int_{-\infty}^{\infty} f(t)g(t)e^{-j\omega t}dt = \frac{1}{2\pi}\int_{-\infty}^{\infty} F(j\nu)G(j\omega - j\nu)d\nu\,. \tag{9.94}$$

For $\omega = 0$, the exponential term on the left side of the Fourier integral disappears and

$$\int\limits_{-\infty}^{\infty} f(t)g(t)dt = \frac{1}{2\pi} \int\limits_{-\infty}^{\infty} F(j\nu)G(-j\nu)d\nu . \qquad (9.95)$$

For the functions of time $f(t)$ and $g(t)$ we will also permit complex functions. We can read from the symmetry property (9.61) and Example 9.7 that for a signal $g(t)$ and the corresponding complex conjugate signal $g^*(t)$ that the relationship between the spectra $G(j\omega)$ and $G^*(j\omega)$ is

$$g(t) \quad \circ\!\!-\!\!\bullet \quad G(j\omega)$$
$$\qquad\qquad\qquad\qquad\qquad\qquad\qquad\qquad (9.96)$$
$$g^*(t) \quad \circ\!\!-\!\!\bullet \quad G^*(-j\omega) .$$

From here we can obtain the general form of Parseval's theorem for complex functions of time

$$\boxed{\int_{-\infty}^{\infty} f(t)g^*(t)dt = \frac{1}{2\pi} \int_{-\infty}^{\infty} F(j\nu) \cdot G^*(j\nu)d\nu .} \qquad (9.97)$$

For $g(t) = f(t)$, we obtain the easy to remember relationship

$$\boxed{\int\limits_{-\infty}^{\infty} |f(t)|^2 dt = \frac{1}{2\pi} \int\limits_{-\infty}^{\infty} |F(j\omega)|^2 d\omega .} \qquad (9.98)$$

To interpret this formula we define the energy of a time signal.

Definition 15: Energy of a time signal

The energy E_f of a signal $f(t)$ is given by

$$E_f = \int\limits_{-\infty}^{\infty} |f(t)|^2 dt .$$

The reason that this integral is called the energy of a signal becomes clear when we imagine $f(t)$, for example, to be a decreasing voltage across an ohmic resistance. The energy converted to heat in the resistance is proportional to the integral over the square of the voltage. Using the magnitude squared permits the use of complex signals.

Parseval's theorem says that the energy of a time signal can be calculated not only in the time-domain, but also in the frequency-domain, by integrating the magnitude squared spectrum. Only the magnitude is involved; the phase clearly has no effect on the energy of the signal.

Example 9.14

As an example of the use of Parseval's theorem, we will calculate the energy E_x of the signal

$$x(t) = \text{si}\left(\frac{t}{2}\right). \tag{9.99}$$

Calculating the integral

$$E_x = \int\limits_{-\infty}^{\infty} x^2(t)dt = \int\limits_{-\infty}^{\infty} \text{si}^2\left(\frac{t}{2}\right) dt = \int\limits_{-\infty}^{\infty} \frac{4 \cdot \sin^2(t/2)}{t^2}dt.$$

directly in the time-domain is possible, but requires some tricks and rearrangements. With the transform pair (9.67)

$$\text{si}\left(\frac{t}{2}\right) \circ\!\!-\!\!\bullet\ 2\pi\,\text{rect}(\omega) \tag{9.100}$$

and Parseval's theorem we can calculate the energy much more easily in the frequency-domain by integrating the rectangle function:

$$E_x = \frac{1}{2\pi} \int_{-\infty}^{\infty} |X(j\omega)|^2\,d\omega = \frac{1}{2\pi} \int_{-\frac{1}{2}}^{\frac{1}{2}} (2\pi)^2 d\omega = 2\pi. \tag{9.101}$$

We have used the fact that the square of a rectangle function is again a rectangle.

■

9.9 Correlation of Deterministic Signals

The concept of correlation actually originates from random signal theory, which we will introduce in Chapters 17 and 18, where it will play a central role. So far we have not encountered random signals, and we would like to emphasise that the following discussion is first of all limited to deterministic (non-random) signals.

Combining two functions of time $f(t)$ und $g(t)$ can be more generally formulated than in (9.97). If we consider the product of $f(t)$ and $g(t)$, shifted in time by τ, the integral expression formed is a function of the separation in time τ. This leads to the definition of the cross-correlation function

9.9.1 Definition

Definition 16: Cross-correlation function

The cross-correlation function *of two deterministic signals $f(t)$ and $g(t)$ is*

$$\varphi_{fg}(\tau) = \int\limits_{-\infty}^{\infty} f(t+\tau)g^*(t)dt \quad . \tag{9.102}$$

Definition 17: Auto-correlation function

For $f(t) = g(t)$ the auto-correlation function *of the function of time $f(t)$*

$$\varphi_{ff}(\tau) = \int\limits_{-\infty}^{\infty} f(t+\tau)f^*(t)dt \; . \tag{9.103}$$

The cross-correlation function describes how two signals are related, while bearing in mind any possible displacement. The auto-correlation function shows how similar the components of a time signal are that occur at different points in time. We will come back to this later, when we have extended (9.102) to cover random signals, but the significant properties are already clear for deterministic signals.

9.9.2 Properties

9.9.2.1 Relationship with Convolution

The integral expression in the definition of cross-correlation (9.102) is certainly related to the convolution integral. In fact, (9.102) can be changed to a convolution with the substitution $t' = -t$

$$\varphi_{fg}(\tau) = \int\limits_{-\infty}^{\infty} f(t+\tau)g^*(t)dt = \int\limits_{-\infty}^{\infty} f(\tau - t')g^*(-t')dt' \; .$$

The cross-correlation function for deterministic signals can therefore also be defined by a convolution

$$\varphi_{fg}(\tau) = \int\limits_{-\infty}^{\infty} f(t+\tau)g^*(t)dt = f(\tau) * g^*(-\tau) \; . \tag{9.104}$$

Likewise for the auto-correlation function

$$\varphi_{ff}(\tau) = \int\limits_{-\infty}^{\infty} f(t + \tau)f^*(t)dt = f(\tau) * f^*(-\tau) \,. \tag{9.105}$$

Example 9.15

What is the cross-correlation $\varphi_{yx}(\tau)$ between the output and input of an LTI-system with impulse response $h(t)$?

Evidently,

$$\begin{aligned}
\varphi_{yx}(\tau) &= y(\tau) * x^*(-\tau) \\
&= h(\tau) * x(\tau) * x^*(-\tau) \\
&= h(\tau) * \varphi_{xx}(\tau)
\end{aligned}$$

The cross-correlation between the output and input of an LTI-system is the same as the convolution of the auto-correlation of the input signal, and the impulse response. We will consider an example, with a delay circuit such that

$$h(t) = \delta(t - t_0) \,,$$

where the cross-correlation is a shifted version of the auto-correlation

$$\varphi_{yx}(\tau) = \delta(\tau - t_0) * \varphi_{xx}(\tau) = \varphi_{xx}(\tau - t_0) \,.$$

$\blacksquare$

9.9.2.2 Symmetry

The question of the symmetry of the cross-correlation function is closely related to interchanging the two functions $f(t)$ and $g(t)$, because a shift of f with respect to g by τ is equivalent to a shift of g with respect to f by $-\tau$. Exchanging f and g in (9.102) and substituting variables yields

$$\varphi_{fg}(\tau) = \varphi_{gf}^*(-\tau) \,. \tag{9.106}$$

Clearly, $\varphi_{fg}(\tau)$ can only be expressed by φ_{gf} and not by φ_{fg} itself. The cross-correlation function therefore has no general symmetry properties.

For $f(t) = g(t)$, it follows from (9.106) for the complex cross-correlation function

$$\varphi_{ff}(\tau) = \varphi_{ff}^*(-\tau) \tag{9.107}$$

and for a real function of time $f(t)$

$$\varphi_{ff}(\tau) = \varphi_{ff}(-\tau) \,. \tag{9.108}$$

The auto-correlation of a complex function of time therefore has conjugate symmetry; the auto-correlation of a real function of time is an even function.

9.9.2.3 Commutivity

While investigating the symmetry properties of the cross-correlation function we found a close relationship with exchanging the time functions. From (9.106) it can be immediately read that in general

$$\varphi_{fg}(\tau) \neq \varphi_{gf}(\tau). \tag{9.109}$$

This means that in contrast to convolution, the cross-correlation is not commutative.

9.9.2.4 The Fourier Transform of the Cross-Correlation Function

The Fourier transform of the cross-correlation function can be most elegantly determined from the convolution relation (9.104). We need for this a special symmetry of the Fourier transform, that as with (9.96), we can read from the general scheme (9.61)

$$\begin{aligned} g(t) &\quad \circ\!\!\!-\!\!\!\bullet \quad G(j\omega) \\[4pt] g^*(-t) &\quad \circ\!\!\!-\!\!\!\bullet \quad G^*(j\omega) \end{aligned} \tag{9.110}$$

With the convolution theorem it follows from (9.104) that

$$\varphi_{fg}(\tau) \qquad\qquad = \qquad\qquad f(\tau) \quad * \quad g^*(-\tau)$$

$$\Phi_{fg}(j\omega) = \mathcal{F}\{\varphi_{fg}(\tau)\} = \int_{-\infty}^{\infty} \varphi_{fg}(\tau) e^{-j\omega\tau} d\tau = F(j\omega) \cdot G^*(j\omega) \tag{9.111}$$

or concisely,

$$\boxed{\varphi_{fg}(\tau) \quad \circ\!\!\!-\!\!\!\bullet \quad F(j\omega)G^*(j\omega) .} \tag{9.112}$$

For the Fourier transform of the auto-correlation function this expression can be simplified to

$$\boxed{\varphi_{ff}(\tau) \quad \circ\!\!\!-\!\!\!\bullet \quad |F(j\omega)|^2 .} \tag{9.113}$$

─── **Example 9.16**

Generalising Example 9.14 we try to calculate the auto-correlation of $x(t) = \text{si}\left(\dfrac{t}{2}\right)$ (9.99). Evaluating the integral

$$\varphi_{xx}(\tau) = \int_{-\infty}^{\infty} x(t+\tau)x(t)dt = \int_{-\infty}^{\infty} \text{si}\left(\frac{t+\tau}{2}\right) \text{si}\left(\frac{t}{2}\right) dt$$

is again so complicated, however, that we would like to find an easier way.

We can read from (9.113) that the Fourier transform of our auto-correlation function is already known from the Fourier transform of the si-function. The auto-correlation function $\varphi_{xx}(\tau)$ then follows by inverse transformation

$$\mathcal{F}\{\varphi_{xx}(\tau)\} \;=\; |X(j\omega)|^2 \;=\; 4\pi^2\mathrm{rect}(\omega) \qquad (9.114)$$

$$\varphi_{xx}(\tau) \;=\; 2\pi\,\mathrm{si}\left(\frac{\tau}{2}\right) \;. \qquad (9.115)$$

To check the results we consider the value at $\tau = 0$. Here, the value of the auto-correlation $\varphi_{xx}(0)$ is the same as the energy E_x of the signal x

$$\varphi_{xx}(0) = \int\limits_{-\infty}^{\infty} x(t)x^*(t)dt = \int\limits_{-\infty}^{\infty} |x(t)|^2dt = E_x \quad .$$

From (9.115) we obtain with $\varphi_{xx}(0) = E_x = 2\pi$ the same value that we had determined in (9.101).

■

9.10   Time-Bandwidth Product

At various times we have seen that there is a close connection between the duration of a signal and the appearance of its spectrum. In Figure 9.18 we made good use of this property to determine how long we needed to measure in the time-domain, to get a clear result in the frequency-domain. This reciprocal connection between the length of the signal in time and the width of the spectrum is clearly not limited to the rectangle function. The similarity theorem (9.69) says that a stretched time signal corresponds to a compressed spectrum (and vice versa), for any time signal.

We now wish to investigate this connection more closely. We are especially interested in the relationship between the period of a signal in time and the width of the corresponding spectrum. The width of the spectrum is known as the *bandwidth*, but we must make it clear what exactly should be understood by the terms period and width.

In Figure 9.7 it is evident what is meant by the duration of a rectangle function. The corresponding spectra are all, however, stretched out infinitely. Despite this, the spectrum of a rectangle function does seem to become wider, in a certain way, when the impulse becomes shorter. There are various possiblities to define the length of a signal and its bandwidth for general signal. We will consider three of them, and will see that they all lead to the same fundamental principle.

9.10.1   Equal-Area Rectangle

The first method for defining the duration of a signal and its bandwidth is shown
in Figure 9.23 for a signal $x(t)$ that is real and symmetrical. Consequently, the
Fourier transform $F\{x(t)\} = X(j\omega)$ has the same properties. We understand the
duration D_1 of a signal $x(t)$ as the duration of a rectangle function that has the
same area and height as $x(t)$ itself. If the time zero is chosen so that it coincides
with the maximum value of $x(t)$ then the height of the rectangle function is equal
to $x(0)$. From the equal-area requirement for the signal $x(t)$ and rectangle of width
D_1 and height $x(0)$

$$\int_{-\infty}^{\infty} x(t)dt = D_1 x(0)$$

we directly obtain the signal duration D_1

$$D_1 = \frac{1}{x(0)} \underbrace{\int_{-\infty}^{\infty} x(t)dt}_{X(0)} \quad . \tag{9.116}$$

The area under the signal $x(t)$ cannot only be determined by integration, but also
by reading the value of the spectrum at $\omega = 0$. This is

$$X(0) = \left[\int_{-\infty}^{\infty} x(t)e^{-j\omega t}dt \right]\Bigg|_{\omega=0} = \int_{-\infty}^{\infty} x(t)dt \; .$$

The signal duration can therefore also be expressed simply by the ratio

$$D_1 = \frac{X(0)}{x(0)} \; .$$

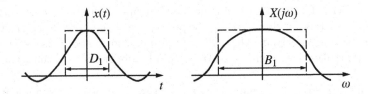

Figure 9.23: Duration D_1 and bandwidth B_1

We understand the width B_1 of a spectrum $X(j\omega)$ as equivalent to the width
of a rectangular spectrum, that has the same height and area as $X(j\omega)$ itself.

Proceeding in the same way as before, we can obtain the bandwidth B_1 as

$$B_1 = \frac{1}{X(0)} \underbrace{\int_{-\infty}^{\infty} X(j\omega)d\omega}_{2\pi\, x(0)} \quad . \tag{9.117}$$

The area under the spectrum $X(j\omega)$ can also be expressed by the value of the signal $x(t)$ at $t = 0$, and it is then

$$x(0) = \left[\frac{1}{2\pi}\int_{-\infty}^{\infty} X(j\omega)e^{j\omega t}d\omega\right]\Bigg|_{t=0} = \frac{1}{2\pi}\int_{-\infty}^{\infty} X(j\omega)d\omega .$$

For the bandwidth B_1 therefore,

$$B_1 = 2\pi\frac{x(0)}{X(0)} .$$

Duration D_1 and bandwidth B_1 clearly have a reciprocal relationship, so for their product

$$D_1\, B_1 \;=\; 2\pi . \tag{9.118}$$

The product $D_1 B_1$ of duration and bandwidth is called the *time-bandwidth product*. From (9.118) we can see that:

> The time-bandwidth product is constant.

This is true for the definitions of duration and bandwidth that we have chosen here: equal-area rectangles for all real and symmetrical time signals.

9.10.2   Tolerance

A different possibilty for defining the duration and bandwidth uses tolerance parameters. Figure 9.24 shows examples for a signal and its spectrum. Outside of the duration D_2, the magnitude of the signal $x(t)$ is always less than q-times $(0 < q < 1)$ its maximum value:

$$|x(t)| \leq q \max|x| \qquad \forall\, t \notin [t_0, t_0 + D_2] . \tag{9.119}$$

The bandwidth B_2 is defined accordingly

$$|X(j\omega)| \leq q \max|X(j\omega)| \qquad \forall\, |\omega| > \frac{B_2}{2} . \tag{9.120}$$

Tolerance parameters of this kind are common in filter design. They permit general statements to be made about a signal or its spectrum, even when the exact behaviour is unkown. From the similarity theorem, the time-bandwidth product $D_2 B_2$ for a certain value of q only depends on the form of the signal $x(t)$. A fixed value or lower limit is unknown in this case.

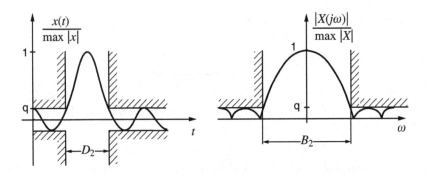

Figure 9.24: Example of tolerance parameters

Example 9.17

Signals of the form

$$x(t) = e^{-\alpha^2 t^2}$$

are called *Gauss impulses.* Their transform pair is (see Appendix Appendix B.3)

$$e^{-\alpha^2 t^2} \circ\!\!-\!\!\bullet \frac{\sqrt{\pi}}{\alpha} e^{-\frac{\omega^2}{4\alpha^2}} . \tag{9.121}$$

From the condition

$$e^{-\alpha^2 (D_2/2)^2} = q$$

we obtain

$$D_2 = \frac{2}{\alpha}\sqrt{-\ln q} .$$

Note that $\ln q$ is negative.

$$\frac{\sqrt{\pi}}{\alpha} e^{-\frac{(B_2/2)^2}{4\alpha^2}} = q\frac{\sqrt{\pi}}{\alpha}$$

$$B_2 = 4\alpha\sqrt{-\ln q}$$

and for the the time-bandwidth product, the value

$$D_2 B_2 = -8\ln q ,$$

is constant for any given q. If q is made smaller, the tolerance parameters become stricter, and the resulting time-bandwidth product $D_2\,B_2$ for Gauss impulses is increased.

■

9.10.3   Second-Order Moments

A third method for defining the signal duration and bandwidth can be obtained by using second-order moments for the magnitude squared of the signal and its spectrum. This definition is motivated by analogies with random signals or mechanical systems.

For simplification, we normalise the signal $x(t)$ that we are considering, so that it has energy $E_x = 1$. We are assuming that

$$\int_{-\infty}^{\infty} |x(t)|^2 dt = \frac{1}{2\pi} \int_{-\infty}^{\infty} |X(j\omega)|^2 d\omega = 1. \tag{9.122}$$

The duration D_3 can then be defined as

$$D_3 = \sqrt{\int_{-\infty}^{\infty} (t - t_s)^2 |x(t)|^2 dt}, \tag{9.123}$$

where

$$t_s = \int_{-\infty}^{\infty} t|x(t)|^2 dt \tag{9.124}$$

is the first moment of $|x(t)|^2$. It corresponds to the *centre of gravity* in mechanics, if the time dependent variable is viewed as corresponding to position (e.g. the distance along a beam), and $|x(t)|^2$ as density. In this analogy, (9.123) is the moment of inertia. If the reader prefers the probability theory analogy, $|x(t)|^2$ can be interpreted as the probability distribution independent of t, (9.124) as the mean and (9.123) as the standard deviation.

The bandwidth B_3 is defined by

$$B_3 = \sqrt{\int_{-\infty}^{\infty} (\omega - \omega_s)^2 , |X(j\omega)|^2 d\omega} \tag{9.125}$$

with

$$\omega_s = \int_{-\infty}^{\infty} \omega |X(j\omega)|^2 d\omega. \tag{9.126}$$

Using Parseval's theorem for $|x(t)|^2$ and $|x'(t)|^2$ and Schwarz's equation, it can be shown [5], that the time-bandwidth product is

$$D_3 B_3 \geq \sqrt{\frac{\pi}{2}}. \tag{9.127}$$

The exact value of $D_3 B_3$ depends on the form of the signal $x(t)$. The minimum is found for a Gauss impulse [5]. Because of formal analogies with quantum mechanics, this equation is also called the *uncertainty relation*.

As a Gauss impulse has a particularly good time-bandwidth product $D_3 B_3$ (9.127), it is employed whenever it is important to pack as much energy as possible into a small frequency band over a small amount of time. This is, for example, a typical demand of a digital transmission system. In short-time spectral analysis, good time and frequency resolution is required at the same time, and Gaussian windows are widely employed.

9.10.4 Summary

From the various definitions of duration and bandwidth and the results obtained we can draw some important conclusions.

Duration and bandwidth of a signal are reciprocal. It is therefore not possible to find a signal that has any desired short duration and at the same time any desired small bandwidth. Shortening the duration of the signal always increases the bandwidth, and vice versa.

This statement is very important for signal transmission and spectral analysis (see Example 9.11). It is formally related to the uncertainty relation from quantum mechanics.

9.11 Exercises

Exercise 9.1

Calculate the Fourier transforms of the following signals with the Fourier integral, as long as it converges. For comparison, give also the Laplace transforms with the regions of convergence.

a) $x(t) = \varepsilon(t)\, e^{-j\omega_0 t}$

b) $x(t) = \mathrm{rect}(0, 1\, t)$

c) $x(t) = \delta(-4t)$

d) $x(t) = \varepsilon(-t)$

e) $x(t) = e^{-j\omega_0 t}$

Note: b) Properties of the Laplace transform are in Chapter 4.   c) Rules for calculations with the delta impulse are in Chapter 8.3.4

Exercise 9.2

For which parts of Exercise 9.1 can the Laplace transform be used to calculate the Fourier transform (see equation (9.9))? Justify your answer.

Exercise 9.3

Give the Fourier transforms of the signals in Exercise 9.1, whose Fourier integrals do not have convergence, using generalised functions. You will find some already known transform pairs and some properties of the Fourier transform useful.

Exercise 9.4

Evaluate $\mathcal{F}\left\{\dfrac{1}{t-a}\right\}$ using the Fourier integral (9.1).

Note: See Section 9.4.4 and use a suitable substitution.

Exercise 9.5

Let $x(t) = \mathrm{si}(\omega_0 t)$.

a) Determine ω_0 so that the zeros of $x(t)$ lie at $t = n \cdot 4\pi$, $n \in \mathbb{Z} \setminus \{0\}$.

b) Calculate $\displaystyle\int_{-\infty}^{\infty} x(t)\, dt$.

c) Sketch $x(t)$ for this choice of ω_0.

Exercise 9.6

Evaluate the Fourier transform of $x(t) = \mathrm{si}(10\pi(t+T))$ and sketch for $T = 0.2$:

a) $|X(j\omega)|$ and $\arg\{X(j\omega)\}$

b) $\mathrm{Re}\{X(j\omega)\}$ and $\mathrm{Im}\{X(j\omega)\}$

Note: displacement rule.

Exercise 9.7

Find the functions of time that correspond to the following Fourier transforms:

$$X_1(j\omega) = \frac{5j\omega + 5}{(j\omega)^2 + 2j\omega + 17}$$

$$X_2(j\omega) = \frac{\sin(2\omega)}{2\omega}$$

$$X_3(j\omega) = \left(\frac{\sin(2\omega)}{2\omega}\right)^2$$

Exercise 9.8

What symmetry does the spectrum have for a) a real signal and b) a purely imaginary signal?

Exercise 9.9

You are given the transform pair $x(t) \circ\!\!-\!\!\bullet X(j\omega)$, $x(t) \in \mathbb{C}$. What are the Fourier transforms of a) $y_a(t) = x(-t)$, b) $y_b(t) = x^*(t)$ and c) $y_c(t) = x^*(-t)$, expressed with $X(j\omega)$? Use the symmetry scheme (9.61).

Exercise 9.10

The signal $x(t)$ has spectrum $X(j\omega)$. With amplitude modulation, $y(t) = (x(t) + m)\sin(\omega_T t)$. Give $Y(j\omega)$ dependent on $X(j\omega)$ and sketch it.

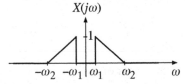

Note: Analyse $\sin(\omega_T t)$ with exponential oscillations and use the modulation theorem.

Exercise 9.11

Calculate $Y(j\omega)$ from Exercise 9.10 with $\mathcal{F}\{\sin \omega_T t\}$ and the multiplication theorem.

Exercise 9.12

Use the duality principle to help calculate

a) $\mathcal{F}\{\pi\delta(t) + \frac{1}{jt}\}$

b) $\mathcal{F}\{\frac{1}{t-ja}\}$

c) $\mathcal{F}\{\text{sign}(t)\}$.

Note: the dual transform pairs can be found in Sections 9.2.2, 9.3 and 9.4.4.

Exercise 9.13

Confirm the transform pair $\dot{\delta}(t) \circ\!\!-\!\!\bullet j\omega$ used in Section 9.7.7 (differentiation theorem) by putting it into the Fourier integral (9.1).

Exercise 9.14

Prove the theorem 'multiplication with t' (9.87) by differentiating the Fourier integral (9.1).

Exercise 9.15

Determine the Fourier transform of the triangular function shown

a) by using the differentiation theorem of the Fourier transform,

b) by multiplying the spectral functions of suitably chosen rectangle impulses, whose convolution gives the triangular function.

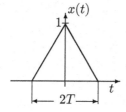

10 Bode Plots

10.1 Introduction

We have already pointed to the close connection between the Laplace transform and the Fourier transform, at the start of Chapter 9. In this chapter we introduce a classical method to determine the magnitude and phase of the Fourier transform of a signal from the pole-zero plot of its Laplace transform. It is mostly used to describe systems so that we can refer to the system function instead of the Laplace transform of the impulse response, and the frequency response instead of the Fourier transform.

Bode plots are used as a fast way of finding the approximate frequency response from the poles and zeros of the system function. At one time, Bode plots were the engineer's standard method for representing a frequency response in graphical form. They would always have logarithmic graph paper ready for this task, but of course, this is no longer necessary, as computers can draw frequency response plots faster and more accurately. Bode plots are, however, a marvellous way of developing an intuitive understanding of how the location of the poles and zeros of the system function affect the frequency response, and it is therefore important for developing and analysing systems.

The fundamental connection between the system function and the frequency response is given by the relationship (9.9), that we can also write here as

$$\mathcal{F}\{h(t)\} = H(j\omega) = H(s)|_{s=j\omega} = \mathcal{L}\{h(t)\}|_{s=j\omega}. \tag{10.1}$$

As stable systems only have decaying oscillations, all poles must lie left of the imaginary axis of the s-plane. The imaginary axis $s = j\omega$ is part of the region of convergence of $\mathcal{L}\{h(t)\}$, so the Fourier and Laplace transforms can be set equal as in (9.9). Anyway, in Chapter 9, we had introduced the Fourier transform with $j\omega$ and not ω, so between the frequency response $H(j\omega) = \mathcal{F}\{h(t)\}$ and the values of the system function $H(s) = \mathcal{L}\{h(t)\}$ for $s = j\omega$, we do not have to make any changes in the notation.

Bode plots represent the frequency response as split into magnitude and phase

$$H(j\omega) = |H(j\omega)| \cdot e^{j\varphi(j\omega)} = |H(j\omega)| \cdot e^{j\,\arg\{H(j\omega)\}}$$

with the following properties

1. The logarithm of the magnitude $|H(j\omega)|$ is plotted against the logarithm of the frequency ω.

2. The phase $\varphi(j\omega) = \arg\{H(j\omega)\}$ is plotted linearly against the logarithm of the frequency.

For transfer functions in rational form, approximate asymptotes for the plots can be drawn very quickly since from

$$H(j\omega) = K \frac{\prod_m (j\omega - s_{0m})}{\prod_n (j\omega - s_{pn})}, \quad K > 0 \tag{10.2}$$

follows

$$\log|H(j\omega)| = \sum_m \log|j\omega - s_{0m}| - \sum_n \log|j\omega - s_{pn}| + \log K$$

$$\varphi(j\omega) = \sum_m \arg(j\omega - s_{0m}) - \sum_n \arg(j\omega - s_{pn}) . \tag{10.3}$$

The logarithm of the magnitude and phase can thus be separately determined for each zero and pole, and the individual magnitudes can be added together when drawn. A logarithmic frequency axis makes it easier to get an overview of multiple decades. Likewise, for reasons of clarity, the ordinates for $\log|H(j\omega)|$ are marked in decibels (dB), i.e., we write $20\log_{10}|H(j\omega)|$.

In the following sections, we will first show the representation of single poles and zeros, and then put the partial results together as complete Bode plots.

10.2 Contribution of Individual Poles and Zeros

We will start with an example, in which we wish to determine the contribution of a real pole at $s = -10$ with the frequency response

$$H(j\omega) = \frac{1}{j\omega + 10} = \left. \frac{1}{s + 10} \right|_{s=j\omega} . \tag{10.4}$$

By substituting different values for ω we obtain the value for $|H(j\omega)|$, $\arg\{H(j\omega)\}$ and $20\log_{10}|H(j\omega)|$, from Table 10.1. The magnitude frequency response is shown in Figure 10.1, in linear and in double logarithmic representation. The linear representation shows neither the behaviour at low frequencies ($0 < \omega < 10$), nor the behaviour at high frequencies ($\omega \gg 10$) particularly clearly. In contrast, the double logarithmic representation makes it clear that the magnitude frequency response can be approximated by two asymptotes: for $\omega \ll 10$ the magnitude frequency response is approximately constant and for $\omega \gg 10$, it decays by 20

Table 10.1: Magnitude, phase and damping of $H(j\omega)$ for various frequencies ω

ω	$|H(j\omega)|$	$\arg\{H(j\omega)\}$	$20\log_{10}|H(j\omega)|$
0	0.1	0	-20dB
1	0,0995	$-5.71°$	-20.04dB
2	0,0981	$-11.31°$	-20.17dB
$\vdots$	$\vdots$	$\vdots$	$\vdots$

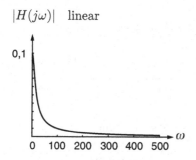 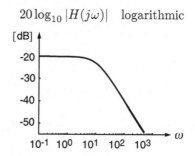

Figure 10.1: Linear and double logarithmic representation of the frequency response

dB/decade. The frequency $\omega = 10$, where both asymptotes meet, corresponds to the magnitude of the pole at $s = -10$ on the real axis. It is also called the *cut-off frequency* of the system.

The exact and the asymptotic behaviour below, at and above the cut-off frequency is summarised in Table 10.2 for a system with a real pole at $s = -a$. The system function is

$$H(s) = \frac{1}{s+a}, \quad a > 0 \quad .$$

The numerical values of the magnitude frequency correspond to the value $a = 10$ from (10.4); the phase angle is correct for any real poles.

At $\omega = 0$ the frequency response is purely real and has the value $H(0) = \frac{1}{a} = \frac{1}{10}$, or in logarithmic form $20\log_{10} H(0) = 20 \times (-1) = -20$ dB. The phase angle is $0°$

At rising frequencies no significant variations from these values are obtained,

Table 10.2: Table of values of $H(j\omega)$, damping and phase

Frequency	$H(j\omega)$	Damping	Phase
$\omega = 0$	$\dfrac{1}{a} = \dfrac{1}{10}$	-20 dB	$0°$
$\omega < 0.1a$			$\approx 0°$
$\omega < a$	$\approx \dfrac{1}{a} = \dfrac{1}{10}$	-20 dB	
$\omega = a$	$\dfrac{1}{ja + a}$	-23 dB	$-45°$
$\omega > a$	$\approx \dfrac{1}{j\omega}$	-20 dB/decade	
$\omega > 10a$			$\approx -90°$

to about $\omega = 0.1a$ At this frequency the phase has the value

$$\varphi(j0.1a) = \arg H(j0.1a) = -\arctan\frac{0.1a}{a} = -\arctan 0.1 = -6°\,.$$

The magnitude of the frequency response first varies significantly from the value at $\omega = 0$ at the cut-off frequency $\omega = a$, and is given by

$$|H(ja)| = \frac{1}{\sqrt{a^2 + a^2}} = \frac{1}{\sqrt{2}} \cdot \frac{1}{a} = \frac{1}{\sqrt{2}}|H(0)|$$

$$20\log_{10}|H(j10)| = -3\,\text{dB} - 20\,\text{dB} = -23\,\text{dB} \quad \text{for} \quad a = 10\,.$$

The phase here is

$$\varphi(ja) = -\arctan\frac{a}{a} = -\arctan 1 = -45°\,.$$

For $\omega > a$ the value of a^2 in the magnitude frequency response

$$|H(j\omega)| = \frac{1}{\sqrt{a^2 + \omega^2}} \approx \frac{1}{\omega}$$

can be neglected compared to ω^2, so that it is then decaying reciprocal to ω.

In the double logarithmic representation, because of $\log|H(j\omega)| = -\log\omega$, a linear relationship between $\log|H(j\omega)|$ and $\log\omega$ is obtained. Increasing the frequency by a factor of 10 (one decade) then leads to a reduction of $\log |H(j\omega)|$ by 20 dB. The magnitude frequency response is said to decay linearly by -20 dB/decade.

The phase reaches the value

$$\varphi(j10a) = \arg H(j10a) = -\arctan\frac{10a}{a} = -\arctan 10 = -84°$$

at $\omega = 10a$. For $\omega \to \infty$ it becomes $-90°$.

It is obvious that the shape of the magnitude frequency response and the phase can be described approximately by a few rules.

- The magnitude frequency response is constant for frequencies below the cut-off frequency $\omega = a$, and afterwards it falls at -20 dB/decade. Its exact path runs underneath the asymptotes and its maximum deviation from them is 3 dB, at the cut-off frequency of $\omega = a$.

- The phase has the constant value $0°$ for $\omega > 10a$. Between $\omega = 0.1a$ and $\omega = 10a$, the phase decreases linearly with the logarithm of frequency. At the cut-off frequency the phase has the value $-45°$ exactly.

The greatest variation between the exact path and the asymptotic approximation is about 6^0 and it occurs at $\omega = 0.1a$ and $\omega = 10a$. Figure 10.2 shows the asymptotic approximation (by solid lines) and the exact path (dashed lines) for $a = 10$. The advantage of this representation is that the exact path can be sketched from the asymptotes and the known deviances at $0.1a$, a and $10a$.

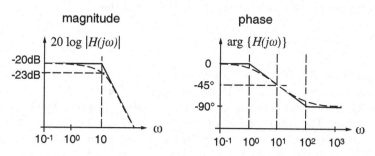

Figure 10.2: Bode plot for $H(s) = \dfrac{1}{s + 10}$

For a real zero, the same rules apply with a few deviations

- The magnitude frequency response climbs above the cut-off frequency at 20 dB/decade.

- The phase climbs between $\omega = 0.1a$ and $\omega = 10a$ from $0°$ to $+90°$.

The plot for a real zero can be obtained from that for a pole by flapping the drawing along the horizontal axis. Figure 10.3 shows the corresponding runs for a zero at $s = -a = -10$, that is for a transfer function of the form $H(s) = s + 10$.

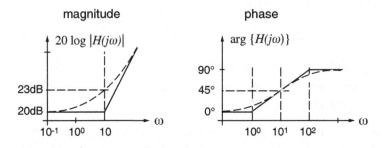

Figure 10.3: Bode plot for $H(s) = s + 10$

10.3 Bode Plots for Multiple Poles and Zeros

After the detailed examination of the contribution of individual real poles and zeros in the last section most of the work required to draw a Bode plot for systems with multiple poles and zeros has already been done. As systems with multiple real poles and zeros (10.2), (10.3) can be written as the sum of the contributions of the individual poles and zeros, we only need to determine the asymptotes for the poles and zeros and put them together to form a complete picture.

We will explain the procedure for the system function

$$H(s) = \frac{s + 1000}{(s + 10)^2} = (s + 1000) \cdot \frac{1}{s + 10} \cdot \frac{1}{s + 10} \, .$$

The double pole will be treated as two simple poles. The cut-off frequencies are $\omega = 1000$ for the zeros, and $\omega = 10$ for both poles. Figure 10.4 above shows the logarithmic representation of the magnitude frequency response from the rules in the last section. The contributions of the poles correspond exactly to the system function (10.4), that we have already dealt with in detail. The contribution of the zero can be given immediately by turning upside down the contribution of the pole above. Because of the greater cut-off frequency of $\omega = 1000$ it is shifted two decades to the right. The value of the constant asymptote can be most easily obtained for $\omega = 0$ to $20 \log_{10} 1000 = 60$ dB.

In Figure 10.4 below, the phase plots are shown. The phases of the poles here are also already known. The phase of the zero can be obtained, again by flapping the plot of the pole as above, and by shifting two decades to the right.

If we add each of the logarithmically represented magnitude frequency responses and the linearly represented phases together, we obtain the complete

20 log |H(jω)|

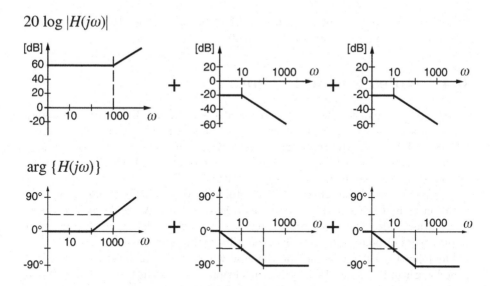

Figure 10.4: Putting together a Bode plot for multiple poles and zeros.

Bode plot, as shown below in Figure 10.5. The double pole as the sum of two simple poles leads to a decay of the magnitude frequency response by $2 \times (-20\,\mathrm{dB/decade}) = -40$ dB/decade, from the cut-off frequency of $\omega = 10$. In the same way the phase decreases between $0.1 \times 10 = 1$ and $10 \times 10 = 100$ by $2 \times 90° = 180°$ from $0°$ to $-180°$.

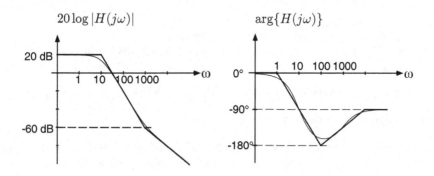

Figure 10.5: Complete Bode plot for $H(s) = \dfrac{s + 1000}{(s + 10)^2}$

   The zero is first noticeable in the Bode plot of the magnitude at the cut-off frequency of $\omega = 1000$. Its slope of 20 dB/decade added to the contribution of the double pole, -40 dB/decade leads to a decay of -20 dB/decade for $\omega > 1000$.

The increase of the phase of 90^0 between $0.1 \times 1000 = 100$ and $10 \times 1000 = 10000$ leads to a total phase of -90° for $\omega > 10000$.

10.4   Rules for Bode Plots

We will now summarise the discussion about determining Bode plots with some simple rules for the magnitude frequency response and the phase. They apply to poles and zeros on the real axis of the left half of the s-plane. The distance separating them should be great enough so that they do not mutually influence each other in the region of the cut-off frequencies . We can see from our previous discussion (see Table 10.2) that a factor of 100 is sufficient difference. The limitation to real poles and zeros has historical reasons. As Bode plots are very easy to draw under these conditions, they are the most widely used. Extension to complex pairs of poles and zeros is covered in further sections.

The following rules also include the case where a pole or a zero lies at $s = 0$. In the logarithmic form of Bode plot, the cut-off frequency is not visible at $\omega = 0$, however, so only the contribution right of the cut-off frequency appears, that is the gradient of ± 20 dB/decade for the magnitude frequency response and $\pm 90^0$ for the phase.

10.4.1   Magnitude Frequency Response

1. Determine the location and order of poles and zeros.

2. Draw axes and plot cut-off frequencies.

3. Start at small ω:
   a) no pole and no zero at $s = 0$ $\rightarrow$ gradient 0
   b) pole at $s = 0$ $\qquad\qquad\qquad\rightarrow$ gradient -20 dB/decade
   c) zero at $s = 0$ $\qquad\qquad\qquad\rightarrow$ gradient +20 dB/decade

   For multiple poles or zeros use multiple gradients.

4. Straight line to the next cut-off frequency.

5. Decrease by 20 dB/decade for each pole, increase by 20 dB/decade for each zero, then continue with step 4 until all cut-off frequencies have been finished with.

6. Label the vertical axis: calculate $|H(j\omega)|$ in a region where the Bode plot is flat.

7. Round off cut-offs by ± 3 dB (or multiples of ± 3 dB for multiple poles or zeros).

10.4.2 Phase

1. Determine the location and order of poles and zeros.

2. Draw axes and plot the cut-off frequencies.

3. Start at small ω:
   a) no pole and no zero at $s = 0$ $\rightarrow$ phase 0
   b) pole at $s = 0$ $\rightarrow$ phase $-90°$
   c) zero at $s = 0$ $\rightarrow$ phase $+90°$

   Multiple poles or zeros have multiple phase angles. Bear in mind that a minus sign before $H(j\omega)$ is a phase offset of $180°$.

4. Straight line to $0.1\times$ the cut-off frequency.

5. Each pole subtracts $90°$, each zero adds $90°$ in a region from $0.1\times$ the cut-off frequency to $10\times$ the cut-off frequency. Continue with step 4 until all cut-off frequencies have been dealt with.

6. Smooth the phase sketch. Round off about $6°$ at $0.1\times$ the cut-off frequency and $10\times$ the cut-off frequency (or the corresponding multiples for multiple poles or zeros).

10.5 Complex Pairs of Poles and Zeros

For complex pairs of poles and zeros there are no simple rules as with real poles and zeros, because the cut-off frequency depends on both the real and imaginary part of the pole or zero.

If the imaginary part is small compared to the real part, a complex pair of poles can be approximated as a double pole on the real axis, because then the cut-off frequency is almost exclusively defined by the real part. The equivalent is also true for pairs of zeros.

If, on the other hand, the real part of a pole is small compared to the imaginary part, the cut-off frequency depends almost exclusively on the imaginary part. The Bode plot with the corresponding cut-off frequency can be drawn as before. As we are dealing with two poles, this amounts to a gradient above the cut-off frequency of -40 dB/decade. The behaviour at the cut-off frequency depends on the relationship between the real part and the imaginary part. A sufficiently large imaginary part in relation to the real part creates a resonance that becomes stronger as the pole nears the imaginary axis. The equivalent is also true for pairs of zeros.

Example 10.1

As an example we will consider the magnitude frequency response and the phase for the system function

$$H(s) = \frac{1}{s^2 + 0.4s + 1.04}$$

with poles at

$$s = -0.2 \pm j .$$

The exact behaviour is depicted in Figure 10.6.

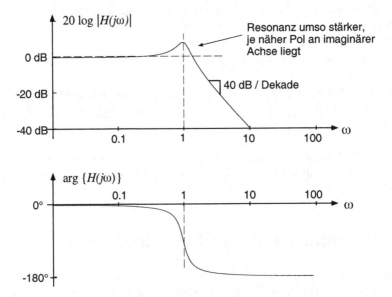

Figure 10.6: Bode plot for a complex pair of poles

The cut-off frequency is in this case mostly determined by the imaginary part and is located at $\omega = 1$. To its left the magnitude frequency response is approximately constant at $-20\log_{10} 1.04 \approx 0$ dB. To the right of the cut-off frequency there is a linear decay of -40 dB/decade

$$20\log_{10}|H(j\omega)| \approx 20\log_{10}\frac{1}{\omega^2} = -40\log_{10}\omega .$$

The form of the resonant peak depends on the distance of the pole from the imaginary axis. The phase decreases in the region of the cut-off frequency from $0°$ to $-180°$, and the exact form of its path near the cut-off frequency must again be determined by the distance of the pole from the imaginary axis.

If either the real or imaginary part may not be neglected in favour of the other, there is no simple method of finding the cut-off frequency and drawing the Bode plot. We can in fact look back at the equations (10.3) and determine the distance of the poles and zeros, and the corresponding angles graphically for each point of interest on the imaginary axis. Figure 10.7 shows the situation for a pair of poles and an arbitrary point on the imaginary axis. The distance to the poles and the corresponding angle can easily be read from the plot.

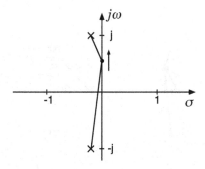

Figure 10.7: Pole-zero diagram for a complex pole pair

Example 10.2

As an example we will examine the system function

$$H(s) = \frac{(s + \frac{1}{2})^2}{(s + \frac{3}{2})(s^2 + \frac{s}{2} + \frac{65}{16})}$$

with the pole-zero diagram from Figure 10.8

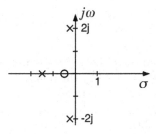

Figure 10.8: Pole-zero diagram for the given transfer function

To estimate the magnitude frequency response we begin by considering the value of the contribution of the system function at $\omega = 0$. From the pole-zero plot we recognise that at $\omega = 2$ there is a resonant point. We thus calculate the magnitude for this point from $H(s)$. As the number of poles is by one greater than the number of zeros, the magnitude frequency response for $\omega \gg 2$ falls at 20 dB/decade. To confirm this we determine the magnitude at $\omega = 10$. In order to approximate the path more accurately, we calculate the value for $\omega = 1$ as well. The distances in equation (10.3) are calculated from the pole-zero plot. As is evident in Figure 10.9, the following approximate values are yielded, if the

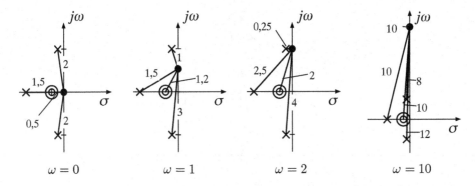

Figure 10.9: Estimating the lengths for various ω

distances of the zeros are multiplied and divided by the distance of the poles:

$$|H(j0)| \approx \frac{\frac{1}{2} \times \frac{1}{2}}{\frac{3}{2} \times 2 \times 2} = \frac{1}{24} \ \widehat{=}\ -27.6 \text{ dB}$$

$$|H(j1)| \approx \frac{1.2 \times 1.2}{1.5 \times 1 \times 3} \approx \frac{1}{3} \ \widehat{=}\ -10 \text{ dB}$$

$$|H(j2)| \approx \frac{4}{2.5 \times 4 \times \frac{1}{4}} = 1.6 \ \widehat{=}\ 4.1 \text{ dB}$$

$$|H(j10)| \approx \frac{10 \times 10}{10 \times 8 \times 12} \approx 0.1 \ \widehat{=}\ -20 \text{ dB}. \tag{10.5}$$

For the phase plot we need the phase at $\omega = 0$ and recognise from the pole-zero diagram that due to the pole $s = -\frac{1}{4} + 2j$ the phase changes sharply around $\omega = 2$. Further, the phase of $H(s)$ for $\omega \gg 2$ takes a value of $-90°$. Considering this, we can estimate the phase at the following points with the pole-zero diagram, as shown in Figure 10.10. We obtain

$$\varphi(j0) = 0$$
$$\varphi(j1) \approx 2 \times 60° - 30° + 90° - 90° = 90°$$
$$\varphi(j2) \approx 2 \times 80° - 50° - 0° - 90° = 20°. \tag{10.6}$$

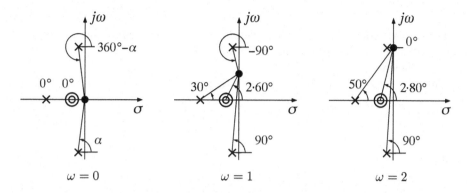

Figure 10.10: Estimation of the phase for various ω

In Figure 10.11 our estimates are displayed with the exact frequency response $|H(j\omega)|$. The estimates are marked as crosses. Likewise, our estimates of the phase as displayed with the exact values for comparison in Figure 10.12.

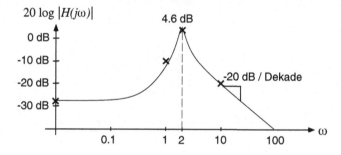

Figure 10.11: Exact magnitude and estimates for various ω

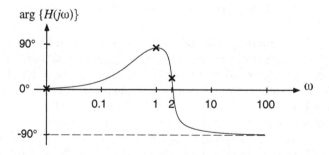

Figure 10.12: Exact phase and estimates for various ω

For second-order systems with a complex pair of poles very close to the imaginary axis, the magnitude response can also be expressed directly by the denominator coefficients.

The pole-zero diagram for system function

$$H(s) = \frac{s^2}{s^2 + 2\alpha s + \omega_0^2}$$

with poles

$$s = -\alpha \pm j\beta, \quad \alpha \ll \beta$$

is shown in Figure 10.13. Under the stated condition $\alpha \ll \beta$ the denominator polynomial takes the form

$$(s + \alpha - j\beta)(s + \alpha + j\beta) = s^2 + 2\alpha s + (\alpha^2 + \beta^2) \approx s^2 + 2\alpha s + \beta^2.$$

The resonant frequency is

$$\omega_0 \approx \beta.$$

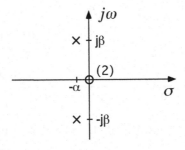

Figure 10.13: Pole-zero diagram with poles very near the imaginary axis

The magnitude response plotted in linear form is shown in Figure 10.14 for $\alpha = 0.1$ and $\beta = 1$. This kind of frequency response occurs for all oscillating systems with low damping and is called a resonance curve . A measure for the resonance is the width of the resonance curve 3 dB below the peak. It can be confirmed that the end points occur at $\omega \approx \beta \pm \alpha$ and the width is therefore $\Delta\omega \approx 2\alpha$. This can also be illustrated easily as in Figure 10.13, if a triangle is drawn between two points $\beta \pm \alpha$ on the imaginary axis and the pole. A dimensionless quantity, known as the Q-factor is obtained by relating $\Delta\omega$ to the resonant frequency ω_0. The Q-factor is given by

$$Q = \frac{\omega_0}{\Delta\omega} \approx \frac{\beta}{2\alpha}.$$

Our example in Figure 10.14 has Q-factor $Q = 5$. The resonant peak lies at Q times the value of $|H(j\omega)|$ for $\omega \to \infty$. This can be clarified by Figure 10.13, where as in (10.5) for $\omega = \beta$ the product of the disctances between the zeros ($= \beta \times \beta$) is divided by the product of the distances between the poles ($\approx \alpha \times 2\beta$).

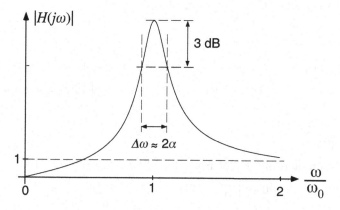

Figure 10.14: Resonance curve of the system

10.6   Exercises

Exercise 10.1

For the following system:

$$u_1 \longrightarrow \boxed{S} \longrightarrow u_2$$

How much amplification does the system have in dB, if

a) $u_2 = 10\,u_1$

b) $u_2 = 10^4\,u_1$

c) $u_2 = \sqrt{2}\,u_1$

d) $u_2 = 0.02\,u_1$

e) $u_2 = -2\,u_1$?

Exercise 10.2

How much amplification in dB does a system have, if the output power is a) 64 times or b) 2 times the input power (the input and output resistances are equal)? First calculate the relationship between input and output voltage.

Exercise 10.3

For $H(s) = \dfrac{1000}{s + 100}$, calculate the the amplification of $H(s)$ in dB and the phase
of $H(s)$ in dB for the frequencies $\omega = 10^k$, $k = 0, 1, \ldots, 5$. The complex values of
$H(s)$ values should be given in the form $a + jb$.

Exercise 10.4

Draw the amplitude and phase diagrams for the Bode plot of $H(s) = \dfrac{s + 1}{s(s + 100)}$.
To find the appropriate scale for the amplitude axis, calculate the amplification at
$\omega = 10$.

Exercise 10.5

Draw a pole-zero and phase plot from the following amplitude plot.

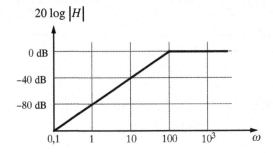

Exercise 10.6

A system has the impulse response $h(t) = -\dfrac{1}{500}\delta(t) - \dfrac{999}{500}e^{-t}\epsilon(t)$.

a) Give $H(s)$ and draw the Bode plot, i.e., amplitude and phase plots with
   appropriately labelled axes.

b) Give the output signal $y(t)$ of the system for the input signal $x(t) = \cos(\omega_0 t)$
   at the normalised frequencies $\omega_0 = 0.01, 1, 10$ and 10^5.

Exercise 10.7

a) Draw the amplitude diagram of the Bode plot for $H(s) = \dfrac{s^2}{(s + 10)^2}$; use the
   frequency $\omega = 100$ for labelling the axis.

b) By what percentage does the exact amplitude differ from the one in the Bode
   plot at $\omega = 100$ and $\omega = 1000$?

c) By how many dB is it rounded off at the cut-off frequency $w = 10$? Exactly how much amplification is yielded at $w = 10$? Give the difference from the rounded value in the Bode plot as a percentage.

Exercise 10.8

Draw the phase plots of $H_1(s) = \dfrac{s^2}{(s+1)^2}$ and $H_2(s) = -\dfrac{s^2}{(s+1)^2}$. Where do the amplitude plots of H_1 and H_2 differ?

Exercise 10.9

A system has the following transfer function:

$$H(s) = 10^4 \cdot \frac{s^2 + 1.1s + 0.1}{s^3 + 1011s^2 + 11010s + 10000}$$

a) Draw the pole-zero plot for this system.

b) Draw the magnitude and phase components of the Bode plot of $H(j\omega)$ with appropriate labelling of axes. What behaviour does $H(s)$ show? Read the maximum gain and the frequency or frequencies where the gain has fallen to a tenth of its maximum.

c) Let the input signal of the system be $x(t) = \varepsilon(t)$. Determine the value of the output signal $y(t)$ for $t \to \infty$ from an amplitude plot.

Exercise 10.10

Mark the cut-off frequencies in the following pole-zero diagram and draw the magnitude component of the Bode plot. Give $H(s)$ and label the $|H|$ axis so that $|H(\omega = 2 \times 10^4)| = 1$.

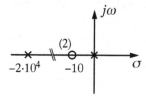

Exercise 10.11

The gain of a system

a) decreases reciprocally

258

10. Bode Plots

b) increases linearly

c) increases quadratically

d) increases with order n

with the frequency. How many $\frac{dB}{decade}$ correspond to this frequency response?

Exercise 10.12

Give the gain $|H|$ of the system from Exercise 10.5 at the frequencies $\omega = 2, \omega = 4, \omega = 8$ and $\omega = 2000$.

Exercise 10.13

Give the gradients $20\frac{dB}{dec}, 40\frac{dB}{dec}$ and $60\frac{dB}{dec}$ in $\frac{dB}{octave}$. Note: octave signifies a doubling of frequency.

Exercise 10.14

Consider a second-order low-pass filter with cut-off frequency $\omega_c = 10^4$ and a DC gain of 20 dB.

a) Draw the amplitude component of the Bode plot with appropriate labelling of axes and give the corresponding transfer function $H_{LP}(s)$.

b) A first-order high-pass filter has been connected to the system, transforming it into a band-pass filter with the following characteristics:

   - max. gain 14 dB
   - gain of 0.5 at $\omega = 10$

   The high-pass filter has cut-off frequency ω_g and for $\omega \to \infty$, gain A. Give $H_{HP}(s)$ and the corresponding amplitude component, then draw the amplitude component of the desired system and determine graphically from it the values of A and ω_g.

Exercise 10.15

The following is known about a system:

- Q-factor = 50,
- maximum gain is 26 dB at $\omega_0 = 10$,
- $H(j\omega)$ increases for $\omega \ll \omega_0$ by $20\frac{dB}{dec}$ and decreases for $\omega \to \infty$ by $20\frac{dB}{dec}$.

a) Draw the amplitude component of the Bode plot. Label the axes as much as possible.

b) Evaluate the transfer function $H(s)$ with the permitted approximations for systems with resonance.

c) Label the axes, using $H(s)$ for $\omega = 1$ and $\omega = 100$.

d) Draw the pole-zero diagram.

e) At about which frequencies is the gain 3 dB below maximum?

f) Under what conditions do we find $|H(j\omega)| \to \infty$? What condition of a real system does this correspond to?

11 Sampling and Periodic Signals

11.1 Introduction

The continuous signals dealt with in previous chapters occur in natural and technical processes, where time (or position) is a continuous variable. To record and process these analog signals with a digital computer we need to convert them to digital signals. This is carried out with an analog-to-digital converter (ADC) in two distinct steps.

- Sampling of a continuous-time signal $x(t)$ at equidistant time points separated by a time T (time quantisation).

$$x[k] = x(kT), \quad k \in \mathbb{Z}.$$

- Storing the sequence of values $x[k]$ in memory cells with a finite number of bits (amplitude quantisation).

The square brackets in $x[k]$ indicate that it is a discrete signal, i.e., a sequence of numbers. The time T between two sequential samples $x(kT)$ is called the *sampling interval*.

We are only concerned here with the first step, the sampling, or time quantisation. Amplitude quantisation is non-linear, because the amplitude is rounded to a finite word length. For computers with a sufficiently great storage range (e.g. 10^{-308} to 10^{308}) and a correspondingly high number of bits (e.g. 64), the approximation is in many cases justifiable so we will forego discussion of it.

The first tool we introduce for describing the sampling process is the impulse train, whose use we will demonstrate on a known problem: representing a continuous periodic signal with a Fourier series.

Both the techniques and the results obtained can be elegantly carried over to deal with sampling. The line spectrum of a periodic signal is closely related to the sample values of a continuous signal by the duality principle of the Fourier transform.

11.2   Delta Impulse Train and Periodic Functions

11.2.1   Delta Impulse Train and its Fourier Transform

We understand *delta impulse train* as the sum of an infinite number of shifted delta impulses, as shown in Figure 11.1. Like the delta impulse itself, it is a generalised function, in the sense of distribution theory. We will be using the symbol $\bot\!\bot\!\bot(t)$, which has the same form as the impulse train, as a short form. Because of its similarity with one of the letters in the cyrillic alphabet, it is also called the 'sha-symbol' . Using this symbol, the impulse train is given by the formula

$$x(t) = \bot\!\bot\!\bot(t) = \sum_{\mu=-\infty}^{\infty} \delta(t-\mu)\,. \qquad (11.1)$$

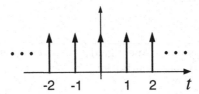

Figure 11.1: Delta impulse train

The Fourier transform of the delta impulse train can be obtained simply by transforming the individual impulses with the displacement theorem,

$$\delta(t-\mu) \quad \circ\!\!-\!\!\bullet \quad e^{-j\mu\omega}$$

to

$$X(j\omega) = \sum_{\mu=-\infty}^{\infty} e^{-j\mu\omega}\,. \qquad (11.2)$$

As each individual exponential term $e^{-j\mu\omega}$ has period 2π, $X(j\omega) = X(j(\omega+2\pi))$. The form of the spectrum $X(j\omega)$ from (11.2) is correct but impractical, because of the infinite sum. In order to derive a more helpful expression, we multiply $X(j\omega)$ with a si-function in the frequency-domain, and obtain

$$Y(j\omega) = X(j\omega) \times \text{si}\left(\frac{\omega}{2}\right)\,. \qquad (11.3)$$

The function $\text{si}(\omega/2)$ has zeros at $\omega = \pm 2\pi$, $\pm 4\pi$, ..., for which the distance 2π corresponds exactly to the period of $X(j\omega)$. Although the product $X(j\omega) \times \text{si}(\omega/2)$

looks complicated, $Y(j\omega)$ is a very simple function, and we will explain this by taking a detour through the time-domain. Inverse Fourier transformation with

$$\text{si}\left(\frac{\omega}{2}\right) \quad \circ\!\!-\!\!\bullet \quad \text{rect}(t)$$

and convolution yields

$$y(t) = \text{Ш}(t) * \text{rect}(t) = 1 \ . \tag{11.4}$$

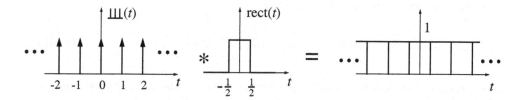

Figure 11.2: Convolution of a delta impulse train with a rectangle function

The time function initially looks more complicated than its spectrum. As $\text{rect}(t)$ has width 1 which corresponds exactly to the interval between the impulses in $\text{Ш}(t)$, however, the convolution $\text{Ш}(t) * \text{rect}(t)$ simply yields the constant 1 (see Figure 11.2). $Y(j\omega)$ is therefore equal to a delta impulse with weight 2π.

$$y(t) = 1 \quad \circ\!\!-\!\!\bullet \quad Y(j\omega) = 2\pi\delta(\omega) \tag{11.5}$$

We could now determine $X(j\omega)$ from (11.3), but since $\text{si}(\omega/2)$ is zero at the points $\omega = \pm 2\pi, \pm 4\pi, \ldots$, and the value of the product is likewise zero – we do not know what the value of $X(j\omega)$ would be. For these points, however, the periodic nature of $X(j\omega = X(j(\omega + 2\pi))$ determined with (11.2) comes into use. By comparing (11.3) and (11.5) it can be confirmed that

$$X(j\omega) = 2\pi\delta(\omega) \quad \text{for } -\pi < \omega \leq \pi . \tag{11.6}$$

For (11.2), periodic continuation yields

$$X(j\omega) = \sum_{\mu=-\infty}^{\infty} e^{-j\mu\omega} = \sum_{\mu=-\infty}^{\infty} 2\pi\delta(\omega - 2\pi\mu) = \text{Ш}\left(\frac{\omega}{2\pi}\right) . \tag{11.7}$$

Here we have used the scaling property (8.19) of the delta impulse

$$2\pi\delta(\omega - 2\pi\mu) = \delta\left(\frac{\omega}{2\pi} - \mu\right) . \tag{11.8}$$

The astonishingly simple result of this can also be expressed as follows: *the Fourier transform of a delta impulse train in the time domain is a delta impulse train in the frequency-domain:*

$$\text{Ш}(t) \circ\!\!-\!\!\bullet \text{Ш}\left(\frac{\omega}{2\pi}\right) . \tag{11.9}$$

In order to keep the derivation as simple as possible, we defined an impulse train with an interval of 1 (11.1). To describe sampling procedures with an arbitrary sampling interval T we use the delta impulse train

$$x(t) = \sum_{\mu=-\infty}^{\infty} \delta(t - \mu T) = \frac{1}{T}\text{Ш}\left(\frac{t}{T}\right) \tag{11.10}$$

We also describe these with the sha-symbol from (11.1), where the scaling property (8.19) is important

$$\delta(t - \mu T) = \frac{1}{T}\delta\left(\frac{t}{T} - \mu\right) \tag{11.11}$$

With the similarity theorem, we can derive from (11.9) the general relation

$$\frac{1}{T}\text{Ш}\left(\frac{t}{T}\right) \circ\!\!-\!\!\bullet \text{Ш}\left(\frac{\omega T}{2\pi}\right) . \tag{11.12}$$

11.2.2   Fourier Transformed Periodic Signals

The connection (11.12) between delta impulse trains in the time-domain and frequency-domain is very elegant and can greatly shorten otherwise complicated calculations. Its use, however, requires some practice, so before we employ it to deal with sampling continuous functions we will try it out on some classical problems, in particular, representing periodic signals in the time- and frequency-domain. Periodic signals have a line spectrum where the distance between the lines is given by the periods in the time-domain. The weighting of individual lines can be determined using Fourier series. We will now use the delta impulse train to derive these.

First we consider a periodic time signal $x(t)$ with period T. It can be represented as convolution of a function $x_0(t)$ with a delta impulse train (see Figure 11.3).

$$x(t) = x_0(t) * \frac{1}{T}\text{Ш}\left(\frac{t}{T}\right) \tag{11.13}$$

The separation T of the impulses is set so that there are no gaps between repetitions of the function $x_0(t)$.

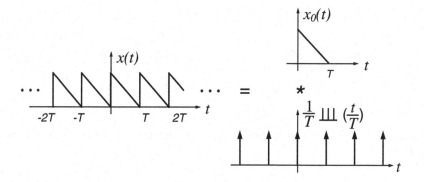

Figure 11.3: Representation of a periodic signal $x(t)$ as the convolution of a period $x_0(t)$ with a delta impulse train

For the Fourier transform $X(j\omega)$ of the periodic signal $x(t)$, we obtain with (11.12) and the convolution theorem

$$X(j\omega) = X_0(j\omega) \text{⊥⊥⊥}\left(\frac{\omega T}{2\pi}\right) = \frac{2\pi}{T}\sum_{\mu}\delta\left(\omega - \frac{2\pi}{T}\mu\right)X_0(j\omega) \qquad (11.14)$$

The Fourier transform $X(j\omega)$ is therefore – as we already know – a line spectrum with a distance of $\omega = \frac{2\pi}{T}$ between the lines. The weights correspond to the respective values of the spectrum $X_0(j\omega)$, multiplied with 2π.

Figure 11.4 shows this relationship between the impulse train and the continuous spectrum $X_0(j\omega)$ of one period. The height of the arrow symbolises the weight of the impulse. The relationship between the periodic time signal and the

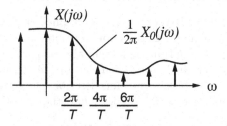

Figure 11.4: The Fourier transform of a periodic function is a line spectrum

individual lines in the spectrum is obtained as the inverse Fourier transform of (11.14). Using the selective property, the integral of the inverse Fourier transform becomes a sum of the individual frequency lines:

$$x(t) \quad = \quad \frac{1}{2\pi} \int_{-\infty}^{\infty} X_0(j\omega) \,\text{Ш}\left(\frac{\omega T}{2\pi}\right) e^{j\omega t} d\omega \qquad (11.15)$$

$$= \quad \frac{1}{T} \sum_{\nu=-\infty}^{\infty} X_0\left(j\frac{2\pi\nu}{T}\right) e^{j\frac{2\pi\nu t}{T}}. \qquad (11.16)$$

Equating this expression with the general form of the Fourier series or a periodic function gives

$$\boxed{x(t) = \sum_{\nu=-\infty}^{\infty} A_\nu e^{j\frac{2\pi\nu t}{T}}.} \qquad (11.17)$$

With (11.16), we obtain

$$\boxed{A_\nu = \frac{1}{T}X_0\left(j\frac{2\pi\nu}{T}\right)} \qquad (11.18)$$

for the complex Fourier coefficients A_ν. The spectrum $X(j\omega)$ of the periodic signal $x(t)$ is therefore a line spectrum where the weight of the individual lines is given by the Fourier coefficients of $x(t)$. The Fourier coefficients are also the values of the spectrum $X_0(j\omega)$ with period $x_0(t)$ at multiples $\omega = \nu\frac{2\pi}{T}$ of the fundamental frequency $\frac{2\pi}{T}$ (see Figure 11.4).

This compact derivation shows that using the impulse train and the sha-symbol saves cumbersome calculation. We will refer back to it when we deal with sampling. The emphasis will then lie on sampling functions of time, instead of finding sample values from a spectrum. Firstly, however, we will deal with convolution of periodic and aperiodic signals.

11.2.3   Convolution of a Periodic and an Aperiodic Signal

The output signal of an LTI-system can be obtained in the time-domain using convolution of the input signal with the impulse response (see Section 8.4.2). If the input signal is periodic, the output signal will also be periodic. We are interested in the relationship between the Fourier coefficients of the input and output signals. The corresponding convolution integral thus describes the convolution of a periodic signal (the input signal) with an aperiodic signal (the impulse response).

─── **Example 11.1**

As an example we consider a periodic input signal $x(t)$ with the Fourier series

$$x(t) = \sum_{k} A_k e^{jk2\pi t} \qquad (11.19)$$

and an LTI-system with impulse response

$$h(t) = e^{-t}\varepsilon(t) \tag{11.20}$$

as shown in Figure 11.5. Comparing (11.19) and (11.17), we recognise that the period is $T = 1$.

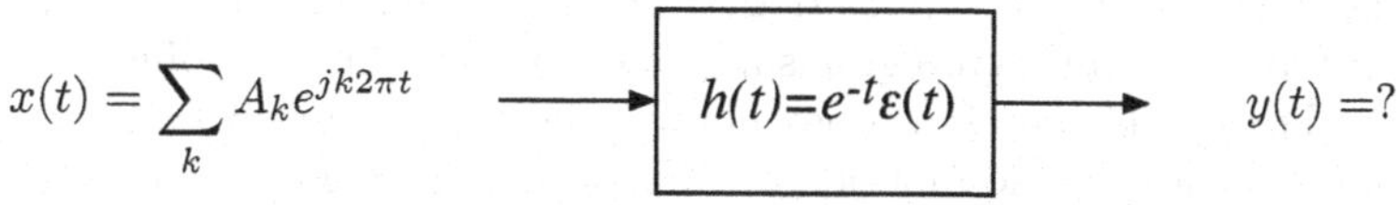

$$x(t) = \sum_k A_k e^{jk2\pi t} \qquad \longrightarrow \qquad \boxed{h(t) = e^{-t}\varepsilon(t)} \qquad \longrightarrow \qquad y(t) =?$$

Figure 11.5: Convolution of a periodic input signal with an aperiodic impulse response

The frequency response of a system is obtained from the system function for $s = j\omega$:

$$H(j\omega) = \frac{1}{j\omega + 1}. \tag{11.21}$$

The Fourier transform of the input signal is a line spectrum:

$$X(j\omega) = \sum_k 2\pi A_k \delta(\omega - 2\pi k). \tag{11.22}$$

The output signal likewise has a line spectrum, as for the product $Y(j\omega) = H(j\omega)X(j\omega)$, we obtain

$$Y(j\omega) = H(j\omega)X(j\omega) = \sum_k \left(\frac{2\pi A_k}{1 + j2\pi k}\right) \delta(\omega - 2\pi k)$$

$$\downarrow \circ \tag{11.23}$$

$$y(t) = \sum_k \left(\frac{A_k}{1 + j2\pi k}\right) e^{j2\pi kt}.$$

The final inverse transform gives the Fourier series of the output signal. Its Fourier coefficients are the product of the Fourier coefficients of the input signal and the values of the system frequency response at the frequencies of the individual spectal lines.

■

Convolution of any periodic signal with period T that can be represented by its Fourier coefficients A_k (11.17), with an aperiodic signal $h(t)$ yields

$$y(t) = \sum_{k=-\infty}^{\infty} C_k e^{j\frac{2\pi kt}{T}} \quad \text{with} \quad C_k = H\left(j\frac{2\pi k}{T}\right) A_k, \tag{11.24}$$

where $H(j\omega) = \mathcal{F}\{h(t)\}$. If $h(t)$ is the impulse response of an LTI-system, $H(j\omega)$ is its frequency response, and (11.24) specifies how the Fourier coefficients are changed by the system.

11.2.4 Periodic Convolution

It seems likely that the results of the last section can be extended to convolution of two periodic signals with the same period, but in fact, the product of both periodic signals under the convolution integral is likewise periodic. The convolution integral of two periodic signals therefore does not converge. Since, however, all of the information about a periodic function is embedded in a single period, it must be possible to define a modified convolution by only integrating over one period.

Definition 18:   Periodic convolution

The periodic convolution *of two periodic signals $x_1(t)$ and $x_2(t)$ with the same period T is given by*

$$y(t) = \int_{\tau_0}^{\tau_0+T} x_1(\tau)\, x_2(t - \tau)\, d\tau = x_1(t) \circledast x_2(t). \qquad (11.25)$$

It is also called cyclic convolution.

The exact location of the integration limits, determined by τ_0 is unimportant as long as the integration is only performed over one period of length T. Periodic convolution corresponds to normal convolution of a periodic signal $x_1(t)$ with a period of $x_2(t)$, or a period of $x_1(t)$ with the periodic signal $x_2(t)$.

The result $y(t)$ of periodic convolution is likewise periodic. We are again interested in the relationship between the coefficients of the Fourier series representation

$$x_1(t) \;=\; \sum_k A_k e^{jk2\pi t/T} \qquad (11.26)$$

$$x_2(t) \;=\; \sum_\ell B_\ell e^{j\ell 2\pi t/T} \qquad (11.27)$$

$$y(t) \;=\; \sum_k C_k e^{jk2\pi t/T} \qquad (11.28)$$

First of all we put (11.26) and (11.27) into (11.25) and re-order the terms

$$y(t) = \sum_k \sum_\ell A_k B_\ell \int_{\tau_0}^{\tau_0+T} e^{j2\pi(k-\ell)\tau/T}\, d\tau \cdot e^{j2\pi\ell t/T} \qquad (11.29)$$

The terms $e^{j2\pi(k-\ell)\tau/T}$ in the integral have magnitude 1 for $k \neq \ell$, and a phase rotating with period $T/(k-\ell)$, and for $k = \ell$ they are constant with the value 1.

Therefore for the integral,

$$\int\limits_{\tau_0}^{\tau_0+T} e^{j2\pi(k-\ell)\,\tau/T}d\tau = \left\{ \begin{array}{ll} T & \ell = k \\ 0 & \ell \neq k \end{array} \right. \tag{11.30}$$

In other words: the basis functions of the Fourier series are orthogonal. Putting this into (11.29) yields

$$y(t) = \sum_k A_k B_k T e^{j2\pi kt/T} \tag{11.31}$$

By comparing (11.28) and (11.31), we find the connection between the Fourier coefficients of $y(t)$, $x_1(t)$ and $x_2(t)$

$$C_k = A_k B_k T\,. \tag{11.32}$$

11.3 Sampling

11.3.1 Ideal Sampling

Now that we are comfortable with the spectra of periodic signals and the delta impulse train, we can use it to describe the sampling of continuous signals. We will first deal with ideal sampling, where we take the precise values of the continuous function at sample points. This idealisation allows us to identify the fundamental principles particularly simply. Later we will see that the concessions necessary for realisation can also be described by simple extensions of the concept of ideal sampling.

To model ideal sampling we start with a continuous signal $\tilde{x}(t)$ and multiply it with an impulse train $\frac{1}{T}\text{⊥⊥⊥}(\frac{t}{T})$, as shown in Figure 11.6. The result $x(t)$ is again a series of impulses. The weight of the individual impulses are the values of the signal at points $t = kT, k \in \mathbb{Z}$. In Figure 11.6, the weights are symbolised by the lengths of the arrows. For the spectrum $X(j\omega)$ of the sampled signal $x(t)$ we use the multiplication theorem (9.75) to obtain

$$x(t) = \tilde{x}(t) \cdot \frac{1}{T}\text{⊥⊥⊥}\left(\frac{t}{T}\right) \circ\!\!-\!\!\bullet \; X(j\omega) = \frac{1}{2\pi}\tilde{X}(j\omega) * \text{⊥⊥⊥}\left(\frac{\omega T}{2\pi}\right). \tag{11.33}$$

Convolution of the spectrum $\tilde{X}(j\omega)$ of the continuous signal with the inpulse train in the frequency-domain $\text{⊥⊥⊥}(\omega T/2\pi)$ gives a periodic continuation of $\tilde{X}(j\omega)$ at multiples of $2\pi/T$, as shown in Fig 11.7.

The frequency $\omega_a = \frac{2\pi}{T}$ is called the *sampling frequency*. The spectrum of the sampled signal in the frequency-domain is clearly periodic with period ω_a, and there is a duality here with the Fourier transform of periodic signals, illustrated by the following scheme.

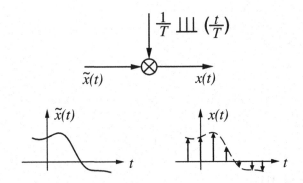

Figure 11.6: Ideal sampling of the signal $\tilde{x}(t)$

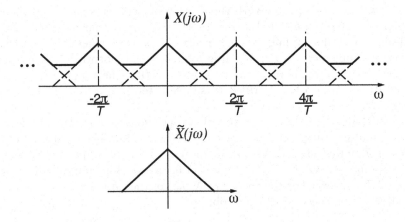

Figure 11.7: Spectrum of the sampled function $\tilde{x}(t)$

11.3.2 Sampling Theorem

Analog signals are sampled so that they can be recorded or processed using digital technology. It is also usually desirable with most applications to reconstruct the analog signal from its sample values, for example, with Compact Discs. The music signals are sampled because there are many technical advantages of storing a digital signal as opposed to an analog signal, but the user of the CD-player is only interested in the analog music signal, and not the sample values. It is therefore worth knowing whether a continuous signal $\tilde{x}(t)$ actually can be reconstructed from the sampled signal $x(t)$. This is exactly what the sampling theorem deals with, which we will be covering in this section.

To begin we introduce the concept of a band-limited signal.

Definition 19:   Band-limited signal

A signal $\tilde{x}(t)$ is band-limited *if its spectrum $\tilde{X}(j\omega)$ is zero*

$$\tilde{X}(j\omega) = \begin{cases} \text{any value} & |\omega| < \omega_g \\ 0 & |\omega| \geq \omega_g, \end{cases}$$

above a frequency ω_g. The frequency ω_g is called the band limit.

Figure 11.8 shows the spectrum of a band-limited signal, which in this case arbitrarily has a triangular form for $|\omega| < \omega_g$. To distinguish between the periodic repetitions by sampling, this is also called the *baseband spectrum*.

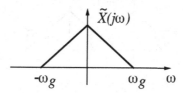

Figure 11.8: Baseband spectrum with limit frequeny ω_g

We can use it to formulate the central principle of the sampling theorem.

If a band-limited signal $\tilde{x}(t)$ is sampled frequently enough, so that the baseband repetitions do not overlap, $\tilde{x}(t)$ can be interpolated from the sampled signal $x(t)$ without error.

The justification for the sampling theorem will be explained by considering it in the frequency-domain. Figure 11.9 shows the spectrum of a sampled signal with the repeated baseband from Figure 11.8 at multiples of the sampling frequency $\omega_a = \frac{2\pi}{T}$. Reconstruction of the original signal requires an interpolation filter $H(j\omega)$, that leaves the baseband unchanged, but suppresses all of the repetitions caused by sampling. Such a filter can always be found if the sampling frequency $\omega_a = \frac{2\pi}{T}$ is greater than or equal to double the bandlimit ω_g of the band-limited signal $\tilde{x}(t)$.

$$\omega_a = \frac{2\pi}{T} \geq 2\omega_g \text{ or } T \leq \frac{\pi}{\omega_g}.$$

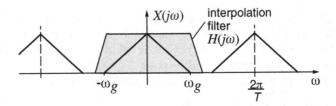

Figure 11.9: For $\omega_g < \frac{\pi}{T}$, the signal $X(j\omega)$ can be reconstructed using the interpolation filter $H(j\omega)$

If this condition is violated, then the repeated baseband overlaps itself, and can no longer be separated by convolution (see Figure 11.10). The overlapping that occurs in the dashed region is also called *aliasing* as an obviously erroneous reconstruction by the interpolation filter leads to high-frequency signal components appearing at other locations in the incorrectly reconstructed baseband. The

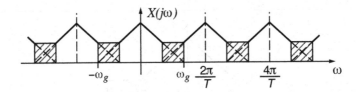

Figure 11.10: Overlapping baseband for $\omega_g > \frac{\pi}{T}$

sampling frequency $\omega_a = \frac{2\pi}{T}$

$$f_a = \frac{1}{T} = \frac{\omega_a}{2\pi} \tag{11.34}$$

is also called the *sampling rate*. It specifies how often the signal will be sampled per unit time. Sometimes f_a is called the sample frequency. To clearly distinguish between f_a and ω_a, the terms 'sampling rate' and 'angular sampling frequency' should be used.

Different choices of the sampling rate cause different cases to occur, shown in Table 11.1. Clearly, critical sampling represents the border case where the sampling rate is as low as possible, but aliasing has still been avoided. In this case the ideal interpolation filter is a rectangle in the frequency-domain with frequency response $H(j\omega)$ and impulse response $h(t)$ (see Figure 11.11):

$$H(j\omega) = T \operatorname{rect}\left(\frac{\omega T}{2\pi}\right) \quad \bullet\!\!-\!\!\circ \quad h(t) = \operatorname{si}\left(\frac{\pi t}{T}\right). \tag{11.35}$$

Table 11.1: Describing different sample rates

$f_a > \dfrac{\omega_g}{\pi}$	oversampling
$f_a = \dfrac{\omega_g}{\pi}$	critical sampling Nyquist-frequency
$f_a < \dfrac{\omega_g}{\pi}$	undersampling $\Rightarrow$ aliasing overlapping spectra

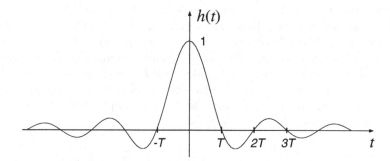

Figure 11.11: Impulse response of an ideal interpolation filter

In the frequency-domain it is clear that the spectrum $\tilde{X}(j\omega)$ of the original signal $\tilde{x}(t)$ can be recovered by multiplication of $X(j\omega)$ with $H(j\omega)$ (11.35). Scaling by T is necessary because the individual impulses of the impulse train $\frac{1}{2\pi}\text{⊥⊥⊥}(\frac{\omega T}{2\pi})$ in (11.33) have a weight of $\frac{1}{T}$.

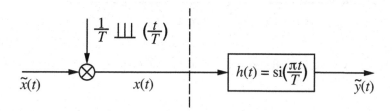

Figure 11.12: Sampling and interpolation with an ideal interpolation filter

In the time-domain we can obtain with some rearrangment, the reconstructed signal $\tilde{y}(t)$ (Figure 11.12)

$$\tilde{y}(t) \;=\; \left[\tilde{x}(t) \cdot \frac{1}{T}\,\text{⊥⊥⊥}\!\left(\frac{t}{T}\right)\right] * \text{si}\!\left(\frac{\pi t}{T}\right) \tag{11.36}$$

$$=\; \left[\sum_{k=-\infty}^{\infty} \tilde{x}(kT)\delta(t-kT)\right] * \text{si}\!\left(\frac{\pi t}{T}\right) \tag{11.37}$$

$$=\; \sum_{k=-\infty}^{\infty} \tilde{x}(kT)\,\text{si}\!\left(\pi\frac{t-kT}{T}\right). \tag{11.38}$$

Perfect reconstruction $\tilde{y}(t) = \tilde{x}(t)$ from the sample values succeeds through inter-polation if the sample values are used as scaling factors of appropriately shifted si-functions (see Figure 11.13). Of course, it must be possible to represent $\tilde{x}(t)$ by shifted and scaled si-functions. From (11.33), with (11.35) – (11.38) we can see that this requirement is equivalent to band-limiting with $\omega_g < \frac{\pi}{T}$. For values $t = kT$ all si-functions but one are zero. This one si-function has its maximum value of 1 at this point. The sample values $\tilde{x}(kT)$ are therefore the weights of the si-functions, and between the sample values, all (infinitely many) of the si-functions are superimposed, such that the nearest sample values have a greater influence on the signal value than those far away.

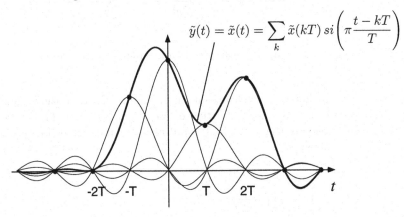

Figure 11.13: A band-limited signal $\tilde{x}(t)$ can be put together from a series of shifted weighted si-functions

The basic idea of the sampling theorem comes from Lagrange (1736-1813), and interpolation with the si-function was described by Whittaker in 1915. The sampling theorem in its current form, however, was introduced by Shannon in 1948.

11.3.3   Sampling Theorem for Complex Band-Pass Signals

If we consider periodic continuation of the spectrum by sampling, for example, Figure 11.9, we see that the baseband spectrum between $-\omega_g$ and ω_g, and the periodic continuation stand next to each other at multiples of the sample frequency. The spectrum of the original signal $\tilde{x}(t)$ could lie just as well between $-\omega_g + \frac{2\pi}{T}$ and $+\omega_g + \frac{2\pi}{T}$, or even between 0 and $\frac{2\pi}{T}$ – the periodic continued spectrum $\tilde{X}(j\omega)$ would be fundamentally indistinguishable. Correct choice of the interpolation filter for reconstruction of $\tilde{x}(t)$ requires knowledge of which region of the frequency axis the signal occupied before sampling. This leads to the sampling theorem for complex band-pass signals.

We start with a signal $\tilde{x}(t)$, with spectrum $\tilde{X}(j\omega)$ (Figure 11.14). As it has no frequency components lower than $(|\omega| < \omega_0)$ or higher than $(|\omega| \geq \omega_0 + \Delta\omega)$, it is known as a *band-pass spectrum*. The corresponding time signal must be complex as the conditions for conjugate symmetry (9.49) are not fulfilled:

$$\tilde{X}(j\omega) \neq \tilde{X}^*(-j\omega) \qquad \rightarrow \ \tilde{x}(t) \ \text{complex}. \qquad (11.39)$$

Complex band-pass signals are of special importance in communications; they are represented by two real signals: one for the real part, and one for the imaginary part.

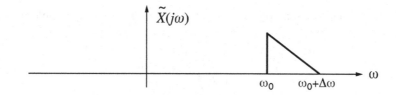

Figure 11.14: Unilateral band-pass spectrum

If we sample this signal with sampling rate

$$f_a = \frac{1}{T} = \frac{\Delta\omega}{2\pi} \ \text{where.} \ \omega_a = \Delta\omega$$

we again obtain a periodic continuation of the spectrum where overlapping is only just avoided; a case of critical sampling. Figure 11.15 shows the spectrum of the sampled signal and the impulse train, that corresponds to the sampling frequency $\omega_a = \Delta\omega$. There is no condition for the relationship between ω_0 and $\Delta\omega$. Both frequencies can be chosen independently. Critical sampling is clearly always possible for complex band-pass signals. We will soon see that for an apparently simple case of a real band-pass signal that it is not.

To recover the original signal, the interpolation filter should be chosen so that it blocks all frequency components not in the spectrum $\tilde{X}(j\omega)$ of the original signal $\tilde{x}(t)$. Its frequency response is a rectangle in the band-pass location:

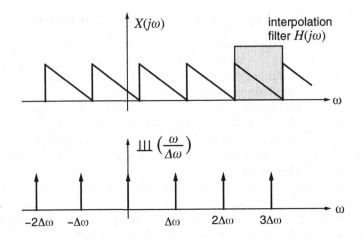

Figure 11.15: Spectrum of the sampled complex signal

$$H(j\omega) = T \operatorname{rect}\left(\left(\omega - \omega_0 - \frac{\Delta\omega}{2}\right)\frac{T}{2\pi}\right). \tag{11.40}$$

There is no conjugate symmetry here either, so the corresponding impulse response is complex:

$$h(t) = \operatorname{si}\left(\frac{\pi t}{T}\right)e^{j(\omega_0 + \frac{\Delta\omega}{2})t}. \tag{11.41}$$

Figure 11.16 shows the real and imaginary parts of $h(t)$ for $\omega_0 = 4\Delta\omega$.

11.3.4   Sampling Theorem for Real Band-Pass Signals

For the apparently simple case of a real band-pass signal, correct choice of sample frequency is not as simple as for the complex band-pass signal. The spectrum of a real band-pass signal is shown in Figure 11.17. The conditions of conjugate symmetry are fulfilled in this case.

Critical sampling requires a sample rate of

$$f_a = \frac{1}{T} = \frac{\Delta\omega}{\pi}, \text{ i.e., } \omega_a = 2\Delta\omega.$$

In contrast to complex band-pass signals, in this case, critical sampling is only possible if ω_0 and $\Delta\omega$ have a particular relationship with each other. Continuation of the band-pass spectrum without gaps requires that the space between $-\omega_0$ and ω_0 in Figure 11.17 can take exactly an even number of half-bands of width $\Delta\omega$.

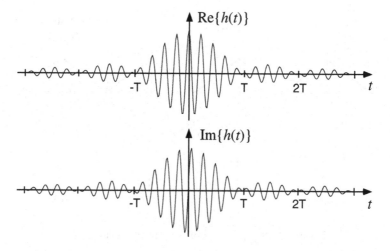

Figure 11.16: Complex impulse response of the interpolation filter

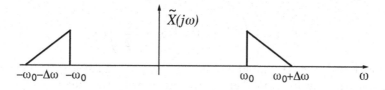

Figure 11.17: Band-pass spectrum of a real signal

The even number condition comes about because of the conjugate symmetry with $\omega = 0$, and therefore

$$\omega_0 = n \cdot \Delta\omega, n \in \mathbb{N} \, . \tag{11.42}$$

Otherwise the space in Figure 11.17 cannot be completely filled with half-bands and critical sampling is impossible. Figure 11.18 shows the spectrum of the critically sampled signal and the corresponding delta impulse train for $n = 2$.

The ideal interpolation filter must take exactly the frequency components that were contained in the original band-pass signal. We can describe it with a rectangular frequency response in the baseband, that is shifted by convolution with two symmetrical delta impulses to the band-pass locations:

$$H(j\omega) = T \operatorname{rect}\left(\frac{\omega T}{\pi}\right) * \left[\delta\left(\omega - \omega_0 - \frac{\Delta\omega}{2}\right) + \delta\left(\omega + \omega_0 + \frac{\Delta\omega}{2}\right)\right] \, . \tag{11.43}$$

The corresponding impulse response is purely real and resembles the impulse response of the interpolation filter (11.35), but because of the band-pass character,

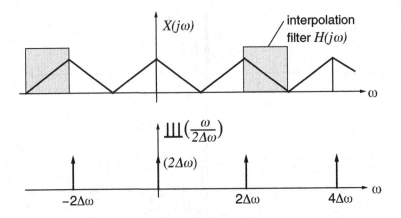

Figure 11.18: Spectrum of the sampled real signal

it is multiplied by another cosine function:

$$h(t) = \text{si}\left(\frac{\pi t}{2T}\right) \cos\left(\left(\omega_0 + \frac{\Delta\omega}{2}\right)t\right) \tag{11.44}$$

Figure 11.19 depicts the impulse response for $\omega_0 = 3 \cdot \Delta\omega$, $(n = 3)$. It can clearly be seen that the envelopes in Figure 11.19 agree with the curve in Figure 11.11 when the time axis is scaled by a factor of 2. (11.44) is also valid for the case $\omega_0 = 0$, in the low-pass case (11.35) (Exercise 11.22).

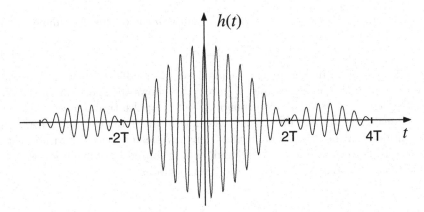

Figure 11.19: Impulse response of an interpolation filter for a critically sampled real band-pass signal with $\omega_0 = 3\Delta\omega$

The sampling frequencies determined for the three cases of critical sampling are shown in Table 11.2: Comparison with Figures 11.8, 11.14 and 11.17 reveals the

Table 11.2: Sampling frequencies for critical sampling

baseband signal	$\omega_a = 2\omega_g$
complex band-pass signal	$\omega_a = \Delta\omega$
real band-pass signal	$\omega_a = 2\Delta\omega$

following simple rule for choosing the sampling frequency: for critical sampling, the required sample frequency ω_a is equal to the complete width of the frequency bands in which $\tilde{X}(j\omega)$ is non-zero. This also requires that the spectra are chosen so that critical sampling is possible.

11.3.5   Non-ideal Sampling

Until now we have not considered physical implementation of the sampling procedure. We have assumed that it is possbible to take the values $\tilde{x}(kT)$ of the signal $\tilde{x}(t)$ at precisely defined points $t = kT$, as depicted in Figure 11.20. In fact, the sampling procedure requires that energy is taken from the signal $\tilde{x}(t)$ for each sample value. For example, an electrical signal can be sampled by taking some charge at each sample point and storing it in a capacitor. The resulting capacitor voltage is a measurement of the sample value. Charging a capacitor requires, however, a certain timespan, so the capacitor voltage cannot be assigned a definite signal value of $\tilde{x}(t)$ at a clearly defined point in time (see Figure 11.20 below).

We can also describe such non-ideal sampling with an impulse train, if we formulate the collected charge over a time τ as integration:

$$\frac{1}{\tau} \int\limits_{t-\tau/2}^{t+\tau/2} \tilde{x}(\nu)d\nu = \tilde{x}(t) * \frac{1}{\tau}\text{rect}\left(\frac{t}{\tau}\right). \tag{11.45}$$

The integration time τ must be smaller than the sampling interval ($\tau < T$). The series of rectangles that describe the sampling procedure is shown in the middle of Figure 11.20. They are each centralised at the desired time point, although strictly speaking, the integrated value is only available at the end of the integration interval. In this manner we avoid having a delay term in the calculation. The sampled signal $x(t)$ is thus

$$x(t) = \left[\tilde{x}(t) * \frac{1}{\tau}\text{rect}\left(\frac{t}{\tau}\right)\right] \cdot \frac{1}{T}\text{Ш}\left(\frac{t}{T}\right). \tag{11.46}$$

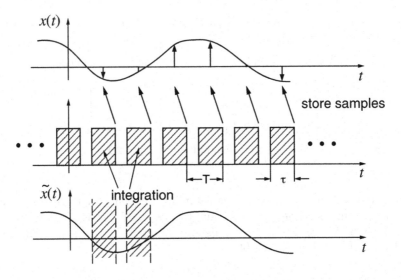

Figure 11.20: Sampling a signal in the real-world

A diagrammatic description of this sampling procedure is shown in Figure 11.21. Compared to the ideal sampling in Figure 11.6, there is an additional system with impulse response $h(t) = \frac{1}{\tau}\text{rect}(\frac{t}{\tau})$

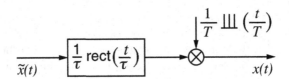

Figure 11.21: Non-ideal sampling of the signals $\tilde{x}(t)$

The description of non-ideal sampling introduced here with the example of an electrical signal also applies to other signals. In every case, measurement of a signal value requires a certain time. A further generalisation is the assumption that the energy transfer is not uniform within the time τ. We can take this into account using any weighting function $a(t)$ with

$$
\begin{aligned}
a(t) &> 0 \quad \text{for } -\frac{\tau}{2} < t < \frac{\tau}{2} \\
a(t) &= 0 \quad \text{otherwise.}
\end{aligned}
\tag{11.47}
$$

instead of the rectangle function. Before the integration, $\tilde{x}(\nu)$ in (11.45) is multiplied by $a(t-\nu)$. In connection with optics, $a(t)$ is also called an *aperture function*.

For the sampled signal,

$$x(t) = \left[\tilde{x}(t) * a(t)\right] \cdot \frac{1}{T} \text{\uparrow\uparrow\uparrow}\left(\frac{t}{T}\right). \tag{11.48}$$

We found the fundamental property with ideal sampling, that the spectrum of the sampled signal consists of the continuation of the spectrum of the non-sampled signal. We will now investigate what influence non-ideal sampling has on the frequency-domain. To simplify this, we will be limited to rectangular apertures in accordance with (11.46). The Fourier transform of (11.46) yields the spectrum of the sampled signal $X(j\omega)$

$$X(j\omega) = \frac{1}{2\pi}\left[\tilde{X}(j\omega) \cdot \text{si}\left(\frac{\omega\tau}{2}\right)\right] * \text{\uparrow\uparrow\uparrow}\left(\frac{\omega T}{2\pi}\right). \tag{11.49}$$

Compared with the spectrum of the ideal sampled signal (11.33), it is striking that the spectrum of the original signal $X(j\omega)$ is weighted with the aperture before periodic continuation. Figure 11.22 shows a possible spectrum of the original signal $\tilde{X}(j\omega)$ (as in Figure 11.8) and the weighting with the Fourier transform of the rectangle impulse. As with the rectangle, the greatest value with $\tau = T$ has been taken, which corresponds to an integration with duration τ over the entire sampling interval T. The first zero of the si-function then lies exactly at the sampling frequency $\omega = 2\pi/T$.

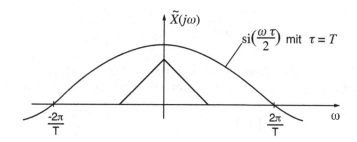

Figure 11.22: Spectrum of the sampled signal

In order to estimate the effect of sampling on the frequency response, we calculate the value of the si-function at the highest frequency that $\tilde{X}(j\omega)$ can contain without causing aliasing errors:

$$\left.\text{si}\left(\frac{\omega\tau}{2}\right)\right|_{\omega=\frac{\pi}{T}} \geq \frac{2}{\pi} \,\hat{=}\, -3.9 \text{ dB for } \tau \leq T. \tag{11.50}$$

The highest frequency component is damped by nearly 4 dB; for all lower frequencies the damping is lower. An integration time τ, that is shorter than the sampling interval likewise reduces the damping effect. In many applications this effect can be tolerated and where appropriate can be balanced out by a suitable post-connected aperture correction filter. For example, most TV cameras have an aperture correction filter that compensates for the filtering effect of the row sampling of the image. In fact, components at frequencies above half the sample frequency are strongly damped by interpolation, but they are not suppressed enough that the resultant aliasing is tolerable. If such frequency components occur they still should be suppressed by a special anti-aliasing pre-filter, that blocks frequencies above half the sample frequency.

11.3.6   Reconstruction

It is not only necessary to consider side-effects from the implementation when sampling, but also during reconstruction of the continuous signals from their sample values. Although the description of reconstruction using an impulse train (11.36 to 11.38) is theoretically very elegant, impulse trains and approximations using short high voltage peaks are not practical for use in electrical circuits. Real digital-analog converters therefore do not use these signals, and use instead sample-and-hold circuits and staircase functions whose step height is the weight of the corresponding delta impulse and so corresponds to the actual sample value.

We still keep the advantages of the delta impulse train, however. The staircase functions can be thought of as a delta impulse train interpolated with a rectangle of width T

$$x'(t) = x(t) * \frac{1}{T} \, \text{rect}\left(\frac{t}{T} - \frac{1}{2}\right). \tag{11.51}$$

Figure 11.23 shows an impulse train and the corresponding rectangle function. The rectangle impulse does not lie symmetrical to the sample points, instead

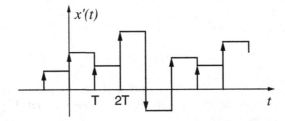

Figure 11.23: Signal interpolated with a sample-and-hold circuit

shifted from the symmetrical position by $T/2$, so because of the shift theorem, a corresponding exponential term appears in the spectrum.

$$X'(j\omega) = X(j\omega) \cdot \text{si}\left(\frac{\omega T}{2}\right) \cdot e^{-j\frac{\omega T}{2}} . \tag{11.52}$$

Figure 11.24 shows the magnitude of the resulting spectrum $X'(j\omega)$. While in Figure 11.22 the baseband spectrum was weighted with a si-function and then periodically continued, here the periodic spectrum $X(j\omega)$ is already provided and is then weighted by multiplying it with the si-function. This damps the reoccurrences of the baseband (shaded in Figure 11.24), but unfortunately does not fully suppress them. The shaded area could, for example, be audible high-frequency noise in an audio signal, or appear in an image as visibly 'blocky' pixels. However, unlike aliasing errors which cannot be separated from the desired signal, these noise components can be fully eliminated by post-connecting a low-pass filter with impulse response $h(t)$ (11.35). The residual reduction in high-frequency components in the baseband can likewise by compensated for using a filter which can be implemented either digitally before the DAC or after as an analog filter.

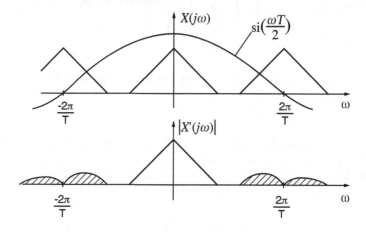

Figure 11.24: Magnitude response $|X'(j\omega)|$ of the signal interpolated with sample-and-hold circuit

<hr>

Example 11.2

The problems described with reconstruction also occur in film projections. Recording and replaying a series of individual images can be considered as sampling of a scene continuously changing in the time-domain. Per second, 24 individual images are taken, so the sampling frequency is

$$f_a = \frac{1}{T} = 24\text{Hz.}$$

Interpolation when the film is played back is very easy to implement when each image is projected for the corresponding length of time, although in fact, they are not there for the entire interval $T = 1/24$s, as between two images, the film has to be moved. There is no practical interpolation filter available for this optical signal, as in Figure 11.9, so the re-occurrences of the spectrum at 24 Hz, 48 Hz, 72 Hz,... are not removed and are perceived by the eyes as flicker. The eyes are sensitive to flicker up to around 60 Hz, but the first re-occurrence at 24Hz creates more visual disturbance. To avoid this problem, a trick can be used: instead of interrupting the projection of an image once for the film transport, it is interrupted twice (see Figure 11.25). The interrupting is done with a mechanical blind which is easily to implement. It doubles the sample rate (to 48 Hz), and pairs of consequential sample values are equal because they come from the same original.

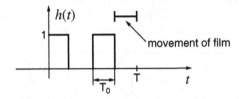

Figure 11.25: Impulse response of the interpolation filter

The effectiveness of the double projection procedure can be explained theoretically. We represent the double projection of an image as the impulse response of an interpolation filter (see Figure 11.25) and find the spectrum:

$$h(t) = \text{rect}\left(\frac{t}{T_0}\right) * \left[\delta\left(t - \frac{T_0}{2}\right) + \delta\left(t - \frac{T}{2} - \frac{T_0}{2}\right)\right] \tag{11.53}$$

$$= \text{rect}\left(\frac{t}{T_0}\right) * \delta\left(t - \frac{T_0}{2} - \frac{T}{4}\right) * \left[\delta\left(t - \frac{T}{4}\right) + \delta\left(t + \frac{T}{4}\right)\right] \tag{11.54}$$

$$H(j\omega) = T_0 \, \text{si}\left(\frac{\omega T_0}{2}\right) e^{-j\omega\left(\frac{T_0}{2} + \frac{T}{4}\right)} \cdot 2\cos\left(\omega \frac{T}{4}\right) \tag{11.55}$$

$$|H(j\omega)| = 2T_0 \, \text{si}\left(\frac{\omega T_0}{2}\right) \cos\left(\omega \frac{T}{4}\right) \tag{11.56}$$

Figure 11.26 shows the periodic spectrum corresponding to 24 images per second and the weighting by the interpolation function $h(t)$. An image signal has a dominant frequency component at $\omega = 0$. The zero of the spectrum at the sampling frequency 24 Hz extensively suppresses the first reoccurrence. The second at 48 Hz is not weakened significantly, but the flickering becomes much less noticeable. Additionally, the flicker reduction only functions when both rectangles that form $h(t)$ (Figure 11.25) are shifted away from each other by exactly $T/2$. Reduction of the components at both 24 Hz and 48 Hz can then be achieved using triple projection (Exercise 11.24).

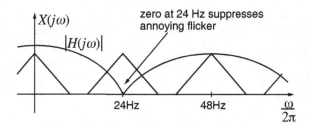

Figure 11.26: Magnitude response of the interpolation filter with a zero at 24 Hz

11.3.7 Sampling in the Frequency-Domain

The sampling theorem states that band-limited signals can be reconstructed from their sample values if the sampling frequency is high enough that no overlapping occurs in the frequency-domain. Because of the duality between the time-domain and frequency-domain, a dual of the sampling theorem can be formulated:

> **Signals of finite duration can be unambiguously reconstructed from a sufficient number of samples from their spectrum. Signals repeated in the time-domain may not overlap.**

We can express this mathematically with the transform pair

$$Y(j\omega) \;=\; \tilde{Y}(j\omega) \cdot \frac{T}{2\pi} \, \text{Ш}\!\left(\frac{\omega T}{2\pi}\right) \tag{11.57}$$

$$y(t) \;=\; \tilde{y}(t) * \frac{1}{2\pi} \, \text{Ш}\!\left(\frac{t}{T}\right). \tag{11.58}$$

Sampling in the frequency-domain is again idealized in (11.57) as multiplication with a delta impulse train. This corresponds to convolution (11.58), that causes periodic continuation of $\tilde{y}(t)$ in the time-domain. If the signal $\tilde{y}(t)$ is limited in time and its duration is less than T, sampling in the frequency-domain does not lead to overlapping in the time-domain. Then $\tilde{y}(t)$ can be recovered from $y(t)$ by multiplication with a time window of length T. In the frequency-domain, this corresponds to an interpolation (convolution) with the spectrum $\mathrm{si}(\omega T/2)$, that transforms the line spectrum into the smooth spectrum belonging to $\tilde{y}(t)$. If overlapping occurs in the time-domain, an error-free reconstruction is impossible. In this context we use the term *time-domain aliasing*.

Sampling in the frequency-domain is always carried out if values of a spectrum are to be stored in the memory of a computer, where continuous functions cannot exist, only a series of numbers. The necessary restriction to time-limited functions is, for example, achieved by splitting the signal into sections with finite duration. The relationship between a series of finite length and their spectra is represented by the *discrete Fourier transform* (DFT).

11.4 Exercises

Exercise 11.1

Write the sketched signal with a sum of delta impulses as well as with the ⊥⊥⊥ symbol.

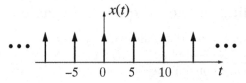

Exercise 11.2

Determine the Fourier transform $X(j\omega)$ of $x(t) = \text{⊥⊥⊥}(at)$ and sketch $x(t)$ and $X(j\omega)$ for $a = \frac{1}{2}$, $a = 1$ and $a = 3$. Note: use transform pair (11.12)

Exercise 11.3

Consider a delta impulse train shifted by t_0:

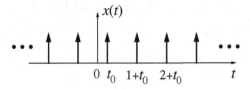

a) Give $x(t)$ and $X(j\omega)$ with ⊥⊥⊥ symbols.

b) What symmetry does $x(t)$ have for $t_0 = 0$, $t_0 = \frac{1}{4}$ and $t_0 = \frac{1}{2}$? Use your answers to derive the symmetry of $X(j\omega)$. Note: see symmetries (9.61).

c) Sketch $X(j\omega)$ for $t_0 = \frac{1}{4}$ and $t_0 = \frac{1}{2}$. Write $X(j\omega)$ as a sum of delta impulses.

Exercise 11.4

Give $X_1(j\omega)$ and $X_2(j\omega)$ as well as $x_1(t) \circ\!\!-\!\!\bullet X_1(j\omega)$ and $x_2(t) \circ\!\!-\!\!\bullet X_2(j\omega)$ using the ⊥⊥⊥- symbol and sketch both functions of time.

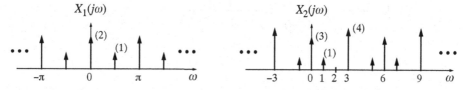

Note: represent $X_\nu(j\omega)$ as a sum of two spectra. For the sketch, represent the result as a sum of delta-impulses.

Exercise 11.5

Calculate the Fourier series of the following periodic functions using a suitable analysis.
Note: evaluating the coefficient formula is not necessary.

a) $x_a(t) = \cos(3\omega_0 t) \times \sin^2(2\omega_0 t)$

b) $x_b(t) = \cos^5(2\omega_0 t) \times \sin(\omega_0 t)$

c) $x_c(t) = \sin^4(3\omega_0 t) \times \cos^2(\omega_0 t)$

Exercise 11.6

Consider the functions
$$
\begin{aligned}
x_1(t) &= \cos(6\omega_0 t) + \cos(9\omega_0 t) \\
x_2(t) &= \sin(\omega_0 t) \times \cos(\sqrt{2}\omega_0 t) \\
x_3(t) &= x_2(t) + \cos(\omega_0 t) \times \sin(\sqrt{2}\omega_0 t) \\
x_4(t) &= \sum_{\nu=1}^{5} \sin(\sqrt{\nu}\omega_0 t) \\
x_5(t) &= \sin\left(\frac{\pi}{2}t\right) + \cos\left(\frac{\pi}{3}t\right)
\end{aligned}
$$

a) Test which are periodic, and give the periods where possible.

b) Give the Fourier transforms $X_1(j\omega)$ ... and $X_5(j\omega)$.

11. Sampling and Periodic Signals

Exercise 11.7

Expand the following periodic signals using the coefficient formula in a complex Fourier series $x(t) = \sum_\mu A_\mu e^{j\omega_0 \mu t}$. Give the fundamental angular frequency ω_0 for each.

a) $x_a(t) = \sin^2 \omega_1 t$

b)

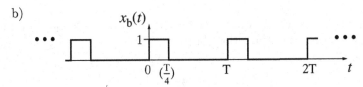

c)

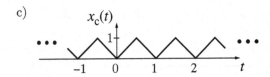

Note: coefficient formula $A_\mu = \dfrac{1}{T} \displaystyle\int\limits_0^T x(t) e^{-j\omega_0 \mu t}\, dt$, $\quad T = \dfrac{2\pi}{\omega_0}$

Exercise 11.8

Show that $X(j\omega) = \sum_\mu e^{-j\mu\omega} = \text{Ш}\left(\dfrac{\omega}{2\pi}\right)$, see (11.7), by expanding the right-hand term into a Fourier series.

Exercise 11.9

Consider $\tilde{x}(t) = \text{si}^2(\pi t)$ and the periodic signal $x(t) = \tilde{x}(t) * \frac{1}{4}\text{Ш}\left(\frac{t}{4}\right)$.

a) Calculate and sketch $\tilde{X}(j\omega)$ and then $X(j\omega)$. How do the weights of the delta impulses of $X(j\omega)$ tie in with $\tilde{X}(j\omega)$?

b) Give the Fourier series expansion of $x(t)$ with (11.18) and sketch the series of the Fourier coefficients A_μ.

Exercise 11.10

Give the Fourier transform of $x(t) = \text{rect}\left(\frac{t}{4}\right)$ and the periodic continuation $x_p(t) = x(t) * \text{Ш}\left(\frac{t}{2}\right)$. Calculate $X_p(j\omega)$

a) using convolution in the time-domain.

b) using the convolution theorem.

Exercise 11.11

(11.22) allows the spectrum of a signal $x(t)$ with period $T = 1$ to be expressed by its Fourier coefficients A_μ. Derive the relationship for general periods T. There are two possible routes to the solution: a) starting with (11.14) and (11.18), and b) using the Fourier transform on (11.17).

Exercise 11.12

Find the Fourier transform of the signal from Exercise 11.7, using the results from Exercise 11.7 and 11.11.

Exercise 11.13

A system with impulse response $h(t) = \sin(t)\,e^{-0.1t}\varepsilon(t)$ is excited by a periodic rectangle signal $x(t)$. Find the output signal $y(t)$ as a Fourier series. See Section 11.2.3.

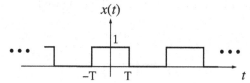

Exercise 11.14

Calculate the cyclic convolution of the signals $f(t)$ and $g(t) = \sin(\omega_0 t)$ using Fourier series. First give the connection between ω_0 and T, so that the cyclic convolution is defined properly.

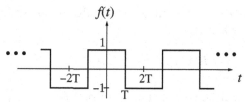

Exercise 11.15

Consider the spectrum $X(j\omega)$ of the signal $x(t)$.

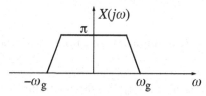

Find $X_a(j\omega) \bullet\!\!-\!\!\circ x(t) \times \frac{1}{T}\,⊥⊥⊥\left(\frac{t}{T}\right)$ and sketch it for $T_1 = \frac{\pi}{2\omega_g}$, $T_2 = \frac{\pi}{\omega_g}$, $T_3 = \frac{2\pi}{\omega_g}$

and $T_4 = \frac{2\pi}{3\omega_g}$. For which cases does aliasing occur? What cases are critically sampled?

Exercise 11.16

A signal $x(t) = \frac{\omega_g}{2\pi} \mathrm{si}^2\left(\frac{\omega_g t}{2}\right)$ is sampled at equidistant points in time νT, $\nu \in \mathbb{Z}$ to form a signal $x_A(t)$ of weighted delta impulses.

a) Give the original spectrum $X(j\omega) = \mathcal{F}\{x(t)\}$ (sketch with labelled axes).

b) Sketch the spectrum $X_A(j\omega) \bullet\!\!-\!\!\circ x_A(t)$ for the case $T = \dfrac{2\pi}{3\omega_g}$.

Exercise 11.17

The sketched rectangular signal $r(t)$ will be investigated.

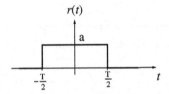

a) What is the spectrum $R(j\omega) = \mathcal{F}\{r(t)\}$ of the signal for both magnitude and phase? Sketch $|R(j\omega)|$.

b) The spectrum $R(j\omega)$ is sampled at equidistant points spaced by ω_0. This creates the new spectrum:

$$R_a(j\omega) = \omega_0 \sum_{\nu=-\infty}^{\infty} R(j\nu\omega_0) \cdot \delta(\omega - \nu\omega_0).$$

Sketch the magnitude of this new spectrum for $|\omega| \leq \frac{4\pi}{T}$ with

    α) $\omega_0 = \frac{2\pi}{4T}$,

    β) $\omega_0 = \frac{2\pi}{2T}$,

    γ) $\omega_0 = \frac{2\pi}{T}$,

    where the weighted delta impulses are represented by arrows of corresponding length.

c) Sketch the functions of time $r_a(t)$, corresponding to the cases α), β), γ) in part b).

Exercise 11.18

The spectrum $X(j\omega)$ is from a unilateral band-pass signal $x(t)$.

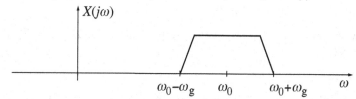

a) Is $x(t)$ real-valued? What symmetry does $x(t)$ have?

b) The signal $x_a(t)$ is created by critical sampling of $x(t)$. Give the sampling frequency f_a and the sample interval T_a for this case. Sketch $X_a(j\omega)$ for $\omega_0 = 9\pi$ and for $\omega_g = 2\pi$.

c) Sketch $X_a(j\omega)$ for sampling with $f_a = \frac{3\omega_g}{2\pi}$, ω_0 and ω_g as above.

Exercise 11.19

The signal $x(t)$ with the sketched spectrum $X(j\omega)$ is modulated with the shown arrangement.

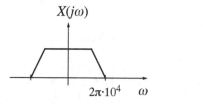

complex modulation

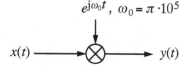

a) Sketch the spectrum $Y(j\omega)$ of the output signal.

b) Draw an arrangement of sampler and reconstruction filter $H(j\omega)$ that can demodulate $y(t)$, so that $x(t)$ is obtained at the output. What are the possible sampling frequencies? Sketch $|H(j\omega)|$ for a suitable reconstruction filter.

Exercise 11.20

A signal with the sketched spectrum is sampled ideally with intervals T.

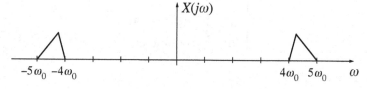

a) Is the signal a high-pass, a low-pass, or a band-pass signal? Is it complex or real?

b) Is critical sampling possible?

c) Draw the spectrum $X_a(j\omega)$ of the sampled signal for $T_1 = \frac{\pi}{\omega_0}$, $T_2 = \frac{2\pi}{3\omega_0}$ and $T_3 = 0.2\frac{\pi}{\omega_0}$.

Exercise 11.21

Can the band-pass signals with the following spectra be critically sampled? For each signal, give the minimum sample frequency at which no aliasing occurs.

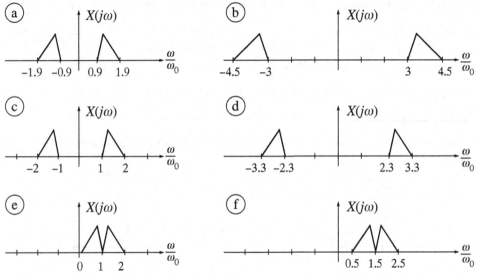

Exercise 11.22

Show that the impulse response of the ideal interpolation filter for critically sampled real band-pass signals (11.44) becomes the low-pass case (11.35) for $\omega_0 = 0$.

Exercise 11.23

A signal $x(t)$, band-limited to $f_g = 20$ kHz is critically sampled at frequency f_a. The non-ideal properties of the sampling are represented by the aperture function

$$a(t) = \begin{cases} 1 + \dfrac{t}{\tau} & \text{for} \quad -\tau < t < 0 \\[2mm] 1 - \dfrac{t}{\tau} & \text{for} \quad 0 \le t < \tau \\[2mm] 0 & \text{otherwise}. \end{cases}$$

For parts a) and b), let $\tau = \dfrac{1}{2 f_a}$.

a) Draw $A(j\omega) \bullet\!\!-\!\!\circ a(t)$ and a suitable example for $X(j\omega)$ in the same sketch.

b) Compare the amplification of the aperture function at the frequency zero and at the band limits of the signal. By how many dB is it less at the band limits?

c) Determine τ so that the aperture function makes a suitable preliminary anti-aliasing filter if sampling is carried out at $f_a = 10$ kHz.

Exercise 11.24

The effect of triple projection on the flickering of cinematic films (see Example 11.2) will be investigated. The interpolation filter takes the form of a mechanical shutter with the sketched impulse response $h(t)$.

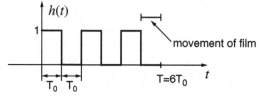

Express $h(t)$ using rect-functions and δ-impulses, as described in (11.53). Then calculate $H(j\omega)$ and sketch $|H(j\omega)|$ for $T = \frac{1}{24\text{Hz}}$. In particular, try to identify the zeros of $H(j\omega)$.

12 The Spectrum of Discrete Signals

In Chapters 1 to 10 we got to know some powerful tools for dealing with continuous signals and systems. We carried out the conversion of real-world continuous signals to sampled signals in Chapter 11, which is necessary for digital processing. The sampled signals were treated as continuous-time variables so that the tools we had learnt, like the Fourier transform, could still be used.

In a computer, we can only work with a sequence of numbers, that are defined over a *discrete* range by a numerical value (the index). Examples of such an index might be the sample number for a digital audio signal, or the pixel address within a digital image. In addition there are discrete signals that have not arisen from sampling a continuous signal. Examples of these are represented in Figures 1.3 and 1.4. We need new tools to describe such sequences, and we will concentrate on them in Chapters 12 - 14. This chapter deals with *discrete signals* and the discrete form of the Fourier transform, the $\mathcal{F}_*$ *transform*. This transform is also referred to as *discrete-time Fourier transform (DTFT)*. The two following chapters deal with *discrete systems* and we will learn the discrete counterpart to the Laplace transform, the *z-transform*.

In Sections 12.1 and 12.2 we will consider discrete signals together with some examples. The discrete-time Fourier transform will also be introduced, which we will use to examine discrete signals in the frequency-domain. We will see that it has similar properties to the Fourier transform for continuous signals. At the end of this chapter we will investigate the relationship between continuous signals and their discrete equivalent as a series of samples.

12.1 Discrete-Time Signals

A discrete-time signal is represented by a sequence of numbers that is called a *time series*. There is no smooth transition between the numbers. Figure 12.1 shows the conventions we will use to represent such signals; in order to distinguish them from continuous-time signals, we put the independent variable in square brackets. In many technical applications, a discrete-time signal arises from the *sampling* of a continuous-time signal $\tilde{x}(t)$, where a sample is taken from $\tilde{x}(t)$ at regular intervals T and is translated as a numerical value $x[k]$:

$$x[k] = \tilde{x}(kT), \quad k \in \mathbb{Z}. \tag{12.1}$$

Here it is essential to distinguish the discrete signal $x[k]$ from the sampled signal

$$\begin{aligned} x_a(t) = \tilde{x}(t) \cdot \frac{1}{T} \text{⊔⊔⊔}\left(\frac{t}{T}\right) &= \sum_k \tilde{x}(kT)\delta(t - kT) \\ &= \sum_k x[k]\,\delta(t - kT), \quad t \in \mathbb{R} \end{aligned} \tag{12.2}$$

introduced in Chapter 11. The dependent variable t is defined for every point in $\mathbb{R}$, although $x_a(t)$ is zero for almost all of these time-points (except for $t = kT$, $k \in \mathbb{Z}$). In contrast, $x[k]$ is only defined where the index k is an integer, so integration of $x[k]$ is not possible, and the Fourier transform from Chapter 9 cannot be used. In the following sections we will deal with discrete signals in depth without being limited to sequences of sampled values, instead we will consider general time-series $x[k]$. We will return to using sampling in Section 12.4. The values $x[k]$

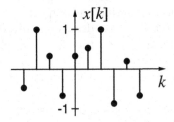

Figure 12.1: Representation of a discrete-time signal $x[k]$

are themselves continuous and in general, they could also be complex: $x[k] \in \mathbb{C}$. Strictly speaking, this is not the case if the series is going to be processed with a computer. The finite word length means that numbers can only be recorded within the limits of the number representation being used, and the values $x[k]$ then become themselves discrete. To distinguish them from the discrete signals with continuous values, we refer to these signals as *digital signals*.

Rounding to discrete values is a non-linear process and is called *quantisation*. Processes involving digital signals therefore cannot be described by LTI-systems. A computer with sufficiently great word length can work with digital signals that are so finely quantised, however, that they can approximate discrete signals with continuous values. We will therefore restrict ourselves to discrete signals from now on.

12.2 Some Simple Sequences

In this section we will be dealing with some simple discrete signals; we already know their continuous-time counterparts. These signals are the impulse, step and exponential series.

12.2.1 Discrete Unit Impulse

The discrete unit impulse is defined by

$$\delta[k] = \begin{cases} 1, & k = 0 \\ 0, & k \in \mathbb{Z} \setminus \{0\} \end{cases}.$$ (12.3)

Figure 12.2 shows this series. The discrete unit impulse has the selective property

$$x[k] = \sum_{\kappa=-\infty}^{\infty} x[\kappa]\delta[k - \kappa],$$ (12.4)

that can be easily checked.

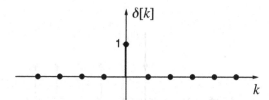

Figure 12.2: Discrete unit impulse

Comparing this with the selective property of the continuous-time delta impulse shows a close relationship between the two:

$$x(t) = \int_{-\infty}^{\infty} x(\tau)\delta(t - \tau)\,d\tau.$$ (12.5)

Integration with respect to the continuous variable τ in (12.5) corresponds to summation with respect to the discrete independent variable κ in (12.4). We will see that replacing integration in the definition of a continuous signal with summation often leads to the equivalent discrete signal.

There is a more significant difference between the discrete unit impulse and the continuous delta impulse: the unit impulse $\delta[k]$ is not a distribution. For $k = 0$ it has the finite value $\delta[0] = 1$, which makes calculations. simple. In contrast to the continuous delta impulse $\delta(t)$, the unit impulse $\delta[k]$ can be directly inserted into a formula and does not require the use of the selective property.

12.2.2   Discrete Step Function

The discrete unit step function from Figure 12.3 is defined by

$$\varepsilon[k] = \left\{ \begin{array}{ll} 1, & k \geq 0 \\ 0, & k < 0 \end{array} \right. . \tag{12.6}$$

The relationship between the unit step function and the unit impulse corresponds to the equivalent for continuous signals:

$$\varepsilon[k] = \sum_{\kappa=-\infty}^{k} \delta[\kappa] \tag{12.7}$$

$$\delta[k] = \varepsilon[k] - \varepsilon[k-1] . \tag{12.8}$$

Integration in the continuous case corresponds to discrete summation and differentiation of continuous signals corresponds to subtracting the neighbouring value in the discrete case.

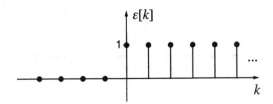

Figure 12.3: Discrete unit step function

12.2.3   Exponential Sequences

Exponential sequences are in general complex sequences of the form

$$x[k] = \hat{X} \, e^{(\Sigma + j\Omega)k} . \tag{12.9}$$

$\hat{X}$ is the complex amplitude, Σ is the damping constant, and Ω is the angular frequency. Discrete exponential sequences are often characterised directly by giving their base, especially if they will be interpreted as the impulse response of a discrete system.

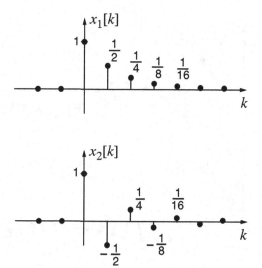

Figure 12.4: Examples of exponential sequences

Example 12.1

Two examples for unilateral exponential sequnces with real values are given by
(12.10) and (12.11), and illustrated in Figure 12.4.

$$x_1[k] = \left(\frac{1}{2} \right)^k \varepsilon[k] \tag{12.10}$$

$$x_2[k] = \left(-\frac{1}{2} \right)^k \varepsilon[k] \tag{12.11}$$

For both exponential series $\hat{X} = 1$ and $\Sigma = -\ln 2$. The angular frequency Ω has
the values $\Omega = 0$ and $\Omega = \pi$ respectively. Both series are the same except the
signs before the odd values.

In contrast to continuous exponential functions, different values of the angular
frequency Ω can lead to the same exponential series. As $e^{j\Omega k}$ is only evaluated for
integers k, adding multiples of 2π to Ω does not change the sequence:

$$e^{j\Omega k} = e^{j(\Omega + 2\pi)k}. \tag{12.12}$$

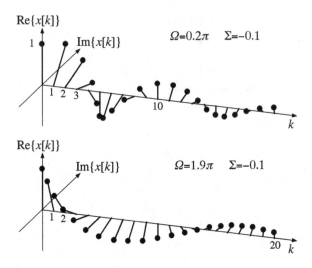

Figure 12.5: Examples of complex exponential series

Example 12.2

The exponential function in Figure 12.5 illustrates the ambiguous character of (12.12): the upper curve has been calculated with an angular frequency of $\Omega = 0.2\pi$. Increasing k by 1 turns the value of the series by 0.2π rad in the mathematically positive direction (from the real part to the imaginary part). For $k = 10$ the same direction as for $k = 0$ is repeated because $10\Omega = 2\pi$; the absolute value is reduced to $\exp(-0.1\times10) = 1/e$, however, because of the damping constant $\Sigma = -0.1$.

In the lower curve the angular frequency is $\Omega = 1.9\pi$. The series makes almost one full turn each time, as Ω is only 0.1π rad less than a whole circle. In comparison to the upper curve, the lower curve does not seem to have a higher frequency, but instead, a lower frequency, turning in the opposite direction. The exponential series with angular frequency $\Omega = 1.9\pi$ is identical to the exponential series with angular frequency $\Omega = 1.9\pi - 2\pi = -0.1\pi$. We already know this phenomenon from western movies, when the wheels of a wagon appear to be turning backwards.

■

The ambiguousness of the frequency of discrete exponential functions is the reason for the occurrence of 'Aliasing' (compare Chapter 11.3.2), where sampling causes different frequencies to overlap. Because of this effect, for a spectral representation of discrete signals, it is necessary to limit the frequency-domain to the width 2π.

12.3 Discrete-Time Fourier Transform

It would be convenient to use the advantages of looking at continuous-time signals in the frequency-domain with discrete-time signals. To this end we will be introducing the discrete counterpart to the Fourier transform, the discrete-time Fourier transform (DTFT). As with the Fourier and Laplace transforms, its use in the frequency-domain requires an inverse transform, transform pairs, theorems and symmetry properties.

12.3.1 Definition of the Discrete-Time Fourier Transform

As a series $x[k]$ is only defined for discrete values of $k \in \mathbb{Z}$, we cannot use the Fourier integral (9.1) introduced in Chapter 9. We therefore define the *discrete-time Fourier transform* or the $\mathcal{F}_*$ *transform* as:

$$\boxed{X(e^{j\Omega}) = \mathcal{F}_*\{x[k]\} = \sum_{k=-\infty}^{\infty} x[k]e^{-jk\Omega}\ .} \qquad (12.13)$$

It transforms a series $x[k]$ into a continuous complex function of a real variable Ω. $X(e^{j\Omega})$ is also called the *spectrum of a series*. In contrast to a continuous signal, it is periodic with 2π, so

$$X(e^{j(\Omega+2\pi)}) = X(e^{j\Omega}). \qquad (12.14)$$

This is easy to see, as each term of the sum in (12.13) contains a 2π-periodic term $e^{jk\Omega}$. In order to see this more clearly, we write $e^{j\Omega}$ as the argument of the $\mathcal{F}_*$ transform and define the Fourier transform over the unity circle of the complex plane. This convention will make the transfer to the z-transform easier.

A *sufficient condition* to show the existence of the spectrum $\mathcal{F}_*\{x[k]\}$ is that the sum of the series $x[k]$ is finite:

$$\sum_k |x[k]| < \infty. \qquad (12.15)$$

12.3.2 Inverse Discrete-Time Fourier Transform

The definition of the spectrum of a sequence from (12.13) represents a Fourier series of $X(e^{j\Omega})$. The period is 2π and the Fourier coefficients are the values $x[k]$. In order to recover the series $x[k]$ from the spectrum $X(e^{j\Omega})$, we have to use the formula for finding Fourier coefficients. It consists of an integration of $X(e^{j\Omega})e^{jk\Omega}$ over one period of the spectrum

$$\boxed{x[k] = \frac{1}{2\pi} \int_{-\pi}^{\pi} X(e^{j\Omega})e^{jk\Omega}d\Omega\ .} \qquad (12.16)$$

This relationship represents the *inverse discrete-time Fourier transform*.

12.3.3   Common Discrete-Time Fourier Transform Pairs

This section contains the Fourier transforms of the simple sequences discussed in Section 12.2. These were the unit impulse, unit step function and bilateral exponential series. As a concluding example we will determine the spectrum of a rectangle function.

12.3.3.1 $\mathcal{F}_*$ Transform of a Discrete Unit Impulse

To calculate the $\mathcal{F}_*$ transform of a unit impulse $x[k] = \delta[k]$ we start with the defining equation of the $\mathcal{F}_*$ transform, and insert the unit impulse. The result can be given immediately, using the selective property of the unit impulse:

$$X(e^{j\Omega}) = \sum_k \delta[k]e^{-j\Omega k} = 1 \ . \tag{12.17}$$

We thus obtain the transform pair

$$\boxed{\delta[k] \ \circ\!\!-\!\!\bullet \ 1 \ .} \tag{12.18}$$

The transform of a shifted unit impulse $x[k] = \delta[k - \kappa]$ leads to a linear change of phase as for a continuous-time signal:

$$X(e^{j\Omega}) = \sum_k \delta[k - \kappa]e^{-j\Omega k} = e^{-j\Omega \kappa} \ , \tag{12.19}$$

and so the transform pair is

$$\boxed{\delta[k - \kappa] \ \circ\!\!-\!\!\bullet \ e^{-j\Omega \kappa} \ .} \tag{12.20}$$

We can see that the transform pair (12.18) is a special case for $\kappa = 0$.

12.3.3.2 $\mathcal{F}_*$ Transform of an Undamped Complex Exponential Series

To find the $\mathcal{F}_*$ transform of an undamped bilateral exponential series $e^{j\Omega_0 k}$, we start with the interesting relationship

$$\sum_k e^{-j\Omega k} = 2\pi \sum_\nu \delta(\Omega - 2\pi\nu) \tag{12.21}$$

that we had derived from equation (11.7) in the last chapter. Now we are using it to determine the spectrum of $x[k] = e^{j\Omega_0 k}$:

$$X(e^{j\Omega}) = \sum_{k=-\infty}^{\infty} e^{-j(\Omega-\Omega_0)k} = 2\pi \sum_{k=-\infty}^{\infty} \delta(\Omega-\Omega_0-2\pi k) \ . \tag{12.22}$$

The transform pair is

$$
e^{j\Omega_0 k} \circ\!\!-\!\!\bullet\ 2\pi \sum_{k=-\infty}^{\infty} \delta(\Omega-\Omega_0-2\pi k) = \bot\!\bot\!\bot\left(\frac{\Omega-\Omega_0}{2\pi}\right) \tag{12.23}
$$

and states that the spectrum of an exponential series $e^{j\Omega_0 k}$ is in fact a delta impulse at $\Omega = \Omega_0$. The ambiguousness we discussed in Section 12.2.3 likewise applies to the spectrum of the delta impulse at $\Omega = \Omega_0 + 2\pi\nu$, $\nu \in \mathbb{Z}$. The resulting impulse train in the frequency-spectrum can be elegantly represented using the sha-symbol we introduced in Chapter 11. The correctness of the transform pair (12.23) can also be confirmed simply by using the inverse $\mathcal{F}_*$ transform on $\bot\!\bot\!\bot\left(\frac{\Omega-\Omega_0}{2\pi}\right)$.

A special case of (12.23) is the spectrum of the sequence $x[k] = 1$. If $\Omega_0 = 0$ is put into (12.23), the transformation yields

$$
x[k] = 1 \circ\!\!-\!\!\bullet\ X(e^{j\Omega}) = 2\pi \sum_{k=-\infty}^{\infty} \delta(\Omega-2\pi k) = \bot\!\bot\!\bot\left(\frac{\Omega}{2\pi}\right). \tag{12.24}
$$

12.3.3.3 $\mathcal{F}_*$ Transform of the Discrete Unit Step Function

The discrete unit step function $\varepsilon[k]$ can be expressed as the sum of a constant term

$$
\varepsilon_1[k] = \frac{1}{2}, \qquad -\infty < k < \infty \tag{12.25}
$$

and a bilateral step series with no middle value

$$
\varepsilon_2[k] = \begin{cases} \frac{1}{2} & k \geq 0 \\[2mm] -\frac{1}{2} & k < 0 \end{cases} \tag{12.26}
$$

$$
\varepsilon[k] = \varepsilon_1[k] + \varepsilon_2[k]. \tag{12.27}
$$

The Fourier transform of the constant term $\varepsilon_1[k]$ can be obtained directly from (12.24)

$$
\mathcal{F}_*\{\varepsilon_1[k]\} = \frac{1}{2}\mathcal{F}_*\{1\} = \frac{1}{2}\bot\!\bot\!\bot\left(\frac{\Omega}{2\pi}\right). \tag{12.28}
$$

To determine the Fourier transform of the second term $\varepsilon_2[k]$ we express the unit impulse $\delta[k]$ by $\varepsilon_2[k]$

$$
\delta[k] = \varepsilon_2[k] - \varepsilon_2[k-1]. \tag{12.29}
$$

Substituting (12.26) confirms this.

Using the laws of linearity and displacement from Section 12.5, we obtain

$$\mathcal{F}_*\{\varepsilon_2[k] - \varepsilon_2[k-1]\} = \mathcal{F}_*\{\varepsilon_2[k]\} - \mathcal{F}_*\{\varepsilon_2[k-1]\} = \mathcal{F}_*\{\varepsilon_2[k]\} - e^{-j\Omega}\mathcal{F}_*\{\varepsilon_2[k]\},$$
(12.30)

together with (12.18) we then obtain from (12.29)

$$\mathcal{F}_*\{\varepsilon_2[k]\} - e^{-j\Omega}\mathcal{F}_*\{\varepsilon_2[k]\} = 1$$
(12.31)

and from here we see that

$$\mathcal{F}_*\{\varepsilon_2[k]\} = \frac{1}{1 - e^{-j\Omega}}.$$
(12.32)

As we cannot divide by zero, (12.32) can only be used for $\Omega \neq \ldots -2\pi, 0, 2\pi, 4\pi, \ldots$. If $\mathcal{F}_*\{\varepsilon_2[k]\}$ would contain delta impulses at these frequencies, they would have to be considered separately. With (12.26), however, $\varepsilon_2[k]$ has zero mean and therefore there can be no delta impulses at $\Omega = 2\pi\nu$, $\nu \in \mathbb{Z}$. It should now be evident why we split $\varepsilon[k]$ up into $\varepsilon_1[k] + \varepsilon_2[k]$. By adding (12.28) and (12.32) together, we finally obtain from (12.27)

$$x[k] = \varepsilon[k] \circ\!\!-\!\!\bullet X(e^{j\Omega}) = \frac{1}{1 - e^{-j\Omega}} + \frac{1}{2}\text{Ш}\left(\frac{\Omega}{2\pi}\right).$$
(12.33)

We now compare this result with the Fourier transform of the unit step function in (9.92). We also found two terms there: one delta impulse and a term for which $s = j\omega$ came from the transfer function of an integrator. With the $\mathcal{F}_*$ transform, the discrete unit step function produces an impulse train instead of an individual impulse. The other term happens to be the transfer function of an accumulator, and we will show that it in fact represents the discrete counterpart of an integrator (see Example 14.5).

12.3.3.4 $\mathcal{F}_*$ Transform of a Unilateral Exponential Sequence

We will now find the $\mathcal{F}_*$ transform of a unilateral exponential sequence $x[k] = a^k\varepsilon[k]$ with $a \in \mathbb{C}$ from the defining equation (12.13):

$$X(e^{j\Omega}) = \sum_{k=0}^{\infty} a^k e^{-j\Omega k} = \sum_{k=0}^{\infty}(a\,e^{-j\Omega})^k.$$
(12.34)

We know that the sum of an infinite geometric series is

$$\sum_{n=0}^{\infty} q^n = \frac{1}{1-q}, \qquad |q| < 1,$$
(12.35)

and from this we can directly obtain

$$x[k] = a^k \varepsilon[k] \circ\!\!-\!\!\bullet\; X(e^{j\Omega}) = \frac{1}{1 - ae^{-j\Omega}}, \qquad |a| < 1. \qquad (12.36)$$

Figure 12.6 shows on the left the first few values of this exponential series for a real a, and the right shows the magnitude of the spectrum. The spectrum is clearly periodic with period 2π. As (12.36) does not converge for $a = 1$, the spectrum of the unit step function (12.33) is not contained in (12.36) as a special case.

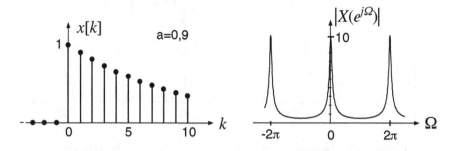

Figure 12.6: Series $x[k]$ and its magnitude spectrum $|X(e^{j\Omega})|$

12.3.3.5 $\mathcal{F}_*$ Transform of a Rectangle Series

We want to find the $\mathcal{F}_*$ transform $X(e^{j\Omega})$ of a rectangle series with length N

$$x[k] = \begin{cases} 1 & \text{for } 0 \le k \le N-1 \\ 0 & \text{otherwise} \end{cases} \qquad (12.37)$$

as shown in Figure 12.7. Inserting this into the defining equation (12.13) leads to

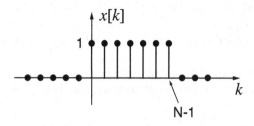

Figure 12.7: Rectangle series of length N

a finite geometric series

$$X(e^{j\Omega}) = \sum_{k=0}^{N-1} e^{-j\Omega k} = \frac{1 - e^{-j\Omega N}}{1 - e^{-j\Omega}}, \qquad (12.38)$$

that can be rearranged as

$$X(e^{j\Omega}) = \frac{e^{-j\Omega\frac{N}{2}}(e^{j\Omega\frac{N}{2}} - e^{-j\Omega\frac{N}{2}})}{e^{-j\frac{\Omega}{2}}(e^{j\frac{\Omega}{2}} - e^{-j\frac{\Omega}{2}})} = e^{-j\Omega\frac{N-1}{2}} \cdot \frac{\sin(\frac{N\Omega}{2})}{\sin(\frac{\Omega}{2})}. \qquad (12.39)$$

The result is shown in Figure 12.8.

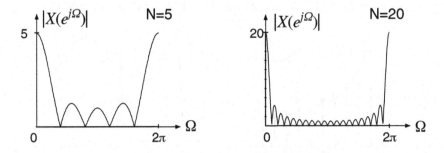

Figure 12.8: Magnitude spectrum of a rectangle series for different N

12.4 Sampling Continuous Signals

Up to now, we have discussed sequences in general, without considering where they come from or what they represent. Now we will examine sequences $x[k]$, that have come about from the sampling of continuous-time signals $\tilde{x}(t)$. The individual members of a series $x[k]$ are the sampled values of the continuous-time signal $\tilde{x}(kT)$, as in (12.1).

According to this simple relationship in the time-domain, a corresponding relationship in the frequency-domain should exist, that means between the Fourier transform of the sampled continuous-time signal and the spectrum of the discrete-time signal.

The signal $x_a(t)$ derived from the continuous-time signal $\tilde{x}(t)$ with ideal sampling (see Figure 12.9) will first of all be represented by an impulse train

$$x_a(t) = \tilde{x}(t) \cdot \frac{1}{T} \perp\!\!\!\perp\!\!\!\perp \left(\frac{1}{T}\right) = \sum_k \tilde{x}(kT)\delta(t - kT). \qquad (12.40)$$

We express the weighting of the delta impulses by the values $x[k]$ of the discrete-time signal (see (12.1))

$$x_a(t) = \sum_k x[k]\delta(t - kT). \qquad (12.41)$$

Transforming $x_a(t)$ with the Fourier integral (9.1) yields

$$
\begin{aligned}
X_a(j\omega) &= \int_{-\infty}^{\infty} \sum_k x[k]\delta(t - kT)e^{-j\omega t}\,dt \qquad (12.42)\\
&= \sum_k x[k]e^{-j\omega kT}.
\end{aligned}
$$

We see from comparison with the definition of the $\mathcal{F}_*$ transform (12.13), that the spectra agree, if $\Omega = \omega T$ is understood to be the normalised angular frequency:

$$
\boxed{X_a(j\omega) = X(e^{j\omega T}).} \qquad (12.43)
$$

The periodic Fourier transform of the sampled continuous-time signal $x_a(t)$ is the same as the spectrum of the discrete signal $x[k]$. The dimensionless angular frequency Ω of $X(e^{j\Omega})$ consists of the angular frequency ω of $X(e^{j\omega})$, normalised with sample interval T. The relationships between the continuous-time signal $\tilde{x}(t)$, the sampled continuous-time signal $x_a(t)$, the series of sampled values $x[k]$ and their spectra are depicted in Figure 12.9.

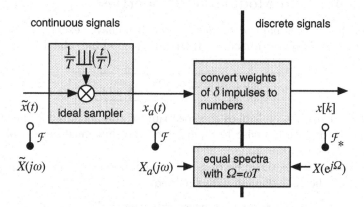

Figure 12.9: Relationship between the $\mathcal{F}$ and $\mathcal{F}_*$ spectrum

The relationship between $\mathcal{F}$ and $\mathcal{F}_*$ spectra gives an important insight that allows us to transfer many important properties and theorems that apply to the spectra of continuous-time signals to spectra of discrete-time signals. In the following section the most important of these are given. It is easy to recognise that some theorems, for example, the similarity theorem, cannot be applied to discrete-time signals because of the sampling process.

12.5 Properties of the $\mathcal{F}_*$ Transform

As the spectrum of a series can be viewed as a spectrum of a continuous-time signal that consists of weighted delta impulses, the similarity, differentiation, and integration theorems described in Chapter 9.7 are true for both $\mathcal{F}$ and $\mathcal{F}_*$. We will discuss the most important properties briefly. A summary of the properties of the $\mathcal{F}$ and $\mathcal{F}_*$ transforms can be found in Appendix Appendix B.8.

12.5.1 Linearity

From the linearity of the summation in (12.13), it follows directly that the principle of superposition still applies to the $\mathcal{F}_*$ transform, and from the linearity property of integration, the same is true for the inverse $\mathcal{F}_*$ transform.

$$
\begin{aligned}
\mathcal{F}_*\{a\,f[k] + b\,g[k]\} &= a\,\mathcal{F}_*\{f[k]\} + b\,\mathcal{F}_*\{g[k]\} \\
\mathcal{F}_*^{-1}\{c\,F(e^{j\Omega}) + d\,G(e^{j\Omega})\} &= c\,\mathcal{F}_*^{-1}\{F(e^{j\Omega})\} + d\,\mathcal{F}_*^{-1}\{G(e^{j\Omega})\}\,.
\end{aligned}
$$

$$(12.44)$$

Here a, b, c and d can be any real or complex constants.

12.5.2 Shift and Modulation Properties

A shift in the time-domain or frequency-domain behaves exactly as with the $\mathcal{F}$ transform or the Laplace transform. Inserting $x[k - \kappa]$ into the defining equation (12.33),

$$
x[k - \kappa] \circ\!\!-\!\!\bullet\, e^{-j\Omega\kappa}\,X(e^{j\Omega})\,,
$$

$$(12.45)$$

is obtained. The shift multiplies the spectrum of the unshifted sequence with a linear-phase term (see Exercise 12.8), although the shift must be a whole number of sample values, i.e., $\kappa \in \mathbb{Z}$.

Likewise, a shift in the spectrum by angular frequency Ω_0 corresponds to a modulation in the time-domain with this frequency:

$$
e^{j\Omega_0 k}\,x[k] \circ\!\!-\!\!\bullet\, X(e^{j(\Omega - \Omega_0)})\,.
$$

$$(12.46)$$

This can also be shown using (12.13) (Exercise 12.8).

12.5.3 Convolution Property of the $\mathcal{F}_*$ Transform

It is useful to know, when finding the response of a discrete system to a discrete-time signal, that convolution in the time-domain corresponds to multiplication of the two $\mathcal{F}_*$ transforms. For a sequence, the *discrete convolution* is defined as

$$
\begin{aligned}
y[k] \;=\; x[k] * h[k] \;&=\; \sum_{\kappa=-\infty}^{\infty} x[\kappa]\, h[k-\kappa] \\
\;=\; h[k] * x[k] \;&=\; \sum_{\kappa=-\infty}^{\infty} h[\kappa]\, x[k-\kappa].
\end{aligned}
\tag{12.47}
$$

This will be more accurately discussed in Chapter 14.6, in connection with discrete LTI-systems. By inserting it into (12.13), it can be shown (see Exercise 12.9), that

$$
\begin{aligned}
y[k] \;&=\; x[k] * h[k] \\
&\circ\!\!-\!\!\bullet \\
Y(e^{j\Omega}) \;&=\; X(e^{j\Omega})\, H(e^{j\Omega}).
\end{aligned}
\tag{12.48}
$$

12.5.4 Multiplication Theorem

Multiplication in the time-domain leads, as expected, to convolution in the frequency-domain. As both convolution terms are periodic, the classic convolution integral would not converge. The multiplication theorem, however, fortunately contains *cyclic convolution*, which we already know from Chapter 11.2.4:

$$
f[k] \cdot g[k] \;\circ\!\!-\!\!\bullet\; \frac{1}{2\pi} F(e^{j\Omega}) \circledast G(e^{j\Omega}) = \frac{1}{2\pi} \int_{-\pi}^{\pi} F(e^{j\eta}) G(e^{j(\Omega-\eta)}) d\eta .
\tag{12.49}
$$

This can be shown by inserting the right side in the definition of the inverse $\mathcal{F}_*$ transform with the help of the modulation theorem (see Exercise 12.10).

── **Example 12.3**

In Example 9.11 we saw that the duration of a signal under measurement determines the resolution of the spectral image of the measured signal. The influence of the finite duration of observation on the measured signal was described by multiplying the signal with a finite window in the time-domain. The spectrum of the measured signal is then obtained from convolution of the original signal with the spectrum of the window. Long measuring windows are advantageous as they correspond to thin spectra (see Figure 9.20).

The discrete case is exactly the same. Trimming a signal to finite length can be described by multiplying it with a rectangle series of the desired length. The spectrum of the observation window approximates the spectrum of the series (12.24), if the rectangle series is very long (see Section 12.3.3.5, Figure 12.8). A longer window of observation leads to a better frequency resolution. ∎

12.5.5   Parseval Relationship

The Parseval relationship can be obtained as a special case of the multiplication property with $g[k] = f^*[k]$:

$$\boxed{\sum_{k=-\infty}^{\infty} |f[k]|^2 = \frac{1}{2\pi} \int_{-\pi}^{\pi} |F(e^{j\Omega})|^2 \, d\Omega \,.}$$

(12.50)

As with continuous-time signals, the Parseval relationship states that the energy of a time signal, defined here by summation of $|f[k]|^2$, can also be found by integrating with respect to $|F(e^{j\Omega})|^2$ in the frequency-domain.

12.5.6   Symmetry Properties of the Discrete-Time Fourier Transform

We define even and odd sequences so that the symmetry axis goes exactly through the element $x[0]$:

$$\text{Even sequences} \qquad x_g[k] = x_g[-k] \tag{12.51}$$
$$\text{Odd sequences} \qquad x_u[k] = -x_u[-k]. \tag{12.52}$$

Accordingly, for odd sequences $x_u[0] = 0$. Every sequence $x[k]$ can be split into an even and an odd part:

$$x_g[k] \;\; = \;\; \frac{1}{2}\left(x[k] + x[-k] \right) \tag{12.53}$$

$$x_u[k] \;\; = \;\; \frac{1}{2}\left(x[k] - x[-k] \right). \tag{12.54}$$

Adding both equations confirms that:

$$x_g[k] + x_u[k] = x[k]. \tag{12.55}$$

Generally, for complex valued signals $x[k]$, the $\mathcal{F}_*$ transform has the same pattern of symmetry as the $\mathcal{F}$ transform (see Chapter 9.5):

$$x[k] \quad = \quad \mathrm{Re}\{x_g[k]\} \quad + \quad \mathrm{Re}\{x_u[k]\} \quad + \quad j\,\mathrm{Im}\{x_g[k]\} \quad + \quad j\,\mathrm{Im}\{x_u[k]\}$$

$$X(e^{j\Omega}) = \mathrm{Re}\{X_g(e^{j\Omega})\} + \mathrm{Re}\{X_u(e^{j\Omega})\} + j\,\mathrm{Im}\{X_g(e^{j\Omega})\} + j\,\mathrm{Im}\{X_u(e^{j\Omega})\}.$$

$$(12.56)$$

Example 12.4

We have already met the transform pair (12.24):

$$x[k] = 1 \;\circ\!\!\!-\!\!\!\bullet\; X(e^{j\Omega}) = \mathrm{III}\left(\frac{\Omega}{2\pi}\right).$$

As $x[k] = 1$ is real and even, we expect a real even $X(e^{j\Omega})$ from scheme (12.56). In fact, $\mathrm{III}\left(\frac{\Omega}{2\pi}\right)$ does turn out to be real and even.

∎

Example 12.5

From (12.56), we expect conjugate symmetrical spectra for discrete-time signals with real values. In particular,

$$\mathrm{Im}\{X(e^{j\Omega})\} = -\mathrm{Im}\{X(e^{-j\Omega})\}.\tag{12.57}$$

At the same time, however, the spectrum is 2π periodic, so

$$\mathrm{Im}\{X(e^{j\Omega})\} = \mathrm{Im}\{X(e^{j(\Omega+2\pi\nu)})\}, \quad \nu \in \mathbb{Z},\tag{12.58}$$

can only fulfill these conditions, for example, if $\Omega = \pi$, for

$$\mathrm{Im}\{X(e^{j\pi})\} = 0.\tag{12.59}$$

The spectrum of a series with real values is therefore real for
$\Omega = \ldots -3\pi, -\pi, \pi, 3\pi, \ldots$. The transform pairs (12.20), (12.33) and (12.39) confirm this.

∎

12.6 Exercises

Exercise 12.1

Regular sampling with interval $T = \dfrac{1}{4}$ of the exponential function $x(t) = e^{(\sigma + j\omega)t}$ gives rise to the exponential series $x[k] = e^{(\Sigma + j\Omega)k}$. Carry out the following steps separately for a) $\omega = 2\pi$ and b) $\omega = 10\pi$ where $\sigma = \ln \dfrac{1}{4}$:

- Give $x_R(t) = \text{Re}\{x(t)\}$ and $x_I(t) = \text{Im}\{x(t)\}$, and sketch both for $t \in [0; 1]$.

- Give Σ and Ω, and mark the sample values $x[k]$ in the sketches of $x(t)$.

Exercise 12.2

Determine the normalised damping constant Σ and normalised angular frequency Ω of the complex exponential series a) e^{-2k}, b) 0.9^k, c) $(-0.9)^k$, d) j^k, e) $\left(\dfrac{1+j}{2}\right)^k$, f) $(-j)^{3k}$, g) j^{5k}, h) j^{9k}. For all parts, choose Ω in the region $[0; 2\pi]$. Which series are the same?

Exercise 12.3

Verify the inverse $\mathcal{F}_*$ transform (12.16) by inserting the defining equation of the $\mathcal{F}_*$ transform (12.13).

Exercise 12.4

Calculate and sketch the $\mathcal{F}_*$ transform of the series $x[k] = \text{si}\left[\dfrac{\pi}{2}k\right]$ using various methods. It will turn out to be the impulse response of an important system.

a) Calculate $\mathcal{F}_*\{x[k]\}$ with the defining equation (12.13).

b) First calculate $X_a(j\omega) = \mathcal{F}\{x_a(t)\}$, with $x_a(t) = \sum \delta(t - kT)\, x[k]$ and then give $\mathcal{F}_*\{x[k]\}$, by using (12.43). Sketch the result.

c) Show that results a) and b) agree, by finding the Fourier series of $X(e^{j\Omega})$. Note that in this case a fundamental period t_0 must be used, instead of a fundamental angular frequency ω_0, as it is a *function of frequency* that you will be representing as a Fourier series.

d) What kind of filter has the impulse response $x[k]$?

Exercise 12.5

A low-pass filter with impulse response $h_1[k]$ has spectrum $H_1(e^{j\Omega})$. One period has been sketched. By reversing every second sign of $h_1[k]$, a new filter is created with $h_2[k] = (-1)^k h_1[k]$. Calculate and sketch $H_2(e^{j\Omega})$.

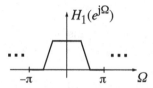

What kind of filter is $h_2[k]$?

Exercise 12.6

In the system shown below, the low-pass filtering of a continuous-time signal is carried out with the help of a discrete low-pass filter. The input signal $x(t)$ is given by its spectrum $X(j\omega)$.

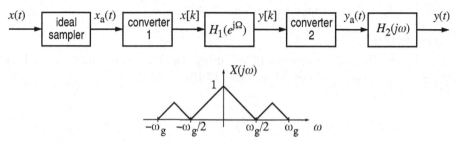

Converter 1 transforms the continuous-time delta impulse train $x_a(t)$ into a discrete-time series $x[k]$, where the values of the series $x[k]$ are the weightings of the delta impulses $(x[k] = x(kT))$. Converter 2 converts the discrete-time series $y[k]$ into the continuous-time signal $y_a(t)$ in the same way.
The signal $x(t)$ is sampled at the Nyquist frequency.

a) Give T and draw the spectrum of $x_a(t) = x(t) \cdot \frac{1}{T}\, \text{⊔⊔⊔}(\frac{t}{T})$.

b) Draw $X(e^{j\Omega}) = \mathcal{F}_*\{x[k]\}$.

c) If $Y(e^{j\Omega}) = \mathcal{F}_*\{y[k]\}$:

$$Y(e^{j\Omega}) = 0 \quad \text{for} \quad \frac{\pi}{2} < |\Omega| \le \pi$$

$$Y(e^{j\Omega}) = T \cdot X(e^{j\Omega}) \quad \text{for} \quad |\Omega| \le \frac{\pi}{2}$$

Determine and draw $H_1(e^{j\Omega})$. Draw $Y_a(j\omega) \bullet\!\!-\!\!\circ y_a(t)$.

d) $H_2(j\omega)$ is a RC-low-pass Nth- order filter with corner frequency $\frac{3}{4}\omega_g$, and the pass section being amplified by a factor of 1. Give $H_2(j\omega)$ and determine N so that spectral repetition of the baseband (where $|\omega| > \omega_g$) is damped by at least 18 dB. Note: see Chapter 10.

Exercise 12.7

Show (12.23) by using the inverse $\mathcal{F}_*$ transform on $\underline{\text{III}}\left(\frac{\Omega-\Omega_0}{2\pi}\right)$.

Exercise 12.8

Show by using the defining equation (12.13) of the $\mathcal{F}_*$ transform,

a) the shift theorem,

b) the modulation theorem.

Exercise 12.9

Show the convolution theorem (12.48). You will need to use the convolution sum for $f[k] * g[k]$ and the defining equation (12.13) of the $\mathcal{F}_*$.

Exercise 12.10

Show the multiplication theorem (12.49), using the cyclic convolution integral and inverse transformation of the result with (12.16).

13 The z-Transform

In Chapter 12 we got to know discrete-time signals $x[k]$ and their spectra $X(e^{j\Omega})$ in the form of the discrete-time Fourier transform $\mathcal{F}_*\{x[k]\}$ from (12.13). The correspondence between the $\mathcal{F}_*$ transform for discrete-time signals and the Fourier transform $\mathcal{F}\{x(t)\}$ for continuous-time signals can be expressed, for example, by the relationship (12.43). For continuous-time signals, however, we also know the Laplace transform $\mathcal{L}\{x(t)\}$, which assigns a function $X(s)\bullet\!\!-\!\!\circ x(t)$ of the complex frequency variable s to the time-signal $x(t)$. A comparable transformation for discrete-time signals is the z-transform. It is (clearly) not named after a famous mathematician, but instead after the letter normally used for its complex frequency variable: z.

Its discussion in this chapter will deal with the same topics as in Chapter 4, when we discussed the Laplace transform. From the definition of the z-transform, we first of all find the relationship between the z-transform and the Fourier transform, and then the relationship between the z-transform and the Laplace transform. After that, we consider convergence, and the properties of the z-transform and inverse z-transform.

13.1 Definition and Examples

13.1.1 Definition of the Bilateral z-Transform

The general definition of the z-transform can be used with a bilateral sequence $x[k]$ where $-\infty < k < \infty$. It is

$$\boxed{X(z) = \mathcal{Z}\{x[k]\} = \sum_{k=-\infty}^{\infty} x[k]z^{-k}; \ z \in \text{ROC} \subset \mathbb{C}.}\tag{13.1}$$

It represents a sequence $x[k]$, which may have complex elements, by a complex function $X(z)$ in the complex z-plane. The infinite sum in (13.1) usually only converges for certain values of z, the region of convergence.

We can think of (13.1) in two ways: by comparison with the Laurent series of a function of a complex argument (see (4.15)), we recognise that the values of the sequence $x[k]$ represent the coefficients of the z-transform's Laurent series at the point $z_0 = 0$. We can therefore take many properties of the z-transform directly from the known properties of the Laurent series.

For the second interpretation we need a result from Chapter 14. There we will see that the sequence z^k is a characteristic sequence of a discrete-time LTI-system. It is the discrete counterpart to the eigenfunctions e^{st} of the continuous-time LTI-system from Chapter 3.2. The z-transform projects a discrete-time signal onto the characteristic series of an LTI-system. We will see that we can therefore form the inverse z-transform by overlapping the characteristic sequences of which the signal is made up.

The transform pairs of a sequence $x[k]$ and its z-transform $X(z)$ is again written with the familiar symbol

$$x[k]\circ\!\!-\!\!\bullet X(z)\,.$$

As we discussed in Chapter 9.2.1, it has no mathematically strict meaning, and is used simply as a typographical symbol.

13.1.2   Examples of the z-Transform

We will now show the properties of the z-transform with some simple examples, paying particular attention to the convergence properties of the sum in (13.1).

———————————————————————————————————— **Example 13.1**

We begin by calculating the z-transform of a sequence of finite length from Figure 13.1:

$$x[k] = \begin{cases} 3 & k = 0 \\ 2 & |k| = 1 \\ 1 & |k| = 2 \\ 0 & \text{otherwise} \end{cases} \tag{13.2}$$

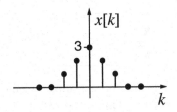

Figure 13.1: Discrete triangular sequence $x[k]$ from Example 13.1

As only a few values of the series are not zero, we can write down the sum in (13.1) directly

$$X(z) = z^2 + 2z + 3 + 2z^{-1} + z^{-2} = \frac{z^4 + 2z^3 + 3z^2 + 2z + 1}{z^2} \quad \text{where} \quad 0 < |z| < \infty.$$
(13.3)

In the rational representation of the function of z, we obtain a denominator polynomial with a double zero at $z = 0$. As the sum in (13.1) only has a finite number of terms, it remains finite for all $0 < |z| < \infty$. The region of convergence of the z-transform $X(z)$ therefore encloses the entire complex plane with the exception of the origin.

∎

Example 13.2

We have already met exponential sequences in Chapter 12.2.3. We will now find the z-transform of a general right-sided exponential sequence

$$x[k] = a^k \cdot \varepsilon[k], \quad a \in \mathbb{C}.$$
(13.4)

Figure 13.2 shows its behaviour for $a = 0.9\, e^{j\pi/6}$ (compare Figure 12.4).

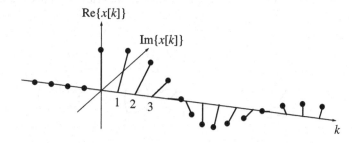

Figure 13.2: Example of a right-sided exponential series

From the defining equation (13.1) we obtain

$$X(z) = \sum_{k=0}^{\infty} a^k z^{-k} = \sum_{k=0}^{\infty} \left(\frac{a}{z}\right)^k = \frac{1}{1 - az^{-1}} = \frac{z}{z - a}.$$
(13.5)

The infinite sum only converges for $|a| < |z|$, which means all values of z in the complex plane that lie outside a circle with radius $|a|$. This region of convergence is the hatched area shown in Fig 13.3. The z-transform has a zero at $z = 0$ and a pole at $z = a$. The circular boundary of the region of convergence is determined by the magnitude of the pole.

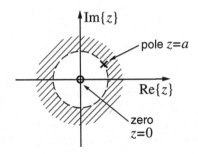

Figure 13.3: Region of convergence of the function $X(z)$ from Example 13.2

For $a = 1$ the sequence $x[k]$ represents the unit step function $\varepsilon[k]$ from (12.6). Its z-transform is

$$\mathcal{Z}\{\varepsilon[k]\} = \frac{z}{z - 1}, \quad |z| > 1 \tag{13.6}$$

■

――――――――――――――――――――――――――――――――――――――― **Example 13.3**

The z-transform of the a left-sided exponential sequence from Figure 13.4

$$x[k] = -a^k \cdot \varepsilon[-k - 1], \quad a \in \mathbb{C} \tag{13.7}$$

is

$$X(z) = \sum_{k=-\infty}^{-1} -a^k z^{-k} = -\sum_{k=-\infty}^{-1} \left(\frac{a}{z}\right)^k = -\sum_{k=1}^{\infty} \left(\frac{z}{a}\right)^k = 1 - \sum_{k=0}^{\infty} \left(\frac{z}{a}\right)^k =$$

$$= 1 - \frac{1}{1 - za^{-1}} = 1 - \frac{a}{a - z} = \frac{z}{z - a}. \tag{13.8}$$

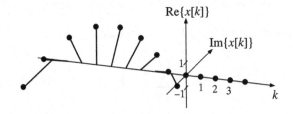

Figure 13.4: Example of a left-sided exponential sequence

This sum converges for $|z| < |a|$, which means all values of z in the complex plane that lie within a circle of radius a (see Figure 13.5). The boundary of the

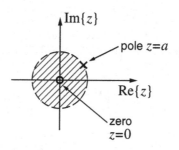

Figure 13.5: Region of convergence for the function $X(z)$ from Example 13.3

region of convergence is circular here too, and its radius is equal to the magnitude of the pole.

In comparison to Example 13.2, we notice that the z-transform of the right-sided exponential sequence (13.4) and the left-sided exponential sequence (13.7) have the same form and only the regions of convergence are different (Figures 13.3 and 13.5). This emphasizes how necessary it is to specify the region of convergence. Without it, a unique inverse transformation is impossible.

We are familiar with this situation from the Laplace transform. In Examples 4.1 and 4.2 we considered left-sided and right-sided continuous-time signals that are likewise only distinguished by the region of convergence (see Figures 4.3 and 4.4).

13.1.3   Illustrative Interpretation of the z-Plane

We can interpret the individual points of the z-plane in a similar way to the s-plane in Chapter 3.1.3. Figure 13.6 shows the corresponding exponential sequence z^k for different values of z.

The values $z = e^{j\Omega}$ on the unit circle correspond to the exponential series $e^{j\Omega k}$ with constant amplitude: $z = 1$ leads to a series with constant values because $e^{j0k} = 1^k = 1$, while $z = -1$ is the highest representable frequency, because $e^{j\pi k} = (-1)^k$. All other values on the unit circle represent complex exponential oscillations of frequency Ω with $-\pi < \Omega < \pi$. Complex conjugate values of z are distinguished by the direction of rotation. Values of $z = re^{j\Omega}$ within the unit circle ($r < 1$) belong to a decaying exponential sequence and values outside the unit circle ($r > 1$) belong to a growing exponential sequence. In Figure 13.6 the exponential series z^k are each only shown for $k \geq 0$. This should not lead to the misinterpretation that we are only dealing with unilateral sequences, since also for $k < 0$, $z^k \neq 0$.

The reader is recommended to memorise the illustration of the z-plane (Figure 13.6). It can be the key to intuitive understanding of the properties of the z-transform and the system function of discrete LTI-systems.

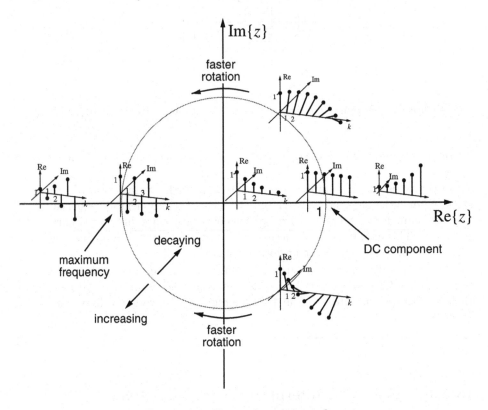

Figure 13.6: Illustration of the z-plane

13.2   Region of Convergence of the z-Transform

We can draw up rules for the region of convergence of the z-transform that are very similar to the corresponding rules for the Laplace transform in Chapter 4.5.3, where we explained the properties of the Laplace transform's region of convergence with a series of examples (Examples 4.1 to 4.5). We could also do this in the case of the z-transform; Examples 13.2 and 13.3 actually correspond to Examples 4.1 and 4.2 respectively. We will, however, forgo continuation of the examples for discrete-time functions, and will just give the rules for the region of convergence in a general form. They each refer to a sequence $x[k]$ and its z-transform $X(z)$ from (13.1).

1. **The region of convergence $X(z)$ in general consists of a circle around the z-plane origin at $z = 0$.**

   As only the magnitude of z is responsible for the convergence of a z-transform, all points on the z-plane with the same magnitude have the same convergence properties. This result is also yielded by the known convergence properties of the Laurent series.

2. **If $x[k]$ is a right-sided signal, the region of convergence lies outside a circle through the singularity furthest from the origin.**

   In Example 13.2 we saw that for a right-sided exponential series, the magnitude of the pole defines the region of convergence. Similarly, for right-sided series with multiple singularities, it is the singularity that lies furthest from the origin. The circle itself is not part of the region of convergence.

3. **If $x[k]$ is a left-sided signal, the region of convergence lies within a circle through the nearest singularity to the origin.**

   The magnitude of the pole also defines the region of convergence for the left-sided exponential series from Example 13.3, although in this case it lies within a circle through the pole. For left-sided series with multiple singularities this is true for the singularity nearest to the origin. The border itself is not part of the region of convergence.

4. **If $x[k]$ is bilateral, i.e. the sum of a left-sided and a right-sided series, then the region of convergence is an annulus between two singularities, if the left-sided and the right-sided regions of convergence overlap.**

   Every bilateral sequence can be formed by putting together a left-sided and a right-sided part. The individual regions of convergence for the z-transforms of each part can be found from rules 2 and 3. For a bilateral sequence the region of convergence consists of the intersection of the left- and right-sided parts. This intersection is an annulus whose inside border is set by the outermost singularity of the right-sided term, and whose outside border is set by the innermost singularity of the left-sided term. If singularities from the right-sided part lie outside singularities from the left-sided part, the intersection is empty, and therefore the z-transform does not converge. In this case we say the the z-transform does not exist.

5. **$X(z)$ is analytic in the entire region of convergence.** As the circular region of convergence is bounded on the inside by the singularities of the right-sided part, and on the outside by the singularities of the left-sided part, it contains no singularities itself. The z-transform is analytic in the entire region of convergence (it is also said to be *regular*, or *holomorphic*).

This means that it can be repeatedly differentiated, and can be interpreted as a Laurent series.

6. **If the sequence $x[k]$ is of finite length, $X(z)$ converges in the entire z-plane, except possibly for $z = 0$ and $z \to \infty$.**

In Example 13.1 we saw that a sum of a finite number of terms yields a finite number. A precondition of this is that all members of the series $x[k]$ are finite: $|x[k]| < \infty$ and $0 < |z| < \infty$. The convergence for $z = 0$ and $z = \infty$ depends on whether the series $x[k]$ contains non-zero values for $k > 0$ and $k < 0$.

13.3 Relationships to Other Transformations

Close relationships exist between the z-transform and other transforms we already know.

- The z-transform and the Fourier transform $\mathcal{F}_*$ of a series are closely connected, like the Laplace transform and the Fourier transform $\mathcal{F}$ of a continuous-time signal (see Chapter 9.3).

- If a sequence $x[k]$ has been created by sampling a continuous-time signal $x(t)$, the values of $x[k]$ can be interpreted as the weightings of individual impulses in an impulse train (see Figure 12.9). Between the Laplace transform $X(s) = \mathcal{L}\{x(t)\}$ and the z-transform $X(z) = \mathcal{Z}\{x[k]\}$ there must likewise be a close connection.

We will now examine more precisely these relationships between the individual transforms.

13.3.1 The z-Transform and Discrete-Time Fourier Transform

When the z-transform of a sequence $x[k]$ is only calculated for values z on the unit circle

$$z = e^{j\Omega}, \qquad \Omega \in \mathbb{R}, \tag{13.9}$$

the Fourier transform of this sequence is immediately obtained. By comparing the instructions for calculating the z-transform and the Fourier transform we recognise that

$$\mathcal{Z}\{x[k]\} = \sum_k x[k]z^{-k} \qquad \mathcal{F}_*\{x[k]\} = \sum_k x[k]e^{-j\Omega k}. \tag{13.10}$$

We have now found the relationship between the two transformations

$$\boxed{\mathcal{F}_*\{x[k]\} = \mathcal{Z}\{x[k]\}\,|_{z=e^{j\Omega}}.} \tag{13.11}$$

This relationship corresponds to the relationship between the Fourier transform of a continuous signal and its Laplace transform (9.9). It also becomes clear why the Fourier transform $X(e^{j\Omega})$ is not simply defined on the real axis Ω, but instead on the unit circle $e^{j\Omega}$. For the discrete-time Fourier transform, the unit circle of the z-plane is the equivalent of the imaginary axis of the s-plane for the continuous-time transform.

This relationship can be generalised if we consider the circle of radius r described by

$$z = re^{j\Omega}, \qquad 0 < r < \infty \tag{13.12}$$

instead of the unit circle on the z-plane. For all values of z in (13.12), the z-transform of $x[k]$ can be given as the Fourier transform of the series $x[k]r^{-k}$:

$$X(z) = X(re^{j\Omega}) = \sum_k x[k](re^{j\Omega})^{-k} = \sum_k x[k]r^{-k}e^{-j\Omega k} . \tag{13.13}$$

This yields therefore

$$\mathcal{F}_*\{x[k]r^{-k}\} = \mathcal{Z}\{x[k]\}\,|_{z=re^{j\Omega}} . \tag{13.14}$$

The relationships (13.11) through to (13.14) only have meaning if, of course, the Fourier transforms of $x[k]$ and $x[k]r^{-k}$ exist, and the unit circle of the z-plane or circle with radius r, belong to the region of convergence of the z-transform. If the unit circle does not belong to the region of convergence of the z-tranform, that does not necessarily mean that the Fourier transform does not exist, instead that the Fourier transform is not analytic. There are actually even sequences that have a Fourier transform but not a z-transform, for example, $x[k] = 1$. If on the other hand, the z-transform convergences on the unit circle, the Fourier transform $X(e^{j\Omega})$ can be analytically continued when $e^{j\Omega}$ is replaced by z. The observations made about Fourier and Laplace transforms in Chapter 9.3 also apply in this case.

Example 13.4

What is the z-transform of the series which has Fourier transform

$$X(e^{j\Omega}) = 1 + \cos\Omega\,?$$

$X(e^{j\Omega})$ can be differentiatied with respect to Ω any number of times, and is therefore analytic, when we replace $e^{j\Omega}$ by z. With

$$X(e^{j\Omega}) = 1 + \cos\Omega = 1 + \frac{1}{2}e^{j\Omega} + \frac{1}{2}e^{-j\Omega}$$

we obtain

$$X(z) = 1 + \frac{1}{2}z + \frac{1}{2}z^{-1} .$$

The region of convergence is the whole z-plane excepting $z = 0$, $z = \infty$, and we must therefore be dealing with a series of finite length. Equating coefficients (13.1) yields directly

$$
x[k] = \begin{cases} \dfrac{1}{2} & k = -1 \\ 1 & k = 0 \\ \dfrac{1}{2} & k = 1 \\ 0 & \text{otherwise} \end{cases}.
$$

■

13.3.2 The z-Transform and Laplace Transform

The relationship between a sampled continuous-time signal $x_a(t)$ and the series of sample values $x[k]$ from (12.2) and Figure 12.9 is

$$
x_a(t) = \sum_k x[k]\delta(t - kT). \tag{13.15}
$$

T is the sampling interval. The Laplace transform is found with the help of the selective property of the delta impulse

$$
\mathcal{L}\{x_a(t)\} = \int_{-\infty}^{\infty} \sum_k x[k]\delta(t - kT)e^{-st}dt = \sum_k x[k]e^{-skT} =
$$
$$
= \sum_k x[k]z^{-k} = \mathcal{Z}\{x[k]\} \quad \text{with} \quad z = e^{sT}, \tag{13.16}
$$

or concisely

$$
X_a(s) = X(e^{sT}). \tag{13.17}
$$

The z-transform $X(z)$ of the sequence $x[k]$ is therefore the same as the Laplace transform of the sampled continuous-time signal $x_a(t)$ for

$$
z = e^{sT}. \tag{13.18}
$$

In order to better understand this relationship between the complex frequency s of a continuous-time signal and the complex frequency z of a series, we will represent it as a projection from the s-plane onto the z-plane. Figure 13.7 illustrates the geometric relationship for a Laplace transform which contains the imaginary axis. The assignment (13.18) represents the:

- imaginary axis $s = j\omega$ of the s-plane as the unit circle $z = e^{j\Omega}$ of the z-plane,

- the left half of the s-plane as the area inside the unit circle on the z-plane,

- the right half of the s-plane as the area outside the unit circle on the z-plane.

If the region of convergence of the Laplace transform contains the imaginary axis of the s-plane (as shown in Figure 13.7), then the region of convergence of the z-plane contains the unit circle.

The way the imaginary axis of the s-plane is projected onto the z-plane as the unit circle is of particular interest. This projection obviously cannot be inverted, as all of the points

$$s_\mu = j\mu\omega_a \text{ with } \omega_a = \frac{2\pi}{T}, \quad \mu \in \mathbb{Z}$$

of the s-plane would be projected onto the point

$$z_\mu = e^{s_\mu T} = e^{j\mu\omega_a T} = e^{j\mu 2\pi} = 1, \quad \forall\,\mu$$

of the z-plane. Correspondingly the points

$$s_\nu = j(\nu\omega_a + \omega_0)$$

would be projected onto

$$z_\nu = e^{s_\nu T} = e^{j(\nu\omega_a T + \omega_0 T)} = e^{j\omega_0 T}, \quad \forall\,\nu.$$

The frequency $\omega_a = 2\pi/T$ is just the angular sampling frequency (see Chapter 11.3.1).

Through the projection (13.18), the imaginary axis $s = j\omega$ therefore becomes transformed to the unit circle of the z-plane, where the circumference of the unit circle corresponds to a section of the imaginary axis which has the same length as the sampling frequency ω_a. The central statement of the sampling theorem can be constructed directly from the definition of the projection (13.18).

The properties of the projection (13.18), projecting a vertical line in the s-plane onto a circle in the z-plane are the same for the imaginary axis as for any straight line parallel to it. From $s = \sigma_0 + j\omega$ with a fixed value of σ_0, it follows with (13.18) that

$$z = e^{(\sigma_0 + j\omega)T} = re^{j\omega T} \qquad \text{where} \quad r = e^{\sigma_0 T}. \tag{13.19}$$

Variation of the values of ω describes z (13.19) as a circle about the origin with radius r. Vertical lines in the left half of the s-plane are projected into the inside of the unit circle of the z-plane (see Figure 13.7) and lines in the right half are projected to the outside of the unit circle.

The Laplace transform of the sampled continuous-time signal (13.15) is periodic with the sampling frequency $\omega_a = 2\pi/T$ in ω, as $e^{-(s+m\omega_a)kT} = e^{-sT}$ for $m \in \mathbb{Z}$ in (13.16). The values of $X_a(s)$ within two adjacent dashed lines in Figure 13.7 are therefore repeated above and below. After the transform $z = e^{sT}$ (13.18) every horizontal strip covers the whole z-plane.

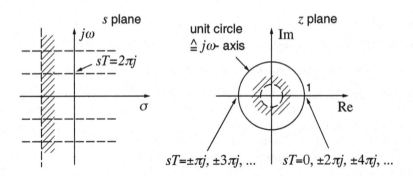

Figure 13.7: The relationship between the z-transform and the Laplace transform of a sampled signal. The s-plane has been projected onto the z-plane with $z = e^{sT}$.

─── **Example 13.5**

The z-transform of the discrete unit step function $\varepsilon[k]$ (13.6) is

$$\mathcal{Z}\{\varepsilon[k]\} = \frac{z}{z-1}\,, \quad |z| > 1\,.$$

We now arrive directly at the Laplace transform of the continuous-time signal

$$x_a(t) = \sum_{k=0}^{\infty} \delta(t-k) = \varepsilon\left(t + \frac{1}{2}\right) \cdot \mathop{\text{⊥⊥⊥}}(t)$$

by setting $z = e^s$.

$$X_a(s) \;=\; \frac{e^s}{e^s - 1}\,, \quad |e^s| > 1$$

$$\text{or} \quad X_a(s) \;=\; \frac{1}{1 - e^{-s}}\,, \quad \text{Re}\{s\} > 0$$

■

13.4   Theorems of the z-Transform

The same holds for practical application of the z-transform as for the Laplace transform, which we have already covered. Calculating the z-transform of a sequence by evaluating the summation formula (13.1) only leads to (13.3) for very simple sequences $x[k]$ (compare Example 13.1). For other cases, we try to refer to these simple series by using some general rules. These general rules will be summarised by a series of theorems that have a lot of similarities with the theorems of the Laplace transform (see Chapter 4.7). That is no coincidence, as we saw in Section 13.3.2 that the z-transform of a series can be thought of as a sampled continuous-time signal consisting purely of delta impulses. It should be clear that many properties of the Laplace transform also apply to the z-transform.

The most important theorems of the z-transform are summarised in Table 13.1. They can also be shown without referring to the Laplace transform by inserting them into (13.1). In contrast to the theorems of the Laplace transform, the independent variable is defined only for integer values. In the square brackets [ ], no real values are permitted, and correspondingly, for the shift theorem, only integer shifts $\kappa \in \mathbb{Z}$ are allowed. As for the similarity theorem (4.24), for sampled signals the only scaling of the time axis that is permitted is $a = -1$, so that becomes the time reversal theorem for the z-transform. Reversing the index of a series of values can be done simply by reading them backwards.

Table 13.1: Theorems of the z-Transform

Theorem	Time-domain	z-domain	New ROC
Linearity	$ax[k] + by[k]$	$aX(z) + bY(z)$	ROC$\supseteq$ ROC$\{x\}\cap$ ROC$\{y\}$
Shift	$x[k - \kappa]$	$z^{-\kappa}X(z)$	ROC$\{x\}$; $z = 0$ and $z \to \infty$ considered separately
Modulation	$a^k x[k]$	$X\left(\dfrac{z}{a}\right)$	ROC $= \left\{z \left\vert \dfrac{z}{a} \in \text{ROC}\{x\}\right.\right\}$
Multiplication by k	$kx[k]$	$-z\dfrac{dX(z)}{dz}$	ROC$\{x\}$; $z = 0$ considered separately
Time reversal	$x[-k]$	$X(z^{-1})$	ROC $= \left\{z \left\vert z^{-1} \in \text{ROC}\{x\}\right.\right\}$

Example 13.6

As an example we will show how the shift theorem of the z-transform can be derived from the shift theorem of the Laplace transform (4.22)

$$\mathcal{L}\{x_a(t-\tau)\} = e^{-s\tau}X_a(s)\,, \quad s \in \mathrm{ROC}\{x\}\,.$$

We make $x_a(t)$ a sampled continuous-time signal, that consists of delta impulses with separation $T = 1$ (13.15). As we want to fix the location of the sample points in time, only displacements of $\tau \in \mathbb{Z}$ are permitted. The z-transform $X(z)$ of the series $x[k]$ that corresponds to $x_a(t)$ can be obtained with (13.17) as $X(z) = X_a(s)|_{s=\ln z}$, and it then follows that

$$\mathcal{Z}\{x[k-\tau]\} = e^{-(\ln z)\tau}X_a(\ln z) = z^{-\tau}X(z)\,.$$

The region of convergence of this z-transform might only change at the points $z = 0$ and $z = \infty$. Of course, the shift theorem can be just as easily shown by putting $x[k-\tau]$ into the definition of the z-transform:

$$\begin{aligned}
\mathcal{Z}\{x[k-\tau]\} &= \sum_k x[k-\tau]z^{-k} &= \sum_\ell x[\ell]z^{-(\ell+\tau)} \\
&= z^{-\tau}\sum_\ell x[\ell]z^{-\ell} &= z^{-\tau}\mathcal{Z}\{x[k]\} \quad.
\end{aligned}$$

Note that delaying a discrete-time signal by one sample corresponds to multiplying by z^{-1}. This makes z^{-1} the system function of a simple delay circuit and the most important and elementary transform for the analysis of discrete-time systems, which we will deal with in Chapter 14.

13.5   Inverse z-Transform

In Section 13.1.1 we interpreted the z-transform $X(z)$ as a Laurent series with coefficients given by the values of the sequence $x[k]$. To invert the z-transform $x[k] = \mathcal{Z}^{-1}\{X(z)\}$, we can therefore use the formula that calculates the coefficients of a Laurent series:

$$x[k] = \frac{1}{2\pi j}\oint X(z)z^{k-1}dz =: \mathcal{Z}^{-1}\{X(z)\}\,. \tag{13.20}$$

The integration follows a closed path around the origin, in a mathematically positive direction. The path of integration must both enclose the origin and stay in the region of convergence of $X(z)$. If the series $x[k]$ has finite length, it is very simple to find (13.20) from the Cauchy integral (5.13) and the resulting residue calculation (Chapter 5.4.1). In the expression

$$X(z)z^{k-1} = \sum_{\kappa} x[\kappa]z^{-\kappa}z^{k-1} = \sum_{\kappa} x[\kappa]z^{-\kappa+k-1}$$

every term within the path of integration is analytic with the exception of poles at $z = 0$. From (5.21) it follows that only the simplest pole $x[\kappa]z^{-1}$ for $\kappa = k$ gives a residuum, and for all other terms the ring integral is zero:

$$\oint X(z)z^{k-1}dz = \oint \sum_{\kappa} x[\kappa]z^{-\kappa}z^{k-1}dz = \sum_{\kappa} \oint x[\kappa]z^{-\kappa+k-1}dz$$

$$= \oint x[k]z^{-1}dz = 2\pi j x[k] \qquad (13.21)$$

For a general bilateral series $x[k]$ with singularities within the path of integration, $x[k]$ must be split into left-sided and right-sided series and with suitable substitution $z \to z^{-1}$ the conditions for (13.21) must be met. Then (13.20) can also be shown generally.

Similar to the inverse form of the Laplace transform (Chapter 4.2), we can interpret (13.20) as a superposition of complex exponential series. In Chapter 14.4 we will show that complex exponential series are characteristic sequences of discrete systems and that the z-transform is therefore particularly suitable for investigating and describing such systems.

For the case where the region of convergence encloses the unit circle of the z-plane, an inverse formula can be created by parametricising the unit circle with

$$z = e^{j\Omega}, \quad -\pi \leq \Omega < \pi \qquad (13.22)$$

which only requires one integration of a real variable. From (13.22), with differentiation, we obtain

$$dz = je^{j\Omega}d\Omega. \qquad (13.23)$$

Substitution of z into (13.20) then yields

$$x[k] = \frac{1}{2\pi j} \oint X(e^{j\Omega})e^{j\Omega(k-1)}dz = \frac{1}{2\pi} \int_{-\pi}^{\pi} X(e^{j\Omega})e^{j\Omega k}d\Omega. \qquad (13.24)$$

This is exactly the same as the inverse formula for the Fourier transform (12.16)! But since we find the Fourier transform of a series on the unit circle of the z-plane, it is of course no longer surprising.

Just as with the inverse Laplace transform, we avoid evaluating an integral expression for the inverse z-transform if there are simpler ways of recovering the series $x[k]$. These methods exist for rational fraction z-transforms, which are the kind we most often encounter, and they are

- for sequences of finite length, assigning each term in the sequence a power of z (compare Example 13.1),

- for sequences of infinite length, decomposing the z-transform into known expressions, for example, with partial fraction expansion.

It would be useful to know from looking at a z-transform, whether it corresponds to a finite or infinite series. We put $X(z)$ in rational fraction form and consider the denominator polynomial:

- If the denominator polynomial of Nth-order is z^N (it only has one zero N times at $z = 0$), then $X(z)$ can be written as a Laurent series with a finite number of terms. The values of $x[k]$ can be read directly.

- If the denominator polynomial is a general polynomial in z, where, apart from z^N there are also lesser powers, then it also has zeros outside the origin of the z-plane. Partial fraction expansion for these zeros yields a sum of terms similar to (13.5), where each has an exponential sequence The exact form of the exponential sequence depends on the region of convergence of $X(z)$ (compare Examples 13.2 and 13.3).

We will deal with both cases with an example.

── **Example 13.7**

The z-transform
$$X(z) = \frac{3z^4 + 5z^2 + 3z - 5}{z^2} \tag{13.25}$$

only has a double pole at $z = 0$ and no others. The corresponding series $x[k]$ can therefore only have a finite number of non-zero values. By dividing we obtain

$$X(z) = \sum_{k=-\infty}^{\infty} x[k]z^{-k} = 3z^2 + 5 + 3z^{-1} - 5z^{-2}. \tag{13.26}$$

The coefficients can be read directly (compare Example 13.1):

$$x[k] = \mathcal{Z}^{-1}\left\{X(z)\right\} = \begin{cases} 3 & k = -2 \\ 5 & k = 0 \\ 3 & k = 1 \\ -5 & k = 2 \\ 0 & \text{otherwise} \end{cases} \tag{13.27}$$

Figure 13.8 shows the series $x[k]$ obtained from $X(z)$. ∎

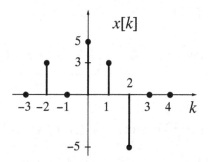

Figure 13.8: Series $x[k]$ obtained from $X(z)$ in Example 13.7

── **Example 13.8**

The z-transform

$$X(z) = \frac{z+1}{z^2 - 2.5z + 1}, \quad |z| > 2 \tag{13.28}$$

has two numerator zeros (poles) outside the origin of the complex plane; $z^2 - 2.5z + 1 = 0$ gives $z_{p1} = \frac{1}{2}$, $z_{p2} = 2$. As the region of convergence lies *outside* a circle around the origin, we expect a right-sided series $x[k] = \mathcal{Z}^{-1}\{X(z)\}$. If it is possible to express $X(z)$ as the sum of two terms (like (13.5)), we can assign each term an exponential oscillation. Direct partial fraction expansion does not lead to the goal as it produces a sum of the form

$$X(z) = A_0 + \frac{A_1}{z - z_{p1}} + \frac{A_2}{z - z_{p2}}, \tag{13.29}$$

while we need an expansion

$$X(z) = B_0 + B_1 \frac{z}{z - z_{p1}} + B_2 \frac{z}{z - z_{p2}} \tag{13.30}$$

to determine the exponential terms. Here it helps to expand $X(z)/z$ into partial fractions, instead of $X(z)$.

$$\frac{X(z)}{z} = \frac{z+1}{z(z - \frac{1}{2})(z - 2)} = \frac{B_0}{z} + \frac{B_1}{z - \frac{1}{2}} + \frac{B_2}{z - 2} \tag{13.31}$$

Determining the partial fraction coefficients, for example, by equating coefficients, leads to $B_0 = 1$, $B_1 = -2$ and $B_2 = 1$. Multiplying by z and term-by-term inversion aided by Example 13.2 yields the series

$$X(z) = 1 - \frac{2z}{z - \frac{1}{2}} + \frac{z}{z - 2} \tag{13.32}$$

$$x[k] \quad = \quad \delta[k] - 2\left(\frac{1}{2}\right)^{k} \varepsilon[k] + 2^{k}\varepsilon[k] . \tag{13.33}$$

13.6   Pole-Zero Diagrams in the z-Plane

As with the Laplace transform of continuous signals, with the z-transform of discrete signals, rational fractions often appear. They correspond to system functions of discrete LTI-systems with a finite number of internal state-variables, which we will discuss in detail in the next chapter. Rational fraction z-transforms are – except for a scaling factor – completely defined by their poles and zeros and can be represented by a pole-zero diagram in the z-plane. As we have already mentioned, the z-transforms of signals of finite duration only have poles at the origin of the z-plane, while signals of infinite duration can have poles at any other points.

Example 13.9

The pole-zero diagram of the z-transform from Example 13.7 has a double pole at the origin and two complex conjugate zeros that are shown in Figure 13.9.

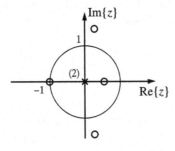

Figure 13.9: Pole-zero diagram for (13.25)

Example 13.10

The pole-zero diagram of the z-transform from Example 13.8 has two real poles and one real zero (Figure 13.10).

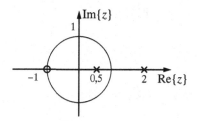

Figure 13.10: Pole-zero diagram for (13.28)

With the rules in Chapter 10 (Bode Diagrams) the magnitude and phase of the Fourier transform can be estimated from the pole-zero diagram of the z-transform after some practice. Unlike Chapter 10, however, the unit circle of the z-plane is used and not the imaginary axis of the s-plane, and it is the angle and distance to the poles and zeros that must be determined. Otherwise the procedure is the same.

Example 13.11

We want to find the frequency response of a system with the pole-zero diagram shown in Figure 13.10. We can find it from the pole-zero diagram up to a factor K.

Ω	$K \cdot \lvert X(e^{j\Omega}) \rvert$
0	$\dfrac{2}{1/2 \times 1} = 4$
$\pm\dfrac{\pi}{4}$	$\dfrac{1.8}{0.7 \times 1.5} \approx 1.7$
$\pm\dfrac{\pi}{2}$	$\dfrac{1.4}{1.1 \times 2.2} \approx 0.6$
$\pm\dfrac{3\pi}{4}$	$\dfrac{0.7}{1.5 \times 3} \approx 0.15$
π	0

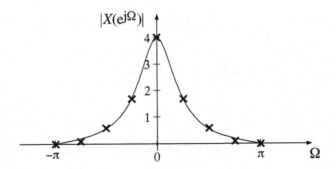

$|X(e^{j\Omega})|$

Figure 13.11: Exact frequency response to (13.28) and the estimate from the pole-zero diagram

Substituting $z = 1$ into (13.28) gives $K = -1$. The estimates are shown in Figure 13.11 with the exact result for comparison. Zeros lying directly on the unit circle force the magnitude of the Fourier transform at this frequency to zero, and poles lying near to the unit circle increase the resonant peak. The nearer to the unit circle that a pole or zero lies, the greater effect it has on the Fourier transform. Poles or zeros at the origin of the z-plane have no influence on the magnitude of the Fourier transform, but they do make a linear contribution towards the phase. ∎

13.7 Exercises

Exercise 13.1

Evaluate the z-transforms of

$x_1[k] = \delta[k-3] - 4\delta[k-2] + 6\delta[k-1] - 4\delta[k] + \delta[k+1]$
$x_2[k] = e^{-ak}\varepsilon[k-2], \quad a \in \mathbb{C}$
$x_3[k] = \begin{cases} (-0.8)^{|k|} & \text{for } |k| \leq 10 \\ 0 & \text{otherwise} \end{cases}$

using the defining equation (13.1).

Exercise 13.2

Evaluate the z-transforms of

$x_1[k] = a^{-k}\varepsilon[k]$
$x_2[k] = -a^{-k}\varepsilon[-k]$
$x_3[k] = 2^{-k}\varepsilon[k] + 0.8^k\varepsilon[-k]$
$x_4[k] = 0.8^{-k}\varepsilon[k] + 0.5^k\varepsilon[-k]$
$x_5[k] = a^{|k|}, \quad a \in \mathbb{R}, a > 0$

and give the region of convergence for each. Note: see Examples 13.2 and 13.3.

Exercise 13.3

$x_\nu(t)$ are functions of time:

$$x_1(t) = e^{(-0.5\frac{t}{T})} \cdot \varepsilon(t)$$

$$x_2(t) = e^{(-0.5\frac{t}{T})} \cos(2\pi\frac{t}{T}) \cdot \varepsilon(t)$$

$$x_3(t) = e^{(-2\frac{t}{T})} \sin(0.5\pi\frac{t}{T}) \cdot \varepsilon(t)$$

$x_\nu(t)$ will be sampled at times $t = kT$ to give the likewise continuous-time signal

$$x_{\nu a}(t) = \sum_{k=0}^{\infty} x_\nu[k] \cdot \delta(t - kT).$$

a) Sketch $x_{\nu a}(t)$

b) Find the Laplace transform $X_{\nu a}(s)$ of the sampled continuous-time signal $x_{\nu a}(t)$.

Note: Determine suitable functions $x_\nu[k]$ and their transforms $X_\nu(z)$.

Exercise 13.4

a) Find the z-transforms of the low-pass impulse responses $x_\nu[k]$.

$$x_1[k] = \begin{cases} 0.5 & \text{for } |k| = 1 \\ 1 & \text{for } k = 0 \\ 0 & \text{otherwise} \end{cases}$$

$$x_2[k] = \sum_{\mu=-1}^{1} \delta[k - \mu]$$

$$x_3[k] = \sum_{\mu=0}^{2} \delta[k - \mu]$$

$$x_4[k] = \begin{cases} \text{si}(0.5\pi k) & \text{for } |k| \leq 4 \\ 0 & \text{otherwise} \end{cases}$$

b) Give the spectra $\mathcal{F}_*\{x_\nu[k]\}$ and sketch the given sequences and magnitudes of the spectra.

Exercise 13.5

Prove the following theorems of the z-transform:

1. using the defining equation (13.1),

2. deriving them from the corresponding Laplace transform theorem:

a) shift theorem

b) modulation theorem

c) the 'multiplication with k' theorem

d) time reversal

Exercise 13.6

Let $x[k] = \delta[k-1] + 2\delta[k] + \delta[k+1]$. Using the modulation theorem, calculate the z-transform of $x_m[k] = e^{j\Omega_0 k}x[k]$ for $\Omega_0 = 0$, $\Omega_0 = \frac{\pi}{2}$ and $\Omega_0 = \pi$. Sketch the corresponding spectra $X_m(e^{j\Omega})$.

Exercise 13.7

Let $X(z)$ be the z-transform of a right-sided discrete signal $x[k]$ where for the poles of $X(z)$, $|z_{p\nu}| \leq 0.5$. Determine the z-transform of the signal and give the region of convergence.

$$\begin{aligned} x_1[k] &= k_0 \cdot x[k] \\ x_2[k] &= x[k-k_0] \cdot \varepsilon[k-k_0] \ . \\ x_3[k] &= (-e)^{\alpha k} \cdot x[k] \end{aligned}$$

Exercise 13.8

Decide whether the following z-transforms belong to a finite or infinite series. Give the region of convergence for each under the assumption that it is a right-sided series being dealt with.

$$H_1(z) = z^3 - 1$$

$$H_2(z) = \frac{1}{z^3} - 1$$

$$H_3(z) = \frac{1}{z^2}$$

$$H_4(z) = 1 + \frac{z+1}{z-0.5}$$

$$H_5(z) = \frac{2z^2 - 4z + 2}{z(z-1)}$$

Exercise 13.9

Determine the inverse z-transforms of the following functions:

$$X_1(z) = \frac{z-1}{z(z+0.5)}, \quad \text{ROC}: |z| > 0.5$$

$$X_2(z) = \frac{z-1}{(z+0.5)^2}, \quad \text{ROC}: |z| > 0.5.$$

Exercise 13.10

Draw the pole-zero diagrams for

a) $H_a(z) = z^3 - 1$

b) $H_b(z) = z^{-3} - 1$

c) $H_c(z) = z^3 + z^{-3}$

Exercise 13.11

The pole-zero diagrams for three similar systems are shown. Calculate and sketch
the impulse responses of the systems (to a constant factor). Where do they differ?

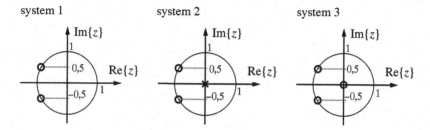

Exercise 13.12

For the sequence $x[k] = \begin{cases} 1 & 0 \le k \le r \\ -1 & r+1 \le k \le 2r+1 \\ 0 & \text{otherwise} \end{cases}$, $r \in \mathbb{N}$

a) find the z-transform $X(z)$

b) find the poles and zeros of $X(z)$

c) sketch a pole-zero diagram for $r = 3$.

14 Discrete-Time LTI-Systems

14.1 Introduction

Now that we have described discrete-time signals in the frequency domain, we can deal with systems which have discrete-time input and output signals. We call these *discrete-time systems*. They have already been briefly mentioned in Chapter 1.2.5, but were left in favour of continuous systems.

The properties of discrete systems have strong parallels with continuous systems. In fact, we can deal with them more quickly by building the corresponding properties to continuous systems. This is in accordance with our general goal, as we want to avoid technical realisations of discrete and continuous systems and deal only with the fundamental relationships.

14.2 Linearity and Time-Invariance

The most general form of a discrete system with one input and one output is shown in Figure 14.1. The system S processes the input series $x[k]$ and produces the output series $y[k]$. The properties of the discrete system S will be restricted so that it is an LTI-system. We can then fall back on Definitions 3 to 6 from Chapter 1, as their general formulation applies to both continuous and discrete systems.

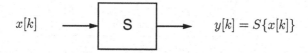

Figure 14.1: Discrete LTI-System

Specialising the superposition principle and time-invariance to the system described by Figure 14.1 yields the following:

- A discrete system S is *linear*, if the superposition principle holds for the response to any two input signals $x_1[k]$ and $x_2[k]$.

$$S\{Ax_1[k] + Bx_2[k]\} = A\,S\{x_1[k]\} + B\,S\{x_2[k]\}. \qquad (14.1)$$

As in (1.3), A and B can be any complex constants.

- A discrete system S is *time-invariant* if for the reaction $y[k]$ to any input signal $x[k]$

$$y[k] = S\{x[k]\} \tag{14.2}$$

the relationship

$$y[k - \kappa] = S\{x[k - \kappa]\}, \quad \forall \kappa \in \mathbb{Z} \tag{14.3}$$

holds. If the index k is instead of time, a discrete position variable, for example, the system S is called *shift-invariant*. In any case, the shift is only defined for integer values of κ, as the difference $k - \kappa$ must give an integer index of the series $x[k]$.

14.3 Linear Difference Equations with Constant Coefficients

14.3.1 Difference Equations and Differential Equations

As linear differential equations with constant coefficients correspond to continuous LTI-systems, linear difference equations with constant coefficients characterize discrete LTI-systems. Examples for such difference equations are

$$y[k] + \frac{1}{2}y[k-1] = x[k] \tag{14.4}$$

$$y[k] - \frac{1}{5}y[k-1] + \frac{1}{3}y[k-2] = x[k] + x[k-1] - \frac{1}{2}x[k-2]. \tag{14.5}$$

The general form of an Nth-order linear difference equation with constant coefficients a_n, b_n is

$$\sum_{n=0}^{N} a_n y[k - n] = \sum_{n=0}^{N} b_n x[k - n]. \tag{14.6}$$

In contrast to the corresponding form (2.3) of differential equation, a shift of the time index k by n values occurs in place of the nth derivatives.

For a given input signal $x[k]$, the difference equation (14.6) has an infinite number of solutions $y[k]$, that can in general be formed from N independent linear components. An unambiguous solution can be obtained with N conditions, given by initial values instead of derivatives, for example, $y[0]$, $y[-1]$, $y[-2]$, $\ldots$, $y[-N + 1]$.

Using the superposition principle (14.1) and the condition for time-invariance (14.3), it can be confirmed that a system described by the difference equation (14.6) is a discrete LTI-system.

14.3.2 Solving Linear Difference Equations

The are two possible methods for solving the linear difference equation (14.6): numerical and analytical.

14.3.2.1 Numerical Solution

By rearranging the difference equation (14.6), we obtain the expression

$$ y[k] = \frac{1}{a_0} \left[\sum_{n=0}^{N} b_n x[k-n] - \sum_{n=1}^{N} a_n y[k-n] \right] , \qquad (14.7) $$

that can be numerically evaluated, beginning with $k = 1, 2, 3, \ldots$. For $k = 1$, the calculation of the initial conditions starts with the known input signal $x[k]$ and with the initial values $y[0]$, $y[-1]$, $\ldots$, $y[-N+1]$. For $k > 1$, the values already known, $y[k-1]$, $y[k-2]$, $\ldots$, are entered into the right-hand side. This method has practical importance, as the right side of (14.7) can be calculated quickly and very accurately using a digital computer. A delay by n can be implemented by shifting the location of the values in the memory of the computer.

14.3.2.2 Analytical Solution

There is also an analytical solution which is possible in much the same way as with the linear differential equation. The fundamental principle is again splitting the solution into an external and internal part

$$ y[k] = y_{\text{ext}}[k] + y_{\text{int}}[k] . \qquad (14.8) $$

Here, $y[k]$ reperesents the solution of the difference equation under the initial conditions, $y_{\text{ext}}[k]$ describes the response to an input signal if the system is at rest and $y_{\text{int}}[k]$ describes the homogenous solution of the initial condition problem without an input signal. The effect of the combined terms is shown in Figure 14.2, which corresponds to Figure 7.1 for continuous initial value problems. The external and internal terms can be calculated in a similar way using the z-transform, just as the Laplace transform was used with continuous systems (see Chapter 7).

14.4 Characteristic Sequences and System Functions of Discrete LTI-Systems

14.4.1 Eigensequences

In Chapter 13.1.1 and 13.5 we mentioned that the inverse z-transform can be interpreted as superimposing eigensequences of discrete LTI-systems. The properties of

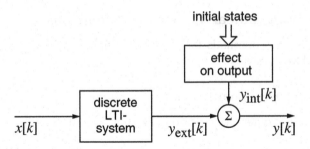

Figure 14.2: Combination of the external and internal terms to form the solution of a differential equation

eigensequences are completely analoguous to those of eigenfunctions of continuous LTI-systems in accordance with Definition 10 in Chapter 3.2. An eigenseries at the input of a discrete LTI-system causes a response at the output that corresponds to the input series with a constant factor (see Figure 14.3). The proof that eigenseries of discrete LTI-systems are exponential series of the form $e[k] = z^k$ will be carried out exactly as for the eigenfunctions in Chapter 3.2.2. We consider a general LTI-

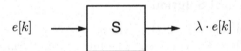

Figure 14.3: System S is excited by the eigenseries $e[k]$

system (Figure 14.1) and only require that it is linear (14.1) and time-invariant (14.2), (14.3). The input signal should be an exponential series $x[k] = z^k$. We want to find the corresponding output signal.

$$y[k] = S\{z^k\}. \qquad (14.9)$$

From the conditions of time-invariance and linearity:

$$y[k - \kappa] = S\{x[k - \kappa]\} = S\{z^{k-\kappa}\}z^{-\kappa}S\{z^k\} = z^{-\kappa}y[k]. \qquad (14.10)$$

These difference equations are only fulfilled at the same time for any κ, if $y[k]$ is a weighted exponential series

$$y[k] = \lambda z^k. \qquad (14.11)$$

Equating $x[k] = z^k$ and $y[k] = \lambda z^k$ yields that every exponential series

$$e[k] = z^k, \quad z \in \mathbb{C} \qquad (14.12)$$

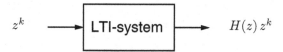

Figure 14.4: z^k are eigenseries of discrete LTI-systems, $H(z)$ is the eigenvalue

is an eigenseries of an LTI-system. The corresponding eigenvalues λ depend in general on z. As with continuous systems we call $\lambda = H(z)$ the *system function* or *transfer function* of a discrete LTI-system (see Figure 14.4).

Like the system function for continuous systems, the system function for discrete systems $H(z)$ has a region of convergence which contains only part of the complex plane. Exciting the system with a complex exponential sequence outside the region of convergence does not lead to an output series with finite amplitude. As with continuous systems, the region of convergence of $H(z)$ is often not explicitly given.

Finally, it should also be mentioned that a unilateral exponential series

$$x[k] = z^k \cdot \varepsilon[k] \tag{14.13}$$

is *not* an eigensequence of a discrete system.

14.4.2   System Function

The system function $H(z)$ can describe the system response to all input signals $x[k]$, and not just eigensequences $e[k] = z^k$. The relationship between a general discrete input sequence $x[k]$ and an exponential sequence is represented by the z-transform, which is given here as the inverse transform of the input signal

$$x[k] = \frac{1}{2\pi j} \oint X(z) z^k \frac{dz}{z} \tag{14.14}$$

and the output signal

$$y[k] = \frac{1}{2\pi j} \oint Y(z) z^k \frac{dz}{z} \tag{14.15}$$

of an LTI-system. The output sequence $y[k]$ is obtained as the system response to $x[k]$

$$y[k] = S\{x[k]\} = \frac{1}{2\pi j} \oint X(z) S\{z^k\} \frac{dz}{z}. \tag{14.16}$$

Inserting in $S\{z^k\} = H(z) z^k$ (see Figure 14.4) and equating with (14.15) yields the relationship between the z-transforms of the input and output signals (see Figure 14.5)

$$\boxed{Y(z) = H(z) X(z).} \tag{14.17}$$

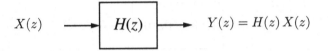

Figure 14.5: System function $H(z)$ of a discrete LTI-system

The system function $H(z)$ (with its region of convergence) is a complete description of the input–output behaviour of a discrete LTI-system. It makes it possible to give the output sequence $y[k]$ for any given input sequence $x[k]$.

14.4.3 Finding System Functions from Difference Equations

A system function can easily be found from a difference equation using the shift theorem of the z-transform (see Table 13.1). We will demonstrate the procedure with a few examples.

———————————————————————————————— **Example 14.1**

A discrete delay circuit delays the input signal. Its 'difference equation is

$$y[k] = x[k-1].\qquad\qquad(14.18)$$

Using the z-transform and its shift theorem (Table 13.1) with $\kappa = 1$ yields

$$Y(z) = \mathcal{Z}\{y[k]\} = \mathcal{Z}\{x[k-1]\} = z^{-1}X(z).\qquad(14.19)$$

A comparison with (14.17) gives the system function of the discrete delay circuit (see Figure 14.6)

$$H(z) = z^{-1}.$$

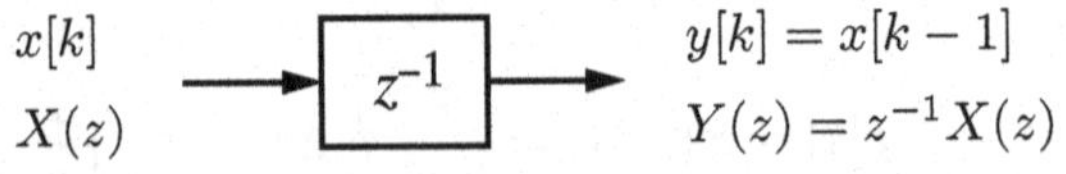

$$
\begin{array}{ccc}
x[k] & & y[k] = x[k-1] \\
X(z) & \boxed{z^{-1}} & Y(z) = z^{-1}X(z)
\end{array}
$$

Figure 14.6: System function of the discrete delay circuit

Example 14.2

We will find the system function of a discrete LTI-system described by the difference equation

$$y[k] - \frac{1}{5}y[k-1] + \frac{1}{3}y[k-2] = x[k] + x[k-1] - \frac{1}{2}x[k-2]. \qquad (14.20)$$

Using the z-transform with its shift theorem

$$Y(z) - \frac{1}{5}Y(z)z^{-1} + \frac{1}{3}Y(z)z^{-2} = X(z) + X(z)z^{-1} - \frac{1}{2}X(z)z^{-2}. \qquad (14.21)$$

From the difference equations, an algebraic equation is formed so that $X(z)$ and $Y(z)$ can be factorized

$$Y(z)\left[1 - \frac{1}{5}z^{-1} + \frac{1}{3}z^{-2}\right] = X(z)\left[1 + z^{-1} - \frac{1}{2}z^{-2}\right]. \qquad (14.22)$$

Sorting the terms yields the system function

$$H(z) = \frac{Y(z)}{X(z)} = \frac{1 + z^{-1} - \frac{1}{2}z^{-2}}{1 - \frac{1}{5}z^{-1} + \frac{1}{3}z^{-2}} = \frac{z^2 + z - \frac{1}{2}}{z^2 - \frac{1}{5}z + \frac{1}{3}}. \qquad (14.23)$$

The system function of a general LTI-system of Nth-degree described by the difference equation (14.6) can be obtained in the same way as the last example. The z-transform and shift theorem yield from (14.6),

$$Y(z) \cdot \sum_{n=0}^{N} a_n z^{-n} = X(z) \cdot \sum_{m=0}^{N} b_m z^{-m} \qquad (14.24)$$

and then the system function

$$H(z) = \frac{Y(z)}{X(z)} = \frac{\sum\limits_{m=0}^{N} b_m z^{-m}}{\sum\limits_{n=0}^{N} a_n z^{-n}} \qquad (14.25)$$

can be found.

With the two examples and with (14.24) und (14.25) we can make some statements with which we are already familiar in principle from dealing with continuous systems and the Laplace transform:

- The z-transform converts a linear difference equation with constant coefficients into an algebraic equation. The algebraic equation can be solved or investigated more easily than the difference equation.

- The system function for a difference equation of finite order is a rational fraction function and can be described by poles and zeros. The frequency response of the system can be estimated easily using the methods described in Chapter 13.6.

- When the system function $H(z)$ is derived from the difference equation, its region of convergence can only be given if additional properties like causality or stability are known. For causal systems the region of convergence lies outside a circle about the origin, through the most distant pole.

14.5   Block Diagrams and State-Space

Like continuous systems, discrete systems can be represented by block diagrams. The parallels between the two kinds of system are particularly noticeable here as instead of an integrator in continuous systems, there is a delay in discrete systems. In the frequency-domain it means that a block with the system function $H(s) = s^{-1}$ (integrator) corresponds to a block with the system function $H(z) = z^{-1}$ (delay circuit).

14.5.1   Direct Form I

The direct form I of a discrete LTI-system as in Figure 14.7 corresponds exactly to the block diagram for continuous sytems in Figure 2.1. The only differences are the delay circuits in place of integrators. Comparison with the system function (14.25) confirms that the block diagram in Figure 14.7 realises a discrete LTI-system with the difference equation (14.6).

14.5.2   Direct Form II

The block diagram in Figure 14.7 can be analysed as two LTI-systems connected in series. The order in which they are connected can be reversed (compare Figure 2.2), but the result is the same, shown in Figure 14.8. As in Chapter 2.2, it is also striking that both vertical series of delays run in parallel and can be replaced by a single series. The result in Figure 14.9 again corresponds to the structure of direct form II for the continuous system shown in Figure 2.3. Direct form III (see Figure 2.5) can be constructed in a similiar way for discrete systems.

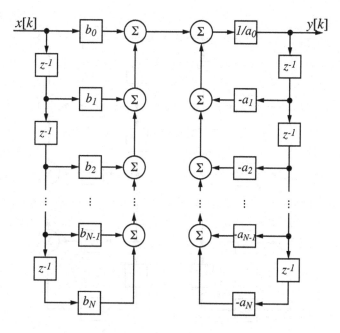

Figure 14.7: Direct form I of a discrete LTI-system

14.5.3   State-Space Description for Discrete LTI-Systems

If a block diagram of an LTI-system is given, a state-space representation of the
form

$$\mathbf{z}[k+1] \quad = \quad \mathbf{A}\mathbf{z}[k] + \mathbf{B}x[k] \tag{14.26}$$

$$y[k] \quad = \quad \mathbf{C}\mathbf{z}[k] + \mathbf{D}x[k]. \tag{14.27}$$

can be formulated.

As state vector $\mathbf{z}[k]$ the values stored in the delay circuits are chosen. We
call (14.26) the system equation, which describes changes in the internal states
depending on the current state $\mathbf{z}[k]$ and the input series $x[k]$. With the form (14.26)
and (14.27) it is easy to characterise systems with multiple inputs and outputs
(Figure 14.10).

Using the z-transform on the state-space description (14.26),(14.27) we obtain:

$$z\mathbf{Z}(z) \quad = \quad \mathbf{A}\mathbf{Z}(z) + \mathbf{B}X(z) \tag{14.28}$$

$$Y(z) \quad = \quad \mathbf{C}\mathbf{Z}(z) + \mathbf{D}X(z). \tag{14.29}$$

These equations are identical to the state-space description of a continuous-time
system in the Laplace-domain except that the complex variable s has in this case

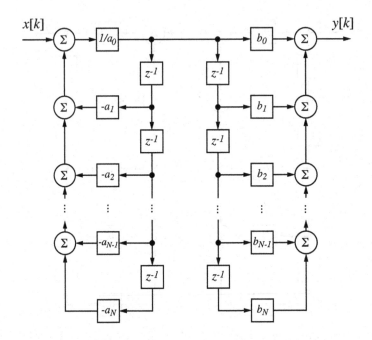

Figure 14.8: Forming direct form II from direct form I

been replaced by the complex variable z. We can therefore also use many of the techniques developed for continuous systems directly on discrete systems. This includes the relationships between the difference/differential equation, block diagram and state-space descriptions (Chapter 2.4), equivalent state-space representations (Chapter 2.5) and controllability and observability (Chapter 2.6). Figure 14.10 shows the block diagram of the state-space description for a discrete LTI-system with state equations (14.26), (14.27), and (14.28), (14.29). Be careful not to confuse the state vector in the time-domain $\mathbf{z}[k]$ with the frequency variable z of the z-transform!

In the case of initial condition problems described by difference equations, the state-space description (14.26) and (14.27) can be extended, where the system is offset to a suitable time k_0 in a suitable initial state $\mathbf{z}[k_0]$ so that the given initial conditions are fulfilled. This can be achieved using a superimposed impulse $\mathbf{z}_0 \delta[k - k_0]$, with which the state-space description is

$$
\begin{aligned}
\mathbf{z}[k+1] &= \mathbf{A}\mathbf{z}[k] + \mathbf{B}x[k] + \mathbf{z}_0 \delta[k+1-k_0] \qquad (14.30) \\
y[k] &= \mathbf{C}\mathbf{z}[k] + \mathbf{D}x[k]
\end{aligned}
$$

If the initial state is given at $k_0 = +1$, the z-transform of the state-space descrip-

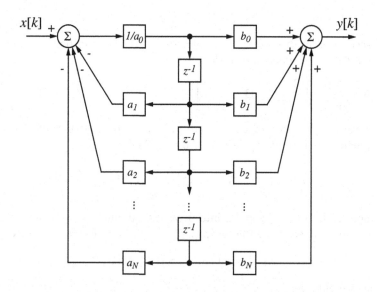

Figure 14.9: Direct form II of a discrete LTI-system

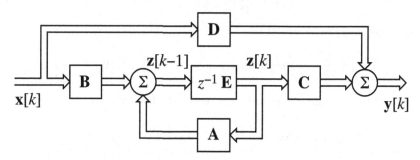

Figure 14.10: Block diagram of the state-space description for a discrete LTI-system

tion is

$$zZ(z) - z_0 = AZ(z) + BX(z) \tag{14.31}$$

$$Y(z) = CZ(z) + DX(z),$$

which corresponds exactly to the initial value problem for continuous systems in the Laplace-domain (7.61), (7.62). The initial state z_0 can be easily reached if the initial conditions are given: $y[1], y[0], y[-1], \ldots, y[-N + 2]$. Starting from (7.64) – (7.66), discrete initial value problems can be solved in the usual way. The procedure will be demonstrated with an example.

Example 14.3

The second-order discrete system in Figure 14.11 is described by the following matrices of the state-space description:

$$\mathbf{A} = \begin{bmatrix} 0 & \frac{1}{4} \\ 1 & 0 \end{bmatrix}, \qquad \mathbf{B} = \begin{bmatrix} 1 \\ 0 \end{bmatrix}, \qquad \mathbf{C} = \begin{bmatrix} 0 & \frac{1}{4} \end{bmatrix}, \qquad D = 1.$$

We calculate the response of the system for $x[k] = 0$, and the initial values $y[0]$ and $y[1]$, using the formula (compare (7.64) – (7.66)):

$$Y(z) = \mathbf{G}(z)\,\mathbf{z}_0, \qquad \mathbf{G}(z) = \mathbf{C}(z\mathbf{E} - \mathbf{A})^{-1} = \frac{\frac{1}{4}}{z^2 - \frac{1}{4}}\begin{bmatrix} 1 & z \end{bmatrix}.$$

The connection between the given initial values $y[0]$ and $y[1]$ and the state vector $\mathbf{z}_0 = \mathbf{z}[1]$, which causes the same output signal, can be read from Figure 14.11

$$z_1[1] = y[0], \qquad z_2[1] = 4y[1].$$

From that we obtain

$$Y(z) = \frac{\frac{1}{4}}{z^2 - \frac{1}{4}}\left(y[0] + z\,4y[1]\right) \qquad (14.32)$$

and splitting into partial fractions

$$Y(z) = \frac{1}{4}z^{-1}\left[\left(y[0] + 2y[1]\right)\frac{z}{z - \frac{1}{2}} - \left(y[0] - 2y[1]\right)\frac{z}{z + \frac{1}{2}}\right].$$

Finally, the output signal is found by inverse z-transformation

$$
\begin{aligned}
y[k] \;=\; & y[0]\frac{1}{4}\left[\left(\frac{1}{2}\right)^{k-1} - \left(-\frac{1}{2}\right)^{k-1}\right]\varepsilon[k-1] + \\
& y[1]\frac{1}{2}\left[\left(\frac{1}{2}\right)^{k-1} + \left(-\frac{1}{2}\right)^{k-1}\right]\varepsilon[k-1].
\end{aligned}
$$

As we had taken the initial state at $k_0 = 1$, the calculated output signal also starts at $k = 1$, and delivers the expected value $y[1]$. This can be confirmed by insertion into the given difference equation. ∎

14.6   Discrete Convolution and Impulse Response

In Chapter 8 we introduced the impulse response as a second important characteristic for continuous systems, in addition to the system function. Its importance can be expressed in three significant properties:

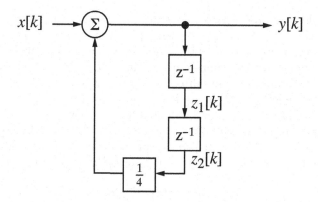

Figure 14.11: Block diagram of the discrete system in Example 14.3

- The impulse response of a system is its reaction to a pulse-shaped input.

- The impulse response is obtained by inverse transforming the system function.

- The output signal of a system is obtained in the time-domain by convolution of the input signal and the impulse response.

The derivation of these properties was not simple, as the pulse-shaped input needed to generate an impulse response is not an ordinary function. The price for the elegance of this system description was generalising functions to distributions. The delta impluse introduced in Chapter 8.3 cannot be characterised simply by its values, only by its effects on other functions, in particular, the selective property.

This system description based on the impulse response can be transferred to discrete systems. We will see that for a discrete system, the impulse response has the same fundamental properties as for a continuous system. In one point the description of discrete systems becomes a lot simpler: the pulse-shaped input signal required to generate the impulse response is now a very simple sequence. It is the unit impulse introduced in Chapter12.2.1, which is a sequence of zeros and a single one. We can deal with these values without the use of generalised functions.

14.6.1 Calculation of the System Response Using Discrete Convolution

To introduce the impulse response of a discrete system we consider Figure 14.12. It shows a discrete system with input signal $x[k]$ and output signal $y[k]$. The response $h[k]$ to a discrete unit impulse (see Chapter 12.2.1) will be called the impulse response of a discrete system.

The impulse response makes it possible to calculate the system response in the time-domain for discrete LTI-systems. As an example we examine the input sequence $x[k]$ in Figure 14.12 with three non-zero values. We can split it into three individual sequences with one non-zero value in each, and then these sequences can be interpreted as a unit impulse shifted by $\kappa = 0, 1, 2$. Each impulse is weighted by the value of the function $x[k]$, so $x[\kappa]\delta[k - \kappa]$ for $\kappa = 0, 1, 2$. At the output of the LTI-system each of the shifted and weighted impulses cause a likewise shifted and weighted impulse response, $x[\kappa]h[k-\kappa]$ for $\kappa = 0, 1, 2$. Because the system is linear, the responses to the individual sequences can be added to give the response to the complete input signal $x[k]$, so $y[k] = \sum_{\kappa=0}^{2} x[\kappa]h[k - \kappa]$. These considerations

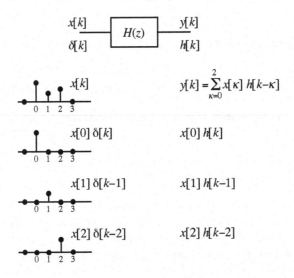

Figure 14.12: Discrete convolution

can also be extended to any input sequence $x[k]$ with more than three values. The summation is performed in the general case from $-\infty$ to ∞. The resulting expression for $y[k]$ is the same as for convolution if the integral is replaced by a summation sign. We therefore refer to this as *discrete convolution*, which will be denoted with $*$:

$$
\begin{aligned}
y[k] &= x[k] * h[k] = \sum_{\kappa=-\infty}^{\infty} x[\kappa]\, h[k - \kappa] \\
&= h[k] * x[k] = \sum_{\kappa=-\infty}^{\infty} h[\kappa]\, x[k - \kappa].
\end{aligned}
\tag{14.33}
$$

The substitution $n = k - \kappa$ quickly shows that discrete convolution is also commutative.

14.6.2   Convolution Theorem of the z-Transform

In the description of the system from Figure 14.12, the relationship between the impulse response $h[k]$ and the system function $H(z)$ remains undefined. To find it, we start with the discrete convolution in (14.33) and use the z-transform on both sides:

$$Y(z) = \sum_{k=-\infty}^{\infty} y[k]z^{-k} = \sum_{k=-\infty}^{\infty} \sum_{\kappa=-\infty}^{\infty} x[\kappa]h[k-\kappa]z^{-k+(\kappa-\kappa)} . \tag{14.34}$$

On the right-hand side we have added in the exponent of z another zero in the form of $\kappa - \kappa = 0$. Interchanging the sums and rearranging terms yields

$$
\begin{aligned}
Y(z) &= \sum_{\kappa=-\infty}^{\infty} x[\kappa]z^{-\kappa} \cdot \sum_{k=-\infty}^{\infty} h[k-\kappa]z^{-(k-\kappa)} \\
&= \sum_{\kappa=-\infty}^{\infty} x[\kappa]z^{-\kappa} \cdot \sum_{n=-\infty}^{\infty} h[n]z^{-n} .
\end{aligned}
\tag{14.35}
$$

In the second row the substitution $n = k - \kappa$ has been used. Both sums are independent of each other and represent the z-transforms of $x[k]$ and $h[k]$ respectively. From this we read:

- Using z-transformation, discrete convolution becomes a multiplication of the corresponding z-transforms.

- The z-transform of the impulse response of a discrete system is the system function.

The first statement identifies the *convolution theorem* of the z-transform, which corresponds to the convolution theorem of the Laplace transform:

$$\boxed{y[k] = x[k] * h[k] \;\circ\!\!-\!\!\bullet\; Y(z) = X(z)\,H(z) \quad \text{ROC}\{y\} \supseteq \text{ROC}\{x\} \cap \text{ROC}\{h\} .}$$

$$\tag{14.36}$$

The region of convergence of $Y(z)$ is at least the intersection between the ROCs of $X(z)$ and $H(z)$. That means that $Y(z)$ definitely converges where both $X(z)$ and $H(z)$ also converge. $Y(z)$ also converges, however, on poles of $H(z)$ if they are at the same time zeros of $X(z)$ (and vice versa). The region of convergence of $Y(z)$ can therefore be greater than the intersection.

The second statement gives the relationship between the impulse response and system function:

$$\boxed{H(z) = \mathcal{Z}\{h[k]\} .} \tag{14.37}$$

As well as the system function is the impulse response a complete description of a discrete LTI-system's input–output behaviour. The system response to any input series can be calculated using convolution with the impulse response. From (14.37), the region of convergence of the system function $H(z)$ can also be found, which in Section 14.4 we assumed was already known.

14.6.3   Systems with Impulse Responses of Finite and Infinite Length

A significant distinguishing feature of discrete systems is the length of the impulse response. Depending on the structure of the system, it can have a finite or infinite number of non-zero values. We will consider two examples for discrete first-order systems.

_____ **Example 14.4**

Figure 14.13 shows the structure of a system with system function

$$H(z) = 1 + bz^{-1}.$$

It consists of a direct path from the input to the output and a parallel path with a delay and a multiplication. The response to an impulse therefore consists of two values and is of finite length.

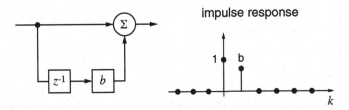

Figure 14.13: System with an impulse response of finite length

A software realisation of this system in the programming language MATLAB could consist, for example, of the following commands:

```
x     = [ 1 0 0 0 0 0];        % input signal: unit impulse
b     = 0.8;                   % multiplier

x_old = 0;                     % memory (initially empty)
y     = zeros(size(x));        % output signal (not yet calculated)

for k=1:6,                     % loop for 6 steps
```

```
    y(k) = x(k) + x_old*b;     % current value of the output signal
      x_old = x(k);            % delaying of the input signal
end
```

The delay is formed by saving the current input value in the variable x_old.   ∎

Example 14.5

Figure 14.14 shows the structure of a system with system function

$$H(z) = \frac{1}{1 - az^{-1}} = \frac{z}{z - a}.$$

It is made up of a direct path and a feedback loop from the output to the input with a delay and a multiplication. The response to an impulse decays steadily for $0 < |a| < 1$, but never quite reaches zero. The impulse response is therefore infinitely long.

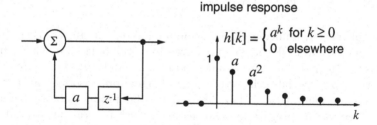

Figure 14.14: System with an impulse response of infinite length

The corresponding program steps in MATLAB are

```
x     = [ 1 0 0 0 0 0];        % input signal: unit impulse
a     = 0.8;                    % multiplier

y_old = 0;                      % memory (initially empty)
y     = zeros(size(x));         % output signal (not yet calculated)

for k=1:6,                      % loop for 6 steps
    y[k] = x[k] + y_old*a;      % current value of the output signal
    y_old = y[k];               % delaying of the output signal
end
```

Here the current output signal is saved in the variable y_old, so that the delay can be implemented.

For $a = 1$ this system represents an accumulator that sums all incoming input values. The impulse response of the accumulator is the discrete unit step (see Figure 14.14 for $a = 1$). We calculated the $\mathcal{F}_*$-transform of the discrete unit step in Chapter 12.3.3.3 and the corresponding z-transform in Example 13.2.

The following terms will be used:

- Systems with finite impulse responses are also called *non-recursive* or *FIR-systems* (FIR – Finite Impulse Response).

- Systems with infinite impulse response are also called *recursive* or *IIR-systems* (IIR – Infinite Impulse Response).

14.6.4 Discrete Convolution

The great practical importance of discrete convolution comes about because the expression

$$y[k] = x[k] * h[k] = \sum_{\kappa=-\infty}^{\infty} x[\kappa]\, h[k - \kappa] \tag{14.38}$$

can be implemented immediately as a computer program. In the usual programming languages, two FOR-loops are required to calculate (14.38), where the outer loop runs via the index k and the inner loop via the index κ. There are, however, processors with a special architecture (digital signal processors), which can carry out the summation with κ in (14.38) as the scalar product of two vectors very quickly. Even for the preparation and testing of such programs it is vital to have mastered discrete convolution on paper. We will therefore deal with it in even greater detail than the calculation of the convolution integral in Chapter 8.4.3.

To begin with we show a method similar to the convolution integral in Chapter 8.4.3:

1. Draw $x[\kappa]$ and $h[\kappa]$ with respect to κ.

2. Reverse $h[\kappa]$: $h[\kappa] \rightarrow h[-\kappa]$.

3. Shift $h[-\kappa]$ by k positions to the *right*: $h[-\kappa] \rightarrow h[k - \kappa]$.

4. Multiplication of $x[\kappa]$ with $h[k - \kappa]$ and summation of the product for all values of κ yields *one* value of $y[k]$.

5. Repeat steps 3 and 4 for all values of k.

Because of the commutativity of convolution, the steps can also be carried out if $x[k]$ and $h[k]$ swap places. The calculation of discrete convolution using these steps will be demonstrated in some examples.

—————————————————————————— **Example 14.6**

We calculate the response of the recursive system from Figure 14.14 with the impulse response

$$h[k] = \varepsilon[k]a^k \qquad 0 < a < 1 \tag{14.39}$$

to an input signal of the form

$$x[k] = \begin{cases} 1 & \text{for } 0 \le k \le K - 1 \\ 0 & \text{otherwise} \end{cases} \tag{14.40}$$

as a result of the convolution

$$y[k] = x[k] * h[k]. \tag{14.41}$$

The impulse response $h[k]$ and the input signal $x[k]$ are shown in Figure 14.15.

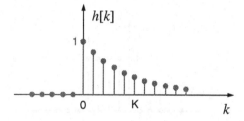

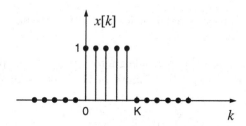

Figure 14.15: Input signal $x[k]$ and impulse response $h[k]$

By reversing and shifting the impulse response, the sequence $h[k - \kappa]$ is obtained, as shown in Fig 14.16 for $k = -2$ and $k = 6$.

For the multiplication of $x[\kappa]$ and $h[k - \kappa]$, and the summation with index κ there are three cases to consider:

- for $k < 0$ $h[k - \kappa]$ and $x[\kappa]$ do not overlap (see Figure 14.16 bottom) and the product is zero for all κ, so

$$y[k] = 0. \tag{14.42}$$

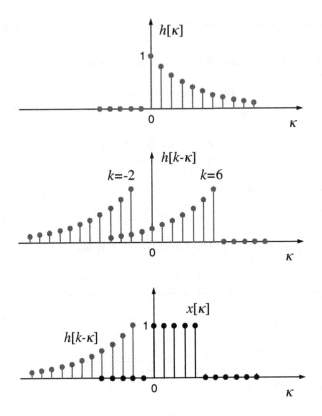

Figure 14.16: Mirrored impulse response does not overlap the input signal $x[k]$ for $k < 0$

- For $0 \leq k < K$, $h[k-\kappa]$ and $x[k]$ begin to overlap, so the product $x[\kappa]h[k-\kappa]$ for $0 \leq \kappa \leq k$ is non-zero (see Figure 14.17). The summation yields

$$y[k] = \sum_{\kappa=0}^{k} a^{k-\kappa} = a^k \sum_{\kappa=0}^{k} (a^{-1})^{\kappa} = \frac{a^k(1 - a^{-k-1})}{1 - a^{-1}} = \frac{1 - a^{k+1}}{1 - a}. \qquad (14.43)$$

- For $K \leq k$ all K values of the input signal $x[k]$ overlap with the reversed and shifted impulse response $h[k - \kappa]$ (see Figure 14.18). It is summed over $0 \leq k < K$ and yields

$$y[k] = \sum_{\kappa=0}^{K-1} a^{k-\kappa} = \frac{a^k(1 - a^{-K})}{1 - a^{-1}} = \frac{a^{k+1}(a^{-K} - 1)}{1 - a}. \qquad (14.44)$$

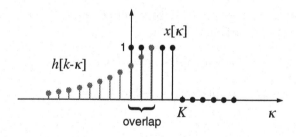

Figure 14.17: For $0 \leq k < K$ the two functions begin to overlap

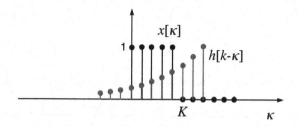

Figure 14.18: Overlap for $K \leq k$

The sum of the values of $y[k]$ from all three component regions is represented in Figure 14.19.

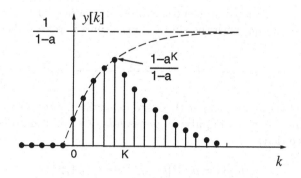

Figure 14.19: End result for $y[k]$ in Example 14.6

Example 14.7

The convolution of a series $x[k]$ with a shifted and scaled unit impulse results in a shifted and scaled variant of $x[k]$:

$$x[k] * A\delta[k - k_0] = Ax[k - k_0] . \qquad (14.45)$$

Figure 14.20 shows an example for $A = -2$ and $k_0 = 3$.

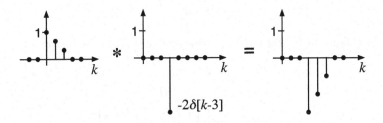

-2δ[k-3]

Figure 14.20: The given series is shifted by 3 and scaled by a factor -2

If $h[k]$ is only a short sequence (e.g. as impulse response of a non-recursive filter), the steps in Figure 14.12 can be carried out to calculate the convolution product $h[k] * x[k]$:

1. Splitting $h[k]$ into shifted and scaled unit impulses $h[\kappa]\delta[k - \kappa]$.

2. Superimposing the component convolution products that are given by the convolution of $x[k]$ with the shifted, scaled unit impulses $h[\kappa]\delta[k - \kappa]$.

Example 14.8

Figure 14.22 shows the use of these steps on the input signal $x[k]$ shown in Figure 14.21 and the short impulse response $h[k]$. Convolution of $x[k]$ with each partial sequence (left)

$$h_\kappa[k] = h[\kappa]\delta[k - \kappa] , \quad \kappa = 0, 1, 2$$

gives the partial product (right)

$$y_\kappa[k] = h[\kappa]x[k - \kappa] , \quad \kappa = 0, 1, 2 \quad ,$$

whose sum over $\kappa = 0, 1, 2$ yields the complete output signal $y[k]$ (bottom).

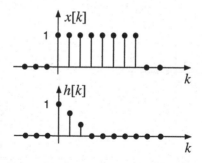

Figure 14.21: Input signal $x[k]$ and impulse response $h[k]$

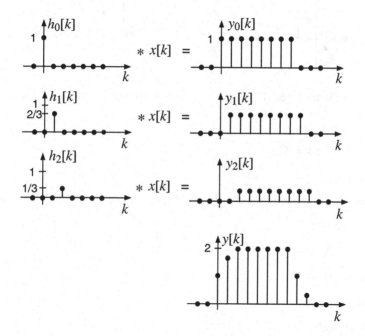

Figure 14.22: Calculating the convolution product of $x[k]$ with the series $h_\kappa[k]$

Convolution of two signals of finite duration results in another finite signal. From Figure 14.23 it can be taken that the first non-zero value of the result $y[k]$ lies at the index which corresponds to the sum of the indices of the first values of $x[k]$ and $h[k]$. For the last value of $y[k]$ the equivalent is true. The duration of $y[k]$ can be obtained from the durations of $x[k]$ and $h[k]$ with

$$\boxed{\text{duration}\{y[k]\} = \text{duration}\{x[k]\} + \text{duration}\{h[k]\} - 1\,.}\qquad(14.46)$$

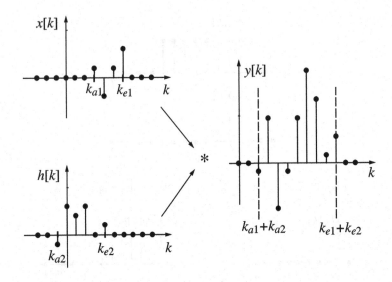

Figure 14.23: Convolution of two finite signals $x[k]$ and $h[k]$

14.7 Exercises

Exercise 14.1

Check the linearity and time-invariance of the following systems $y[k] = S\{x[k]\}$:

a) $y[k] = a\,x[k]$

b) $y[k] = x[k - 5]$

c) $y[k] = a + x[k]$

d) $y[k] = a^k\,x[k]$

e) $y[k] = x[k] - x[k - 1]$

f) $y[k] = \displaystyle\sum_{\mu=0}^{k} x[\mu]$

g) $y[k] = \displaystyle\sum_{\mu=-\infty}^{k} x[\mu]$

h) $y[k] = c \cdot y[k - 1] + x[k]$

i) $y[k] = \dfrac{1}{k} x[k]$

j) $y[k] = a^{x[k]}$

Exercise 14.2

Are the following systems shift-invariant? Justify your answer.

a) Downsampler by a factor of 2: $y[k] = x[2\,k]$

b) Upsampler by a factor of 2: $y[k] = \begin{cases} x\left[\frac{1}{2}k\right] & k \text{ even} \\ 0 & k \text{ odd} \end{cases}$

c) Downsampler followed by a upsampler:

$$x[k] \longrightarrow \boxed{\downarrow 2} \longrightarrow \boxed{\uparrow 2} \longrightarrow y[k]$$

   Investigate especially a delay by one step and by two steps.

Exercise 14.3

A system is given by the difference equation

$$y[k] = x[k] - 2y[k-1] - y[k-2]\,.$$

Determine the reaction to a) $x[k] = \delta[k]$ and b) $x[k] = \varepsilon[k]$ numerically for $k \geq -1$.

Exercise 14.4

Solve the following discrete initial value problem numerically for $k \in [0;5]$:

difference equation: $y[k] + y[k-2] = x[k]$

initial conditions: $y[0] = 0$

   $y[1] = 5$

excitation: $x[k] = (-1)^k \varepsilon[k]$

Exercise 14.5

Compute numerically the output signal of the following system for $k \in [0;3]$. Give $y_{\text{ext}}[k]$, $y_{\text{int}}[k]$ and the initial condition $y[0]$, if the output of the delay element has the initial state $z[0] = 2$ and $x[k] = \varepsilon[k]$.

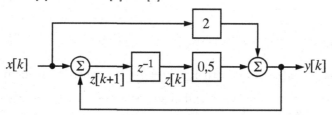

Exercise 14.6

a) Transform the block diagram from Exercise 14.5 into direct form I, II and III. Determine for each case the initial states of the delay elements in order to fulfill the initial conditions. For which form is the initial state unique?

b) Derive a difference equation from the block diagram from exercise 14.5 and compare the coefficients with a).

Exercise 14.7

Give the direct forms II and III for the system from Exercise 14.4. Determine the initial state for each so that the initial conditions are fulfilled.

Exercise 14.8

A system is given by the difference equation $y[k] - \frac{1}{2}y[k-1] + \frac{1}{4}y[k-2] = x[k]$.

Calculate the response to $x[k] = \left(\frac{1}{2}\right)^k \varepsilon[k]$ with the inverse z-transform.

Exercise 14.9

a) Calculate the impulse response $h_1[k]$ of the discrete system

$$H_1(z) = \frac{a\,z^2}{z^2 - z + 0,5}, \quad |z| > c.$$

Are we dealing with an FIR or an IIR system (justification)?

b) $H_1(z)$ will be approximated with an FIR system $H_2(z)$. The impulse responses of $h_1[k]$ and $h_2[k]$ should agree for $k \leq 5$. Find $H_2(z)$ and produce a block diagram realisation.

Exercise 14.10

Solve the initial value problem in Exercise 14.5 analytically using the z-transform:

$y[k] - 0.5y[k-1] = 2x[k] + 0.5x[k-1]$

$y[0] = 3$

$x[k] = \varepsilon[k]$

a) Calculate $y_{\text{ext}}[k] \circ\!\!-\!\!\bullet Y_{\text{ext}}(z) = H(z) \cdot X(z)$.

b) Calculate $y_{\text{int}}[k]$. Start with $Y_{\text{int}}(z) = \dfrac{Az}{z - z_p}$, where z_p is the pole of $H(z)$.

Then determine A in the time-domain using the initial conditions. Note: see Chapter 7.3.3.

c) Give $y[k]$. Compare the results with Exercise 14.5.

Exercise 14.11

Solve the initial value problem from Exercise 14.4 analytically with the z-transform. Calculate the external and internal terms as in Exercise 14.10. Verify the result with Exercise 14.4.

Exercise 14.12

Using the convolution sum (14.33), calculate $c[k] = a[k] * b[k]$, for

a) $a[k] = \varepsilon[k] - \varepsilon[k-3]$; $b[k] = \delta[k] + 2\delta[k-1] - \delta[k-2]$

b) $a[k] = 0.8^k \varepsilon[k]$; $b[k] = \delta[k+2] + 0.8\delta[k+1]$,   with sketches.

Exercise 14.13

Sketch the convolution product $y[k] = x[k] * h[k]$:

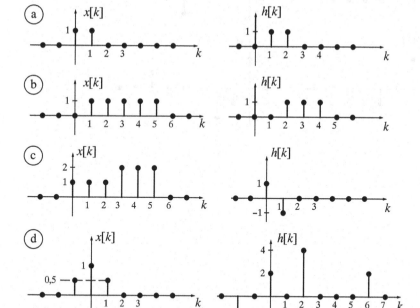

Exercise 14.14

An input and output signal for a FIR system are given. Determine its impulse response using the idea of discrete convolution.

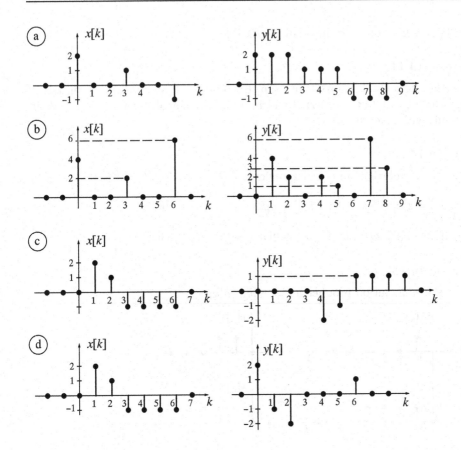

15 Causality and the Hilbert Transform

Our examination of systems has so far given little thought to whether or not they can really be implemented. Although in some of the examples we started from an implementation (e.g., an electrical network), and found system descriptions like the impulse response or transfer function, we have not yet investigated the reverse: implementation of a given system function. We will correct this in this chapter and the next. In accordance with our stated goals at the start of the book, we will not be concerned with implementations using a specific technology. Instead, we will be looking for general criteria that have to be fulfilled for implementation to be possible. These criteria can be simplified into two concepts: causality and stability. In this chapter we will deal with questions connected to the causality of a system. The problem of system stability will be dealt with in the next chapter.

Causality in general signifies a causal relationship between two or more processes, for example, between the input and output of a system. In the language of signals and systems this means more than just a logical connection (*if* input, *then* output), it also means that an event cannot happen before it has been caused by another (*first* input, *then* output).

In this chapter we will be learning a simple time-domain characterisation using the impulse response. Its extension to the frequency-domain then leads to the idea of the Hilbert transform. These new tools can also be useful when the time and frequency-domains are exchanged. The so-called analytical signal follows from this use of the duality principle.

The examination of causality is closely related for continuous and discrete-time systems, so we will be dealing with both kinds of system in parallel.

15.1 Causal Systems

We will be introducing the concept of causality in steps, beginning with systems in general and specialising at first into linear systems and then finally LTI systems.

15.1.1 General Systems

First of all we will be considering the general continuous system from Figure 15.1, which has no special properties that we can use. A causal relationship between the input and output signal exists, if two input signals $x_1(t)$ and $x_2(t)$ are the same up to time t_0, then they will give rise to the same output signals $y_1(t)$ and $y_2(t)$:

$$x_1(t) = x_2(t) \quad \text{for} \quad t < t_0 \quad \Longrightarrow \quad y_1(t) = y_2(t) \quad \text{for} \quad t < t_0. \quad (15.1)$$

We could also say that if two input signals $x_1(t)$ and $x_2(t)$ are not different up to a time point t_0, the output signals may not differ until this time point is reached. As neither linearity nor time-invariance are pre-conditions, the condition (15.1)

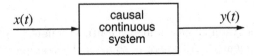

Figure 15.1: Definition of causality for continuous systems

must be fulfilled for all possible pairs of input signals $x_1(t)$ and $x_2(t)$ and for all time points t_0.

For discrete systems, Figure 15.2 holds the corresponding condition for the discrete point in time k_0

$$x_1[k] = x_2[k] \quad \text{for} \quad k < k_0 \quad \Longrightarrow \quad y_1[k] = y_2[k] \quad \text{for} \quad k < k_0. \quad (15.2)$$

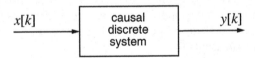

Figure 15.2: Definition of causality for discrete systems

15.1.2 Linear Systems

With linear systems, the difference $x(t) = x_1(t) - x_2(t)$ of the input signals from (15.1) corresponds to the difference $y(t) = y_1(t) - y_2(t)$ of the output signals. The condition for causality is thus more simply given by

$$x(t) = 0 \quad \text{for } t < t_0 \quad \Longrightarrow \quad y(t) = 0 \quad \text{for } t < t_0. \quad (15.3)$$

A linear system is causal if an input signal $x(t)$ that is zero until time t_0 causes an output signal $y(t)$, that is also zero until time t_0. This condition must be true for all time points t_0. The corresponding causality condition for linear discrete systems is simply

$$x[k] = 0 \quad \text{for } k < k_0 \quad \Longrightarrow \quad y[k] = 0 \quad \text{for } k < k_0. \quad (15.4)$$

15.1.3 LTI-Systems

LTI-systems are time-invariant as well as being linear, and so the conditions (15.3) and (15.4) only need to be formulated for one point in time, for example, $t_0 = 0$ or $k_0 = 0$, obtaining for continuous and discrete systems:

$$x(t) = 0 \quad \text{for } t < 0 \qquad \Longrightarrow \qquad y(t) = 0 \quad \text{for } t < 0 \qquad (15.5)$$
$$x[k] = 0 \quad \text{for } k < 0 \qquad \Longrightarrow \qquad y[k] = 0 \quad \text{for } k < 0. \qquad (15.6)$$

With help from the convolution relationships (8.39) and (14.33),

$$y(t) = h(t) * x(t) = \int_{-\infty}^{\infty} h(\tau)x(t - \tau)d\tau \qquad (15.7)$$

$$y[k] = h[k] * x[k] = \sum_{\kappa=-\infty}^{\infty} h[\kappa]x[k - \kappa] \qquad (15.8)$$

we can express the conditions (15.5) and (15.6) significantly more concisely using the impulse response.

For continuous systems it follows from (15.7) with $x(t) = 0$ for $t < 0$ that

$$y(t) = \int_{-\infty}^{t} h(\tau)x(t - \tau)d\tau. \qquad (15.9)$$

The causality condition $y(t) = 0$ for $t < 0$ is then fulfilled when the impulse response $h(t)$ disappears for $t < 0$:

$$h(t) = 0 \quad \text{for} \quad t < 0. \qquad (15.10)$$

In the same way, for discrete systems it follows from (15.8) that

$$h[k] = 0 \quad \text{for} \quad k < 0. \qquad (15.11)$$

Figures 8.18 and 14.16 to 14.18 show calculations of convolution with the impulse response of causal systems (continuous and discrete). We have already learnt from these examples that the convolution result is non-zero only for times later than $t_a = t_{a1} + t_{a2}$, where t_{a1} and t_{a2} indicate the beginning of the signals to be convoluted (see Figures 8.30 and 14.23). In the case investigated here, $t_a = t_{a1} = t_{a2}$ is zero.

The condition for causality of LTI-systems can also be expressed concisely in words:

> An LTI-system is causal if its impulse response is a right-sided function and it disappears for negative time.

This statement is equally true for both continuous and discrete systems.

From the properties of the ROC of the Laplace transform from Chapter 4.5.3 (or the z-transform from Chapter 13.2), we can derive a general property for the ROC of the transfer function of a causal LTI-system. Given separately for continuous and discrete systems:

- The transfer function of a continuous causal LTI-system is the Laplace trans- for of a right-sided function of time. It converges to the right of a vertical line in the complex s-plane, i.e., for $\operatorname{Re}\{s\} > \sigma$.

- The transfer function of a discrete causal LTI-system is the z-transform of a right-sided series. It converges outside a circle in the complex z-plane, i.e., for $|z| > a$.

The transfer function of a causal LTI-system can be identified significantly more precisely than only by the location of the ROC. We will learn about this property using the spectra of causal signals in the next sections.

15.2   Causal Signals

15.2.1   Time-Domain

We have already seen that the impulse response of a causal LTI-system must be a right-sided signal. The following investigations can be further generalised, if we extend them to cover all right-sided signals, irrespective of whether the signal is an impulse response of an LTI-system. For this reason, we have to position the zero point of the time axis so that the right-sided signal begins no earlier than at zero. Because of the close connection with causal systems, right-sided signals that begin no earlier than at $t = 0$ or $k = 0$ are also known as *causal signals*:

Definition 20:   Causal Signals

Causal signals are signals that fulfill the conditions set out in (15.10) and (15.11), for the impulse responses of continuous or discrete systems, so

$$x(t) = 0 \quad for \quad t < 0 \qquad and \qquad x[k] = 0 \quad for \quad k < 0. \tag{15.12}$$

Example 15.1

The signal $x(t) = \varepsilon(t+1)e^{-t}$ is a right-sided signal, but is not causal. By shifting the zero of the time axis, we can obtain an equivalent signal $u(t)$, that is causal:

$$u(t) = x(t-1) = \varepsilon(t)e^{-t+1}.$$

■

15.2.2   Spectra of Causal Signals

The causality of a signal is a property that must also be expressed in the spectrum of a signal. We will be investigating the characteristic properties of the spectrum of causal signals in the next section. The results will be general for causal signals and their spectra. As every causal signal can also be an impulse response for an LTI-system, however, these spectral properties are also valid for the frequency response of causal LTI-systems.

15.2.2.1   Continuous-Time Signals

We will start with the construction of the spectral properties of continuous-time signals. If a signal $h(t)$ is causal and does not contain any Dirac impulse at $t = 0$, multiplying it with a step funtion $\varepsilon(t)$ does not change the signal

$$h(t) = h(t) \cdot \varepsilon(t) \,. \tag{15.13}$$

Taking the Fourier transform, and using the multiplication property (9.75), together with the spectrum of the step function (9.92), we obtain

$$H(j\omega) = \frac{1}{2\pi} H(j\omega) * \left[\pi\delta(\omega) + \frac{1}{j\omega} \right] \,. \tag{15.14}$$

We proceed using the selectivity property of the Delta impulse $\delta(\omega)$

$$H(j\omega) = \frac{1}{2} H(j\omega) + \frac{1}{2\pi} H(j\omega) * \frac{1}{j\omega} \,, \tag{15.15}$$

and from there it follows immediately that

$$H(j\omega) = \frac{1}{\pi} H(j\omega) * \frac{1}{j\omega} \,. \tag{15.16}$$

This result means first of all that the spectrum of a causal signal is not changed by convolution with $\frac{1}{j\omega}$ and division by π. In order to illustrate this better, we split (15.16) into real and imaginary parts:

$$\mathrm{Re}\{H(j\omega)\} = \frac{1}{\pi}\mathrm{Im}\{H(j\omega)\} * \frac{1}{\omega} \tag{15.17}$$

$$\mathrm{Im}\{H(j\omega)\} = -\frac{1}{\pi}\mathrm{Re}\{H(j\omega)\} * \frac{1}{\omega} \,. \tag{15.18}$$

As $\frac{1}{j\omega}$ is purely imaginary, it is the real part of the spectrum $\mathrm{Re}\{H(j\omega)\}$ that determines the imaginary part of the convolution product and vice versa. We can recognise that the real and imaginary parts can be transformed into each other by an operation which is identical apart from its sign. This operation is called the *Hilbert transform*, and is defined by

$$\mathcal{H}\{X(j\omega)\} = \frac{1}{\pi} X(j\omega) * \frac{1}{\omega} = \frac{1}{\pi} \int_{-\infty}^{\infty} \frac{X(j\eta)}{\omega - \eta} \, d\eta \,. \tag{15.19}$$

The Hilbert transform consists of a convolution with $\frac{1}{\omega}$ and a division by π, and is also an LTI-system, although here it is defined in the frequency-domain. In contrast to the Laplace and Fourier transforms the Hilbert transform is not a transformation between the time-domain and frequency-domain, rather it assigns a function of variables (here frequency) to another function of the same variables, which is then the *Hilbert transform*. Calculating the Hilbert transform by evaluating the integral in (15.19) demands some care, as the denominator becomes zero for $\eta = \omega$. The correct procedure is shown in detail in [19, 20], for example.

With the Hilbert transform we can concisely formulate the relationship between the real and imaginary parts of the spectrum of causal signals:

$$\begin{aligned} \mathrm{Re}\{H(j\omega)\} &= \mathcal{H}\{\mathrm{Im}\{H(j\omega)\}\} \tag{15.20} \\ \mathrm{Im}\{H(j\omega)\} &= -\mathcal{H}\{\mathrm{Re}\{H(j\omega)\}\} \,. \tag{15.21} \end{aligned}$$

We can now reach the conclusion that the spectrum of a causal signal has the characteristic that its real and imaginary parts can be found from each other by the Hilbert transform (15.20, 15.21). As the impulse response of a causal system is a causal function, this is true for the frequency response of causal LTI-systems. Clearly the real and imaginary parts of the frequency response of an LTI-system, and likewise the magnitude and phase, cannot be given independently from each other.

The result derived in this section can be easily extended to signals $h(t)$ that contain a delta impulse $h_0\delta(t)$ at the origin, which we had initially excluded. We replace $h(t) \circ\!\!-\!\!\bullet H(j = \omega)$ in (15.13) – (15.18) by $h(t) - h_0\delta(t) \circ\!\!-\!\!\bullet H(j\omega) - h_0$. In place of (15.20) and (15.21), this yields

$$\begin{aligned} \mathrm{Re}\{H(j\omega)\} &= \mathrm{Re}\{h_0\} + \mathcal{H}\{\mathrm{Im}\{H(j\omega)\}\} \tag{15.22} \\ \mathrm{Im}\{H(j\omega)\} &= \mathrm{Im}\{h_0\} - \mathcal{H}\{\mathrm{Re}\{H(j\omega)\}\} \,. \tag{15.23} \end{aligned}$$

15.2.2.2 Discrete Signals

The case for discrete causal signals is very similar to that which we have just examined for continuous signals. We start with a causal discrete signal $h[k]$, that does not change when multiplied with the the unit step function:

$$h[k] = h[k] \cdot \varepsilon[k] \,. \tag{15.24}$$

The spectrum of this signal is obtained with the multiplication property (12.49) and the spectrum of the unit step function (12.33)

$$H(e^{j\Omega}) = \frac{1}{2\pi} H(e^{j\Omega}) \circledast \left[\frac{1}{1 - e^{-j\Omega}} + \frac{1}{2} \text{⊥⊥⊥}\left(\frac{\Omega}{2\pi}\right) \right] \tag{15.25}$$

and using the selectivity property

$$H(e^{j\Omega}) = \frac{1}{2} H(e^{j\Omega}) + \frac{1}{2\pi} H(e^{j\Omega}) \circledast \frac{1}{1 - e^{-j\Omega}} \,. \tag{15.26}$$

Finally it follows

$$H(e^{j\Omega}) = \frac{1}{\pi} H(e^{j\Omega}) \circledast \frac{1}{1 - e^{-j\Omega}} \,. \tag{15.27}$$

In order to find a similar relationship between (15.17) and (15.18), we have to deal with the value $h[0]$ at $k = 0$. Using elementary trigonometric equations we write

$$\frac{1}{1 - e^{-j\Omega}} = \frac{1}{2} + \frac{1}{2j \tan \frac{\Omega}{2}} \,. \tag{15.28}$$

Carrying out cyclic convolution $\circledast$ of $H(e^{j\Omega})$ with the constant $\frac{1}{2\pi}$ yields the value $h[0]$, so that instead of (15.27), we can use

$$H(e^{j\Omega}) = h[0] + H(e^{j\Omega}) \circledast \frac{1}{2\pi j \tan \frac{\Omega}{2}} \tag{15.29}$$

as well. Re-writing as real and imaginary parts gives

$$\text{Re}\{H(e^{j\Omega})\} = \text{Re}\{h[0]\} + \text{Im}\{H(e^{j\Omega})\} \circledast \frac{1}{2\pi \tan \frac{\Omega}{2}} \tag{15.30}$$

$$\text{Im}\{H(e^{j\Omega})\} = \text{Im}\{h[0]\} - \text{Re}\{H(e^{j\Omega})\} \circledast \frac{1}{2\pi \tan \frac{\Omega}{2}} \,. \tag{15.31}$$

Here again are the relationships between real and imaginary parts that we can simplify with a suitable definition of the Hilbert transform for periodic spectra. This is

$$\boxed{\mathcal{H}_*\{X(e^{j\Omega})\} = X(e^{j\Omega}) \circledast \frac{1}{2\pi \tan \frac{\Omega}{2}} = \frac{1}{2\pi} \int_{-\pi}^{\pi} \frac{X(e^{j\eta})}{\tan(\frac{\Omega-\eta}{2})} \, d\eta \,.} \tag{15.32}$$

The Hilbert transform $\mathcal{H}_*$ is also a linear, time-invariant system. Thus we obtain the relation between the real and the imaginary part of the spectrum of a causal discrete signal

$$\text{Re}\{H(e^{j\Omega})\} = \text{Re}\{h[0]\} + \mathcal{H}_*\{\text{Im}\{H(e^{j\Omega})\}\} \tag{15.33}$$

$$\text{Im}\{H(e^{j\Omega})\} = \text{Im}\{h[0]\} - \mathcal{H}_*\{\text{Re}\{H(e^{j\Omega})\}\}. \tag{15.34}$$

These relationships are very similar to their continuous-time counterparts (15.22, 15.23), and it is also the case that with discrete causal LTI-systems, the real and imaginary parts (likewise magnitude and phase of the frequency response) cannot be given independently from each other.

15.3   Signals with a One-Sided Spectrum

The properties of causal signals and their spectra can also be extended to right-sided spectra and their corresponding functions of time, using the duality between the time and frequency-domains. We thus obtain results that are closely related to those found in Section 15.2. Because of the duality, the time-domain and frequency-domain are simply interchanged.

We start with a right-sided spectrum that contains no negative frequency components

$$H(j\omega) = 0 \quad \text{for} \quad \omega < 0 \tag{15.35}$$

and are interested in the special properties of the time signal $h(t) = \mathcal{F}^{-1}\{H(j\omega)\}$, that is limited by the unilateral spectrum. First of all we establish that $H(j\omega)$ does not exhibit conjugated symmetry (compare (9.49)). Then it follows that $h(t)\circ\!\!-\!\!\bullet H(j\omega)$ is not a real function, but in fact consists of both real and imaginary parts. Starting with

$$H(j\omega) = H(j\omega) \cdot \varepsilon(\omega) \tag{15.36}$$

we can proceed in the same way as for equations (15.13) to (15.21), in order to obtain a dual result like (15.20), (15.21):

$$\text{Re}\{h(t)\} = -\mathcal{H}\{\text{Im}\{h(t)\}\} \tag{15.37}$$

$$\text{Im}\{h(t)\} = \mathcal{H}\{\text{Re}\{h(t)\}\}. \tag{15.38}$$

This result can also be extended to signals with non-zero mean having a Dirac impulse $2\pi H_0\delta(\omega)\bullet\!\!-\!\!\circ H_0$ in their spectrum

$$\text{Re}\{h(t)\} = \text{Re}\{H_0\} - \mathcal{H}\{\text{Im}\{h(t)\}\} \tag{15.39}$$

$$\text{Im}\{h(t)\} = \text{Im}\{H_0\} + \mathcal{H}\{\text{Re}\{h(t)\}\} \tag{15.40}$$

For time signals, the same definition of the Hilbert transform as in (15.19) is true:

$$\mathcal{H}\{x(t)\} = \frac{1}{\pi}x(t) * \frac{1}{t} = \frac{1}{\pi} \int\limits_{-\infty}^{\infty} \frac{x(\tau)}{t-\tau}d\tau. \qquad (15.41)$$

Here the independent variable is in contrast to (15.19) just called t and not ω. A shorter and significantly clearer way of carrying out the Hilbert transform by calculating the convolution integral (15.41) can be obtained by switching to the frequency-domain. Using the convolution property (9.70) on (15.41) yields

$$\mathcal{H}\{x(t)\} = \mathcal{F}^{-1}\left\{\frac{1}{\pi}X(j\omega) \cdot \mathcal{F}\left\{\frac{1}{t}\right\}\right\}. \qquad (15.42)$$

The transform pair (9.39) follows

$$\mathcal{H}\{x(t)\} = \mathcal{F}^{-1}\left\{-jX(j\omega)\,\text{sign}\,(\omega)\right\}. \qquad (15.43)$$

The Hilbert transform $\hat{x}(t)$ of a time signal $x(t)$ is obtained if a new spectrum is formed from the spectrum $X(j\omega)$ by inverting the sign for $\omega < 0$ and multiplying by $-j$. The function of time $\hat{x}(t)$ corresponding to this spectrum is the Hilbert transform of $x(t)$.

This procedure is already well known, if we think of the Hilbert transform of a time signal as the effect of a system $\mathcal{H}$ on the input signal $x(t)$ (Fig 15.3). The function $1/(\pi t)$ is then the impulse response of the system $\mathcal{H}$ and (15.41) describes the convolution of the input signal $x(t)$ with the impulse response. The alternative calculation in the frequency-domain is the multiplication of the input signal's spectrum $X(j\omega)$ with the transfer function of the system $\mathcal{H}$

$$H_{\mathcal{H}}(j\omega) = -j\text{sign}(\omega). \qquad (15.44)$$

This method using the frequency-domain is significantly simpler and safer to use than evaluating the convolution integral (15.41). Furthermore, exploiting duality this method can also be used on (15.19) and likewise on discrete causal systems and unilateral periodic spectra.

From (15.40), it can be seen that a certain real part $\text{Re}\{h(t)\}$, also defines the corresponding imaginary part apart from an additive constant, if the spectrum has no contribution for $\omega < 0$. Therefore, from a real signal $x(t)$ a new signal $x_1(t)$ can be derived by

$$x_1(t) = x(t) + j\mathcal{H}\{x(t)\}. \qquad (15.45)$$

The relations (15.39) and(15.40) hold for its real and imaginary components. It therefore has a one-sided spectrum $X_1(j\omega)$. The mean value of the imaginary part has been arbitrarily set to $\text{Im}\{H_0\} = 0$; it could have been any other value.

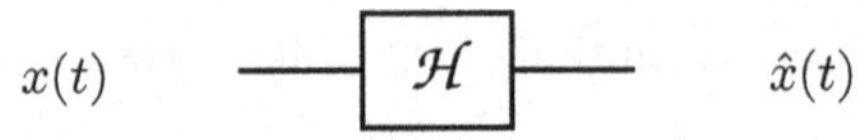

$$x(t) \quad \boxed{\mathcal{H}} \quad \hat{x}(t)$$

$$x(t) \quad * \quad \frac{1}{\pi t} \quad = \quad \hat{x}(t)$$

$$X(j\omega) \quad \cdot \quad \frac{1}{\pi}\left[-j\pi\,\text{sign}(\omega)\right] \quad = \quad \hat{X}(j\omega)$$

Figure 15.3: Hilbert-transform in the time-domain and frequency-domain

The complex signal of time $x_1(t)$ is called the *analytical signal* corresponding to the real signal $x(t)$. The analytical signal allows a simple description of important communications and signal processing systems like modulation, sampling of band-pass signals, filter banks and others. The following example shows a typical application.

─── **Example 15.2**

Figure 15.4 shows an arrangement for transmission of a real signal $x(t)$. Its spectrum $X(j\omega)$ has a band-pass characteristic with limit frequencies ω_1 and ω_2. In general, $X(j\omega)$ is complex, but in Figure 15.4 only a real valued spectrum is shown, for the sake of simplicity. $X(j\omega)$ has conjugate symmetry (9.49), as $x(t)$ is real, and because of this symmetry, the left sideband is a reflection of the right and contains no further information.

It is advantageous to transmit only one of the two sidebands, so that only half the bandwidth is needed. The missing sideband can be reconstructed by the receiver as the symmetry is known. It is also useful to reduce the bandwidth if the signal is going to be sampled (see Chapter 11.3.2). From Table 11.2 we can see that a complex band-pass signal with a unilateral spectrum (shown in Figure11.14) requires only half the sampling frequency that a real band-pass signal needs. It now remains to be shown how a complex band-pass signal with a unilateral spectrum is formed from a real band-pass signal.

This is where the Hilbert transform comes in. The signal $x_1(t)$ (15.45) has the real and imaginary parts

$$\text{Re}\{x_1(t)\} = x(t)\,, \qquad \text{Im}\{x_1(t)\} = \mathcal{H}\{x(t)\} \qquad (15.46)$$

and the desired right-sided spectrum $X_1(j\omega)$. It can be critically sampled with $\omega_a = \omega_2 - \omega_1$, regardless of the specific values of ω_1 and ω_2. The resultant signal $x_2(t)$ has a periodic spectrum that does not overlap. The original signal $x(t)$ can

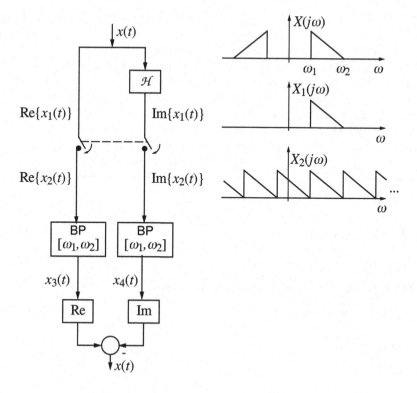

Figure 15.4: Signal transmission with the Hilbert transformator

be obtained by filtering both of the real signals $\text{Re}\{x_2(t)\}$ and $\text{Im}\{x_2(t)\}$ using a complex band-pass BP, with transfer function (11.40)

$$H_{\text{BP}}(j\omega) = \begin{cases} T & \omega_1 < \omega < \omega_2 \\ 0 & \text{otherwise} \end{cases}.$$

and the sampling interval $T = \dfrac{2\pi}{\omega_2 - \omega_1}$. The signals $x_3(t)$ and $x_4(t)$ produced are both complex signals, and their combination

$$x_3(t) + j x_4(t) = x_1(t)$$

yields the signal $x_1(t)$, which has a one-sided spectrum. The real part $\text{Re}\{x_1(t)\}$ is the real original signal $x(t)$, and $x_3(t)$ and $x_4(t)$ must be combined accordingly

$$\text{Re}\{x_1(t)\} = \text{Re}\{x_3(t) + j x_4(t)\} = \text{Re}\{x_3(t)\} - \text{Im}\{x_4(t)\}$$

for $x(t)$ to be recovered. As the input signal of both complex band-pass filters is purely real, it is clear that in one branch only the real part of the impulse response must be considered, and in the other only the imaginary part.

Compared to critical sampling of real valued band-pass signals (see Chapter 11.3.4), the sampling rate is reduced by a factor of two. However, consider that in our example, each sampling value has a real and an imaginary part, while in Chapter 11.3.4 all sampling values were real. The required number of real sampling values per time is thus the same. However, we may use analog-to-digital converters with half the sampling rate and we need not pay attention to the relation between the band edges ω_1 and ω_2 for critical sampling.

■

15.4   Exercises

Exercise 15.1

Which of the following discrete systems are causal?

a) $y[k] = c_1 x[k+1] + c_0 x[k]$

b) $y[k] = a[k]x[k]$

c) $y[k] = a[k+1]x[k]$

d) $y[k] = \sin(\pi \cdot x[k])$

e) $y[k] = x[2k]$

Exercise 15.2

For $\omega_0 > 0$, calculate the Hilbert transform of a)$e^{j\omega_0 t}$, b) $\sin \omega_0 t$, c) $\cos \omega_0 t$ and d) $\cos 2\omega_0 t$ in the frequency-domain. Note: see Figure15.3. For each part, give the phase shift between the input and output signal caused by the Hilbert transform.

Exercise 15.3

Consider the real signal $x(t)$ with spectrum $X(j\omega)$. Some properties of the Hilbert transform $y(t) = \mathcal{H}\{x(t)\} = \dfrac{1}{\pi t} * x(t)$ are to be investigated. The impulse response of the Hilbert transformator is indicated by $h(t)$.

a) Give $H(j\omega) \bullet\!\!-\!\!\circ h(t)$ and sketch the magnitude and phase of $H(j\omega)$. How does the Hilbert transform affect $|X(j\omega)|$ and $\arg\{X(j\omega)\}$?

b) Is $y(t)$ real, imaginary or complex? What symmetry does the Hilbert transform of the even part $x_e(t)$ and the odd part $x_o(t)$ have?

c) Calculate the cross-correlation function of the (deterministic) signals $x(t)$ and $y(t)$ for $\tau = 0$, i.e. $\varphi_{xy}(0) = \int_{-\infty}^{\infty} x(t)y(t)\, dt$. Which property of the Hilbert transform of real signals can be derived from this result? Note: see Chapter 9.8 and 9.9.

d) Compute the Hilbert transform of the Fourier series

$$x_F(t) = \frac{A_0}{2} + \sum_{\nu=1}^{\infty} [a_\nu \cos(\omega_0 \nu t) + b_\nu \sin(\omega_0 \nu t)], \text{ with } \omega_0 > 0.$$

Exercise 15.4

Given the signal $x(t) = \dfrac{\omega_g}{\pi} \cdot \dfrac{1 - \cos \omega_g t}{\omega_g t}$ and its Hilbert transform $\hat{x}(t) = \mathcal{H}\{x(t)\}$.

a) Calculate and sketch $X(j\omega) \bullet\!\!-\!\!\circ x(t)$.
   Note: Use the multiplication theorem of the Fourier transform, Chapter 9.7.5.

b) Determine $\hat{X}(j\omega) \bullet\!\!-\!\!\circ \hat{x}(t)$ graphically and calculate $\hat{x}(t)$. Note: consider Figure 15.3.

c) Let $x_a(t) = x(t) + j\hat{x}(t)$. Which special property exhibits $X_a(j\omega) \bullet\!\!-\!\!\circ x_a(t)$?

d) Calculate $X_a(j\omega)$ formally in dependence on $X(j\omega)$.

Exercise 15.5

What is the connection between the energy of a signal $x(t)$ and the Hilbert transformed signal $\hat{x}(t) = \mathcal{H}\{x(t)\}$? Assume that the signal has no DC component. Note: refer to the Parseval theorem in Chapter 9.8.

Exercise 15.6

Let $X_1(j\omega)$ be the spectrum of the real signal $x_1(t)$ and $X_2(j\omega) = (1 + \text{sign}(\omega))X_1(j\omega)$ the one-sided spectrum of a signal $x_2(t)$.

a) Determine $x_2(t)$ in dependence on $x_1(t)$.
   Note: Consider which transform corresponds to the required convolution.

b) Determine the relation of the signal energies of $x_1(t)$ und $x_2(t)$. State both energies directly in the frequency domain.

c) Determine the relation of the signal energies of $x_1(t)$ und $x_2(t)$ by a consideration in the time domain. Use the result of Exercise 15.5.

Exercise 15.7

A sketch of the real function $P(j\omega)$ is shown, which is part of the transfer function

$H(j\omega) = P(j\omega) + jQ(j\omega)$ of a causal system.

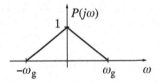

a) Determine the corresponding impulse response $h(t) \circ\!\!-\!\!\bullet H(j\omega)$.   Note:
   $\varepsilon(0) = \dfrac{1}{2}$.

b) Find the imaginary function $Q(j\omega)$ with (15.18).

c) What symmetry do $p(t) \circ\!\!-\!\!\bullet P(j\omega)$, $q(t) \circ\!\!-\!\!\bullet Q(j\omega)$ and $Q(j\omega)$ show?

Exercise 15.8

An audio signal with the spectrum $S(j\omega)$ is to be transmitted. In order to save bandwidth, the signal spectrum is 0 for $|\omega| > \omega_g = 2\pi \times 4$ kHz.

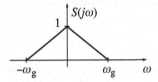

a) The signal $s(t)$ is modulated by $\cos(\omega_0 t)$.

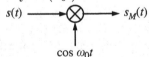

   Give the spectrum of $s_M(t)$.

b) How much bandwidth does $S_M(j\omega)$ use compared to $S(j\omega)$?

c) To remove the disadvantage found in part b), the signal will be transmitted according to the following representation.

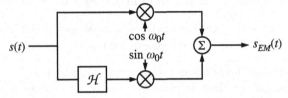

   This method, known as single sideband modulation (SSB) takes advantage of the redundancy present in the spectrum of a real signal. Draw the spectrum of $S_{EM}(j\omega)$.

d) How much bandwidth does $S_{EM}(j\omega)$ use compared to $S(j\omega)$?

16 Stability and Feedback Systems

As well as causality, another important criterion that must be met for a system to be implemented is stability. If a system is stable, and the input signal is bounded, then the output signal cannot grow beyond limit. This condition must be fulfilled, if continuous systems are to be realised based on the fundamental physical principle of conservation of energy, for example, electrical, optical, mechanical, hydraulic or pneumatic systems. For a discrete system the signal amplitude must stay within certain limits as defined by the permitted numerical range of the computer being used. The first thing we will do in this chapter is investigate the connections between stability, the frequency response and the impulse response for general LTI-systems. Then we will restrict ourselves to causal systems and consider stability tests using the pole-zero diagram. Finally we will discuss some typical uses of systems with feedback. We will again be dealing with continuous and discrete systems at the same time.

16.1 BIBO, Impulse Response and Frequency Response Curve

There are various possible ways of defining the stability of a system. Among them, BIBO-stability is particularly useful for LTI-systems. First of all we will introduce the concepts *bounded function* and *bounded series*.

Definition 21: Bounded function, bounded series

A function is said to be bounded when its magnitude is less than a fixed limit for all time t:

$$|x(t)| < M_1 < \infty, \ \forall\, t\,. \tag{16.1}$$

Correspondingly, the condition

$$|x[k]| < M_2 < \infty, \ \forall\, k\,. \tag{16.2}$$

is sufficient for a series to be bounded.

With these definitions we can define stability:

Definition 22:   Stability

*A continuous-time (discrete-time) LTI-system is stable if it reacts to a bounded
input function $x(t)$ (input series $x[k]$) with a bounded output function $y(t)$
(output series $y[k]$).*

This can be neatly expressed as:

$$\text{Bounded Input} \quad \longrightarrow \quad \text{Bounded Output.}$$

This stability definition is therefore called BIBO-stability.

As with causality, stability conditions can be given for the impulse response
of LTI-systems. From these, we can also find connections with the frequency
response and the transfer function. We will derive this separately for continuous
and discrete systems.

16.1.1   Continuous LTI-Systems

From the general definition of BIBO-stability, the following condition for the stability of continuous LTI-systems can be derived:

A continuous LTI-system is stable if and only if its impulse response can
be absolutely integrated:

$$\int_{-\infty}^{\infty} |h(t)|\, dt < M_3 < \infty. \qquad (16.3)$$

We will first show that this condition is sufficient for BIBO-stability. To do this
we form the magnitude of the output signal $|y(t)|$ using the convolution integral.
Taking a bounded input signal $x(t)$ (16.1) and an impulse response $h(t)$ that can
be absolutely integrated, we can directly find an upper bound for $|y(t)|$:

$$
\begin{aligned}
|y(t)| &= \left| \int_{-\infty}^{\infty} x(\tau) h(t-\tau) d\tau \right| \leq \int_{-\infty}^{\infty} |x(\tau)|\, |h(t-\tau)| d\tau \\
&< \int_{-\infty}^{\infty} M_1 |h(t-\tau)| d\tau \quad = \quad M_1 \int_{-\infty}^{\infty} |h(t)| dt \\
&< \quad M_1 M_3 < \infty.
\end{aligned}
\qquad (16.4)
$$

Note that the impulse response *must* be abolutely integratable. In order to
show this, we consider the rather unpleasant bounded input signal

$$x(t) = \frac{h^*(-t)}{|h(-t)|}, \qquad (16.5)$$

and calculate the value

$$y(0) = \int_{-\infty}^{\infty} x(\tau)h(-\tau)d\tau = \int_{-\infty}^{\infty} \frac{h^*(\tau)h(\tau)}{|h(\tau)|}d\tau = \int_{-\infty}^{\infty} |h(\tau)|d\tau. \qquad (16.6)$$

The result is only finite when the impulse response can be absolutely integrated, otherwise $|y(t)|$ would not be bounded.

As we discovered in Chapter 9.2.2, if the impulse response can be absolutely integrated, the Fourier transform $H(j\omega) = \mathcal{F}\{h(t)\}$, – the frequency response of the stable system – must exist. Because the Fourier integral is bounded (see (9.5)), it can be analytically continued, and is the same as the Laplace transform on the imaginary axis $s = j\omega$. That means that for stable systems, the imaginary axis of the s-plane is part of the region of convergence of the system function. The frequency response of a stable system cannot have any singularities or discontinuities.

16.1.2 Discrete Systems

For discrete LTI-systems:

> A discrete LTI-system is stable if and only if its impulse response is summable:
>
> $$\sum_{k=-\infty}^{\infty} |h[k]| < M_4 < \infty. \qquad (16.7)$$

The proof is exactly as for continuous systems.

It can be shown in the same way how the existence of the frequency response as the Fourier transform $H(e^{j\Omega}) = \mathcal{F}_*\{h[k]\}$ of the impulse response $h[k]$ agrees with the transfer function $H(z) = \mathcal{Z}\{h[k]\}$ on the unit circle of the z-plane. Correspondingly, the frequency response of a stable system must not have any discontinuities or singularities.

16.1.3 Examples

We will clarify the use of the stability criteria (16.3) for continuous systems with a few examples. Stability is shown in the same way for discrete systems, using (16.7).

── **Example 16.1**

For the simplest example we consider a system with impulse response

$$h(t) = e^{-at}\varepsilon(t) \qquad a \in \mathbb{R}. \qquad (16.8)$$

To determine the stability from the criteria in(16.3), we investigate whether the impulse response can be absolutely integrated:

$$\int_{-\infty}^{\infty} |e^{-at}\varepsilon(t)|\, dt = \int_{0}^{\infty} e^{-at} dt = \begin{cases} \dfrac{1}{a} & \text{for } a > 0 \\[2ex] \infty & \text{otherwise} \end{cases} \tag{16.9}$$

and obtain the result: the system is stable for $a > 0$.

For $a = 0$, the impulse response takes the form of the step function $h(t) = \varepsilon(t)$. We recognise the impulse response of an integrator from Chapter 8.4.4.1, and with (16.9) we can say the following about the system stability: an integrator is not BIBO-stable. This is not surprising when we recall the response of an integrator to a step function $\varepsilon(t)$. The integral over this bounded input signal grows with increasing time over all limits and thus violates the condition for BIBO-stability. ∎

Example 16.2

When we dealt with the sampling theorem in Chapter 11.3.2, we used a system with a rectangular frequency response and a sinc function impulse response (11.35) as an interpolation filter for sampling. This is also called an *ideal low-pass filter*. We now want to investigate whether such a system can be realised.

Considering the impulse response in Figure 11.11, it can be immediately seen that the impulse response of the ideal low-pass is a bilateral signal. The ideal low-pass is therefore not causal.

As the frequency response of a stable system cannot have any discontinuities, we suspect that the ideal low-pass is also unstable. In order to confirm this, the unit of time in (11.35) is chosen so that $T = \pi$ for simplification, and the impulse response is

$$h(t) = \mathrm{si}(t)\,. \tag{16.10}$$

We now have to determine whether the integral

$$\int_{-\infty}^{\infty} |h(t)|\, dt = \int_{-\infty}^{\infty} \frac{|\sin(t)|}{|t|}\, dt = 2\int_{0}^{\infty} \frac{|\sin(t)|}{|t|}\, dt \tag{16.11}$$

has a finite value or grows beyond limit. We do this by estimating the area under $|h(t)|$ with a series of triangles whose areas are all less than the area of the individual sidelobes in Figure 16.1. The baselines of the triangles each have width π and heights decreasing by $1/t$. The infinite sum of these triangle area is a lower bound for the area under the magnitude component of the impulse response $|h(t)|$:

$$\sum_{k=0}^{\infty} \frac{1}{2} \cdot \pi \cdot \frac{1}{\pi k + \frac{3\pi}{2}} = \frac{1}{2} \sum_{k=0}^{\infty} \frac{1}{k + \frac{3}{2}} < \frac{1}{2} \int_{-\infty}^{\infty} |h(t)|\, dt\,. \tag{16.12}$$

The sum itself is a known harmonic series, and does not converge, which shows that the impulse response of the ideal low-pass cannot be absolutely integrated. The ideal low-pass is therefore neither causal nor stable.

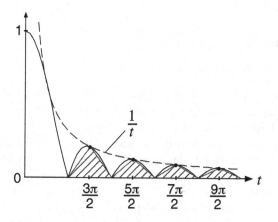

Figure 16.1: Estimating the area under the |si|-function

After this example we have to ask whether it is worth examining such idealised systems, when they are impossible to implement. The answer is typical from the point of view of systems theory:

- The ideal low-pass filter with an entirely real rectangular frequency plot is a very simple concept. It allows many operations to be considered in the simplest form, for example, limiting spectra, and reconstructing continuous signals from samples.

- The ideal low-pass cannot be precisely realised but can be approximated if some concessions can be accepted. An example for this could be:

  - Noncritical sampling.
    Oversampling allows the steepness of the edges to be reduced (see Figure 11.9). The impulse response then decays more quickly and therefore can be absolutely integrated and stable behaviour has been achieved.

  - Permitting a time-delay.
    The impulse response of the ideal low-pass filter can be approximated well by a causal system if it is shifted far enough to the right.

16.2   Causal Stable LTI-Systems

LTI-systems that are causal and stable at the same time can be recognised by
the location of their singularities in the complex s-plane. We will first give their
properties in a general form and then specialise the result for LTI-systems with
rational transfer functions.

16.2.1   General Properties

16.2.1.1   Continuous Systems

The ROC for the transfer function of a continuous LTI-system gives us important
information about the causality and stability of the system. We will look back at
the ROC of the Laplace transform in Chapter 4.5.3.

Causality means that the impulse response is a right-sided signal. The ROC
of the Laplace transform for right-sided signals lies to the right of a line in the
s-plane, which is parallel to the imaginary axis. Therefore, the transfer function
of a causal LTI-system converges in a region of the s-plane that lies to the right
of the singularity with the greatest real part.

Now we look again at the property of stability. It implies that it must be possible to absolutely integrate the impulse response. We have already said that this
means the imaginary axis of the s-plane must be part of the region of convergence
of the transfer function. As this must lie to the right of a border parallel to the
imaginary axis, however, this means that the border must lie to the left of the
imaginary axis, in the left half of the s-plane (see Figure 4.5).

Where then, do the singularities of the transfer function lie? Since the ROC is
free from singularities, they must lie to the left of the line, in the left half of the
s-plane.

We can summarise this insight with the following statement.

All of the singularities of the transfer function $H(s)$ for a causal, stable,
continuous LTI-system must lie in the left half of the s-plane.

16.2.1.2   Discrete Systems

For discrete systems we look at the properties of the ROC of the z-transform, to
find similar results.

For a causal discrete LTI-system, the ROC of the transfer function $H(z)$ lies
outside a circle around the origin of the z-plane. Because the unit circle must be
part of the ROC for a stable system, the radius of the border must be less than 1,
and since the singularities cannot lie in the ROC, they must be inside this circle.
Therefore:

> All of the singularities of the system function $H(z)$ for a causal, stable, discrete LTI-system must lie within the unit circle of the z-plane.

16.2.2 LTI-Systems with Rational Transfer Functions

The general statements can be made more precise for systems for which the transfer function is rational. Their singularities are simple or multiple poles that define the characteristic frequencies of the system. We can describe these systems with pole-zero diagrams.

16.2.2.1 Continuous Systems

For continuous LTI-systems a stability criterion follows immediately.

> The poles of the system function $H(s)$ for a causal and stable continuous LTI-system lie in the left half of the s-plane.

To illustrate this property we must recall the internal term $y_{\text{int}}(t)$ of the output signal from Chapter 7.3.3. It describes the part of the output signal that is caused by the initial state of the system. The Laplace transform of the internal term can be represented (see (7.92)) by partial fractions with poles s_i. In the time-domain this corresponds to the sum of the system's characteristic frequencies, which we have only written out here for N simple poles:

$$Y_{\text{int}}(s) = \sum_{i=1}^{N} \frac{A_i}{s - s_i}$$

$$y_{\text{int}}(t) = \sum_{i=1}^{N} A_i e^{s_i t} \varepsilon(t). \tag{16.13}$$

For pole $s_i = \sigma_i + j\omega_i$, the real part $\sigma_i < 0$ is in the left half of the plane, so the corresponding characteristic frequency decays

$$\lim_{t \to \infty} e^{s_i t} = \lim_{t \to \infty} e^{\sigma_i t} \cdot e^{j\omega_i t} = 0. \tag{16.14}$$

The condition for stability, that all poles lie in the left half of the s-plane, means that in the time-domain, the response to the initial conditions decays with time, just what we expect for a stable system. This is clear from Figure 3.3 which shows that only poles in the left half of the s-plane correspond to decaying exponential oscillations. The decay of the internal term means that the inital state of a system does not define the system behaviour, as long as one is prepared to wait long enough.

The location of the zeros in the complex plane does not influence the characteristic frequencies and thus has no influence on the stability of the system.

16.2.2.2   Discrete Systems

A corresponding result holds for discrete systems:

> The poles of the system function $H(z)$ of a causal and stable discrete
> LTI-system lie within the unit circle of the z-plane.

Here we can also see the connection between the location of the poles in the
complex frequency plane and the system's characteristic frequencies in the time-
domain (see Exercise 16.5). The stability conditions can also be illustrated to aid
understanding, as in Figure 13.6. The internal term of the system response here
only decays if the corresponding poles lie within the unit circle. Again, the initial
states have no influence if one waits long enough.

─── **Example 16.3**

Figure 16.2 shows four pole-zero diagrams of both continuous (left) and discrete
(right) systems. We can see immediately from the location of the poles that only
the first and third continuous system (likewise discrete system) are stable. If,
however, bilateral impulse responses are permitted – systems which are not causal
– the first three continuous systems and the first, second and fourth discrete system
are stable. The ROC must be chosen so that it includes the imaginary axis of the
s-plane, or the unit circle of the z-plane. When poles lie directly on the imaginary
axis or unit circle, this becomes impossible, and the system will always be unstable.

─── ■

16.2.3   Stability Criteria

If the transfer function of an LTI-system in rational form is determined from a
differential or difference equation, only the numerator and denominator coefficients
are obtained at first. In order to check the stability with the location of the poles,
the zeros of the denominator polynomial must be determined. This can be given
in closed form for polynomials up to the third degree, but for higher-order systems
iterative procedures are necessary. Once the digital computers needed to carry out
these procedures were unavailable, but modern computers can do this easily. To
avoid having to perform the numerical search for the zeros by hand, a series of easy
to perform stability tests were developed. Instead of calculating the individual pole
locations, they just determine whether all poles lie in the left half of the s-plane
(or unit circle of the z-plane). We will only be considering one test for continuous
and discrete systems because the tests all have essentially the same effect.

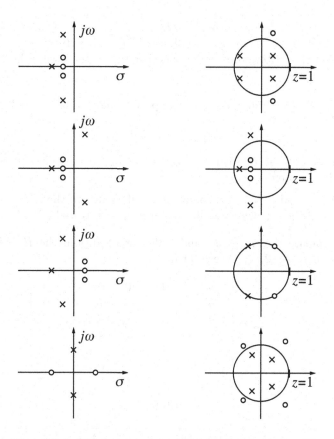

Figure 16.2: Pole-zero plots for stable and unstable systems

16.2.3.1 Continuous Systems

For continuous systems we will describe Hurwitz's stability criteria. It does not involve searching for zeros, and directly uses the coefficients of the denominator polynomial in the transfer function.

To use it we put the transfer function in the form

$$H(s) = \frac{P(s)}{Q(s)}, \tag{16.15}$$

so that the numerator polynomial

$$Q(s) = s^N + a_1 s^{N-1} + a_2 s^{N-2} + \ldots + a_{N-1} s + a_N \tag{16.16}$$

has a one as the coefficient of the highest order.

A polynomial is a *Hurwitz polynomial* if all of its zeros have a negative real part. The system is stable if its numerator polynomial $Q(s)$ is a Hurwitz polynomial, and Hurwitz's stability criteria determine whether this is the case. The test consists of two parts:

- A necessary condition for a Hurwitz polynomial is that all coefficients a_n are positive:

$$a_n > 0, \quad n = 1, \ldots, N. \qquad (16.17)$$

It can be only be fulfilled if all orders s^n in $Q(s)$ exist, as otherwise there would be a coefficient $a_n = 0$.

For $N = 1$ and $N = 2$ this condition is sufficient and the test is finished, but for $N > 2$ the following conditions must also be tested.

- To formulate the necessary and sufficient condition, the *Hurwitz determinants* for $\mu = 1, 2, \ldots, N$ are set up:

$$\Delta_\mu = \begin{vmatrix} a_1 & 1 & 0 & 0 & \cdots & 0 \\ a_3 & a_2 & a_1 & 1 & \cdots & 0 \\ a_5 & a_4 & a_3 & a_2 & \cdots & 0 \\ \vdots & \vdots & \vdots & \vdots & & \vdots \\ a_{2\mu-1} & a_{2\mu-2} & & & \cdots & a_\mu \end{vmatrix} \qquad (16.18)$$

with

$$a_\nu = 0 \text{ for } \nu > N. \qquad (16.19)$$

$Q(s)$ is a Hurwitz polynomial if all Hurwitz determinants are positive.

$$\Delta_\mu > 0 \text{ for } \mu = 1, 2, \ldots N \qquad (16.20)$$

That concludes the Hurwitz test. We illustrate how it is carried out in Example 16.4. A related procedure – Routh's stability test – can be found in [23, 19].

16.2.3.2   Discrete Systems

For discrete systems, it must be determined whether all poles lie within the unit circle. We will therefore be using a procedure which maps the complex z-plane onto the complex s-plane, where we can carry out a stability test for continuous systems.

The *bilinear transform* is a suitable way of doing this:

$$z = \frac{s+1}{s-1}, \quad s = \frac{z+1}{z-1}. \qquad (16.21)$$

It maps the inside of the unit circle $|z| < 1$ onto the left half-plane $\text{Re}\{s\} < 0$. We can show this by considering the magnitude of $s = \sigma + j\omega$

$$|z| = \sqrt{\frac{(\sigma + 1)^2 + \omega^2}{(\sigma - 1)^2 + \omega^2}} \cdot \tag{16.22}$$

For $|z| < 1$ follows $\sigma < 0$.

The stability of the discrete system

$$H(z) = \frac{P(z)}{Q(z)} \tag{16.23}$$

is guaranteed if the zeros of the denominator polynomial $Q(z)$ lie in the unit circle of the z-plane. To check this we form the the rational function $\tilde{Q}(s)$ from $Q(z)$, using the bilinear transform

$$\tilde{Q}(s) = Q\left(\frac{s+1}{s-1}\right). \tag{16.24}$$

Zeros of $\tilde{Q}(s)$ (and so also of $Q(z)$) can only arise from the numerator polynomial of $\tilde{Q}(s)$, so we test whether the numerator of $\tilde{Q}(s)$ is a Hurwitz polynomial. If it is, the zeros must lie in the left half of the s-plane and the zeros of $Q(z)$ must lie in the unit circle of the z-plane.

The bilinear transformation (16.21) in comparison with other transformations that map the area within the unit circle onto the left half of the s-plane has the advantage that $\tilde{Q}(s)$ is a rational function and we can use the Hurwitz test, for example.

On closer inspection, however, our procedure is still flawed: the point $z = 1$ is projected to the point $s = \infty$ (see (16.21)), so it is not included in the Hurwitz test. We have to test this point $Q(1)$ as well.

The stability criterion for a discrete system with transfer function $H(z)$ is now:

> A discrete system with transfer function $H(z) = \dfrac{P(z)}{Q(z)}$ is stable if the numerator polynomial of $\tilde{Q}(s) = Q\left(\dfrac{s+1}{s-1}\right)$ is a Hurwitz polynomial and if $Q(1) \neq 0$.

We will now demonstrate the stability test with an example.

Example 16.4

Figure 16.3 shows a fourth-order recursive discrete system. In order to determine its stability we first read from the block diagram:

$$Y(z) = Y(z) \left(\frac{3}{2} z^{-1} + \frac{3}{4} z^{-2} + \frac{3}{8} z^{-3} + \frac{1}{4} z^{-4} \right) + X(z) \tag{16.25}$$

and from here obtain the system function

$$H(z) = \frac{Y(z)}{X(z)} = \frac{z^4}{z^4 - \frac{3}{2} z^3 - \frac{3}{4} z^2 - \frac{3}{8} z - \frac{1}{4}} = \frac{z^4}{Q(z)}. \tag{16.26}$$

Using the bilinear transformation, the denominator $Q(z)$ becomes the partial fraction $\tilde{Q}(s)$:

$$\tilde{Q}(s) = Q \left(\frac{s+1}{s-1} \right) = \frac{-1.875 \left(s^4 - 1.4667 s^3 - 3.2 s^2 - 3.8667 s - 1 \right)}{(s-1)^4}. \tag{16.27}$$

We must now test the polynomial in parenthesis in the numerator of $\tilde{Q}(s)$ to ascertain whether or not it is a Hurwitz polynomial. The necessary condition for a Hurwitz polynomial (16.17) is not fulfilled as not all of the coefficients are positive. We can now make the following deductions:

⇒ $\tilde{Q}(s)$ has zeros with $\mathrm{Re}\{s\} \geq 0$.

⇒ $Q(z)$ has zeros with $|z| \geq 1$.

⇒ $H(z)$ has poles with $|z| \geq 1$.

⇒ The system is unstable.

Of course, the zeros of the denominator polynomial $Q(z)$ can be found using a computer. The result is

$$Q(z) = (z - 2) \left(z + \frac{1}{2} \right) \left(z^2 + \frac{1}{4} \right).$$

There is a zero at $z = 2$ which is outside the unit circle, and this confirms the result we obtained using the bilinear transform and the Hurwitz test.

■

16.3 Feedback Systems

The discussion of stability so far has been limited to systems that were exclusively described by their transfer function. Now we will consider feedback systems that we are already familiar with from Chapter 6.6.3. Of course, an LTI-system with

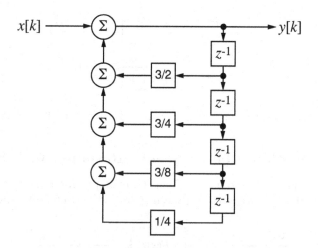

Figure 16.3: Fourth-order recursive discrete-time system

feedback *can* be described by a transfer function, as shown in Figure 6.15, but the information about the individual transfer functions of the forward and feedback paths is lost (in Figure 6.15 these are $F(s)$ and $G(s)$). The feedback principle is used in many areas, both natural and technical. We will show three typical problems that can be solved using feedback.

16.3.1   Inverting a System Using Feedback

In Example 6.9 we saw that feedback can invert the transfer function of a system. Figure 16.4 shows the situation again, but in contrast to Figure 6.15, the feedback path contains another change of sign. The transfer function of the closed loop is then

$$H(s) = \frac{Y(s)}{X(s)} = \frac{K}{1 + KG(s)} \approx \frac{1}{G(s)} \text{ for } K\,|G(s)| \gg 1. \qquad (16.28)$$

The poles of the whole feedback system $H(s)$ are the zeros of $G(s)$. They must lie in the left half of the s-plane, so that stability can be assured. The poles of $G(s)$, however, have no influence on the stability of the feedback system.

Continuous systems that have no zeros in the right s-plane are called *minimal phase systems*. Correspondingly, discrete systems with no zeros outside of the unit circle are also called minimal phase systems. For the inverse system $\frac{1}{G(s)}$ (or $\frac{1}{G(z)}$) to be stable, $G(s)$ (or $G(z)$) has to be minimal phase and additionally may not have any zeros on the imaginary axis (or on the unit circle) of the complex frequency plane.

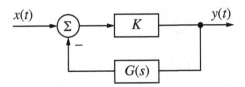

Figure 16.4: Structure of the system $G(s)$ with feedback loop

16.3.2 Smoothing the Frequency Response with Feedback

When constructing an electronic amplifier, for example, for audio use, the principle of negative feedback (illustrated in Figure 16.5) is very useful. An amplifier with a

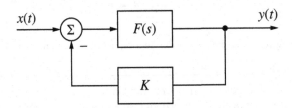

Figure 16.5: Smoothing the frequency response with feedback

very high factor of amplification $|F(j\omega)| \gg 1$ can be built easily, if strong variations in the frequency response $F(j\omega)$ can be tolerated. Some of the amplification is then sacrificed, by connecting the output back to the input with a gain factor K. The resulting frequency response

$$H(j\omega) = \frac{F(j\omega)}{1 + KF(j\omega)} \approx \frac{1}{K} \quad \text{for} \quad |K \cdot F(j\omega)| \gg 1,$$

is nearly constant. It is important, of course, that $|F(j\omega)|$ is large enough that the inequality $|K \cdot F(j\omega)| \gg 1$ is also true for small values of K, so that the negative feedback system has a sufficiently high amplification $1/K$.

16.3.3 Using Feedback to Stabilise a System

In many control system applications, the forward path (the 'plant') is a large technical installation that cannot be changed by simple means. The system behaviour can only be modified by coupling a second system to it, usually as a backwards path, creating a closed feeback loop. This second system is also called the *controller*, and is constructed so that its properties can be changed without too much effort. We will now be considering whether an unstable plant can be coupled with a feedback system, to yield a complete system that is stable.

We will start with the first-order plant from Figure 16.6. The forward path of the system has the first-order transfer function

$$H(s) = \frac{b}{s-a}, \quad a > 0, \quad b > 0$$

and a pole at $s = a > 0$ in the right half of the s-plane. It is therefore unstable. We choose for the feedback path, a P-circuit (shown in Figure 8.24) with amplification K. The transfer function of the feedback system is

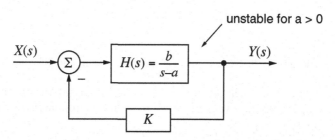

Figure 16.6: First-order plant with P-control

$$Q(s) = \frac{Y(s)}{X(s)} = \frac{H(s)}{1 + KH(s)} = \frac{b}{s - a + K \cdot b} \tag{16.29}$$

and it has a pole at $s = a - Kb$. The location of the pole can now be influenced by the amplification factor K of the P-circuit. Figure 16.7 shows the possible pole locations dependent on amplification K. For $K = 0$ the feedback has no effect and the pole lies at $s = a$. Growing values of K move the pole to the left and when $K > a/b$, it is in the left half of the plane. Sufficiently large amplification can make the overall system stable. The path on the s-plane that the pole follows, dependent of the amplification, is called the *root locus*.

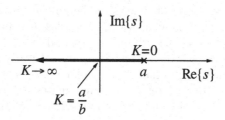

Figure 16.7: Root locus of the control circuit from Figure 16.6

Feedback can also have the opposite effect: a stable plant can be destabilised by feedback. Here we start again with Figure 16.6, but this time choose $a < 0$,

$b > 0$ and $K < 0$. The pole of the system without feedback now lies in the left half of the s-plane. For $K < 0$ it moves to the right and at $K = a/b$ it reaches the right half of the plane. Figure 16.8 shows the corresponding root locus.

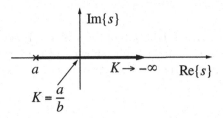

Figure 16.8: Root locus of Figure 16.6 for $a < 0, K < 0$

The influence of the P-circuit and the sign appearing at the summation node in the feedback path on the system as a whole can be summarised as follows:

• Positive feedback destabilises a system.

• Negative feedback stabilises a system.

Unfortunately, for higher-order systems the relationships are not so simple. We can show this with a second-order plant depicted in Figure 16.9. The transfer function in the forward path

$$H(s) = \frac{b}{s^2 + a} \quad , \quad a \in \mathbb{R} \tag{16.30}$$

has two poles at $s = \pm\sqrt{-a}$ and is therefore unstable for all a. Using a P-circuit for feedback the overall transfer function

$$Q(s) = \frac{Y(s)}{X(s)} = \frac{H(s)}{1 + KH(s)} = \frac{b}{s^2 + a + Kb} \tag{16.31}$$

is obtained. Figure 16.10 shows the root locus for $a < 0$.

For $K = 0$ all of the poles of the system lie on the real axis, one in the right half-plane. For $K < 0$ nothing changes, as the poles just move further away. For $K > 0$ the poles move along the imaginary axis and for $K > (-a)/b$ they form a complex conjugated pole pair on the imaginary axis. Clearly it is not possible here to move both poles into the left half of the s-plane using P-circuit feedback. This can be confirmed by a Hurwitz test. The denominator polynomial $s^2 + a + Kb$ in (16.31) is not a Hurwitz polynomial, as the coefficient of the linear term in s is equal to zero.

The system can be successfully stabilised with proportional-differential feedback as in Figure 16.11. This leads to an overall transfer function for the system

$$Q(s) = \frac{Y(s)}{X(s)} = \frac{H(s)}{1 + G(s)H(s)} = \frac{b}{s^2 + bK_2 s + (a + K_1 b)}, \tag{16.32}$$

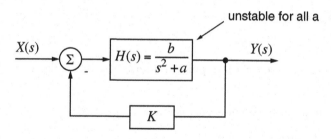

Figure 16.9: Second-order plant with proportional control

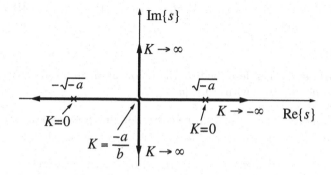

Figure 16.10: Root locus of the control circuit from Figure 16.9

for which the denominator polynomial is a Hurwitz polynomial, if the constant K_1 and K_2 are chosen so that

$$bK_2 > 0 \quad \text{and} \quad a + K_1b > 0. \tag{16.33}$$

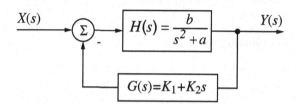

Figure 16.11: Second-order plant with proportional-differential control

16.4   Exercises

Exercise 16.1

The stability of a system with impulse response $h(t) = e^{-0.1t}\sin(5\pi t)\varepsilon(t)$ is to be investigated.

a) Can the impulse response be absolutely integrated?

b) Check the location of the poles.

Exercise 16.2

Show that (16.7) is a necessary and sufficient condition for the BIBO-stability of a discrete LTI-system.

Exercise 16.3

A discrete LTI-system is described by the following differential equation:
$y[k] = x[k] - a^6 x[k-6]$ with $0 < |a| < \infty$.

a) Determine $H_1(z) = \frac{Y(z)}{X(z)}$ and the corresponding pole-zero diagram. Give the ROC of $H_1(z)$. For what values of a is the system stable?

b) Determine the transfer function $H_2(z)$ of a second discrete system so that $H_1(z) \cdot H_2(z) = 1$. Give both possible ROCs of $H_2(z)$ and determine for each whether it is stable.

Exercise 16.4

A system is given by the differential equation $y[k] - \frac{1}{2}y[k-1] + \frac{1}{4}y[k-2] = x[k]$.

a) Give the transfer function $H(z) = \frac{Y(z)}{X(z)}$ of the system.

b) Is the system stable? This should be determined

α) with the pole-zero diagram,

β) with the bilinear transformation $z = \frac{s+1}{s-1}$ and a stability test for continuous systems.

c) Is the system minimal phase?

d) Is the system causal?

Exercise 16.5

Using the internal component of the output signal, give a motivation that a discrete system is stable if its poles lie within the unit circle of the complex z-plane (see Section 16.2.2).

Exercise 16.6

Test the stability of the systems

a) $H(s) = \dfrac{s^3 + 1}{-2.5s^4 - 11.25s^3 - 20s^2 - 17.5s - 5}$

b) $H(s) = \dfrac{s^3 - 0.125}{s^4 - 4.5s^3 + 8s^2 - 7s + 2}$

c) $H(s) = \dfrac{s + 2}{s^3 + s + 2}$,

without explicity finding the poles. Are the systems minimal phase?

Exercise 16.7

Where is the left half of the unit circle on the z-plane projected to on the s-plane when transformed by the bilinear transform (16.21)?

Exercise 16.8

A control loop with $E(s) = s$ and $G(s) = \dfrac{2V_0(s^2 + 1)}{s}$ is given.

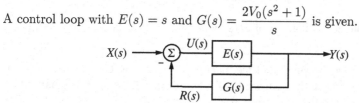

For the open loop the transfer function is $H_0(s) = \frac{R(s)}{U(s)}$ and for the closed loop, $H(s) = \frac{Y(s)}{X(s)}$.

a) Determine $H_0(s)$ and $H(s)$.

b) Is $H(s)$ stable? Where are the poles of $H(s)$?

c) Can the system be stabilised by a suitable choice of V_0?

d) $V_0 = \frac{1}{70}$. At what frequency ω_0 is the system unstable?

e) Calculate $r(t)$ for $u(t) = \sin(6t)$ and $V_0 = \frac{1}{70}$.

Exercise 16.9

Consider a system S_1 with transfer function $H(s) = \dfrac{1}{s^2 + 1}$.

a) Is $H(s)$ stable?

b) S_1 will now be incorporated into a feedback loop.

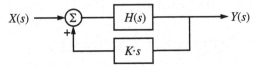

Give the new transfer function $H_r(s) = \frac{Y(s)}{X(s)}$.

c) For what values of K is the feedback system stable?

d) Test the stability for $K < 0$ with $h_r(t)$.

Exercise 16.10

A system has transfer function $H(s) = \frac{Y(s)}{X(s)}$ with $F(s) = \dfrac{s}{s^2 - 2s + 5}$.

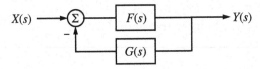

a) Draw the pole-zero diagram of $H(s)$ when $G(s) = 0$. Is $H(s)$ stable in this case?

b) To stabilise $H(s)$ proportional feedback is provided by $G(s)$ (with real amplification K). Draw the root locus of $H(s)$ for $0 < K < \infty$. For what values of K does the stabilisation succeed?

Exercise 16.11

$H(s)$ is as in Exercise 16.10, although $F(s) = \dfrac{1}{s^2 - 2s + 5}$.

a) Draw the root locus for $G(s) = K$. Can the system be stabilised?

b) If $G(s)$ is changed to provide differential feedback with K real, can the system then be stabilised? If yes, for what values of K?
   Note: consider the Hurwitz test.

17 Describing Random Signals

17.1 Introduction

All of the continuous and discrete signals that we have considered so far in the time-domain and frequency-domain were signals that could be described by mathematical functions. We could calculate the values of the signal, add and subtract signals, delay signals, and form derivatives and integrals. We found integration very useful, for convolution, for Fourier and Laplace transforms, and we also used complex integration for the inverse Laplace and inverse z-transforms. For discrete signals summation is used instead of integration. This was all possible because we had assumed that every signal had one definite value at every point in time, and that every signal could be described by a mathematical formula, however complicated that formula might be.

Many signals that occur in practise do not conform to this assumption. It would be theoretically possible to describe the speech signal from Figure 1.1 with the properties of the human vocal tract by superimposing various waves, but this would not lead to a technically realistic solution. It is completely impossible to assign functions to noise signals, or signals made up of chaotic oscillations. A new concept must be found to represent such irregular processes. Just understanding that a signal waveform can have an unpredictable value and is therefore random, does not actually help much. To deal with system inputs and outputs in the way we are used to, random signals must be described by non-random, or 'deterministic' quantities. This can be performed by the so-called expected values, which are introduced in the next section. Then we will deal with stationary and ergodic random processes, for which a significantly simpler calculation is possible with expected values. An important class of expected values are correlation functions, which will also be discussed. All of these forms for describing random signals will be introduced for continuous signals, and the chapter concludes by extending the concepts to discrete random signals.

17.1.1 What Are Random Signals?

The signals that we have been working with so far are called *deterministic signals*, which means that a signal has a known unambiguous value at every point in time. A signal can also be deterministic when it cannot be described by simple mathematical functions but instead, for example, by an infinite Fourier series.

Signals that have unknown behaviour are called *non-deterministic signals*, *stochastic signals* or *random signals*. Examples of random signals in electronics are interference signals like antenna noise, amplifier distortion or thermal resistance noise, but useful signals can also be stochastic. In communications, it is pointless transmitting a signal already known to the receiver. In fact, for a receiver, the less content of a message that can be predicted, the greater the information content in the message.

17.1.2   How Can Random Signals Be Described?

To describe random signals we can first of all try to start with the signals themselves. When the signal form itself cannot be mathematically described, it can still be measured and a graph can be obtained, for example, in Figures 17.1 and 17.2. It is not known whether Fourier, Laplace or convolution integrals of the random signals exist, or whether the methods used to calculate the spectrum or system function are even defined. Even when the existence of an integral is certain, the resulting spectrum or output signal is itself again a random variable whose behaviour we cannot make any general statements about.

We can, for example, interpret the signal in Figure 17.1 as the noise of an amplifier and calculate the response of a post-connected system. The knowledge of this output signal, however, cannot be transferred to other situations, as another amplifier of the same kind would produce another noise signal, $x_2(t)$. The first amplifier would also never repeat the noise signal $x_1(t)$, so we can do very little with an output signal calculated from it.

The solution to this problem is found not by considering individual random signals, but instead by analysing the process that produces the signals. In our example it means that we should derive general statements about the noise behaviour. Of course, it is impossible to know the noise signals in individual amplifiers in advance. Instead, the typical noise power can be given, for example. This has two significant advantages:

- the given noise power is a deterministic property which can be calculated in a normal way,

- it is the same for all amplifiers of a particular model.

We need to introduce some new terms to extend our discussion to general random signals. A process that produces random signals will be called a *random process*. The entirety of all random signals that it can produce is called an *ensemble* of random signals. Individual random signals (e.g., $x_1(t)$, $x_2(t)$, $x_i(t)$ in Figure 17.1) are called *sample functions* or *realisations* of a random process. We will concentrate on the random process that produces the signals as only it gives information on all its sample functions.

To characterise random signals deterministically, we use *statistical averages* also called just *averages*. They can be classified into characteristics of random processes which hold for a complete ensemble of random signals (*expected value*), and *time-averages*, which are found by averaging one sample function along the time-axis. In the next section we will deal with expected values and time-averages.

17.2 Expected Values

17.2.1 Expected Value and Ensemble Mean

Several different sample functions of a process are represented in Figure 17.1. We can imagine that they are noise signals, that are measured at the same time on various amplifiers of the same model. As *expected value* (also *ensemble mean*) we define the mean value that is obtained at the same time from all sample functions of the same process:

$$\mathrm{E}\{x(t_1)\} = \lim_{N \to \infty} \frac{1}{N} \sum_{i=1}^{N} x_i(t_1).\tag{17.1}$$

As we can obtain different means at different times, the expected value is in general time-dependent:

$$\mathrm{E}\{x(t_1)\} \neq \mathrm{E}\{x(t_2)\}.\tag{17.2}$$

In Figure 17.1 it can be observed that the mean of the functions $x_1(t), x_2(t), \cdots, x_i(t)$ takes another value at time t_2 than at time t_1.

Since the averaging in Figure 17.1 runs in the direction of the dashed lines, the expected value is an average *across the process*. In contrast, the time-average is taken in the direction of the time-axis, and is an average *along the process*.

The definition of the expected value in (17.1) should be understood as a formal description and not as a method for its calculation. It says that it should be determined from *all* sample functions of a process, which is in practice an impossible task. The expected value identifies the whole process, not just selected sample functions. If we wish to actually evelute an expected value, there are three available methods.

- From precise knowledge of the process the expected value can be calculated without averaging sample functions. We need tools from mathematical probability, however, that we did not want as pre-requisites.

- Averaging a finite number of sample functions can give an approximation of the expected value from (17.1). This is equivalent to the limit in (17.1) being only partially carried out. The approximation becomes more accurate as more sample functions are included.

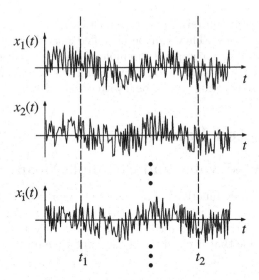

Figure 17.1: Example of sample functions $x_i(t)$

- Under certain pre-conditions, that we will deal with later in more detail, the ensemble average can also be expressed by the time-average for a sample function.

─── **Example 17.1**

We use the example of a die to show the three methods for finding an exact or approximate expected value:

- From knowledge of the process, which is the complete symmetry of the die, we can say that every number has the same probability of appearing, so the expected value is:

$$E\{x(t)\} = \frac{1 + 2 + 3 + 4 + 5 + 6}{6} = 3.5\,.$$

As the die's symmetry does not change, the expected value is the same for any time t.

- If we repeat the procedure with many dice whose symmetry properties we do not know precisely, we can also approximately calculate the ensemble average by averaging all of the numbers on the die's faces. As the expected values can in general change over time, we should take the ensemble mean for all dice at the same time.

- Changing statistical properties over time is not a concern for normal dice. We can further say that the average for many dice at the same time will yield approximately the same results as one die thrown repeatedly. The numbers that come up are the discrete values of a sample function along the time axis. In this case, the expected value can also be expressed by the time-average.

■

Before proceeding with the relationships between the ensemble-average and time-average, we will discuss some more general forms of expected values.

17.2.2 First-Order Expected Values

The expected value $E\{x(t)\}$ tells us what value to expect on average from a random process, but it does not fully characterise the process.

Figure 17.2 shows sample functions of two random processes that have the same (time-variable) average, but they clearly differ in other properties. The sample functions of random process B vary much more around the average than those of process A. In order to describe such properties we introduce the general *first-order expected value*:

$$E\{f(x(t))\} = \lim_{N\to\infty} \frac{1}{N} \sum_{i=1}^{N} f(x_i(t)). \tag{17.3}$$

In contrast to (17.1), $x(t)$ is here replaced by a function $f(x(t))$. By choosing different functions f, different first-order averages are obtained. The reason why we use the term *first*-order expected values is because they only take into account the amplitudes of the sample functions at *one* point in time. We will soon learn about higher-order expected values, where the values at more than one time are combined together. The mean according to (17.1) is contained in (17.3) for $f(x) = x$. It is also called the *linear average* and is denoted by $E\{x(t)\} = \mu_x(t)$.

For $f(x) = x^2$ we obtain the *quadratic average*:

$$E\{x^2(t)\} = \lim_{N\to\infty} \frac{1}{N} \sum_{i=1}^{N} x_i^2(t). \tag{17.4}$$

We can use it to describe the average power of a random process, for example, the noise power of an amplifier without load.

The square of the deviation from the linear average is also important. It is obtained from (17.3) for $f(x) = (x - \mu_x)^2$ and is called the *variance*:

$$E\left\{(x(t) - \mu_x(t))^2\right\} = \sigma_x^2(t). \tag{17.5}$$

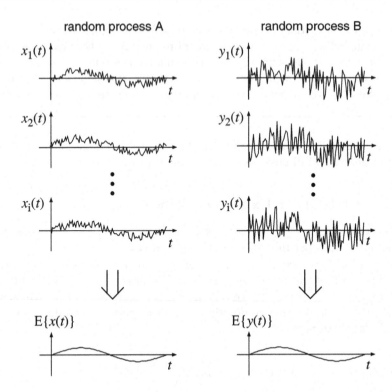

Figure 17.2: Random processes A and B with a differing distribution between the individual sample functions

Taking the positive square root of the variance gives the *standard deviation* $\sigma_x(t)$. Variance and standard deviation are measures for the spread of the amplitude around the linear average for a random signal. The random process A in Figure 17.2 clearly has a lower variance than random process B.

The linear average and the variance are by far the most frequently used first-order expected values. When combined with second-order expected values they are sufficient to characterise many common random signals. For general random signals it is possible to define more first-order expected values by choosing a different $f(x)$ in (17.3).

—————————————————————————————————— **Example 17.2**

A general characterisation with first-order expected values is obtained by putting
$$f(x) = \varepsilon(\Theta - x)$$
into (17.3) and determining the expected value dependent on the threshold Θ. We

thus obtain the *distribution function*

$$P_{x(t)}(\Theta) = \mathrm{E}\{\varepsilon(\Theta - x(t))\}.$$

It returns the probability that $x(t)$ is less than the threshold Θ. It is depicted in Figure 17.3 for the die from Example 17.1. The derivative

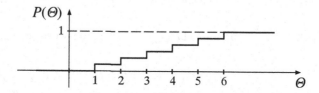

Figure 17.3: Distribution function for the numbers on a die

$$\frac{dP_{x(t)}(\Theta)}{d\Theta} =: p_{x(t)}(\Theta)$$

is called the *probability density function* of the random signal x at time t. From $P_{x(t)}(\Theta)$ or $p_{x(t)}(\Theta)$, all first-order expected values can be calculated:

$$\mathrm{E}\{f(x(t))\} = \int_{-\infty}^{\infty} f(\Theta)p_{x(t)}(\Theta)d\Theta.$$

For the die from Figure 17.3, this yields a linear average $\mu_x(t) = 3.5$, as in Example 17.1.

∎

17.2.3 Calculating with Expected Values

Expected values can be dealt with in a similar way to functions, if the corresponding calculation rules are observed. There are actually only two simple rules to remember:

- The expected value $\mathrm{E}\{\cdot\}$ is a linear operator for which the principle of superposition holds

$$\boxed{\mathrm{E}\{ax(t) + by(t)\} = a\mathrm{E}\{x(t)\} + b\mathrm{E}\{y(t)\}.} \qquad (17.6)$$

Here a and b can also be deterministic functions of time $a(t)$ and $b(t)$.

- The expected value of a deterministic signal $d(t)$ is the value of the signal itself, as $d(t)$ can be thought of as a realisation of a random process with identical sample functions:

$$\boxed{E\{d(t)\} = d(t)\,.} \tag{17.7}$$

Example 17.3

We use both rules (17.6) and (17.7) to express the variance with the linear average and square mean:

$$
\begin{aligned}
\sigma_x{}^2(t) &= E\{(x(t) - \mu_x(t))^2\} &= E\{x^2(t) - 2x(t)\mu_x(t) + \mu_x{}^2(t)\} \\
&= E\{x^2(t)\} - 2\mu_x(t)E\{x(t)\} + \mu_x{}^2(t) \\
&= E\{x^2(t)\} - \mu_x{}^2(t) \tag{17.8}
\end{aligned}
$$

It is therefore sufficient to know only two of the three quantities – the linear average, the square average and the variance – and then the third can be calculated. The relationship between them is often used in practice to calculate the variance. If N sample functions are available, $x_i(t)$ and $x_i^2(t)$ are summed and the result is divided by N. The quantity $x_i(t) - \mu_x(t)$ cannot be averaged because in the first pass, the linear average $\mu_x(t)$ is not yet available. A second pass can be avoided by calculating the variance in accordance with (17.8). ∎

17.2.4 Second-Order Expected Values

First-order expected values hold for a certain point in time, and therefore they cannot register the statistical dependencies that exist between different points in a signal. With *second-order expected values*, however, this is possible. They link the signal at two different points:

$$E\{f(x(t_1), x(t_2))\} = \lim_{N \to \infty} \frac{1}{N} \sum_{i=1}^{N} f(x_i(t_1), x_i(t_2))\,. \tag{17.9}$$

The auto-correlation function (ACF) is an important second-order expected value, which is obtained from (17.9) for $f(\mu, \nu) = \mu\nu$:

$$\boxed{\varphi_{xx}(t_1, t_2) = E\{x(t_1)\, x(t_2)\}\,.} \tag{17.10}$$

It describes the relationship between the values of a random signal at times t_1 and t_2. High values for the auto-correlation function indicate that $x(t)$ at times t_1 and t_2 take similar values. For $t_1 = t_2$, the auto-correlation function becomes the mean square

$$\varphi_{xx}(t_1, t_1) = E\{x^2(t_1)\}\,. \tag{17.11}$$

Example 17.4

Figure 17.4 shows which properties of a random signal are represented by second-order expected values. The random processes A and B have identical first-order expected values, in particular their distribution function $P_{x(t)}(\Theta) = P_{y(t)}(\Theta)$ (Example 17.2). The sample functions of random process A change much more slowly with time, however, than those of process B. We can therefore expect a much greater value of auto-correlation function $\varphi_{xx}(t_1, t_2)$ for A's neighbouring values t_1 and t_2 than for the auto-correlation function $\varphi_{yy}(t_1, t_2)$ for B. This is confirmed by the measured ACFs shown in Figure 17.5.

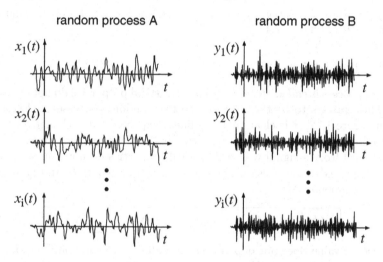

Figure 17.4: Illustration of two random processes A and B with identical first-order expected values and differing second-order expected values

Expected values of first- and second-order are by far the most important in practical applications. Many descriptions of stochastic functions rely solely on these two. However, advanced models of complicated random processes resort also to higher-order expected values. This emerging branch of signal analysis is therefore called *higher-order statistics*.

17.3 Stationary Random Processes

In Section 17.2.1 we considered various possible methods for calculating expected values. One of these was expressing the ensemble mean with the time-average. The conditions under which this is possible will be explained in this section.

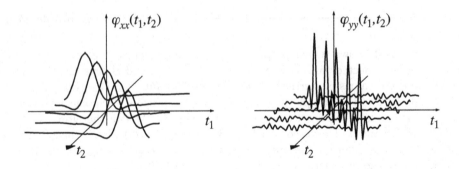

Figure 17.5: Auto-Correlation functions $\varphi_{xx}(t_1, t_2)$ and $\varphi_{yy}(t_1, t_2)$ of the random processes A and B in Figure 17.4

17.3.1 Definition

A random process is called *stationary* if its statistical properties do not change with time. This appears to be the case for the two random processes in Figure 17.4, while in Figure 17.2, it is clear that the linear expected value changes with time.

For a precise definition we start with a second-order expected value as in (17.9). If it is formed from a signal whose statistical properties do not change with time, then the expected value does not change if both time points t_1 and t_2 are shifted by the same amount Δt:

$$\mathrm{E}\{f(x(t_1), x(t_2))\} = \mathrm{E}\{f(x(t_1 + \Delta t), x(t_2 + \Delta t))\}. \qquad (17.12)$$

The expected value does not depend on the individual time points t_1 and t_2, but instead on their difference. We can use this property to define stationary.

Definition 23:   Stationary

A random process is stationary *if its second-order expected values only depend on the difference of observed time points $\tau = t_1 - t_2$.*

In (17.7) we considered deterministic signals as a special case of random signals, for which the linear expected value is equal to the current function value. This means that deterministic signals can only be stationary if they are constant in time. Any deterministic signal that changes with time is therefore not stationary. The same holds for finite random signals, as a signal that is zero before or after a certain point in time changes its statistical properties and cannot be stationary.

From this definition for second-order expected values we can derive the properties of first order expected values. As first-order expected values for $f(x(t_1), x(t_2)) = g(x(t_1))$ are a special case of second-order expect values we find:

$$\mathrm{E}\{g(x(t_1))\} = \mathrm{E}\{g(x(t_1 + \Delta t))\}. \qquad (17.13)$$

That means that for stationary random processes, the first-order expected values do not depend on time. In particular, for the linear average and the variance:

$$\mu_x(t) = \mu_x, \quad \sigma_x{}^2(t) = \sigma_x{}^2. \tag{17.14}$$

The auto-correlation function can be expressed more simply with $\tau = t_1 - t_2$:

$$E\{x(t_1)x(t_2)\} = E\{x(t_1)x(t_1 - \tau)\} = E\{x(t_2 + \tau)x(t_2)\} = \varphi_{xx}(\tau). \tag{17.15}$$

The auto-correlation function $\varphi_{xx}(\tau)$, that we introduced in (17.15) as a one-dimensional function is not the same function as $\varphi_{xx}(t_1, t_2)$ in (17.10), which is a function of two variables, t_1 and t_2. The two φ_{xx} are linked for stationary random signals (17.15). If the stationary condition only holds for

$$f(x(t_1), x(t_2)) = x(t_1)\,x(t_2) \tag{17.16}$$
$$f(x(t_1), x(t_2)) = x(t_1), \tag{17.17}$$

however, and not for general functions $f(\cdot, \cdot)$, then the random process is called *weak stationary*. Also for the linear average, the variance and the auto-correlation function of weak stationary processes (17.13), (17.14), and (17.15) hold. We can thus obtain the following statement: for a weak stationary process the linear average and the correlation properties contained in the auto-correlation function are constant with time. The concept of weak stationary processes is usually used in the context of modelling and analysis of random procesess, where it is often a pre-condition. When only the linear average and auto-correlation function are considered, weak stationary is a less limiting condition that that of (strictly) stationary processes for all expected values of first- and second-order.

Example 17.5

We take a random process with linear average $E\{x(t)\} = 0$ and the ACF

$$\varphi_{xx}(t_1, t_2) = \mathrm{si}(t_1 - t_2).$$

This random process is definitely weakly stationary, as we can write its ACF as

$$\varphi_{xx}(\tau) = \mathrm{si}(\tau)$$

and its linear average $\mu_x = 0$ is constant. Thus (17.12) is fulfilled for the functions (17.16), (17.17). The variance of the signal is

$$\sigma_x^2 = E\{x^2\} = \varphi_{xx}(0) = 1.$$

With the given information we cannot tell whether or not the random process is stationary in the strict sense.

                                                              Example 17.6

The weak stationary random signal $x(t)$ from Example 17.5 is modulated by a deterministic signal $m(t) = \sin \omega t$, so a new random signal $y(t)$ is created:

$$y(t) = m(t)x(t) = \sin(\omega t)x(t).$$

The linear average of $y(t)$ in accordance with (17.6), is

$$\mathrm{E}\{y(t)\} = m(t) \cdot \mathrm{E}\{x(t)\} = 0.$$

The ACF is

$$\varphi_{yy}(t_1, t_2) = \mathrm{E}\{y(t_1)y(t_2)\} = \mathrm{E}\{m(t_1)x(t_1)m(t_2)x(t_2)\}$$

$$= m(t_1)m(t_2)\varphi_{xx}(t_1, t_2).$$

$\varphi_{xx}(t_1, t_2)$ only depends on the difference between the observation times $\tau = t_1 - t_2$ (see Example 17.5), but this is not true for

$$m(t_1)m(t_2) = \sin \omega t_1 \sin \omega t_2.$$

Therefore $y(t)$ is neither stationary nor weak stationary. For example, the mean square at time $t = \frac{\pi}{2\omega}$ is

$$\varphi_{yy}\left(\frac{\pi}{2\omega}, \frac{\pi}{2\omega}\right) = \sin \frac{\pi}{2} \sin \frac{\pi}{2} \cdot \mathrm{si}(0) = 1.$$

but at time $t = 0$, it is

$$\varphi_{yy}(0, 0) = \sin(0) \sin(0) \cdot \mathrm{si}(0) = 0.$$

---   ■

17.3.2 Ergodic Random Processes

We will now return to the question of the conditions under which we can express the ensemble averages by the time average. We first define the first-order time average

$$\overline{f(x_i(t))} = \lim_{T \to \infty} \frac{1}{2T} \int_{-T}^{T} f(x_i(t)) \, dt \tag{17.18}$$

and the second-order time average

$$\overline{f(x_i(t), x_i(t - \tau))} = \lim_{T \to \infty} \frac{1}{2T} \int_{-T}^{T} f(x_i(t), x_i(t - \tau)) \, dt. \tag{17.19}$$

The apperently cumbersome limit $T \rightarrow \infty$ is necessary because the integral $\int\limits_{-\infty}^{\infty} f(\cdot)dt$ does not exist for a function f of a stationary random signal. The average is therefore formed from a section of the signal with length $2T$ and this length is then extended to infinity.

If the time-averages (17.18), (17.19) agree for *all* sample functions of a random process and are also equal to the ensemble average, we can then express it with the time average of any sample function. Random processes of this kind are called *ergodic*.

Definition 24: Ergodic

A stationary random process for which the time-averages of each function are the same as the ensemble average is called an ergodic random process.

It must be proven for individual cases whether or nor a stationary random process is ergodic. Often this proof cannot be exactly carried out, and in these cases it can be assumed that the process is ergodic, as long as no indications to the contrary occur. The big advantage of ergodic processes is that knowledge of an individual sample function is sufficient to calculate expected values with the time-average.

Similar to stationary processes, there is also a restriction for certain random processes. If the ergodicity conditions only hold for

$$f(x(t_1), x(t_2)) = x(t_1)\,x(t_2) \tag{17.20}$$
$$f(x(t_1), x(t_2)) = x(t_1) \tag{17.21}$$

but not for general functions $f(\cdot, \cdot)$, the random process is then *weak ergodic*. Like the idea of weak stationary, it is used for modelling and analysis of random processes with minimal restriction.

_____ **Example 17.7**

A random process produces sample functions

$$x_i(t) = \sin(\omega_0 t + \varphi_i),$$

where ω_0 is a fixed quantity, but the phase φ_i is completely random. All phase angles are equally likely.

The process is stationary because the second-order expected values (17.12) only depend on the difference between the observation times. The ACF, for example, is

$$\varphi_{xx}(t_1, t_2) \quad = \mathrm{E}\{\sin(\omega_0 t_1 + \varphi)\sin(\omega_0 t_2 + \varphi)\}$$

$$= \frac{1}{2\pi} \int\limits_{0}^{2\pi} \sin(\omega_0 t_1 + \varphi)\sin(\omega_0 t_2 + \varphi)d\varphi$$

$$= \frac{1}{2}\cos(\omega_0 t_1 - \omega_0 t_2) = \frac{1}{2}\cos\omega_0\tau \quad \mathrm{mit} \quad \tau = t_1 - t_2\,.$$

The integral averages all phase angles between 0 and 2π.

The process is also ergodic because the time-average (17.19) agrees with the ensemble average. We can verify this with the example of the ACF, where

$$\overline{x_i(t)x_i(t-\tau)} \quad = \lim_{T\to\infty} \frac{1}{2T} \int\limits_{-T}^{T} \sin(\omega_0 t + \varphi_i)\sin(\omega_0(t-\tau) + \varphi_i)dt$$

$$= \lim_{N\to\infty} \frac{1}{2N+1} \sum_{\mu=-N}^{N} \int\limits_{(-\pi+2\pi\mu)/\omega}^{(\pi+2\pi\mu)/\omega} \sin(\omega_0 t + \varphi_i)\sin(\omega_0(t-\tau) + \varphi_i)dt$$

$$= \frac{1}{2}\cos\omega_0\tau\,.$$

If the random process also has a random peak value $\hat{x}$ in addition to the random phase φ, so that

$$x(t) = \hat{x}\sin(\omega_0 t + \varphi)\,,$$

it is indeed stationary, but not ergodic. For example, while $\mathrm{E}\{x^2(t)\} = \frac{1}{2}\mathrm{E}\{\hat{x}^2\}$, the time-average of the square of a certain sample function i is, however, $\overline{x_i^2(t)} = \frac{1}{2}\hat{x}_i^2$. ∎

17.4 Correlation Functions

Now that we are comfortable with expected values and have learnt ways of calculating them, we can return to the task from Section 17.1 at the beginning of this chapter. We want to describe the properties of random signals with deterministic variables. This will now be done with first- and second-order expected values. We start with ergodic random processes if needed, so that the expected values can be expressed and evaluated with time averages.

The most important expected values are the linear average, the auto-correlation function and a generalisation derived from it, the cross-correlation function.

For the linear expected value, in accordance with (17.1), (17.14) and (17.18) we obtain

$$E\{x(t)\} = \mu_x = \overline{x(t)} = \lim_{T\to\infty} \frac{1}{2T} \int_{-T}^{T} x(t)dt . \qquad (17.22)$$

As ergodic functions are stationary, the linear average μ_x is (like all other first order expected values) independent of time. From Definition 24 it is equal to the time average $\overline{x(t)}$. As the time average can in this case be formed from any sample function, the sample function in (17.22) is no longer marked by an index (compare (17.18)).

We will get to know the other expected values better in the following sections, in particular the auto-correlation and cross-correlation functions. As we only use second-order expected values in the form of (17.20), weak ergodicity will be a sufficient pre-requisite.

First we consider correlation functions of purely real signals. The results will later be extended to complex signals.

17.4.1 Correlation Functions of Real Signals

17.4.1.1 Auto-Correlation Function

We have already seen that the auto-correlation function of a weak stationary signal can be expressed by the time difference of the two multiplied signal values (compare (17.15)):

$$\varphi_{xx}(\tau) = E\{x(t)x(t-\tau)\} . \qquad (17.23)$$

The product $x(t)x(t-\tau)$ can in general take positive as well as negative values. Depending on which of the two predominate, the expected value is positive or negative. For $\tau = 0$, however, $x^2(t)$ is never negative and must also be greater than for any other value of τ. That means that the auto-correlation function $\varphi_{xx}(\tau)$ has a maximum value at $\tau = 0$. We can show this easily, by considering

$$E\{(x(t) - x(t-\tau))^2\} = \varphi_{xx}(0) - 2\varphi_{xx}(\tau) + \varphi_{xx}(0) \geq 0 \qquad (17.24)$$

which is always positive, and which leads directly to $\varphi_{xx}(\tau) \leq \varphi_{xx}(0)$. From

$$E\{(x(t) + x(t-\tau))^2\} = \varphi_{xx}(0) + 2\varphi_{xx}(\tau) + \varphi_{xx}(0) \geq 0 \qquad (17.25)$$

we know that there is also a lower bound $\varphi_{xx}(\tau) \geq -\varphi_{xx}(0)$. This lower bound can be made narrower for signals with non-zero means, by considering

$$\tilde{x}(t) = x(t) - \mu_x \qquad (17.26)$$

and the ACF

$$\varphi_{\tilde{x}\tilde{x}}(\tau) = E\{(x(t) - \mu_x)(x(t-\tau) - \mu_x)\} = \varphi_{xx}(\tau) - \mu_x^2 . \qquad (17.27)$$

If we use (17.25) on $\varphi_{\tilde{x}\tilde{x}}(\tau)$, the ACF of the signal is $\varphi_{xx}(\tau) \geq -\varphi_{\tilde{x}\tilde{x}}(0) + \mu_x^2 = \varphi_{xx}(0) + 2\mu_x^2$. The value $\varphi_{xx}(0)$ can also be expressed by the variance σ_x^2 and the linear mean μ_x, in accordance with (17.4):

$$-\varphi_{xx}(0) + 2\mu_x^2 = -\sigma_x^2 + \mu_x^2 \leq \varphi_{xx}(\tau) \leq \varphi_{xx}(0) = \sigma_x^2 + \mu_x^2 \,. \tag{17.28}$$

In the relationship $\varphi_{xx}(\tau) \leq \varphi_{xx}(0)$, the equals sign represents the case where $x(t)$ is a periodic function of time. If the shift by τ is exactly equal to a shift by one period or a multiple, then $x(t)x(t+\tau) = x^2(t)$ and also $\varphi_{xx}(\tau) = \varphi_{xx}(0)$. There is no shift, however, between $x(t)$ and $x(t+\tau)$ for which the expected value becomes $E\{x(t)x(t+\tau)\} > E\{x^2(t)\}$. If such an effect is observed when measuring an ACF, this means that the pre-conditions for a weak stationary process are not there.

A further property of the auto-correlation function is it symmetry with respect to $\tau = 0$. As the value of $\varphi_{xx}(\tau)$ only depends on the displacement between the two functions in the product $x(t)x(t+\tau)$ we can substitute $t' = t + \tau$ and this yields

$$E\{x(t)x(t+\tau)\} = E\{x(t'-\tau)x(t')\} = E\{x(t')x(t'-\tau)\} \,. \tag{17.29}$$

The symmetry property

$$\varphi_{xx}(\tau) = \varphi_{xx}(-\tau) \tag{17.30}$$

is then immediately obtained.

For the behaviour of the auto-correlation function for $\tau \to \infty$, no general statements can be made. In many cases, there is no relationship between distantly separated values. These values are then said to be *uncorrelated*.

This property is expressed in the expected values of the signal, such that the second-order expected value is decomposed into the product of two first-order expected values:

$$E\{x(t)x(t-\tau)\} = E\{x(t)\} \cdot E\{x(t-\tau)\} \qquad |\tau| \to \infty \,. \tag{17.31}$$

As the linear expected value of a stationary signal does not depend on time,

$$E\{x(t)\} = E\{x(t-\tau)\} = \mu_x \tag{17.32}$$

and therefore for the auto-correlation function

$$\varphi_{xx}(\tau) = \mu_x^2 \qquad |\tau| \to \infty \,. \tag{17.33}$$

For signals whose values are uncorrelated if far apart, the only relationship between them is the (time-independent) linear average μ_x.

Figure 17.6 shows a typical auto-correlation function with the properties just discussed:

$$
\begin{array}{rll}
\text{maximum value:} & \varphi_{xx}(0) \;=\; \sigma_x^2 + \mu_x^2 \;\geq\; \varphi_{xx}(\tau) & (17.34) \\
\text{lower bound:} & \varphi_{xx}(\tau) \;\geq\; -\sigma_x^2 + \mu_x^2 & (17.35) \\
\text{symmetry:} & \varphi_{xx}(\tau) \;=\; \varphi_{xx}(-\tau) & (17.36) \\
\text{uncorrelated for } |\tau| \to \infty : & \lim_{|\tau|\to\infty} \varphi_{xx}(\tau) \;=\; \mu_x^2 & (17.37)
\end{array}
$$

The negative values of the auto-correlation function in Figure 17.6 show that for these values the shift τ of the expected value $E\{x(t+\tau)x(t)\}$ is negative. It means that with a separation τ, $x(t+\tau)$ and $x(t)$ are expected to have different signs. Like large positive values of the auto-correlation function, large negative values also indicate a strong relationship between the values of $x(t)$. The lower bound $-\sigma_x^2 + \mu_x^2$ is not reached in this example. The first three properties (17.34)

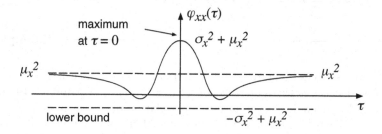

Figure 17.6: Example of an auto-correlation function

– (17.36) hold for all stationary random signals $x(t)$, but (17.37) only holds when distantly separated values do not correlate. In Example 17.7, (17.37) does not hold, for example.

Example 17.8

In Example 17.4 we considered two random processes A and B whose sample functions clearly have statistical interdependencies. Under the assumption that these processes are also stationary for the section shown, we can characterise them with their auto-correlation functions $\varphi_{xx}(\tau)$ and $\varphi_{yy}(\tau)$, shown in Figure 17.7. They are illustrated in Figure 17.7, and are found simply as cross-sections through $\varphi_{xx}(t_1, t_2)$ and $\varphi_{yy}(t_1, t_2)$ in Figure 17.5, along the t_1 or t_2 axis. The auto-correlation function $\varphi_{xx}(\tau)$ in particular is very similar to the auto-correlation function from Figure 17.6. In this case, however, the linear average is zero, which is also visible from the sample functions in Figure 17.4. The significantly faster changes in the time-behaviour of random process B in Figure 17.4 are also expressed by a faster decay of the autocorrleation function $\varphi_{yy}(\tau)$ from its maximum value in Figure 17.7. The maximum value $\varphi_{yy}(0)$ itself is equal to the corresponding value $\varphi_{xx}(0)$ of random process A, as it was already assumed that the first-order expected values are equal. Additionally, $y(t)$ clearly has a zero mean. Both ACFs do not reach the lower limit $-\sigma_x^2 + \mu_x^2 = -\sigma_y^2 + \mu_y^2$.

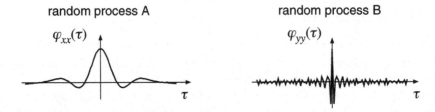

Figure 17.7: Auto-correlation function of the random processes A and B from Figure 17.4

─── ■

─── **Example 17.9**

Figure 17.8 shows the auto-correlation function of the speech signal for a male voice. The distinct negative correlation at a shift of $\tau \approx 5$ms points towards the fundamental frequency. As the negative correlation occurs at a shift of one half-wave, this indicates a fundamental of about 100 Hz. For larger shifts, the ACF does not approach a constant value. In fact, vowels are periodic over a large length of time. For very large τ, however, we find $\varphi_{xx}(\tau) = 0$ because the speech signal has a zero mean.

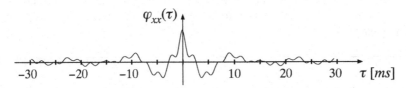

Figure 17.8: Auto-Correlation function of a speech signal

─── ■

17.4.1.2   Auto-Covariance Function

Introducing expected values should enable the use of deterministic functions to replace the random signals themselves in the system description. That means we could carry out the same operations that we did on deterministic input and output signals, the most important being the Fourier transform. If the function of time can be absolutely integrated (see Chapter 9.2.2), this is sufficient condition for the existence of the deterministic functions. This property is not given for auto-correlation functions, however, if the linear average μ_x is non-zero.

In order to overcome the difficulty stated above, the linear average can be removed from the outset and instead of the signal $x(t)$, the zero mean signal $(x(t) - \mu_x)$ can be considered. Its auto correlation function is called the *auto-covariance function* of $x(t)$ and is denoted by $\psi_{xx}(\tau)$:

$$\psi_{xx}(\tau) = \mathrm{E}\{(x(t) - \mu_x)(x(t - \tau) - \mu_x)\}. \tag{17.38}$$

Using the calculation rules from Section 17.2.3 we obtain

$$\psi_{xx}(\tau) = \varphi_{xx}(\tau) - \mu_x^2, \tag{17.39}$$

just as in (17.8). The properties of the auto-covariance function correspond to those of the auto-correlation function for zero mean signals.

maximal value:	$\sigma_x^2 = \psi_{xx}(0) \geq \psi_{xx}(\tau)$	(17.40)
lower bound:	$\psi(\tau) \geq -\sigma_x^2 = -\psi(0)$	(17.41)
symmetry:	$\psi_{xx}(\tau) = \psi_{xx}(-\tau)$	(17.42)
uncorrelation for $\|\tau\| \to \infty$:	$\lim_{\|\tau\| \to \infty} \psi_{xx}(\tau) = 0$	(17.43)

17.4.1.3 Cross-Correlation Function

The auto-correlation function is given by the expected value of two signal values that are taken from *one* random process at two different times. This idea can be extended to signal values from different random processes. The corresponding expected value is called the *cross-correlation function*. To represent its properties correctly we have to extend the earlier defintions of second-order expected values, stationary and ergodic random processes to deal with two random processes. A *second-order joint expected value* is the expected value of a function $f(x(t_1), y(t_2))$, formed with signals from two different random processes:

$$E\{f(x(t_1), y(t_2))\} = \lim_{N \to \infty} \frac{1}{N} \sum_{i=1}^{N} f(x_i(t_1), y_i(t_2)). \tag{17.44}$$

For the cross-correlation function $\varphi_{xy}(t_1, t_2)$, it holds in general that (compare (17.10)):

$$\varphi_{xy}(t_1, t_2) = E\{x(t_1) \cdot y(t_2)\}. \tag{17.45}$$

The cross-correlation function is denoted like the auto-correlation function, but the second random process is indicated by another letter in the index.

Next we extend the idea of stationary from Definition 23 to cover two random processes. We call two random processes *joint stationary*, if their joint second-order expected values only depend on the difference $\tau = t_1 - t_2$. For joint stationary random processes the cross-correlation function then takes a form similar to (17.15):

$$\varphi_{xy}(\tau) = E\left\{x(t+\tau) \cdot y(t)\right\} = E\left\{x(t) \cdot y(t-\tau)\right\}. \qquad (17.46)$$

Finally we introduce the *second-order joint time-average*

$$\overline{f(x_i(t), y_i(t-\tau))} = \lim_{T\to\infty} \frac{1}{2T} \int_{-T}^{T} f(x_i(t), y_i(t-\tau))dt \qquad (17.47)$$

and call two random processes for which the joint expected values agree with the joint time-averages *joint ergodic*. There are also weak forms for joint stationary and joint ergodic random processes, where the corresponding conditions are only fulfilled for $f(x(t_1), y(t_2)) = x(t_1)y(t_2)$, $f(x(t_1), y(t_2)) = x(t_1)$ and $f(x(t_1), y(t_2)) = y(t_2)$.

The cross-correlation function performs a similar function for two random processes that the auto-correlation function does for one random process. It is a measure for the relationship of values from the two random process at two times separated by τ. The extension to two random processes causes some differences to the auto-correlation function.

First of all, two random processes can be uncorrelated not only for large timespans but also for all values of τ. Their cross-correlation function is then the product of the linear expected values μ_x and μ_y of the individual random processes:

$$\varphi_{xy}(\tau) = \mu_x\,\mu_y \quad \forall\,\tau. \qquad (17.48)$$

There is also the case that two random processes are not uncorrelated for all values of τ, but at least for $|\tau| \to \infty$:

$$\varphi_{xy}(\tau) = \mu_x\,\mu_y \ \text{ for } |\tau| \to \infty. \qquad (17.49)$$

Furthermore, the cross correlation function does not have the even symmetry of the auto-correlation function, as from (17.46) and swapping x and y, we only obtain

$$\varphi_{xy}(\tau) = \varphi_{yx}(-\tau) \neq \varphi_{xy}(-\tau). \qquad (17.50)$$

The auto-correlation function $\varphi_{xx}(\tau)$ can be obtained from the cross-correlation function $\varphi_{xy}(\tau)$ as a special case $y(t) = x(t)$. Then using (17.50), we can find the symmetry property (17.36) of the auto-correlation function.

17.4.1.4 Cross-Covariance Function

The cross-correlation function can also be formed for zero-mean signals $(x(t) - \mu_x)$ and $(y(t) - \mu_y)$ and this leads to the *cross-covariance function* (compare Section 17.4.1.2):

$$\psi_{xy}(\tau) = \mathrm{E}\{(x(t) - \mu_x)(y(t - \tau) - \mu_y)\}. \tag{17.51}$$

As with the auto-covariance function in (17.39), the calculation rules from Section 17.2.3 yield the relationship between cross-covariance function $\psi_{xy}(\tau)$ and cross-correlation function $\varphi_{xy}(\tau)$:

$$\psi_{xy}(\tau) = \varphi_{xy}(\tau) - \mu_x \mu_y. \tag{17.52}$$

17.4.2 Correlation Functions of Complex Signals

When we introduced the various correlation functions in Section 17.4.1 we stuck to real signals for the sake of simplicity. In many applications, however, complex signals will appear as in the signal transmission example (15.2). In this section we will therefore be extending the use of correlation functions to complex random processes. These are random processes that produce complex sample functions.

To introduce the correlation functions for complex signals we proceed differently to Section 17.4.1. There we started with the auto-correlation function and introduced the cross-correlation function as a generalization that contained other correlation functions (cross-covariance, auto-correlation, auto-covariance) as special cases.

Here we start with the cross-correlation function for complex signals and derive the other correlation functions from it. To do this we must assume that $x(t)$ and $y(t)$ represent complex random processes that are joint weak stationary.

17.4.2.1 Cross-Correlation Function

There are several possibilities for extending the cross-correlation function to cover complex random processes. We will choose a definition that allows a particularly straight forward interpretation of the crosspower spectrum. There are different definitions in other books (for example [19]). According to (17.46), we define the cross-correlation function for complex random processes as

$$\boxed{\varphi_{xy}(\tau) = \mathrm{E}\{x(t + \tau)\,y^*(t)\}.} \tag{17.53}$$

The only difference to the definition for real random processes is that the conjugate complex function of time $y^*(t)$ is used. For real random processes (17.53) becomes (17.46).

In general, the cross-correlation function for complex random processes is also neither symmetrical nor commutative:

$$\varphi_{xy}(\tau) = \varphi_{yx}^*(-\tau) \neq \varphi_{yx}(\tau). \tag{17.54}$$

For uncorrelated random process we obtain

$$\varphi_{xy}(\tau) = E\left\{x(t+\tau)\right\} E\left\{y^*(t)\right\} = \mu_x \mu_y^* \quad \forall \tau. \tag{17.55}$$

17.4.2.2 Auto-Correlation Function

The auto-correlation function of a complex random process can be obtained from the cross-correlation function as in (17.53), for $y(t) = x(t)$:

$$\boxed{\varphi_{xx}(\tau) = \mathrm{E}\{x(t+\tau)\,x^*(t)\}.} \tag{17.56}$$

As in the real case, the auto-correlation function consists of a symmetry relation at the transition from τ to $-\tau$. We can obtain it directly from (17.56) by substituting $t' = t + \tau$ and by using the calculation rules for conjugate complex quantities:

$$\begin{aligned}
\varphi_{xx}(\tau) &= E\left\{x(t+\tau)x^*(t)\right\} = E\left\{x(t')x^*(t'-\tau)\right\} = E\left\{[x(t'-\tau)x^*(t')]^*\right\} \\
&= [E\left\{x(t'-\tau)x^*(t')\right\}]^* = \varphi_{xx}^*(-\tau) \quad .
\end{aligned} \tag{17.57}$$

or more concisely

$$\boxed{\varphi_{xx}(\tau) = \varphi_{xx}^*(-\tau).} \tag{17.58}$$

The conjugate symmetry here can be recognised from (9.47), and is expressed as an even real part and an odd imaginary part of $\varphi_{xx}(\tau)$. For real random processes the imaginary part of $\varphi_{xx}(\tau)$ is zero and the even symmetry holds according to (17.36). In any case, the odd imaginary part disappears at $\tau = 0$, so for complex random processes, $\varphi_{xx}(0)$ is also purely real. With the same reasoning as in Section 17.4.1.1, it holds that the magnitude of $\varphi_{xx}(\tau)$ is maximal at $\tau = 0$, and it can be expressed by the variance σ_x^2 and the mean μ_x:

$$\varphi_{xx}(\tau) \leq \varphi_{xx}(0) = E\left\{x(t)x^*(t)\right\} = \sigma_x^2 + \mu_x \mu_x^* = \sigma_x^2 + |\mu_x|^2 . \tag{17.59}$$

While the mean μ_x is in general complex for a complex random process, the variance is always a real quantity

$$\sigma_x^2 = E\left\{[x(t) - \mu_x][x(t) - \mu_x]^*\right\} = E\left\{|\,x(t) - \mu_x\,|^2\right\} , \tag{17.60}$$

as the square expected value is formed in this case with the magnitude squared.

17.4.2.3   Cross-Covariance and Auto-Covariance Function

From the cross-correlation function and the auto-correlation function we again use conversion to zero-mean signals $(x(t) - \mu_x)$ and $(y(t) - \mu_y)$ to obtain the complex cross-covariance and complex auto-covariance functions. The cross-covariance function is given by

$$\psi_{xy}(\tau) = \mathrm{E}\{(x(t+\tau) - \mu_x)(y(t) - \mu_y)^*\}\,, \qquad (17.61)$$

where the relationship

$$\psi_{xy}(\tau) = \varphi_{xy}(\tau) - \mu_x \mu_y^* \qquad (17.62)$$

with the cross-correlation function holds. The cross-covariance function is also neither symmetrical nor commutative:

$$\psi_{xy}(\tau) = \psi_{yx}^*(-\tau) \neq \psi_{yx}(\tau)\,. \qquad (17.63)$$

If both random processes are not correlated for large timespans τ, the cross-covariance function approaches zero for $|\tau| \to \infty$:

$$\psi_{xy}(\tau) = 0 \quad \text{f'ur} \quad |\tau| \to \infty\,. \qquad (17.64)$$

The auto-covariance function $\psi_{xx}(\tau)$ comes from the cross-covariance function $\psi_{xy}(\tau)$ where $y(t) = x(t)$. Its relationship with the auto-correlation function is:

$$\psi_{xx}(\tau) = \varphi_{xx}(\tau) - \mu_x \mu_x^* = \varphi_{xx}(\tau) - |\mu_x|^2 \qquad (17.65)$$

with the conjugate symmetry

$$\psi_{xx}(\tau) = \psi_{xx}^*(-\tau)\,. \qquad (17.66)$$

The maximum value at $\tau = 0$ is equal to the variance σ_x^2:

$$\psi_{xx}(0) = \sigma_x^2\,. \qquad (17.67)$$

The auto-covariance function also disappears for $|\tau| \to \infty$ if the random process no longer correlates when τ is large:

$$\psi_{xx}(\tau) = 0 \quad \text{for} \quad |\tau| \to \infty\,. \qquad (17.68)$$

17.5   Power Density Spectra

Describing random signals with expected values and in particular with correlation functions showed that the properties of random signals can be expressed by deterministic qunatities. Instead of a random time signal there would be, for example, the deterministic auto-correlation function which is likewise a deterministic quantity. In the past chapters we saw that describing signals in the frequency-domain is particularly useful, and often leads to elegant methods for analysing LTI-systems. We would therefore like to have a deterministic description of random signals in the frequency-domain as well.

The idea that first comes to mind is describing sample functions $x_i(t)$ of a random process as expected values instead of considering them as random signals, but this idea is actually unsuitable. Stationary random signals can never be integrated absolutely (9.4), as they do not decay for $|t| \to \infty$. Therefore the Laplace integral cannot exist and the Fourier transform can only exist in special cases. Instead of transforming into the frequency-domain and then forming expected values, we form the expected values in the time-domain and then transfer the deterministic quantities to the frequency-domain. This idea is the basis for the definition of the power density spectrum.

17.5.1 Definition

We start with the auto-correlation function or the auto-covariance function of a weak stationary random process and form its Fourier transform:

$$\boxed{\Phi_{xx}(j\omega) = \mathcal{F}\{\varphi_{xx}(\tau)\}.}$$ (17.69)

It is also called the *power density spectrum* of the random process. The power density spectrum characterises statistical dependencies of the signal amplitude at two different points in time. Correspondingly, the Fourier transform of a cross-correlation function can be formed giving the *cross-power density spectrum*:

$$\boxed{\Phi_{xy}(j\omega) = \mathcal{F}\{\varphi_{xy}(\tau)\}.}$$ (17.70)

$\Phi_{xy}(j\omega)$ is also called the *cross-spectrum*.

─── **Example 17.10**

In Example 17.7 we examined a stationary random process which produced sinusoidal sample functions $x_i(t) = \sin(\omega_0 t + \varphi_i)$ with random phase φ_i. For these sample functions it is sometimes even possible to give the Fourier transform

$$X_i(j\omega) = [\pi\delta(\omega - \omega_0) - \pi\delta(\omega + \omega_0)]e^{j\omega/\omega_0 \varphi_i}.$$

Its linear average

$$\mathrm{E}\{X(j\omega)\} = 0$$

does not say much, however, and the mean square

$$\mathrm{E}\{X_i(j\omega)X_i^*(j\omega)\}$$

cannot be given because of the delta impulse. In the best case we can form

$$E\{|X_i(j\omega)|\} = \pi\delta(\omega - \omega_0) + \pi\delta(\omega + \omega_0)$$

as a deterministic description of the random process. Forming the expected values first in the time-domain and then transforming them yields the power density spectrum

$$\Phi_{xx}(j\omega) = \frac{\pi}{2}\delta(\omega - \omega_0) + \frac{\pi}{2}\delta(\omega + \omega_0) \quad \bullet\!\!-\!\!\circ \quad \varphi_{xx}(\tau) = \frac{1}{2}\cos\omega_0\tau$$

as the Fourier transform of the ACF. It is similar to $E\{|X_i(j\omega)|\}$ and likewise indicates that the random signal only contains frequency components at $\pm\omega_0$.

17.5.2 Power Density Spectrum and Mean Square

The mean square of a random process can also be calculated directly from the power density spectrum. First, the auto-correlation function is expressed as the inverse Fourier transform of the power density spectrum:

$$\varphi_{xx}(\tau) = \mathcal{F}^{-1}\{\Phi_{xx}(j\omega)\} = \frac{1}{2\pi}\int_{-\infty}^{\infty}\Phi_{xx}(j\omega)e^{j\omega\tau}d\omega. \qquad (17.71)$$

As the mean square is equal to the value of the auto-correlation function at $\tau = 0$, we can obtain the relationship between the power density spectrum and mean square by putting $\tau = 0$ into (17.71):

$$\boxed{E\{|x(t)|^2\} = \varphi_{xx}(0) = \frac{1}{2\pi}\int_{-\infty}^{\infty}\Phi_{xx}(j\omega)d\omega.} \qquad (17.72)$$

The mean square is therefore equal to the integral of the power density spectrum, multiplied by a factor $1/2\pi$.

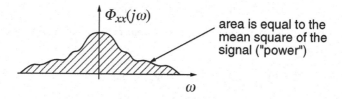

Figure 17.9: The area under the power density spectrum $\Phi_{xx}(j\omega)$ of a signal $x(t)$ is proportional to the mean square of the signal

We saw by introducing the mean square into (17.4) that it can also be used as a measure of the power of a random process, for example, as noise power of random interference. Relation (17.72) suggests the interpretation that the power density spectrum describes the power distribution over various frequencies for a random process. We will return to this idea in Chapter 18.2.6 where it will be refined and made more precise.

17.5.3 Symmetry Properties of the Power Density Spectrum

The symmetry properties of the power density spectrum come directly from the conjugate symmetry of the auto-correlation function (17.58). As the real part of the auto-correlation function is an even function, we know from the symmetry properties of the Fourier transform (9.61) that the power density spectrum must be purely real:

$$\text{Im}\{\Phi_{xx}(j\omega)\} = 0. \tag{17.73}$$

It also holds if $x(t)$ describes a random processs. In Chapter 18.2.6 we will show that

$$\Phi_{xx}(j\omega) \geq 0$$

holds in the general case.

The cross-power density spectrum and the cross-correlation function show no particularly interesting symmetries, just

$$\Phi_{xy}(j\omega) = \Phi_{yx}^*(j\omega) \tag{17.74}$$

from (17.54).

For real random processes a further symmetry property can be obtained, however. The auto-correlation function is then a real and even function

$$\varphi_{xx}(\tau) = \varphi_{xx}(-\tau), \tag{17.75}$$

in accordance with (17.30). Its Fourier transform (9.61) is likewise real and even, and the power density spectrum of a real random process is therefore an even function:

$$\Phi_{xx}(j\omega) = \Phi_{xx}(-j\omega). \tag{17.76}$$

For the cross-power density spectrum of two real random processes we obtain the conjugate symmetry

$$\Phi_{xy}(j\omega) = \Phi_{xy}^*(-j\omega) = \Phi_{yx}(-j\omega) = \Phi_{yx}^*(j\omega) \tag{17.77}$$

from (17.74) and (9.61). As the cross-correlation function is in general not symmetrical, the cross-power density spectrum of a real random process is also not real. Just like the CCF, the cross-power density spectrum is not commutative.

As an example we will consider the power density spectrum in Figure 17.10 for the speech signal from Example 17.9. As expected, it is real and positive, and as the speech signal is, of course, itself real-valued, the power density spectrum is also an even function. In Figure 17.10 therefore, only the positive frequencies are shown. The concentration of power density in the 100 Hz region is clearly recogniseable. This result agrees with the rough estimation of the fundamental speech frequency in Example 17.9.

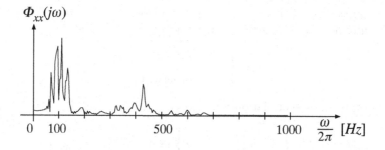

Figure 17.10: Power density spectrum of the speech signal from Figure 17.10

17.5.4 White Noise

Many interference sources, like amplifier noise or radio interference, can be described by random processes with a power density spectrum that is almost constant over a large range of frequencies. Such random processes are often approximated by an idealised process, whose power density spectrum is not frequency dependent:

$$\Phi_{nn}(j\omega) = N_0 . \tag{17.78}$$

Since all frequencies occur evenly, the idealised process is called *white noise*. The name is derived from the term 'white light', which represents a uniform mixture of every colour in the spectrum.

The auto-correlation function of white noise can only be represented by a distribution, the delta impulse:

$$\varphi_{nn}(\tau) = \mathcal{F}^{-1}\{N_0\} = N_0\,\delta(\tau) . \tag{17.79}$$

This means that samples of white noise taken at different times do not correlate.

From the representation of white noise in the time-domain (17.79) and in the frequency-domain (17.78), we see that a white noise signal must have infinite power:

$$\varphi_{nn}(0) = \frac{1}{2\pi} \int\limits_{-\infty}^{\infty} N_0 \, d\omega \to \infty \; . \tag{17.80}$$

White noise is therefore an idealisation that cannot actually be realised. Despite this, it has extremely simple forms of power density spectrum and auto-correlation function that are very useful, and the idealisation can be justified if the white noise signal is high- or low-pass filtered, and the highest frequency components suppressed. This leads to a refined model for random processes: band-limited white noise with a rectangular power density spectrum

$$\Phi_{nn}(j\omega) = \begin{cases} N_0 & \text{for } |\omega| < \omega_{\max} \\ 0 & \text{otherwise} \end{cases} , \tag{17.81}$$

with a si-shaped auto-correlation function

$$\varphi_{nn}(\tau) = \mathcal{F}^{-1}\left\{ N_0 \operatorname{rect}\left(\frac{\omega}{2\omega_{\max}} \right) \right\} = N_0 \frac{\omega_{\max}}{\pi} \operatorname{si}(\tau \omega_{\max}) \tag{17.82}$$

and finite power

$$\varphi_{nn}(0) = N_0 \frac{\omega_{\max}}{\pi} \; . \tag{17.83}$$

It characterises noise processes whose power is evenly distributed below a band limit $\omega_{\max}$. Figure 17.11 shows the power density spectrum and the auto-correlation function.

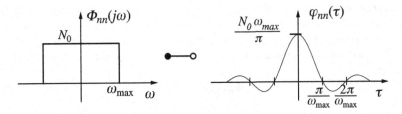

Figure 17.11: Band-limited white noise in the frequency-domain and time-domain

17.6   Describing Discrete Random Signals

The concepts discussed so far for continuous random signals can easily be extended to random sequences. As most of the reasoning and derivation is very similar, we will forego examining them in detail. However, we need to be aware of the definitions of stationary and ergodicity as they apply to random sequences.

Stationary means in this case that the second-order expected values $E\left\{f(x[k_1], y[k_2])\right\}$ only depend on integer differences $\kappa = k_1 - k_2$ between the discrete time variables k_1 and k_2.

The time-average, necessary for the definition of ergodicity, is given for sample sequences of a discrete random process by

$$\overline{f(x_i[k])} = \lim_{K\to\infty} \frac{1}{2K+1} \sum_{k=-K}^{K} f(x_i[k]).\qquad(17.84)$$

From here, the auto-correlation, cross-correlation and covariances can be considered in the same way as for continuous random processes. Instead of auto-correlation and cross-correlation functions, we will be using auto-correlation and cross-correlation sequences.

For the cross-power density spectrum of two discrete weak stationary random processes, we obtain from the discrete-time Fourier transform the spectrum

$$\Phi_{xy}(e^{j\Omega}) = \mathcal{F}_*\{\varphi_{xy}[\kappa]\},\qquad(17.85)$$

periodic in Ω. The power density spectrum of a discrete random process corresponds to the case where $y = x$.

─── **Example 17.12**

A weak stationary continuous random process $\tilde{x}(t)$ is characterised by an exponentially decaying ACF

$$\varphi_{\tilde{x}\tilde{x}}(\tau) = e^{-\omega_0|\tau|}.$$

Samples are taken at intervals T

$$x[k] = \tilde{x}(kT) \quad k \in \mathbb{Z}.$$

The auto-correlation sequence is

$$\varphi_{xx}[\kappa] = E\{x[k+\kappa]x[k]\} = E\{\tilde{x}(kT+\kappa T)\tilde{x}(kT)\} = \varphi_{\tilde{x}\tilde{x}}(\kappa T) = e^{-\omega_0 T|\kappa|}.$$

The power density spectrum of the discrete random process is

$$\Phi_{xx}(e^{j\Omega}) = \mathcal{F}_*\{\varphi_{xx}[\kappa]\} = \frac{1 - e^{-2\omega_0 T}}{1 + e^{-2\omega_0 T} - 2e^{-\omega_0 T}\cos\Omega}$$

and is shown in Figure 17.12. It is positive, real and even, like the power density spectrum of a continuous random process with real values. In addition, it is 2π periodic. As the auto-correlation sequences $\varphi_{xx}[\kappa]$ is yielded simply by sampling

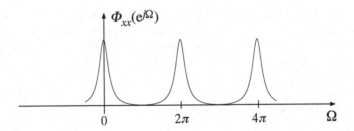

Figure 17.12: Power density spectrum of the discrete random process φ_{xx}

the auto-correlation sequences $\varphi_{\tilde{x}\tilde{x}}(t)$, the power density spectrum of the discrete random process $\Phi_{xx}(e^{j\Omega})$ is simply the periodic continuation of the power density spectrum $\Phi_{\tilde{x}\tilde{x}}(j\omega)$, in accordance with (11.33)

$$\Phi_{xx}(e^{j\Omega}) = \frac{1}{2\pi}\Phi_{\tilde{x}\tilde{x}}\left(\frac{j\Omega}{T}\right) * \underline{\sqcup}\sqcup\left(\frac{\Omega}{2\pi}\right).$$

■

17.7   Exercises

Exercise 17.1

For which of the following ensembles might hold:

a) $\lim\limits_{N\to\infty} \dfrac{1}{N}\sum\limits_{i=1}^{N} x_i(t) = \lim\limits_{T\to\infty} \dfrac{1}{2T}\int_{-T}^{T} x_i(t)\,dt$, for any i

b) $\lim\limits_{N\to\infty} \dfrac{1}{N}\sum\limits_{i=1}^{N} x_i{}^2(t) = \lim\limits_{T\to\infty} \dfrac{1}{2T}\int_{-T}^{T} x_i{}^2(t)\,dt$, for any i

Start by verbally formulating the conditions.

Ensemble 1

$x_1(t)$

$x_2(t)$

$x_3(t)$

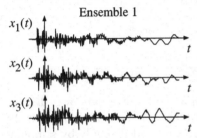

Ensemble 2

$x_1(t)$

$x_2(t)$

$x_3(t)$

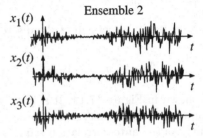

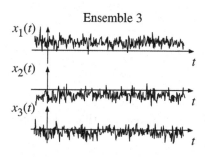

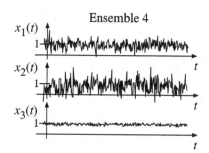

Ensemble 5: from Figure 17.1

Exercise 17.2

Discuss which of the ensembles from Exercise 17.1 could be:
a) weak stationary,
b) weak ergodic.

Exercise 17.3

Consider a random process $x(t)$ at the output of a CD-player where the sample functions are produced by someone continually repeating a ten second section from a rock CD. Assume that it is the ideal case in which this has been happening for an infinite time and the CD-player will go on playing infinitely.

a) How large are $E\{x(t)\}$ and $\overline{x_i(t)}$ under the assumption that the output has no DC component?

b) Using the first-order expected values, discuss whether the process could be stationary.

c) Using the first-order expected values, discuss whether the process could be ergodic.

Exercise 17.4

Take two uncorrelated random processes $x_1(t)$ and $x_2(t)$, where $\mu_{x_1} = 2$, $\mu_{x_2} = 0$, $E\{x_1^2(t)\} = 5$ and $E\{x_2^2(t)\} = 2$. Calculate the averages $\mu_y(t)$, $\sigma_y^2(t)$, $E\{y^2(t)\}$ for the random process $y(t) = x_1(t) + x_2(t)$.

Exercise 17.5

Find the variance of $v(t) = x(t) + y(t)$, where $x(t)$ is a random signal with variance $\sigma_x^2 = 10$ and $y(t)$ is any deterministic signal.

Exercise 17.6

The ergodic random process $x(t)$ has $\mu_x = 1$ and $\sigma_x^2 = 4$. Calculate $\mu_y(t)$, $\sigma_y^2(t)$, $E\{y^2(t)\}$ and $\overline{y_i(t)}$ for

a) $y(t) = x(t) + K$

b) $y(t) = x(t) + \sin t$

c) $y(t) = x(t) + \varepsilon(t)$

d) $y(t) = x(t) \cdot 5\varepsilon(t)$

For each part, say whether $y(t)$ is ergodic. Assume in part d) that all sample functions $x_i(t)$ are even.

Exercise 17.7

Give $\mu_x(t)$, $E\{x^2(t)\}$, $\sigma_x^2(t)$ and $\overline{x_i(t)}$ for the deterministic signal $x(t) = e^{-0.1t}\varepsilon(t)$.

Exercise 17.8

Consider the discrete random process 'throwing a die', where a) $x[k]$ is the number thrown and b) $x[k]$ is the square of the number thrown. Find $\mu_x[k]$, $\sigma_x[k]$ and $E\{x^2[k]\}$. Are the processes ergodic?

Exercise 17.9

Consider the discrete random process 'throwing a loaded die' where six always appears at times $k = 3N$, $N \in \mathbb{Z}$, but the numbers thrown at other times are distributed equally, and where $y[k]$ is the number thrown. Find $\overline{y_i[k]}$ and $\overline{y_i{}^2[k]}$, and also $\mu_y[k]$, $\sigma_y[k]$ and $E\{y^2[k]\}$. Is the process stationary and/or ergodic?

Exercise 17.10

Calculate the ACF of deterministic signal $x(t) = K$, $K \in \mathbb{C}$ with (17.56).

Exercise 17.11

What peculiarities does the ACF have for the random process from Exercise 17.3? Does it only depend on the difference of the averaging points? Give $\varphi_{xx}(t_0, t_0 + 10s)$.

Exercise 17.12

Let $x(t)$ and $y(t)$ be two real random signals.

a) Show that $E\{x(t)y(t)\} = 0$ entails from $E\{(x(t) + y(t))^2\} = E\{x^2(t)\} + E\{y^2(t)\}$.

b) How can $E\{x(t)y(t)\}$ be simplified if $x(t)$ and $y(t)$ are uncorrelated?

c) Give the conditions for uncorrelated $x(t)$ and $y(t)$ so that $E\{x(t)y(t)\} = 0$.

Exercise 17.13

What are the formulae for power, DC component, effective value and AC power for a deterministic real signal $d(t)$?

Exercise 17.14

The AC component power of a deterministic real signal $d(t)$ with DC component μ can be calculated in two ways: $P = \overline{(d(t) - \mu)^2}$ or $P = \overline{d^2(t)} - \mu^2$. Show that both formulae produce the same result.

Exercise 17.15

Derive the equation $\psi_{xy}(\tau) = \varphi_{xy}(\tau) - \mu_x \mu_y$ which holds for all real stationary random processes. Start with the definition (17.51).

Exercise 17.16

The relationship between two stationary random processes $x(t)$ and $y(t)$ is expressed by $\varphi_{xy}(\tau) = \dfrac{4\tau^2 + 10}{1 + \tau^2}$. Find μ_y, $\varphi_{yx}(\tau)$, $\psi_{xy}(\tau)$ and $\psi_{yx}(\tau)$, if $\mu_x = 1$.

Exercise 17.17

A stationary random process $v(t)$ with $\varphi_{vv}(\tau) = e^{-|\tau|}$ is used to create a further random process $u(t)$ by delaying $v(t)$ by $t_0 = 10$. Determine

a) μ_v, μ_u, $\varphi_{uv}(\tau)$, $\varphi_{vu}(\tau)$ and $\varphi_{uu}(\tau)$

b) $\Phi_{vv}(j\omega)$, $\Phi_{uv}(j\omega)$ and $\Phi_{vu}(j\omega)$.

c) What symmetry properties do $\Phi_{vv}(j\omega)$, $\Phi_{uv}(j\omega)$ and $\Phi_{vu}(j\omega)$ have? Is the random process complex?

Exercise 17.18

Determine the auto-correlation sequence $\varphi_{xx}[\kappa]$, the auto-covariance sequence $\psi_{xx}[\kappa]$ and the power density spectrum $\Phi_{xx}(e^{j\Omega})$ of the random process from Exercise 17.8a).

Exercise 17.19

Consider the ergodic discrete random process 'throwing a tetrahedral die', which has 0 marked on one side and 1 marked on the other three; $x[k]$ is the number lying underneath. Determine μ_x, σ_x^2, $\varphi_{xx}[\kappa]$, $\psi_{xx}[\kappa]$ and $\Phi_{xx}(e^{j\Omega})$.

Exercise 17.20

The ensemble $x[k]$ is produced as in Exercise 17.19, and a further ensemble is produced by throwing a second tetrahedron whose sides are marked with the numbers 1,2,3 and 4; $y[k]$ is the number lying underneath. Determine μ_y, σ_y^2, $\varphi_{yy}[\kappa]$, $\varphi_{xy}[\kappa]$ and $\varphi_{yx}[\kappa]$.

Exercise 17.21

The sketched power density spectrum of a random process $x(t)$ is given. Determine the power and auto-correlation function of $x(t)$.

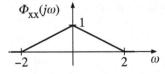

18 Random Signals and LTI-Systems

Now that we have described random signals with correlation functions and power density spectra, we can use these new tools to investigate the response of LTI-systems to random signals. Rather than being interested in the precise behaviour of the output signal when a particular sample function of a random process appears at the input, we want to describe the system output as a random process and are looking for the expected values of the output process dependent on the expected values of the input process.

We will see that the relationships between the expected values at the input and output of an LTI-system take a similar form to the relationships between deterministic input and output signals.

To begin the chapter we will deal with random signals that are the result of multiplying a random signal with a constant or by adding random signals together. Then we will describe the response of LTI-systems to random signals. Finally we will consider applications of the theory.

18.1 Combining Random Signals

With expected values, correlation functions and power density spectra, we have learnt the most significant forms for describing random signals in the time-domain and frequency-domain. Now we are interested in the connections between random signals and what influence they have on the description forms just mentioned. The discussion will be restricted to multiplication of a random signal with a constant and adding two random signals together, as shown in Figures 18.1 and 18.2. These are two significant elements of block diagrams as we found in Chapter 2.2. In Section 18.2 we will consider general LTI-systems.

18.1.1 Multiplication of a Random Signal with a Factor

Multiplying a random signal $x(t)$ with a complex, constant factor K, in accordance with Figure 18.1 creates the new random signal

$$y(t) = Kx(t).$$

(18.1)

We now want to express the auto-correlation function of the output signal $\varphi_{yy}(\tau)$ and the cross-correlation function $\varphi_{yx}(\tau)$ of the output signal and input signal, with the auto-correlation function of the input signal $\varphi_{xx}(\tau)$.

For the auto-correlation function of the output signal,

$$\varphi_{yy}(\tau) = E\{y(t+\tau)y^*(t)\} = E\{Kx(t+\tau)K^*x^*(t)\} = |K|^2\varphi_{xx}(\tau) \qquad (18.2)$$

is immediately obtained. Likewise the cross-correlation function between $x(t)$ and $y(t)$ is

$$\varphi_{yx}(\tau) \;=\; E\{y(t+\tau)x^*(t)\} = E\{Kx(t+\tau)x^*(t)\} = K\varphi_{xx}(\tau) \qquad (18.3)$$
$$\varphi_{xy}(\tau) \;=\; E\{x(t+\tau)y^*(t)\} = E\{x(t+\tau)K^*x^*(t)\} = K^*\varphi_{xx}(\tau). \qquad (18.4)$$

For the (cross-)power density spectra, corresponding relationships are obtained, because the Fourier transform is linear.

These results can be summarised in a clearer form.

$$
\begin{array}{lll}
\varphi_{xx}(\tau) & \circ\!\!-\!\!\bullet & \Phi_{xx}(j\omega) & (18.5) \\
\varphi_{yx}(\tau) = K\varphi_{xx}(\tau) & \circ\!\!-\!\!\bullet & \Phi_{xx}(j\omega)K = \Phi_{yx}(j\omega) & (18.6) \\
\varphi_{xy}(\tau) = K^*\varphi_{xx}(\tau) & \circ\!\!-\!\!\bullet & \Phi_{xx}(j\omega)K^* = \Phi_{xy}(j\omega) & (18.7) \\
\varphi_{yy}(\tau) = |K|^2\varphi_{xx}(\tau) & \circ\!\!-\!\!\bullet & \Phi_{xx}(j\omega)|K|^2 = \Phi_{yy}(j\omega) & (18.8)
\end{array}
$$

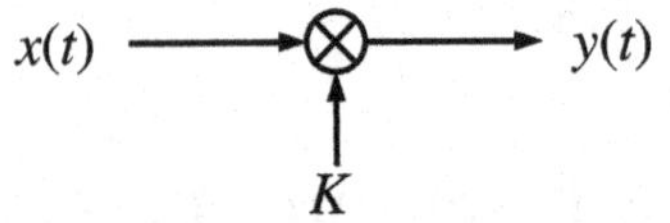

Figure 18.1: Multiplication of a random signal with a constant factor K

18.1.2   Addition of Random Signals

Figure 18.2 shows two random signals $f(t)$ and $g(t)$ being added to form a new random signal $y(t)$. The properties of both random process $f(t)$ and $g(t)$ are known, but the properties of the summed process $y(t)$ must be determined. From the random processes $f(t)$ and $g(t)$ we only expect enough information to form auto- and cross-correlation functions, in particular, that $f(t)$ and $g(t)$ are joint weak stationary random processes. Complex sample functions are permitted.

The addition of random signals is the simplest and most frequently used model for describing noise-like interference. We can, for example, represent the output signal of an amplifier $y(t)$ as the sum of the ideal (noise-free) amplified signal $f(t)$ and the noise $g(t)$. The ideal signal $f(t)$ can be a speech or music signal and

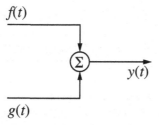

Figure 18.2: Addition of two random signals $f(t)$ and $g(t)$

is here described as a random process. Even when this model does not exactly reproduce the interference in a multi-stage amplifier, in many cases it still delivers a good approximation of the actual relationships. Its greatest advantage is the simple analysis that we are about to carry out for the auto-correlation function and the power density spectrum, and then for the cross-correlation function and crosspower density spectrum.

18.1.2.1 Auto-Correlation Function and Power Density Spectrum

The summed signal $y(t)$ from Figure 18.2 is simply

$$y(t) = f(t) + g(t). \tag{18.9}$$

Its auto-correlation function (17.56)

$$\varphi_{yy} = E\{y(t+\tau)y^*(t)\} \tag{18.10}$$

can be used to work back to the auto-correlation and cross-correlation functions of $f(t)$ and $g(t)$, with the calculation rules from Chapter 17.2.3. Substituting (18.9) into (18.10) yields

$$\begin{aligned} \varphi_{yy}(\tau) &= E\{(f(t+\tau)+g(t+\tau))(f^*(t)+g^*(t))\} & (18.11) \\ &= \varphi_{ff}(\tau)+\varphi_{fg}(\tau)+\varphi_{gf}(\tau)+\varphi_{gg}(\tau) \quad. & (18.12) \end{aligned}$$

The auto-correlation function $\varphi_{yy}(\tau)$ is therefore formed from the auto-correlation functions of the signals $f(t)$ and $g(t)$ and their cross-correlation functions.

The power density spectrum at the output is found by Fourier transforming (18.12):

$$\Phi_{yy}(j\omega) = \Phi_{ff}(j\omega) + \Phi_{fg}(j\omega) + \Phi_{gf}(j\omega) + \Phi_{gg}(j\omega). \tag{18.13}$$

Using (17.74) we can summarise both crosspower densities as a real variable

$$\Phi_{fg}(j\omega) + \Phi_{gf}(j\omega) = \Phi_{fg}(j\omega) + \Phi_{fg}^*(j\omega) = 2\mathrm{Re}\{\Phi_{fg}(j\omega)\} \tag{18.14}$$

and obtain for the power density spectrum $\Phi_{yy}(j\omega)$, the expression

$$\boxed{\Phi_{yy}(j\omega) = \Phi_{ff}(j\omega) + 2\mathrm{Re}\{\Phi_{fg}(j\omega)\} + \Phi_{gg}(j\omega)} \qquad (18.15)$$

which is obviously real.

In addition, the power density spectrum $\Phi_{yy}(j\omega)$ is made up of the spectra of the signals $f(t)$ and $g(t)$ and of their mutual crosspower density spectrum.

The results so far hold for the addition of any two random processes $f(t)$ and $g(t)$. Possible correlation between the random processes is taken into account by the cross-correlation function $\varphi_{fg}(\tau)$ and crosspower density spectrum $\Phi_{fg}(j\omega)$.

These relationships become much simpler if both random processes are uncorrelated and at least one has a zero mean. Then, from (17.55) we obtain

$$\varphi_{fg}(\tau) = 0 \quad \circ\!\!-\!\!\bullet \quad \Phi_{fg}(j\omega) = 0 \qquad (18.16)$$

and the relations (18.12) and (18.15) simplify to

$$\boxed{\begin{aligned} \varphi_{yy}(\tau) &= \varphi_{ff}(\tau) + \varphi_{gg}(\tau) \\ \Phi_{yy}(j\omega) &= \Phi_{ff}(j\omega) + \Phi_{gg}(j\omega). \end{aligned}} \qquad \begin{aligned} (18.17) \\ (18.18) \end{aligned}$$

Assuming that the processes are uncorrelated holds in many cases where $g(t)$ represents a signal interfering with the useful signal $f(t)$. Examples are amplifier noise independent of the input signal or atmospheric effects on radio transmission that is likewise independent of the transmitted signal. As this kind of interference also usually has a zero mean, the simple relationships (18.17) and (18.18) also hold. They say that when two uncorrelated random processes are added, and as long as at least one has a zero mean, the auto-correlation functions and power density spectra are also added together to form the respective functions for the complete signal model.

18.1.2.2 Cross-Correlation Function and CrossPower Density Spectrum

We next consider the cross-correlation function between $f(t)$ and the sum $y(t) = f(t) + g(t)$ as well as the corresponding crosspower density spectrum. Correlation between $f(t)$ and $g(t)$ is now permitted. From (17.53) and (18.9) the cross-correlation function $\varphi_{fy}(\tau)$ is obtained:

$$\boxed{\varphi_{fy}(\tau) = \mathrm{E}\{f(t+\tau)(f^*(t) + g^*(t))\} = \varphi_{ff}(\tau) + \varphi_{fg}(\tau).} \qquad (18.19)$$

The cross-correlation function between a sum term and the summed signal is the sum of the auto-correlation function of the corresponding sum term and the cross-correlation function between both sum terms.

For the cross spectrum the same rule applies using the respective power spectra of $f(t)$ and $g(t)$

$$\Phi_{fy}(j\omega) = \Phi_{ff}(j\omega) + \Phi_{fg}(j\omega) \,. \tag{18.20}$$

These results also become particularly simple if both random processes are uncorrelated and at least one has a zero mean. The cross-correlation function $\varphi_{fg}(\tau)$ and the crosspower density spectrum $\Phi_{fg}(j\omega)$ then disappear and we are left with:

$$\begin{aligned}
\varphi_{fy}(\tau) &= \varphi_{ff}(\tau) \tag{18.21}\\
\Phi_{fy}(j\omega) &= \Phi_{ff}(j\omega) \,. \tag{18.22}
\end{aligned}$$

All results in this section hold correspondingly for the cross-correlation function and the crosspower density spectrum between $g(t)$ and $y(t)$.

18.2   Response of LTI-Systems to Random Signals

Now that we have defined the statistical description of input and output processes that arise by addition or multiplication of each other, we can consider the corresponding relationships for input and output signals of LTI-systems. As description forms for LTI-systems, we will choose the impulse response and frequency response. No assumptions are made about the inner structure of the system. Next it must be clarified whether the a stationary or ergodic input signal brings about the same properties in the output signal. To do this we first derive the connections between the different averages at the input and output of LTI-systems in detail.

18.2.1   Stationarity and Ergodicity

We start with an LTI-system as in Figure 18.3 and consider, if the input process is stationary or an ergodic random process, then does the output process also have these properties? If that is the case, we can also use the correlation function and power density description that was introduced in Chapter 17 on the output process, under the condition of weak stationarity.

If the input process is stationary then the second-order expected values do not change when the input signal is shifted by time Δt (compare (17.12)):

$$E\{f(x(t_1), x(t_2))\} = E\{f(x(t_1 + \Delta t), x(t_2 + \Delta t))\}\,. \tag{18.23}$$

Because the system time-invariant, for the output signal $y(t) = S\{x(t)\}$,

$$y(t_i + \Delta t) = S\{x(t_i + \Delta t)\}\,, \tag{18.24}$$

holds and from (18.23) we obtain

$$E\{g(y(t_1), y(t_2))\} = E\{g(y(t_1 + \Delta t), y(t_2 + \Delta t))\}\,, \tag{18.25}$$

as $g(S\{x(t_1)\}, S\{x(t_2)\})$ also represents a time-invariant system. The output process is therefore likewise stationary. Corresponding statements about weak stationary and ergodic processes can be obtained in the same way.

Now we know that for an input signal which is

a) weak stationary or stationary,

b) weak ergodic or ergodic,

the output signal also has the same properties. Likewise, input and output processes show the same joint properties. The different relationships are illustrated in Figure 18.3.

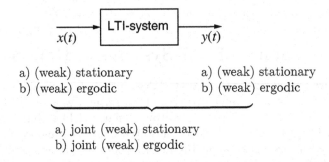

a) joint (weak) stationary
b) joint (weak) ergodic

Figure 18.3: LTI-system with input signal $x(t)$ and output signal $y(t)$

18.2.2   Linear Mean at the Output of an LTI-System

To determine the linear mean at the output of an LTI-system with impulse response $h(t)$ as in Figure 18.4, we start with the convolution (8.39)

$$y(t) = x(t) * h(t) \tag{18.26}$$

and form the linear mean $E\{y(t)\} = \mu_y(t)$ of the output signal:

$$\mu_y(t) = E\{y(t)\} = E\{x(t) * h(t)\} = E\{x(t)\} * h(t) = \mu_x(t) * h(t). \tag{18.27}$$

The input signal $x(t)$ is a random variable, but the impulse response $h(t)$ is not, so the expected value formed with $x(t)$ and $\mu_y(t)$ can be expressed by $\mu_x(t)$ and $h(t)$. Here, $x(t)$ does not have to be a stationary process and this means $\mu_x(t)$ and $\mu_y(t)$ are therefore dependent on time. Comparing (18.26) and (18.27), we see that the relationship between the determined linear expected values $\mu_x(t)$ and $\mu_y(t)$ has the same form as the convolution between the input signal $x(t)$ and the output signal $y(t)$.

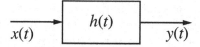

Figure 18.4: LTI-system with impulse response $h(t)$

If the input signal $x(t)$ is stationary, then the linear mean $E\{x(t)\} = \mu_x$ is constant in time and the convolution relationship simplifies to

$$\mu_y = E\{y(t)\} = \int_{-\infty}^{\infty} \mu_x h(\tau)\, d\tau = \mu_x \cdot H(0). \tag{18.28}$$

The constant linear average μ_x can be brought outside of the integral, and the remaining integral over $h(\tau)$ can be written as the value $H(0)$ of the Fourier integral (9.1) at $\omega = 0$. The mean of a stationary random signal is therefore transferred like the DC component of a deterministic signal. Thus we obtain the simple formula

$$\boxed{\mu_y = \mu_x \cdot H(0)\,.} \tag{18.29}$$

18.2.3   Auto-Correlation Function at the Output of an LTI-System

We will calculate the auto-correlation function at the system output for a stationary input signal. The auto-correlation function

$$\varphi_{yy}(\tau) = E\{y(t + \tau)\, y^*(t)\} \tag{18.30}$$

is obtained from the auto-correlation function $\varphi_{xx}(\tau)$ of the input signal and the impulse response $h(t)$, by using convolution (18.26) and some rearrangements. The aim of the rearrangements is to move the expectation operator into the resulting integrals and apply it to the input signal.

In the first step we express the output signal $y(t)$ in (18.30) with the convolution integral:

$$\varphi_{yy}(\tau) = E\left\{ \int_{\mu} h(\mu)x(t + \tau - \mu)\, d\mu \int_{\nu} h^*(\nu)x^*(t - \nu)\, d\nu \right\}. \tag{18.31}$$

Bringing both convolutions together to form a double integral, sorting the terms within the integral and using the expected values on the resulting product of the input signals yields

$$\varphi_{yy}(\tau) = \int_{\nu} \int_{\mu} h(\mu)\, h^*(\nu)\, E\left\{ x(t + \tau - \mu)\, x^*(t - \nu) \right\} d\mu\, d\nu\,. \tag{18.32}$$

The expected value is now formed from the product of the values of the input signal at different time points and therefore represents the auto-correlation function of the input signal. Because the input signal is stationary, the auto-correlation function depends on the difference between the time points $t + \tau - \mu$ and $t - \nu$:

$$\varphi_{yy}(\tau) = \int_\nu \int_\mu h(\mu)\, h^*(\nu)\, \varphi_{xx}(\tau - \mu + \nu)\, d\mu\, d\nu. \qquad (18.33)$$

We have now found the first connection between the auto-correlation functions at the input $\varphi_{xx}(\tau)$ and at the output $\varphi_{yy}(\tau)$.

We can further simplify the rather involved expression (18.33), by recalling the correlation of deterministic signals introduced in Chapter 9.9. To use the definition of the auto-correlation function for deterministic signals, given in (9.103), we first substitute $\theta = \mu - \nu$ into (18.33) and obtain

$$\varphi_{yy}(\tau) = \int_\theta \int_\mu h(\mu)\, h^*(\mu - \theta)\, d\mu\ \varphi_{xx}(\tau - \theta)\, d\theta. \qquad (18.34)$$

After another substitution $\lambda = \mu - \theta$, we recognise in the inner integral the auto-correlation function $\varphi_{hh}(\theta)$ as in (9.103):

$$\varphi_{hh}(\theta) = \int_\mu h(\mu)\, h^*(\mu - \theta)\, d\mu = \int_\lambda h(\lambda + \theta)\, h^*(\lambda)\, d\lambda = h(\theta) * h^*(-\theta). \qquad (18.35)$$

Despite having the same notation, the auto-correlation function $\varphi_{hh}(\theta)$ does not represent an expected value of a random process, as the the impulse response $h(t)$ is a deterministic function. We can, however, interpret $\varphi_{hh}(\theta)$ as an expected value of $h(\lambda + \theta)\, h^*(\lambda)$, like we did with the random processes, if we define the expected value for deterministic functions with the integration in (18.35). As the auto-correlation function $\varphi_{hh}(\theta)$ describes the LTI-system, it is also called a *filter auto-correlation function (filter ACF)*. Describing deterministic LTI-systems using characteristics which are similar to those for random signals has the advantage that it allows the initially cumbersome expression (18.33) to be represented in a simpler and more easily remembered form. To derive such an expression we put (18.35) into (18.34) and obtain

$$\varphi_{yy}(\tau) = \int_\theta \varphi_{hh}(\theta)\ \varphi_{xx}(\tau - \theta)\, d\theta = \varphi_{hh}(\tau) * \varphi_{xx}(\tau). \qquad (18.36)$$

The auto-correlation function at the output $\varphi_{yy}(\tau)$ is now obtained as convolution of the auto-correlation function at the input $\varphi_{xx}(\tau)$ with the filter auto-correlation function $\varphi_{hh}(\tau)$. The filter auto-correlation function is itself the convolution of the impulse response $h(t)$ with $h^*(-t)$. In this way the relationship between the auto-correlation functions at the input and output of an LTI-system and its impulse response can be summarised with two simple equations:

$$\varphi_{yy}(\tau) = \varphi_{hh}(\tau) * \varphi_{xx}(\tau) \tag{18.37}$$

$$\varphi_{hh}(\tau) = h(\tau) * h^*(-\tau). \tag{18.38}$$

$$\tag{18.39}$$

The mean square as a measure for the power of the output signal is obtained by evaluating the convolution integral at $\tau = 0$:

$$\mathrm{E}\{|y(t)|^2\} = \varphi_{yy}(0) = \int_{\tau} \varphi_{hh}(\tau)\,\varphi_{xx}(-\tau)\,d\tau. \tag{18.40}$$

Example 18.1

We will consider an ideal delay circuit with the impulse response

$$h(t) = \delta(t - t_0).$$

With (18.38) and (18.39) we obtain

$$\varphi_{hh}(\tau) = \delta(\tau - t_0) * \delta(-\tau - t_0) = \delta(\tau)$$

$$\varphi_{yy}(\tau) = \delta(\tau) * \varphi_{xx}(\tau) = \varphi_{xx}(\tau).$$

The ACF is not changed by the delay circuit.

∎

For discrete systems with impulse response $h[k]$ there are similar relationships for the auto-correlation sequences of the output signal. With the definition of discrete convolution (12.48) as a sum, the formulae for discrete systems look exactly like the expressions (18.38), (18.39) for continuous systems:

$$\varphi_{yy}[\kappa] = \varphi_{hh}[\kappa] * \varphi_{xx}[\kappa] \tag{18.41}$$

$$\varphi_{hh}[\kappa] = h[\kappa] * h^*[-\kappa]. \tag{18.42}$$

18.2.4 Cross-Correlation Function Between the Input and Output of an LTI-System

The cross-correlation function between the input and the output of an LTI-system can be obtained with a similar procedure to that for the auto-correlation function. For a stationary input signal, starting with

$$\varphi_{xy}(\tau) = \mathrm{E}\{x(t + \tau)\,y^*(t)\} \tag{18.43}$$

the convolution integral is inserted

$$\varphi_{xy}(\tau) = \mathrm{E}\left\{ x(t+\tau) \int_\mu h^*(\mu) x^*(t-\mu)\,d\mu \right\}, \qquad (18.44)$$

then interchanging the sequence of integration and expected values yields

$$\varphi_{xy}(\tau) = \int_\mu h^*(\mu) \mathrm{E}\{x(t+\tau)\, x^*(t-\mu)\}\,d\mu. \qquad (18.45)$$

In the expected value we again recognise the auto-correlation function of the input signal $x(t)$:

$$\varphi_{xy}(\tau) = \int_\mu h^*(\mu)\, \varphi_{xx}(\tau+\mu)\,d\mu. \qquad (18.46)$$

By substituting $\nu = -\mu$ we finally reach the desired expression for the cross-correlation function of the input and output of an LTI-system and its impulse response:

$$\varphi_{xy}(\tau) = \int_\nu h^*(-\nu)\, \varphi_{xx}(\tau-\nu)\,d\nu = h^*(-\tau) * \varphi_{xx}(\tau). \qquad (18.47)$$

The formula obtained for the cross-correlation function $\varphi_{yx}(\tau)$ between the input and output is easier to remember. To derive it from (18.47) we first make use of the symmetry relationships for correlation functions from Chapter 17.4.2. From (17.54) in combination with (18.46) we obtain

$$\varphi_{yx}(\tau) = \varphi_{xy}^*(-\tau) = \int_\mu h(\mu)\, \varphi_{xx}^*(-\tau+\mu)\,d\mu. \qquad (18.48)$$

The conjugate symmetry of the auto-correlation function (17.58) yields

$$\varphi_{yx}(\tau) = \int_\mu h(\mu)\, \varphi_{xx}(\tau-\mu)\,d\mu = h(\tau) * \varphi_{xx}(\tau). \qquad (18.49)$$

Both possible cross-correlation functions between the input and output of an LTI-system are easily obtained by convolution of the auto-correlation function at the input with the impulse response $h(\tau)$ and $h^*(-\tau)$:

$$\begin{aligned}
\varphi_{xy}(\tau) &= h^*(-\tau) * \varphi_{xx}(\tau) & (18.50) \\
\varphi_{yx}(\tau) &= h(\tau) * \varphi_{xx}(\tau). & (18.51) \\
& & (18.52)
\end{aligned}$$

--- **Example 18.2**

We again consider the ideal delay circuit from Example 18.1, and with (18.52) find

$$\varphi_{yx}(\tau) = \delta(\tau - t_0) * \varphi_{xx}(\tau) = \varphi_{xx}(\tau - t_0).$$

The cross-correlation function between the input and output of a delay circuit is simply a shifted version of the auto-correlation function. This simple insight is used in many engineering systems to measure signal delay, for example, radar or sonar.

--- ■

According to the continuous case, for a discrete LTI-system with impulse response $h[k]$ holds

$$\varphi_{xy}[\kappa] \quad = \quad h^*[-\kappa] * \varphi_{xx}[\kappa] \qquad (18.53)$$

$$(18.54)$$

$$\boxed{\varphi_{yx}[\kappa] \quad = \quad h[\kappa] * \varphi_{xx}[\kappa].} \qquad (18.55)$$

$$(18.56)$$

18.2.5 Power Density Spectrum and LTI-Systems

We have now derived the relationships between the correlation functions at the input and output of LTI-systems and the impulse response. They turned out to be very similar to convolution of a deterministic input signal and impulse response. An even simpler description can be obtained in the frequency-domain by multiplying the input spectrum with the frequency response of the LTI-system. As we have already met the power density spectrum as a frequency-domain description of random signals, it is reasonable to assume that there are some similar relationships between (cross-)power density spectra at the input and output of a system and its frequency response.

We will start with the cross-correlation function $\varphi_{yx}(\tau)$ as in (18.52). Fourier transformation $\Phi_{xx}(j\omega) = \mathcal{F}\{\varphi_{xx}(\tau)\}$ of the ACF (17.69) and the CCF (17.70), $\Phi_{yx}(j\omega) = \mathcal{F}\{\varphi_{yx}(\tau)\}$, and the convolution theorem (9.70) yields

$$\Phi_{yx}(j\omega) = \Phi_{xx}(j\omega)H(j\omega). \qquad (18.57)$$

The cross-power density spectrum $\Phi_{yx}(j\omega)$ is obtained from the power density spectrum $\Phi_{xx}(j\omega)$ of the input signal by multiplying it with the frequency response $H(j\omega)$ of the LTI-system.

The cross-power density spectrum $\Phi_{xy}(j\omega)$ could also be derived in the same way from (18.51), using Fourier transformation. We can obtain it more quickly from (18.57), however, with (17.74) as

$$\Phi_{xy}(j\omega) = \Phi_{xx}(j\omega)H^*(j\omega). \qquad (18.58)$$

Here we have used the fact that the power density spectrum $\Phi_{xx}(j\omega)$ is real (17.73).

For the power density spectrum of the output signal $\Phi_{yy}(j\omega)$ a simple relationship can be found by Fourier transforming (18.38). The Fourier transform of the ACF $\varphi_{hh}(\tau)$ can, after the results of Chapter 9.9.2.4 (Correlation of Deterministic Signals), be expressed by the magnitude squared frequency response $H(j\omega)$:

$$\varphi_{hh}(\tau)\circ\!\!\!-\!\!\!\bullet\Phi_{hh}(j\omega) = H(j\omega)H^*(j\omega) = |H(j\omega)|^2 \qquad (18.59)$$

yielding

$$\Phi_{yy}(j\omega) = \Phi_{xx}(j\omega)|H(j\omega)|^2 \,. \qquad (18.60)$$

The significant relationships for correlation functions and power density can now be summarised in a clearer form:

$$
\begin{array}{rcl}
\varphi_{xx}(\tau) & \circ\!\!-\!\!\bullet & \Phi_{xx}(j\omega) \hfill (18.61)\\
\varphi_{yx}(\tau) = h(\tau) * \varphi_{xx}(\tau) & \circ\!\!-\!\!\bullet & \Phi_{xx}(j\omega)H(j\omega) = \Phi_{yx}(j\omega) \hfill (18.62)\\
\varphi_{xy}(\tau) = h^*(-\tau) * \varphi_{xx}(\tau) & \circ\!\!-\!\!\bullet & \Phi_{xx}(j\omega)H^*(j\omega) = \Phi_{xy}(j\omega) \hfill (18.63)\\
\varphi_{yy}(\tau) = \varphi_{hh}(\tau) * \varphi_{xx}(\tau) & \circ\!\!-\!\!\bullet & \Phi_{xx}(j\omega)|H(j\omega)|^2 = \Phi_{yy}(j\omega)\,. \hfill (18.64)
\end{array}
$$

Equations (18.61) to (18.64) have the same form as the corresponding equations (18.5) to (18.8) for multiplication with a complex constant K. The expressions obtained are a special case of (18.61) to (18.64) for $H(j\omega) = K \;\bullet\!\!-\!\!\circ\; h(\tau) = K\,\delta(\tau)$.

The results obtained in the preceding sections can be made even clearer and more general. Instead of the auto-correlation function $\varphi_{yy}(\tau)$ or the cross-correlation function $\varphi_{yx}(\tau)$, we could have used the cross-correlation function $\varphi_{yr}(\tau)$ with any other random signal $r(t)$, as shown in Figure 18.5. The same

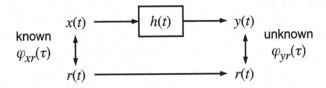

Figure 18.5: CCF of a signal with the input and output of an LTI-system

steps as in the derivation of $\varphi_{yx}(\tau)$ in Section 18.2.4 would then have lead to

$$\varphi_{yr}(\tau) = \varphi_{xr}(\tau) * h(\tau) \ , \qquad (18.65)$$

and the corresponding representation in the frequency-domain, in accordance
with (18.57) and (18.58)

$$
\begin{array}{rcll}
\Phi_{yr}(j\omega) &=& \Phi_{xr}(j\omega)\,H(j\omega) & (18.66)\\
\Phi_{ry}(j\omega) &=& \Phi_{rx}(j\omega)\,H^{*}(j\omega)\,. & (18.67)
\end{array}
$$

Using a suitable substitution for r, equations (18.62) to (18.64) can be found
from the last two expressions. From (18.66) we let $r = x$ and $r = y$, and obtain

$$
\begin{array}{rcll}
\Phi_{yx}(j\omega) &=& \Phi_{xx}(j\omega)\,H(j\omega) & (18.68)\\
\Phi_{yy}(j\omega) &=& \Phi_{xy}(j\omega)\,H(j\omega) & (18.69)
\end{array}
$$

and from (18.67) for $r = x$

$$
\Phi_{xy}(j\omega) = \Phi_{xx}(j\omega)\,H^{*}(j\omega)\,. \tag{18.70}
$$

Finally, (18.69) and (18.70) yield

$$
\Phi_{yy}(j\omega) = \Phi_{xy}(j\omega)\,H(j\omega) = \Phi_{xx}(j\omega)\,H^{*}(j\omega)H(j\omega) = \Phi_{xx}(j\omega)|H(j\omega)|^{2}\,. \tag{18.71}
$$

It is therefore sufficient to remember formulae (18.66) and (18.67); all of the important relations between the (cross-)power density spectra at the input and output
of LTI-systems can be derived from them.

─── **Example 18.3**

A typical stock exchange index $x(t)$ has a power density spectrum of the form

$$
\Phi_{xx}(j\omega) = \frac{2a}{\omega^{2} + a^{2}}\,,
$$

where a is very small. The change of the index with time conveniently produces a
differentiator with the transfer function

$$
H(j\omega) = j\omega\,.
$$

The resulting derivative $y(t) = \dot{x}(t)$ has the power density spectrum

$$
\Phi_{yy}(j\omega) = \omega^{2}\Phi_{xx}(j\omega) = \frac{2a\omega^{2}}{\omega^{2} + a^{2}}\,.
$$

For $\omega^{2} \gg a^{2}$,

$$
\Phi_{yy}(j\omega) \approx 2a\,.
$$

The corresponding ACF is

$$
\varphi_{yy}(\tau) \approx 2a\delta(\tau)\,.
$$

Sequential changes of the stock exchange index are therefore completely uncorrelated.

─── ■

18.2.6 Interpretation of the Power Density Spectrum

We can explain the interpretation of the power density spectrum (which should actually be called the autopower density spectrum) with a thought experiment. We start with a random signal $x(t)$ which has power density spectrum $\Phi_{xx}(j\omega)$. It will be filtered through an ideal band-pass filter with frequency response

$$H(j\omega) = \begin{cases} 1 & \text{for } \omega_0 \leq \omega < \omega_0 + \Delta\omega \\ 0 & \text{otherwise} \end{cases} . \qquad (18.72)$$

We have already encountered band-pass filters of this kind when we looked at sampling complex band-pass signals in Chapter 11.3.3. Because they do not have conjugate symmetry they have complex impulse responses, and they produce complex output signals from real input signals. However, this is not necessarily a problem; we will imagine that $\Delta\omega$ is very small, and the band-pass therefore lets through only a very narrow frequency band.

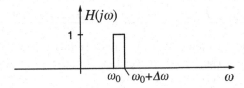

Figure 18.6: Band-pass which has been shifted to the right, and is therefore complex

To determine the power of the output signal, we calculate the quadratic mean of the output signal $y(t) = h(t) * x(t)$. We use the inverse Fourier transform of $\Phi_{yy}(j\omega)$ from (18.64), instead of the convolution integral as in (18.40):

$$E\{|y(t)|^2\} = \frac{1}{2\pi} \int_{-\infty}^{\infty} \Phi_{xx}(j\omega)\, |H(j\omega)|^2\, d\omega . \qquad (18.73)$$

As $H(j\omega)$ has a narrow band character, the integral with respect to ω only covers a narrow band of the power density spectrum $\Phi_{xx}(j\omega)$, of width $\Delta\omega$:

$$E\{|y(t)|^2\} = \frac{1}{2\pi} \int_{\omega_0}^{\omega_0 + \Delta\omega} \Phi_{xx}(j\omega)\, d\omega \approx \frac{1}{2\pi} \Phi_{xx}(j\omega_0)\Delta\omega . \qquad (18.74)$$

The power density spectrum represents the distribution of a random signal's power onto an infinite number of infinitesimal frequency bands of width $\Delta\omega$. This justifies the name power density and confirms the interpretation from Chapter 17.5.2 (see Figure 17.9). At the same time we can obtain

$$\Phi_{xx}(j\omega) \geq 0 \quad \forall \omega , \qquad (18.75)$$

which we gave in Chapter 17.5.2 without proof.

The interpretation of the cross-power density spectrum $\Phi_{xr}(j\omega)$ between a signal $x(t)$ and a reference signal $r(t)$ can be carried out similarly, in principle, even though the result is not quite so descriptive. Again having in mind the analysis of $x(t)$ into many narrow band pass signals according to (18.66), $\Phi_{xr}(j\omega_0)$ turns out to be the cross spectrum $\Phi_{yr}(j\omega_0)$ of the frequency component $y(t) = x(t) * h(t)$ with frequency ω_0.

If the reference signal $r(t)$ is also filtered with a corresponding band-pass filter, the cross-power spectrum does not change. Clearly only components of the same frequency contribute to the cross-correlation between two stationary signals. The correlation between these band-pass components is recorded in the cross-power density spectrum. If there is a fixed amplitude and phase relationship between $x(t)$ and $r(t)$, it will appear in the cross spectrum. If the frequency components do not correlate, $\Phi_{xr}(j\omega) = 0$.

18.2.7 Measuring the Transfer Behaviour of an LTI-System

The transfer behaviour of an LTI-system is completely described by its impulse response or transfer function. There are various ways of measuring them. Theoretically it is possible to excite the system with a delta impulse and measure the output response but, in practice, the high amplitudes required for an approximation of the delta impulse cause many problems. Another way is to excite the system with a sine wave with varying frequency and measure the frequency response, with the amplitude and phase of the output signal. A prerequisite is that apart from its response to the sinusoidal input signal, the output signal may not contain any interference, because otherwise it would falsify the measurement. Beyond that, the frequency of the sinusoidal signal can only be changed slowly enough that no undesired transients appear.

A better, more modern method uses a wideband noise signal and statistical analysis of the output signal. One corresponding measurement setup is shown in Figure 18.7. The wideband measurement signal at the input is described as white noise. This approximation is always justified if the bandwidth of the noise source is much greater than the bandwidth of the system under investigation. The unknown transfer behaviour is obtained from the cross-correlation between the input and output signals.

For white noise with noise power $N_0 = 1$ (see Chapter 17.5.4):

$$\varphi_{xx}(\tau) = \delta(\tau) \qquad \circ\!\!-\!\!\bullet \qquad \Phi_{xx}(j\omega) = 1 \quad . \tag{18.76}$$

For the cross-correlation function $\varphi_{yx}(\tau)$ between the output and input, and for the cross-power density spectrum $\Phi_{yx}(j\omega)$ follows from (18.62):

$$\varphi_{yx}(\tau) \quad = \quad \varphi_{xx}(\tau) * h(\tau) = \delta(\tau) * h(\tau) = h(\tau) \tag{18.77}$$

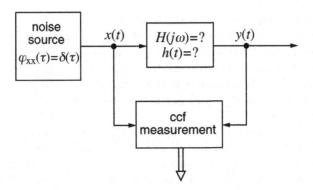

Figure 18.7: Analysing an LTI-system with white noise

$$\Phi_{yx}(j\omega) \;\; = \;\; H(j\omega)\,.\tag{18.78}$$

The cross-correlation function $\varphi_{yx}(\tau)$ and cross-power density spectrum $\Phi_{yx}(j\omega)$ immediately give the impulse response and transfer function of an unknown LTI-system. If the signal $y(t)$ is overlapped by a noise signal $n(t)$, the result remains unchanged, as long as $x(t)$ and $n(t)$ do not correlate (Exercise 18.17).

18.3 Signal Estimation Using the Wiener Filter

The field of signal estimation concerns reconstruction of a signal that has been corrupted by various influences. The original signal is corrected as much as possible using knowledge of the statistical properties of the corrupting influence. If we are dealing with weak stationary random processes and LTI-systems, then we will be able to put the theory we developed in Section 18.2 to good use.

The problem is depicted in Figure 18.8. The original signal $s(t)$ is not directly accessible; only the signal $x(t)$ can be observed. The influence which corrupts the original signal before it is observed will be described with the properties of an LTI-system and additionally, an interference signal (noise) with a random character. No details about the structure of the LTI-system and the noise signal are known. We will assume that we only know the cross-power density spectrum $\Phi_{xs}(j\omega)$ between the original signal $s(t)$ and the observed signal $x(t)$, for example, from measurement with a known original signal. The cross-power density spectrum is a mathematical formulation of the influence for the corrupting influence on the original signal. The power density spectrum $\Phi_{xx}(j\omega)$ of the observed signal can be determined by measurement at any time, and is thus also known.

The filter $H(j\omega)$ will be developed so that its output signal $y(t)$ comes as close to the original signal as possible. The task of the filter is therefore to eliminate

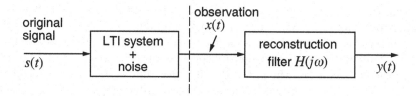

Figure 18.8: Estimation of the original signal $s(t)$

the influence on the original signal. In the following sections we will develop a method that provides a suitable transfer function $H(j\omega)$ from the known quantities $\Phi_{xx}(j\omega)$ and $\Phi_{xs}(j\omega)$ in a systematic way. The filter is called an *optimal* or *Wiener* filter.

Estimation of this kind is necessary in many different applications as these examples show.

Measurement   In this case $s(t)$ is a physical quantity whose time-behaviour is measured. As each measurement affects the process being measured, the measuring device gives an inaccurate observed version $x(t)$ of the original signal $s(t)$. Additionally, errors in measurement can be modelled using noise signals.

Signal transmission   If a signal $s(t)$ is to be transmitted to another location, it will be affected by the non-ideal properties of the transmission chain and may also be subject to interference, so the received signal $x(t)$ no longer corresponds to the transmitted signal $s(t)$.

Recording   In order to store a signal $s(t)$, it must be changed so that it complies with the requirements of the target medium. The signal $x(t)$ read at a later time contains influences from the recording and reading equipment, and the interference here can also be modelled as noise.

18.3.1   Deriving the Transfer Function of the Wiener Filter

To derive the optimal transfer function $H(j\omega)$ we first of all need to find a mathematical approach which should precisely formulate the requirement that the output signal $y(t)$ is 'as similar as possible' to the original signal $s(t)$. To achieve this, we will introduce a measure for the difference between $s(t)$ and $y(t)$, called the error power or the estimation error

$$e(t) = y(t) - s(t). \tag{18.79}$$

We represent the error power using the expected value

$$E\{|e(t)|^2\} = E\{|y(t) - s(t)|^2\}. \tag{18.80}$$

Although we do not know the original signal $s(t)$ or the reconstruction error $e(t)$, we can express its power with known statistical quantities.

We know from Chapter 17.5.2 that the mean square $E\{|e(t)|^2\}$ can also be calculated by integrating the power density spectrum $\Phi_{ee}(j\omega)$ (compare Figure 17.9). We will therefore consider the power density spectrum $\Phi_{ee}(j\omega)$ of the reconstruction error $e(t)$, and try to make it as small as possible. $\Phi_{ee}(j\omega)$ is expressed by the power density spectra $\Phi_{yy}(j\omega)$ of the output signal and $\Phi_{ss}(j\omega)$ of the original signal. From (18.79), with expression (18.13), we obtain

$$\Phi_{ee} = \Phi_{yy} - \Phi_{ys} - \Phi_{sy} + \Phi_{ss} \qquad (18.81)$$

for the addition of random variables. Here the argument $(j\omega)$ is left out to simplify the notation.

To determine the transfer function H of the estimation filter, we need an expression for the power density spectrum Φ_{ee}, dependent on H. Using (18.64) and the general relationships (18.66) and (18.67) with $r = s$, the correlation with y can be expressed by the correlation with x:

$$\Phi_{ee} = \Phi_{xx}\, H\, H^* - \Phi_{xs}\, H - \Phi_{sx}\, H^* + \Phi_{ss} \qquad (18.82)$$

The estimation filter H that leads to the smallest error power is obtained from (18.82) by differentiating Φ_{ee} with respect to H. It should be noted here that the variables in (18.82) and H are both complex. Therefore we write (18.82) dependent on the magnitude $|H|$ and phase ϕ of the frequency response

$$H = |H|\, e^{j\phi} \qquad (18.83)$$

and obtain

$$\Phi_{ee} = \Phi_{xx}\, |H|^2 - \Phi_{xs}\, |H|e^{j\phi} - \Phi_{sx}\, |H|e^{-j\phi} + \Phi_{ss}\,. \qquad (18.84)$$

We can now differentiate with respect to the real variable $|H|$

$$\frac{d\Phi_{ee}}{d|H|} = 2\,|H|\,\Phi_{xx} - \Phi_{xs}\, e^{j\phi} - \Phi_{xs}^*\, e^{-j\phi} = 2\,|H|\,\Phi_{xx} - 2\,\mathrm{Re}\{\Phi_{xs}e^{j\phi}\} \overset{!}{=} 0 \quad (18.85)$$

The optimal magnitude response $|H|$ is determined from the requirement that the derivative of Φ_{ee} with respect to $|H|$ is zero:

$$|H| = \frac{\mathrm{Re}\{\Phi_{xs}\, e^{j\phi}\}}{\Phi_{xx}} \qquad (18.86)$$

The phase ϕ is determined in the same way by differentiating Φ_{ee} with respect to ϕ

$$\frac{d\Phi_{ee}}{d\phi} = -j\,|H|\,e^{j\phi}\,\Phi_{xs} + j\,|H|\,e^{-j\phi}\,\Phi_{xs}^* \overset{!}{=} 0\,. \qquad (18.87)$$

The optimal phase ϕ can be derived from

$$\text{Im}\{\Phi_{xs}\, e^{j\phi}\} = \text{Im}\{|\Phi_{xs}|\, e^{j(\arg\{\Phi_{xs}\}+\phi)}\} = 0 \qquad (18.88)$$

where the imaginary part becomes zero when the phase is

$$\phi = -\arg\{\Phi_{xs}\}. \qquad (18.89)$$

The phase response of the estimation filter is therefore chosen to be equal to the negative phase response of the cross-correlation function Φ_{xs}. The magnitude $|H|$ is now also completely defined, because from (18.86), and (18.89), we find

$$|H| = \frac{\text{Re}\{\Phi_{xs}\, e^{j\phi}\}}{\Phi_{xx}} = \frac{\text{Re}\{|\Phi_{xs}|\, e^{j(\arg\{\Phi_{xs}\}+\phi)}\}}{\Phi_{xx}} = \frac{|\Phi_{xs}|}{\Phi_{xx}}. \qquad (18.90)$$

With (18.83), the complex frequency response of the estimation filter can be obtained from (18.89) and (18.90).

$$\boxed{H(j\omega) = \frac{\Phi_{xs}^*(j\omega)}{\Phi_{xx}(j\omega)} = \frac{\Phi_{sx}(j\omega)}{\Phi_{xx}(j\omega)}} \qquad (18.91)$$

This estimation filter corrects the observed signal $x(t)$ such that the deviation of the result $y(t)$ from the original signal $s(t)$ exhibits the smallest error power possible. We derived this result without knowing the original signal $s(t)$. Anyway, the description of random processes by mean values and power spectra enables the construction of an optimal estimation filter. The estimation filter given by (18.91) is called a Wiener filter, according to Norbert Wiener (1894-1964), a pioneer of estimation theory.

18.3.1.1 Linear Distortion and Additive Noise

The derivation of the Wiener filter as in (18.91) is general because we required no knowledge of the kind of signal interference. The result can be illustrated by using a common model for signal interference from deterministic and random sources.

Figure 18.9 shows the same set-up as in Figure 18.8. The interference source is in this case more accurately modelled by an LTI-system with frequency response $G(j\omega)$ and an additive noise source $n(t)$. The frequency response $G(j\omega)$ can, for example, stand for the frequency response of an amplifier, a transmission cable, or a radio channel. All external interference sources are summarized by the additive noise signal $n(t)$. The noise signal may not correlate with the original signal $s(t)$.

We derive the frequency response for this set-up of the Wiener filter from the general case in (18.91), using (18.64) and (18.66). First, we obtain

$$\Phi_{vx} = \Phi_{sx}G \qquad (18.92)$$

from (18.66). For the cross-spectrum Φ_{vx}, we use (18.22) and (18.64) to obtain

$$\Phi_{vx} = \Phi_{vv} = \Phi_{ss}GG^* \,. \tag{18.93}$$

From (18.92) and (18.93), we finally find

$$\Phi_{sx} = \Phi_{ss}G^* \,. \tag{18.94}$$

The power density spectrum is obtained from (18.18) and (18.64)

$$\Phi_{xx} = \Phi_{vv} + \Phi_{nn} = \Phi_{ss}|G|^2 + \Phi_{nn} \,. \tag{18.95}$$

If we now put equations (18.94) and (18.95) into (18.91), we obtain the Wiener filter for the set-up in Figure 18.9

$$H(j\omega) = \frac{\Phi_{ss}(j\omega)G^*(j\omega)}{\Phi_{ss}(j\omega)|G(j\omega)|^2 + \Phi_{nn}(j\omega)} \,. \tag{18.96}$$

Knowing the frequency response $G(j\omega)$ makes it possible in this case to find from the cross-power density spectrum $\Phi_{sx}(j\omega)$, the power density spectrum $\Phi_{ss}(j\omega)$ of the original signal and $\Phi_{nn}(j\omega)$ of the noise signal.

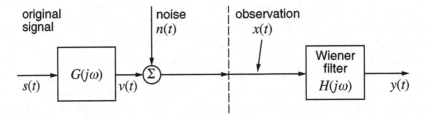

Figure 18.9: Reconstruction of a signal with linear distortion from additive noise

18.3.1.2   Ideal Transmission and Additive Noise

From the form of the Wiener filter in (18.96), we can derive two important special cases, whose mode of operation we can easily see. First we assume an ideal transfer, with $G(j\omega) = 1$, so that the original signal is only distorted by the additive noise (see Figure 18.10). The task of the Wiener filter is then reduced to finding the best possible suppression of the noise signal $n(t)$. The corresponding frequency response is

$$H(j\omega) = \frac{\Phi_{ss}(j\omega)}{\Phi_{ss}(j\omega) + \Phi_{nn}(j\omega)} \tag{18.97}$$

The mode of operation of the filter can be derived directly from the frequency response (18.97):

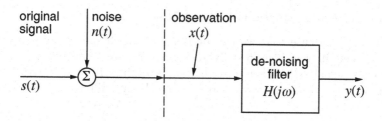

Figure 18.10: Filter for removing additive noise from a signal

- For $\Phi_{ss}(j\omega) \gg \Phi_{nn}(j\omega)$ the signal power dominates, and the filter allows the signal and the low-power noise to pass through unhindered: $H(j\omega) \approx 1$.

- For $\Phi_{ss}(j\omega) \ll \Phi_{nn}(j\omega)$ the noise dominates, and the filter bars both the noise and the smaller signal component: $H(j\omega) \approx 0$.

Figure 18.11 shows an example of an original signal with a strongly frequency-dependent power density spectrum and a noise source that produces white noise. The filter lets the frequency components pass where the original signal outweighs the noise, and stops the frequencies where the original signal makes no contribution. If both power density spectra are of the same order of magnitude, the frequency response of the filter takes a value between 0 and 1.

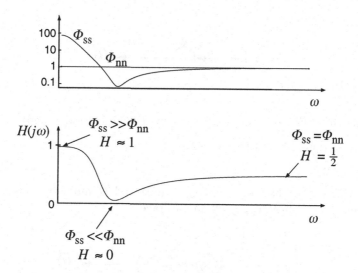

Figure 18.11: Power density spectrum $\Phi_{ss}(j\omega)$ of the original signal, $\Phi_{nn}(j\omega)$ of the noise, and frequency response $H(j\omega)$ of the Wiener filter.

18.3.1.3 Linear Distortion Without Noise

The other special case that we can derive from the frequency response (18.96) of the Wiener filter is identified by removing the noise source: $\Phi_{nn}(j\omega) = 0$. The original signal $s(t)$ is then only corrupted by the LTI-system with frequency response $G(j\omega)$. The frequency response of the Wiener filter then takes the form

$$H(j\omega) = \frac{\Phi_{ss}(j\omega)G^*(j\omega)}{\Phi_{ss}(j\omega)|G(j\omega)|^2} = \frac{1}{G(j\omega)} \qquad (18.98)$$

which means that it is trying to remove the effects of the distorting system $G(j\omega)$. As long as $|G(j\omega)| \neq 0$, this will be successful, and is even independent of the power density spectrum $\Phi_{ss}(j\omega)$ of the original signal $s(t)$. We have already discussed this case in Chapter 8.5.2 'de-convolution', but because the noise was disregarded, it was not quite complete. In addition, we have to bear in mind the stability of $H(j\omega)$ (Chapter 17.3.1).

18.4   Exercises

Exercise 18.1

A complex constant C will be added to a stationary random signal $x(t)$ with average μ_x and ACF $\varphi_{xx}(\tau)$.

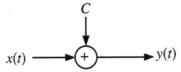

Find $\varphi_{xy}(\tau)$ and $\varphi_{yy}(\tau)$

  a) using the definitions of the CCF and ACF, (17.53) and (17.56)

  b) with Sections 18.1.2.1 and 18.1.2.2.

Exercise 18.2

The following system has complex constants A and B and a stationary complex input signal $x(t)$ with ACF $\varphi_{xx}(\tau)$ and average μ_x.

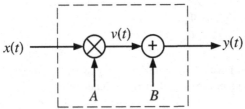

With Section 18.1, find

a) $\varphi_{vv}(\tau)$

b) $\varphi_{yy}(\tau)$

c) $\varphi_{xv}(\tau)$

Note: use the answer from Exercise 17.10.

Exercise 18.3

For the system from Exercise 18.2, find $\varphi_{xy}(\tau)$.

a) Use the definition of the CCF (17.53).

b) Can the problem also be solved with (18.61)-(18.64)? Justify your answer.

Exercise 18.4

Consider the following system with 2 inputs and 2 outputs and complex constants
A and B. The random processes $u(t)$ and $v(t)$ are stationary, have zero mean and
are complex.

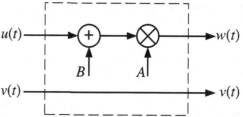

From the given correlation properties $\varphi_{uu}(\tau)$, $\varphi_{vv}(\tau)$ and $\varphi_{vu}(\tau)$, find

a) the ACF of $w(t)$

b) the CCF between v and w

c) the CCF between w and v

d) the CCF between u and w

e) the CCF between w and u

Use only the definition of the CCF for complex signals (17.53) and the expression
(17.54). Finally, verify results a), d) and e) with Section 18.1.

Exercise 18.5

The linear combination $z(t) = Ax(t) + By(t)$ is formed from two independent
random processes $x(t)$ and $y(t)$. $A, B \in \mathbb{R}$, and $x(t)$ has a zero mean and $y(t)$ has
mean μ_y. Find $\varphi_{xz}(\tau)$, $\varphi_{yz}(\tau)$ and $\varphi_{zz}(\tau)$.

Exercise 18.6

A linear, time-invariant system is described by its transfer function

$$H(s) = \frac{s^2 - 2s + 2}{(s+2)(s^2 + 2s + 2)}$$

Its impulse response is $h(t)$.

a) Calculate the filter ACF $\varphi_{hh}(\tau)$ and sketch its response.

b) White noise with power density $N_0 = 1$ is applied to the system input. Give the auto-correlation function $\varphi_{yy}(\tau)$ and the power P_y of the output signal $y(t)$.

Exercise 18.7

A system has the impulse response $h(t)$, an input signal $x(t)$ and output signal $y(t)$.
$x(t)$ is stationary and has a zero-mean, $\varphi_{xx}(\tau) = \delta(\tau)$ and $h(t) = \text{si}(t)$.
Find

a) the power density spectrum of $x(t)$

b) μ_x and μ_y

c) $\varphi_{hh}(\tau)$

d) $\varphi_{yy}(\tau)$

e) $\varphi_{xy}(\tau)$

f) Power and variance of $x(t)$ and $y(t)$.

Exercise 18.8

A system has the transfer function

$$H(j\omega) = \cos(\frac{\pi}{2\omega_g}\omega)\text{rect}(\frac{\omega}{2\omega_g})$$

a) Determine the impulse response $h(t)$.

b) Find the filter ACF $\varphi_{hh}(\tau)$

c) At the system input a signal is applied with power density $\Phi_{xx}(j\omega) = N_0 + m\delta(\omega)$. Find the ACF of the input signal $\varphi_{xx}(\tau)$ and the mean μ_x.

d) For the output signal $y(t)$, find the mean μ_y, the auto-correlation function $\varphi_{yy}(\tau)$, the output signal power P_y and the power density spectrum $\Phi_{yy}(j\omega)$.

Exercise 18.9

A system with the transfer function

$$H(s) = \frac{s-1}{s^2 + 3s + 2}$$

is excited by white noise with power density N_0. Determine the auto-correlation function, the mean and the variance of the output variable $y(t)$.

Exercise 18.10

A system with transfer function $H(s)$ is driven by a white noise signal $x(t)$ with power density N_0. The auto-correlation of the output signal is measured $\varphi_{yy}(\tau) = N_0 \frac{\alpha}{2} e^{-\alpha|\tau|}$, where $\alpha > 0$.

a) Determine a possible transfer function $H(s)$ for the system.

b) Can the transfer function be unambiguously determined?

Exercise 18.11

Two causal LTI-systems are described with their impulse responses $h_1(t)$ and $h_2(t)$. They both have a stochastic input signal $x(t)$ with auto-correlation function $\varphi_{xx}(\tau)$. Two stochastic output signals $y_1(t)$ and $y_2(t)$ are produced.

a) Express the auto-correlation functions $\varphi_{y_1 y_1}(\tau)$ and $\varphi_{y_2 y_2}(\tau)$ of the output signals with the given signal and system descriptions.

b) Determine the cross-correlation functions $\varphi_{y_1 x}(\tau)$ and $\varphi_{y_2 x}(\tau)$.

c) Give the corresponding cross-correlation function $\varphi_{y_1 y_2}(\tau)$.

Exercise 18.12

Derive (18.54) from the CCF of complex random series $\varphi_{xy}[\kappa] = E\{x[k+\kappa]x^*[k]\}$.

Exercise 18.13

Derive (18.41) from the ACF of complex random series $\varphi_{xx}[\kappa] = E\{x[k+\kappa]x^*[\kappa]\}$.

Exercise 18.14

The following set-up is often used to generate a signal with the spectral properties of a discrete speech signal, where $\varphi_{nn}[k] = \delta[k]$.

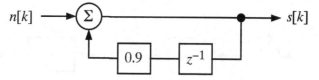

a) Calculate the frequency response $H(e^{j\Omega})$ of the system with output $n[k]$ and output $s[k]$.

b) Calculate $\Phi_{ss}(e^{j\Omega})$ and sketch it (e.g., with MATLAB).

Exercise 18.15

A transmission channel is characterised by the following system with real constants a and b:

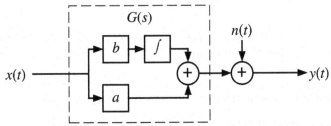

The interference $n(t)$ does not correlate with $x(t)$. Both signals are white noise with power densities $\Phi_{nn}(j\omega) = N_0$ and $\Phi_{xx}(j\omega) = 1$.
Determine the transfer function $H(j\omega)$ of a system that reconstructs a signal $x(t)$ from $y(t)$ with minimal mean square error. The reconstructed signal at the output of $H(j\omega)$ should be denoted by $\tilde{x}(t)$.

a) Which statistical signal properties must be known to solve the problem? Give $H(j\omega)$ dependent on these variables. Is the solution optimal with respect to the task set?

b) Find $H(j\omega)$ as well as $\Phi_{\tilde{x}\tilde{x}}(j\omega)$ for the transmission channel shown above for $a = 1$ and $b = 100$. Set $N_0 = 0$. Sketch $\Phi_{xx}(j\omega)$, $\Phi_{yy}(j\omega)$, $|H(j\omega)|$ and $\Phi_{\tilde{x}\tilde{x}}(j\omega)$ in logarithmic form for $10^{-2} < \omega < 10^4$.

c) Solve b) for $N_0 = 9999$. Instead of $\Phi_{yy}(j\omega)$, sketch the behaviour of $\Phi_{xx}(j\omega)\cdot$ $|G(j\omega)|^2$ and for comparison, also plot $\Phi_{nn}(j\omega)$ in the same diagram.

Note: treat the sketches in the same way as Bode diagrams, or plot the curves using a computer.

Exercise 18.16

The transmission of a signal $s(t)$ is subject to interference from noise $n(t)$ and distortion from $G(s)$, as shown in the illustration.

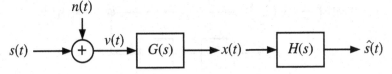

The signals $s(t)$ and $n(t)$ are uncorrelated.

Determine the Wiener filter $H(j\omega)$ that will try to reconstruct the original signal $s(t)$ from $x(t)$:

a) generally, depending on $\Phi_{nn}(j\omega)$, $\Phi_{ss}(j\omega)$ and $G(j\omega)$,

b) for $\Phi_{nn}(j\omega) = 1$, $\Phi_{ss}(j\omega) = \text{rect}(\dfrac{\omega}{2\omega_g})$ and $G(s) = \dfrac{s}{s+10}$.

Exercise 18.17

The frequency response $H(j\omega)$ under the influence of interference $n(t)$ is measured with the arrangement from Figure 18.7.

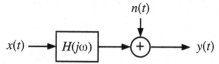

Show that $\Phi_{yx}(j\omega) = H(j\omega)$ holds when $\varphi_{xx}(\tau) = \delta(\tau)$ and when $n(t)$ and $x(t)$ do not correlate.

Appendix A   Solutions to the Exercises

Solution 1.1

    a) discrete-amplitude, discrete-time, i.e., digital

    b) continuous-amplitude, discrete-time, i.e., not digital

    c) continuous-amplitude, continuous-time, not digital

    d) discrete-amplitude, continuous-time, not digital

    e) continuous-amplitude, discrete-time

    f) continuous-amplitude, continuous-time

Solution 1.2

If a hard disk is considered a "black box", a digital signal is saved on it as a series of ones and zeros.

If, however, the inner workings are considered, it must be more precisely defined:

The bit-stream to be written is digital. The write voltage has discrete amplitude but is continuous-time, so it is neither an analog nor a digital signal. The magnetic field strength in the disk and the read voltage are continuous-amplitude and continuous-time signals, i.e., analog. The read is converted by recovering the clock signal and a hard decision into a series of ones and zeros, and is a digital signal.

Solution 1.3

    a) x_1 : analog, as it is continuous in time and amplitude

        x_2 : analog (although x_2 only changes its value at certain points in time, it is defined at every point)

        x_3 : discrete-time, discrete-amplitude, digital

    b) System 1: linear, time-variant, analog, with memory, causal

        System 2: non-linear, time-invariant, neither analog nor digital, memoryless, causal

Solution 1.4

a) linear, time-invariant, memoryless and therefore also causal

b) non-linear, time-invariant, memoryless, causal

c) linear
   time-invariant, as the response to an input signal shifted by τ is the same as the output signal shifted by τ: $S\{x(t-\tau)\} = x(t-\tau-T) = y(t-\tau)$
   causal, as the output does not depend on future input signals
   with memory, as the delay requires that the signal is saved

d) linear
   time-invariant, see c)
   not causal as the output is equal to the input signal in the future[1]
   with memory, as the response to the input signal also depends on other points in time

e) linear
   time-invariant as $S\{x(t-\tau)\} = \dfrac{dx(t-\tau)}{dt} = y(t-\tau)$
   with memory, causal

f) linear
   time-invariant as

$$S\{x(t-\tau)\} = \frac{1}{T} \int\limits_{t-T}^{t} x(t'-\tau)\,dt' \overset{\eta=t'-\tau}{=} \frac{1}{T} \int\limits_{t-T-\tau}^{t-\tau} x(\eta)\,d\eta = y(t-\tau)$$

   causal
   with memory as all past values from $t=0$ are saved

g) linear, time-invariant, causal, with memory

h) linear
   time-variant as $S\{x(t-\tau)\} = x(t-\tau-T(t)) \neq y(t-\tau) = x(t-\tau-T(t-\tau))$
   causal, with memory

i) linear, time-variant, not causal, with memory

---
[1]cannot be realised

Solution 1.5

a) S_1 is linear as the response to a linear combination of multiple inputs

$$S_1\{Ax_a(t) + Bx_b(t)\} = m \cdot (Ax_a(t) + Bx_b(t)) \cdot \cos(\omega_T t)$$

is equal to the linear combination of the individual response

$$A\,S_1\{x_a(t)\} + B\,S_1\{x_B(t)\} = m \cdot Ax_a(t) \cdot \cos(\omega_T t) + m \cdot Bx_b(t) \cdot \cos(\omega_T t).$$

S_2 is non-linear, as

$$S_2\{Ax_a(t) + Bx_b(t)\} = [1 + m(Ax_a(t) + Bx_b(t))] \cdot \cos(\omega_T t) \neq$$
$$A\,S_2\{x_a(t)\} + B S_2\{x_b(t)\} = [A + Amx_a(t) + B + Bmx_b(t)] \cdot \cos(\omega_T t).$$

b) S_1 varies with time because

$$S_1\{x(t-T)\} = mx(t-T)\cdot\cos(\omega_T t) \neq y_1(t-T) = mx(t-T)\cdot\cos(\omega_T(t-T)),$$

the same holds for S_2.

c) S_1 and S_2 are real as from $x(t) \in \mathbb{R}$ we obtain: $y_{1,2}(t) \in \mathbb{R}$.

d) S_1 and S_2 are memoryless, as the input signal is only used at time t.

Solution 1.6

We know from 3. that the system is not linear and time-invariant at the same time. It could, however, be either one of the two, and it is not possible to say which.

Solution 2.1

a) Show that:

$$\sum_{i=0}^{N} \alpha_i \frac{d^i y(t-\tau)}{dt^i} = \sum_{k=0}^{M} \beta_k \frac{d^k x(t-\tau)}{dt^k}.$$

Substituting variables $t' = t - \tau \Rightarrow \dfrac{dt'}{dt} = \dfrac{d(t-\tau)}{dt} = 1 \Rightarrow dt = dt'$, i.e., the substitution leads to the above equation.

b) If y_1 is the system response to x_1 and y_2 the response to x_2:

$$\sum_{k=0}^{M} \beta_k \frac{d^k(Ax_1 + Bx_2)}{dt^k} = A \sum_k \beta_k \frac{d^k(x_1)}{dt^k} + B \sum_k \beta_k \frac{d^k(x_2)}{dt^k}$$

$$= A \sum_{i=0}^{N} \alpha_i \frac{d^i(y_1)}{dt^i} + B \sum_{i=0}^{N} \alpha_i \frac{d^i(y_2)}{dt^i} = \sum_{i=0}^{N} \alpha_i \frac{d^i(Ay_1 + By_2)}{dt^i}$$

Solution 2.2

3rd order system. Direct form II is canonical as it uses the minimum number of energy stores (integrators). By setting the coefficients $a_0 = 0.5$; $a_1 = 0$; $a_2 = -3$; $a_3 = 1$; $b_{0.1} = 0$; $b_2 = 0.1$ and $b_3 = 1$, we find the block diagrams shown in Figs. 2.1 and 2.3.

Solution 2.3

$$1.25 \frac{d^3 y}{dt^3} - \frac{d^2 y}{dt^2} + 2y = x$$

Solution 2.4

a) Block diagram

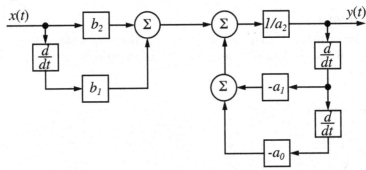

b) A canonical form is found by interchanging the left and right sides and using a common differentiator. It does not have to fulfill any conditions as differentiators with the same input signals have the same output signals.

Solution 2.5

a) The state-space representation is given by

$$\dot{z} = \begin{bmatrix} -50 & 0 & -0.5 \\ 0 & -20 & -0.2 \\ 100 & 100 & 0 \end{bmatrix} z + \begin{bmatrix} 0.5 \\ 0 \\ 0 \end{bmatrix} x$$

$$y = \begin{bmatrix} 0 & -100 & 0 \end{bmatrix} z.$$

If every state is put at an integrator output and an adder is connected to every integrator input, a signal flow graph can always be derived from the state-space representation.

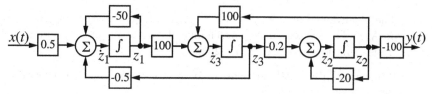

b) $0.1\dfrac{d^3y}{dt} + 7\dfrac{d^2y}{dt} + 107\dfrac{dy}{dt} + 200y = 100x$

c) Insert the coefficients $a_0 = -0.1$; $a_1 = -7$; $a_2 = -107$; $a_3 = -200$ and $b_3 = 100$ into Fig. 2.3.

d) Both are canonical.

Solution 2.6

a) States at the outputs of the integrators. We choose: z_1 at the output of the left integrator, z_2 at the output of the right integrator.

b) $\dot{\mathbf{z}} = \begin{bmatrix} e & 0 \\ d & f \end{bmatrix} \mathbf{z} + \begin{bmatrix} a \\ b \end{bmatrix} \mathbf{x}$

c) $\begin{bmatrix} y_1 \\ y_2 \end{bmatrix} = \mathbf{y} = \begin{bmatrix} 0 & 1 \\ 1 & 0 \end{bmatrix} \mathbf{z} + \begin{bmatrix} c \\ 0 \end{bmatrix} \mathbf{x}$

d) Parallel form

e) $a \neq 0$ and $b \neq 0$

f) Not completely observable, not observable from any of the outputs as every row of $\hat{\mathbf{C}}$ has a zero.

Solution 2.7

a) $\mathbf{A} = \begin{bmatrix} 0 & 1 \\ -5 & -4 \end{bmatrix}$ $\mathbf{b} = \begin{bmatrix} 0 \\ 1 \end{bmatrix}$

$\mathbf{c} = \begin{bmatrix} -3 & -8 \end{bmatrix}$ $d = 2$

b) The transformation will be carried out with a modal form of $\mathbf{A}$ that is not in general unambiguous. With $\mathbf{T} = \begin{bmatrix} 1 & 1 \\ -2+j & -2-j \end{bmatrix}$ we obtain:

$$\hat{\mathbf{A}} = \begin{bmatrix} -2+j & 0 \\ 0 & -2-j \end{bmatrix} \quad \hat{\mathbf{b}} = \begin{bmatrix} -0.5j \\ 0.5j \end{bmatrix}$$

$$\hat{\mathbf{c}} = \begin{bmatrix} 13-8j & 13+8j \end{bmatrix} \quad \hat{d} = 2.$$

The eigenvalues of the system are $-2 \pm j$ (equal to the diagonal elements of $\hat{\mathbf{A}}$). The system does not change its input-output behavior with the transformation from Eq. 2.47 - 2.50.

c) It is controllable as all elements of $\hat{\mathbf{b}}$ are non-zero, and it is observable as all elements of $\hat{\mathbf{c}}$ are non-zero.

Solution 2.8

a) $$\mathbf{T} = \mathbf{T}^{-1} = \begin{bmatrix} 0 & 0 & \cdots & 0 & 1 \\ 0 & 0 & \cdots & 1 & 0 \\ & \vdots & & & \vdots \\ 0 & 1 & \cdots & 0 & 0 \\ 1 & 0 & \cdots & 0 & 0 \end{bmatrix}$$

b) The proof is easy if one recognises that left multiplication with $\mathbf{T}^{-1}$ can be interpreted as horizontally mirroring the matrix elements, and right multiplication with $\mathbf{T}$ as vertical mirroring.

Solution 3.1

$$x_a(t) = \frac{1}{j2} e^{(-2+j5)t} - \frac{1}{j2} e^{(-2-j5)t} + e^{-2t}$$

$$\Rightarrow s_1 = -2, \quad s_2 = -2 - j5, \quad s_3 = -2 + j5$$

$$x_b(t) : s_{1,2} = \pm j\omega_0, \quad s_{3,4} = \pm 2j\omega_0$$

$$x_c(t) : s_{1,2} = \pm j\omega_0$$

Solution 3.2

a) yes, b) no, c) no, d) no, e) yes, as $H(s = j2) = 0$

Solution 3.3

a) Consistent normalisation for the components:

$$R : \frac{1V}{1A} = 1\Omega; \quad L : \frac{1V \cdot 1s}{1A} = 1H; \quad C : \frac{1A \cdot 1s}{1V} = 1F.$$

This yields the normalised component values: $R = 1$; $C = \frac{1}{6}$; $L = 1.2$

b) $\dfrac{U_2(s)}{U_1(s)} = \dfrac{\frac{R}{sRC+1}}{\frac{R}{sRC+1} + sL} = \dfrac{1}{LC \cdot s^2 + \frac{L}{R} \cdot s + 1}$

$H(s) = \dfrac{1}{0.2\,s^2 + 1.2\,s + 1}$

c) Normalising to 1s and 1V: $u_1(t) = e^{-3t}\cos(-4t) = \dfrac{1}{2}e^{s_1 t} + \dfrac{1}{2}e^{s_1{}^* t}$ with $s_1 = -3 + j4$

$H(s_1) = \dfrac{1}{0.2(-7 - j24) + 1.2(-3 + j4) + 1} = -0.25$

$H(s_1{}^*) = H(s_1)^* = -0.25$

d) $u_2(t) = \dfrac{1}{2}H(s_1)\,e^{s_1 t} + \dfrac{1}{2}H(s_1)^*\,e^{s_1{}^* t} = -\dfrac{1}{4}e^{-3t}\cos(4t)$

De-normalising with t in s and u in V: $u_2(t) = -\dfrac{1}{4}\,V e^{-\frac{3}{s}t}\cos(\dfrac{4}{s}t)$

Solution 3.4

Where $t < 0$: $\quad y(t) = 0 = \lambda \cdot x(t)$

Where $t \geq 0$: $\quad y(t) = \displaystyle\int_0^t e^{s\tau}d\tau = \dfrac{1}{s}(e^{st} - 1) \neq \lambda \cdot e^{st}$

Since in the second case there is no λ to satisfy the condition, $x(t)$ is not an eigenfunction of the integrator.

Solution 4.1

a) $X(s) = \dfrac{1}{j2}\underbrace{\displaystyle\int_0^{\infty} e^{(j-s)t}\,dt}_{\text{ROC}:Re\{s\} > 0} - \dfrac{1}{j2}\underbrace{\displaystyle\int_0^{\infty} e^{(-j-s)t}\,dt}_{\text{ROC}:Re\{s\} > 0} = \dfrac{1}{s^2 + 1}$

b) $X(s) = \displaystyle\int_{-\infty}^{\infty} \dfrac{1}{j2}(e^{jt} - e^{-jt})\,e^{-st}\,dt$

$= \dfrac{1}{j2}\underbrace{\displaystyle\int_{-\infty}^{\infty} e^{(j-s)t}\,dt}_{\text{no ROC}} - \dfrac{1}{j2}\underbrace{\displaystyle\int_{-\infty}^{\infty} e^{(-j-s)t}\,dt}_{\text{no ROC}}$

c) $X(s) = \underbrace{\int\limits_{T}^{\infty} e^{(2-s)t}\, dt}_{\text{ROC:}Re\{s\} > 2} = \dfrac{1}{s-2} e^{(2-s)T}$

d) $X(s) = \underbrace{\int\limits_{0}^{\infty} t e^{(2-s)t}\, dt}_{\text{ROC:}Re\{s\} > 2} = \left[t \cdot \dfrac{1}{2-s} e^{(2-s)t} \right]_0^{\infty} - \int\limits_{0}^{\infty} 1 \cdot \dfrac{1}{2-s} e^{(2-s)t}\, dt$

$$= 0 - \left(\dfrac{1}{2-s} \right)^2 (-1) = \dfrac{1}{(s-2)^2}$$

e) $X(s) = \int\limits_{-\infty}^{0} \dfrac{1}{2}(e^{2t} - e^{-2t})e^{-st}\, dt$

$$= \dfrac{1}{2} \underbrace{\int\limits_{-\infty}^{0} e^{(2-s)t}\, dt}_{\text{ROC:}Re\{s\} < 2} - \dfrac{1}{2} \underbrace{\int\limits_{-\infty}^{0} e^{(-2-s)t}\, dt}_{\text{ROC:}Re\{s\} < -2} = -\dfrac{2}{s^2 - 4}$$

Solution 4.2

Right-sided functions have exponential order, if M, C and T can found such that:
$|x(t)| \le Me^{Ct}$ for $t \ge T$.

a) yes, e.g. with $M = 1$, $C = 1$, $T = 0$

b) yes, e.g. with $M = 7$, $C = 5$, $T = 0$

c) yes, e.g. with $M = 1$, $C = 5$, $T = 0$

d) yes, e.g. with $M = 1$, $C = 6$, $T = 0$

e) no

f) yes, e.g. with $M = 1$, $C = 0$, $T = 0$

Solution 4.3

First determine M, C, D and T so that $|x(t)| \le Me^{Ct}$ for $t \ge T$ and $|x(t)| \le Me^{Dt}$ for $t \le -T$. The bilateral Laplace transform exists if the region of convergence is not empty, i.e. if $D > C$.

a) $M = 1, C = 0, D = 0, T = 0 \Rightarrow ROC = \{\ \}$

b) $M = 1, C = 0, D = $ any value, $T = 0 \Rightarrow ROC : 0 < \mathrm{Re}\{s\}$

c) $M = 1, C > 0, D < 0, T = 0 \Rightarrow ROC = \{\,\}$

d) $M = 5, C = 2, D = 2, T = 0 \Rightarrow ROC = \{\,\}$

Solution 4.4

Eq. 4.1: $\mathcal{L}\{x(t)\} = X(s) = \displaystyle\int_{-\infty}^{\infty} x(t)e^{-st}\,dt$

$$\mathcal{L}\{A\,f(t) + B\,g(t)\} = \int_{-\infty}^{\infty} [A\,f(t) + B\,g(t)]e^{-st}\,dt$$

$$A\int_{-\infty}^{\infty} f(t)e^{-st}\,dt + B\int_{-\infty}^{\infty} g(t)e^{-st}\,dt = A\,\mathcal{L}\{f(t)\} + B\,\mathcal{L}\{g(t)\}$$

Solution 4.5

a) $F(s) = \dfrac{2s + 3}{(s+2)(s+1)}$ $ROC : \mathrm{Re}\{s\} > -1$

b) $G(s) = \dfrac{3s + 1}{(s+3)(s+1)}$ $ROC : \mathrm{Re}\{s\} > -1$

c) $F(s) + G(s) = \dfrac{5s^2 + 16s + 11}{(s+3)(s+2)(s+1)} = \dfrac{5s + 11}{(s+2)(s+3)}$ $ROC : \mathrm{Re}\{s\} > -2$

Solution 4.6

Eq. 4.1: $\mathcal{L}\{x(t)\} = X(s) = \displaystyle\int_{-\infty}^{\infty} x(t)e^{-st}\,dt$

a) subst. $t = t' - \tau$, $\dfrac{dt}{dt'} = 1 \Rightarrow dt = dt'$

$$X(s) = \int_{\tau-\infty}^{\tau+\infty} x(t' - \tau)e^{-s(t'-\tau)}\,dt' = e^{s\tau}\int_{-\infty}^{\infty} x(t' - \tau)e^{-st'}\,dt'$$

$$e^{-s\tau} X(s) = \int_{-\infty}^{\infty} x(t' - \tau)e^{-st'}\,dt' = \mathcal{L}\{x(t' - \tau)\}$$

b) subst. $s = s' - \alpha$

$$X(s' - \alpha) = \int_{-\infty}^{\infty} x(t)e^{-(s'-\alpha)t}\, dt = \int_{-\infty}^{\infty} e^{\alpha t}x(t)e^{-s't}\, dt = \mathcal{L}\{e^{\alpha t}x(t)\}$$

Solution 4.7

Gl. 4.1: $\mathcal{L}\{x(t)\} = X(s) = \displaystyle\int_{-\infty}^{\infty} x(t)e^{-st}\, dt$

subst: $t = at'$, $\dfrac{dt}{dt'} = a \Rightarrow dt = a\, dt'$, $a \neq 0$

$$X(s) = \int_{-\frac{1}{a}\cdot\infty}^{\frac{1}{a}\cdot\infty} x(at')e^{-sat'} \cdot a\, dt' = \underbrace{a \cdot \operatorname{sign}(a)}_{|a|} \cdot \int_{-\infty}^{\infty} x(at')e^{-sat'}\, dt'$$

subst: $s = \dfrac{s'}{a}$

$$X\left(\frac{s'}{a}\right) = |a| \int_{-\infty}^{\infty} x(at')e^{-s't'}\, dt' = |a|\mathcal{L}\{x(at')\}$$

Solution 4.8

I. $x(t) = \varepsilon(t) \circ\!\!-\!\!\bullet X(s) = \dfrac{1}{s}$, $\operatorname{Re}\{s\} > 0$

II.   with the shift theorem:
$$e^{-at}\varepsilon(t) \circ\!\!-\!\!\bullet X(s+a) = \frac{1}{s+a}, \quad \operatorname{Re}\{s\} > \operatorname{Re}\{-a\}$$

III.   first scale transform pair I with $a = -1$ in time:
$$\varepsilon(-t) \circ\!\!-\!\!\bullet \frac{1}{|-1|} \cdot X(-s) = \frac{-1}{s}, \quad \operatorname{Re}\{s\} < 0$$

$$-\varepsilon(-t) \circ\!\!-\!\!\bullet \frac{1}{s} = X(s)\operatorname{Re}\{s\} < 0,$$

then use the shift theorem:
$$-e^{-at}\varepsilon(-t) \circ\!\!-\!\!\bullet X(s) = \frac{1}{s+a}, \quad \operatorname{Re}\{s\} < \operatorname{Re}\{-a\}$$

IV.   see Eq. 4.48

V.   see Eq. 4.49

VI. $\qquad \sin(\omega_0 t)\,\varepsilon(t) \qquad\qquad = \quad \dfrac{1}{j2}(e^{j\omega_0 t} - e^{-j\omega_0 t})\varepsilon(t)$

$$\circ\!\!\!\bullet$$

$$\dfrac{1}{j2}\left(\mathcal{L}\{e^{j\omega_0 t}\varepsilon(t)\} - \mathcal{L}\{e^{-j\omega_0 t}\,\varepsilon(t)\}\right) \quad = \quad \dfrac{1}{j2}\left(\dfrac{1}{s - j\omega_0} - \dfrac{1}{s + j\omega_0}\right)$$

$$= \quad \dfrac{\omega_0}{s^2 + \omega_0{}^2}, \qquad \mathrm{Re}\{s\} > 0$$

VII. similar to VI.

Solution 5.1

$$\oint_W \dfrac{F(s)}{s - s_0}\,ds = \int_0^1 \dfrac{F(s(\nu))}{s(\nu) - s_0}\left(\dfrac{ds}{d\nu}\right)d\nu = \int_0^1 \dfrac{F(s_0 + \delta\,e^{j2\pi\nu})}{\delta\,e^{j2\pi\nu}}\cdot \delta\cdot 2\pi j\,e^{j2\pi\nu}\,d\nu$$

$$= 2\pi j \int_0^1 F(s_0 + \delta\,e^{j2\pi\nu})\,d\nu = 2\pi j\,F(s_0)$$

Solution 5.2

$$F(s) = \dfrac{A}{s+1} + \dfrac{B}{s+2} + \dfrac{C}{s+5}$$

$$A = \lim_{s\to -1}[F(s)\,(s+1)] = \dfrac{2 - 2s}{(s+2)(s+5)}\Big|_{s=-1} = 1$$

$$B = \lim_{s\to -2}[F(s)\,(s+2)] = \dfrac{2 - 2s}{(s+1)(s+5)}\Big|_{s=-2} = -2$$

$$C = \lim_{s\to -5}[F(s)\,(s+5)] = \dfrac{2 - 2s}{(s+1)(s+2)}\Big|_{s=-5} = 1$$

$$F(s) \quad = \quad \dfrac{1}{s+1} - \dfrac{2}{s+2} + \dfrac{1}{s+5}$$

$$\bullet\!\!\!\circ$$

$$f(t) \quad = \quad [e^{-t} - 2e^{-2t} + e^{-5t}]\,\varepsilon(t)$$

Solution 5.3

a) $F(s) = \dfrac{A_1}{s+1} + \dfrac{A_2}{(s+1)^2} + \dfrac{A_3}{(s+1)^3} + \dfrac{B}{s+4}$

$$B = \lim_{s \to -4}[F(s)(s+4)] = \frac{2s-1}{(s+1)^3}\bigg|_{s=-4} = \frac{1}{3}$$

$$A_3 = \frac{1}{0!}\lim_{s \to -1}[F(s)(s+1)^3] = \frac{2s-1}{(s+4)}\bigg|_{s=-1} = -1$$

$$A_2 = \frac{1}{1!}\lim_{s \to -1}\frac{d}{ds}[F(s)(s+1)^3] = \frac{2(s+4)-(2s-1)}{(s+4)^2}\bigg|_{s=-1}$$

$$= \frac{9}{(s+4)^2}\bigg|_{s=-1} = 1$$

$$A_1 = \frac{1}{2!}\lim_{s \to -1}\frac{d^2}{ds^2}[F(s)(s+1)^3] = \frac{1}{2}\frac{0\cdot(s+4)^2 - 2(s+4)\cdot 9}{(s+4)^4}\bigg|_{s=-1}$$

$$= \frac{1}{2}\frac{-18}{(s+4)^3}\bigg|_{s=-1} = -\frac{1}{3}$$

$$F(s) = -\frac{1}{3}\frac{1}{s+1} + \frac{1}{(s+1)^2} - \frac{1}{(s+1)^3} + \frac{1}{3}\frac{1}{s+4}$$

$$f(t) = \left[-\tfrac{1}{3}e^{-t} + te^{-t} - \tfrac{1}{2}t^2e^{-t} + \tfrac{1}{3}e^{-4t}\right]\varepsilon(t)$$

b) $F(s) = \dfrac{A_1}{s+1} + \dfrac{A_2}{(s+1)^2} + \dfrac{A_3}{(s+1)^3} + \dfrac{B}{s+4}$

$B = \dfrac{1}{3}$ and $A_3 = -1$ can be calculated as in a).

Equating coefficients is then possible with only two more equations.

$$2s - 1 = A_1(s+1)^2(s+4) + A_2(s+1)(s+4) + A_3(s+4) + B(s+1)^3$$

$s^3:\quad 0 = A_1 + B \quad \Rightarrow \quad A_1 = -B = -\dfrac{1}{3}$

$s^0:\quad -1 = 4A_1 + 4A_2 + 4A_3 + B \quad \Rightarrow \quad A_2 = \dfrac{1}{4}(-1 - 4A_1 - 4A_3 - B) = 1$

c) With $F(s)$ as in a) the equation system

$$\begin{aligned}
0 &= A_1 + B \\
0 &= 6A_1 + A_2 + 3B \\
2 &= 9A_1 + 5A_2 + A_3 + 3B \\
-1 &= 4A_1 + 4A_2 + 4A_3 + B
\end{aligned}$$

must be solved.

Solution 5.4

a) $F(s) = \dfrac{A}{s+1+j2} + \dfrac{A^*}{s+1-j2}$

$$A = \lim_{s \to -1+j2} [F(s)\,(s+1+j2)] = \left. \frac{s+3}{s+1-j2} \right|_{s=-1-j2} = \frac{-j2+2}{j4} = \frac{1}{2} + \frac{j}{2}$$

$$f(t) = [A\,e^{-(1+j2)t} + A^*\,e^{-(1-j2)t}]\,\varepsilon(t) = e^{-t}\,[\cos(2t) + \sin(2t)]\,\varepsilon(t)$$

b) Since there are no poles apart from the conjugated complex pair, equating coefficients is trivial.

$$F(s) = \frac{As+B}{s^2+2s+5} = \frac{s+3}{s^2+2s+5} = \frac{s+3}{(s^2+2s+1)+4} = \frac{(s+1)+2}{(s+1)^2+2^2}$$

$$F(s) = \frac{(s+1)}{(s+1)^2+2^2} + \frac{2}{(s+1)^2+2^2}$$

$$f(t) = [e^{-t}\cos(2t) + e^{-t}\sin(2t)]\,\varepsilon(t)$$

Solution 5.5

$$F(s) = \frac{A}{s+2} + \frac{Bs}{s^2+w_0^2} + \frac{C}{s^2+w_0^2}$$

$$A = \lim_{s \to -2}[F(s)\,(s+2)] = \left. \frac{s}{s^2+w_0^2} \right|_{s=-2} = \frac{-2}{4+w_0^2}$$

B and C must be determined by equating coefficients:

$$(Bs+C)(s+2) + A\,(s^2+w_0^2) = s$$

$$s^2: \quad B+A = 0 \quad \Rightarrow \quad B = \frac{2}{4+w_0^2}$$

$$s^0: \quad 2C + Aw_0^2 = 0 \quad \Rightarrow \quad C = -\frac{1}{2}Aw_0^2 = \frac{w_0^2}{4+w_0^2}$$

$$F(s) = \frac{1}{4+w_0^2}\left[-2\,\frac{1}{s+2} + 2\,\frac{s}{s^2+w_0^2} + w_0\,\frac{w_0}{s^2+w_0^2}\right]$$

$$f(t) = \frac{1}{4+w_0^2}\left[-2e^{-2t} + 2\cos(w_0t) + w_0\sin(w_0t)\right]\varepsilon(t)$$

Solution 5.6

$$F(s) = \frac{(s+1)(s-2)}{s^2(s+2)(s+3)(s-1)} = \frac{A}{s^2} + \frac{B}{s} + \frac{C}{s+2} + \frac{D}{s+3} + \frac{E}{s-1}$$

$$A = s^2 F(s)\Big|_{s=0} = \frac{-2}{-6} = \frac{1}{3}$$

$$C = (s+2)F(s)\Big|_{s=-2} = \frac{4}{-12} = -\frac{1}{3}$$

$$D = (s+3)F(s)\Big|_{s=-3} = \frac{10}{36} = \frac{5}{18}$$

$$E = (s-1)F(s)\Big|_{s=1} = \frac{-2}{12} = -\frac{1}{6}$$

$$F(s) = \frac{1/3}{s^2} + \frac{B}{s} + \frac{-1/3}{s+2} + \frac{5/18}{s+3} + \frac{-1/6}{s-1}$$

Determine B by evaluating $F(s)$ at the zero $s = -1$.

$$F(-1) = \frac{1}{3} - B - \frac{1}{3} + \frac{5}{36} + \frac{1}{12} = -B + \frac{8}{36} \overset{!}{=} 0 \Rightarrow B = \frac{2}{9}$$

Then:　$$F(s) = \frac{1/3}{s^2} + \frac{2/9}{s} - \frac{1/3}{s+2} + \frac{5/18}{s+3} - \frac{1/6}{s-1}$$

Inverse transformation using Table 4.1 yields:

$$f(t) = \left(\frac{1}{3}t + \frac{2}{9} - \frac{1}{3}e^{-2t} + \frac{5}{18}e^{-3t} - \frac{1}{6}e^{t}\right)\varepsilon(t)$$

Solution 5.7

$$F(s) = \frac{-3s^3 - 12s^2 - 16s - 5}{(s+1)^2(s+2)(s+3)} = \frac{A}{(s+1)^2} + \frac{B}{s+1} + \frac{C}{s+2} + \frac{D}{s+3}$$

$$A = (s+1)^2 F(s)\Big|_{s=-1} = \frac{2}{2} = 1$$

$$C = (s+2)F(s)\Big|_{s=-2} = \frac{3}{1} = 3$$

$$D = (s+3)F(s)\Big|_{s=-3} = \frac{16}{-4} = -4$$

$$F(s) = \frac{1}{(s+1)^2} + \frac{B}{s+1} + \frac{3}{s+2} - \frac{4}{s+3}$$

A trick to calculate B is putting some value of s (but not a pole) into $F(s)$. In this case $F(s)$ is evaluated at $s = 0$.

$$F(0) = 1 + B + \frac{3}{2} - \frac{4}{3} = B + \frac{7}{6} \overset{!}{=} -\frac{5}{6} \Rightarrow B = -2$$

Then: $F(s) = \dfrac{1}{(s+1)^2} - \dfrac{2}{s+1} + \dfrac{3}{s+2} - \dfrac{4}{s+3}$

Inverse transformation using Table 4.1 yields:

$f(t) = \left((t-2)\,e^{-t} + 3\,e^{-2t} - 4\,e^{-3t} \right) \varepsilon(t)$

Solution 6.1

$$H(s) \;=\; \frac{\frac{1}{sC}}{R + sL + \frac{1}{sC}} = \frac{1}{sRC + s^2 LC + 1}$$

$$\;=\; \frac{5}{s^2 + 2s + 5} = \frac{5}{(s - s_p)(s - s_p{}^{*})}, \qquad s_p = -1 + j2$$

With $\varepsilon(t) \circ\!\!\!-\!\!\bullet\ \dfrac{1}{s}$ we obtain:

$$Y(s) = \frac{1}{s} \cdot H(s) = \frac{5}{s\,(s - s_p)(s - s_p{}^{*})} = \frac{A}{s} + \frac{B}{s - s_p} + \frac{B^{*}}{s - s_p{}^{*}}$$

Calculating the partial fraction coefficients:

$$A = \left.\frac{5}{(s - s_p)(s - s_p{}^{*})}\right|_{s=0} = \frac{5}{s_p\, s_p{}^{*}} = 1$$

$$B = \left.\frac{5}{(s - s_p{}^{*})\,s}\right|_{s=s_p} = \frac{5}{(s_p - s_p{}^{*})\,s_p} = -0.5 + j0.25 = |B|\,e^{j\Theta}$$

with $|B| = 0.25\sqrt{5}$ and $\Theta = \pi + \arctan(\tfrac{1}{2})$

Using Table 4.1 for inverse transformation gives the step response

$$y(t) \;=\; [A + B\,e^{s_p t} + B^{*}\,e^{s_p^{*}}]\,\varepsilon(t)$$

$$\;=\; [A + |B|\,e^{-t}\,(e^{(j\Theta + j2t)} + e^{(-j\Theta - j2t)})]\,\varepsilon(t)$$

$$\;=\; [A + |B|\,e^{-t} \cdot 2\,\cos(2t + \Theta)]\,\varepsilon(t)$$

Solution 6.2

The gradient of the tangent at $u(0) = 1$ is $\left.\dfrac{du(t)}{dt}\right|_{t=0} = -\dfrac{1}{T}$. This yields an intersection $x = T$.

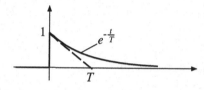

Solution 6.3

a) $H(s) = K\dfrac{(s-1+j)(s-1-j)}{(s+1+j)(s+1-j)} = K\dfrac{s^2 - 2s + 2}{s^2 + 2s + 2}$,   second order system

b) $H(s) = K\dfrac{s-1}{(s-1)(s-j2)(s+j2)} = K\dfrac{1}{s^2 + 4}$,   second order system

c) $H(s) = K\dfrac{s+1}{(s-2)(s+2)} = K\dfrac{s+1}{s^2 - 4}$,   second order system

d) $H(s) = K\dfrac{s-4}{s(s+3)}$,   second order system

Solution 6.4

a) $H(s) = \dfrac{s^2 + 4s + 4}{s^2 + 2s + 1} = \dfrac{(s+2)^2}{(s+1)^2}$

Pole-zero diagram of the transfer function:

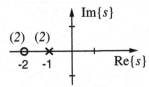

b) $H(s) = \dfrac{s^2 + 4s - 21}{s^3 + 3s^2 + 25s + 75} = \dfrac{(s-3)(s+7)}{(s+5j)(s-5j)(s+3)}$

Pole-zero diagram of the transfer function:

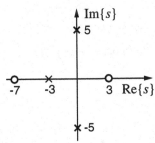

Solution 6.5

From the pole-zero diagram we see that:

$$H(s) = K\dfrac{s-1}{(s+1-5j)(s+1+5j)(s+2)} = K\dfrac{s-1}{s^3 + 4s^2 + 30s + 52}$$

Ascertainment of K:

$$H(0) = K\frac{-1}{52} \overset{!}{=} +1 \quad \Rightarrow \quad K = -52$$

From $H(s) = \dfrac{-52s + 52}{s^3 + 4s^2 + 30s + 52}$ we determine the system differential equation

$$\frac{d^3y}{dt^3} + 4\frac{d^2y}{dt^2} + 30\frac{dy}{dt} + 52y = -52\frac{dx}{dt} + 52x$$

Solution 7.1

a) The signal is also already known before $t = 0$, so it is not an initial value problem.

b) It is $Y(s) = H(s) \cdot X(s)$ with

$$H(s) = \frac{1}{s+1}, \qquad \operatorname{Re}\{s\} > -1$$

$$X(s) = \frac{\omega_0}{s^2 + \omega_0^2} + \frac{1}{(s+2)^2}, \qquad -2 < \operatorname{Re}\{s\} < 0$$

For $-1 < \operatorname{Re}\{s\} < 0$ it follows

$$Y(s) = \frac{1}{s+1}\left(\frac{\omega_0}{s^2 + \omega_0^2} + \frac{1}{(s+2)^2}\right)$$

$$= \frac{\omega_0}{(s+1)(s^2 + \omega_0^2)} + \frac{1}{(s+1)(s+2)^2}$$

$$= \frac{A}{s+1} + \frac{B}{s+j\omega_0} + \frac{B^*}{s - j\omega_0} + \frac{C}{s+1} + \frac{D}{s+2} + \frac{E}{(s+2)^2}$$

Finding the partial fraction coefficients:

$$A = (s+1)\frac{\omega_0}{(s+1)(s^2 + \omega_0^2)}\bigg|_{s=-1} = \frac{\omega_0}{1 + \omega_0^2}$$

$$B = (s + j\omega_0)\frac{\omega_0}{(s+1)(s^2 + \omega_0^2)}\bigg|_{s=-j\omega_0} = \frac{1}{-2\omega_0 - 2j} = \frac{-\omega_0 + j}{2\omega_0^2 + 2}$$

$$C = (s+1)\frac{1}{(s+1)(s+2)^2}\bigg|_{s=-1} = 1$$

$$D = \frac{d}{ds}\left[(s+2)^2\frac{1}{(s+1)(s+2)^2}\right]\bigg|_{s=-2} = -\frac{1}{(s+1)^2}\bigg|_{s=-2} = -1$$

$$E = (s+2)^2\frac{1}{(s+1)(s+2)^2}\bigg|_{s=-2} = -1$$

Inserting the coefficients gives:

$$Y(s) = \frac{\omega_0}{1+\omega_0^2} \cdot \frac{1}{s+1} + \frac{(-\omega_0+j)(s-j\omega_0)+(-\omega_0-j)(s+j\omega_0)}{(s^2+\omega_0^2)(2\omega_0^2+2)} +$$

$$+ \frac{1}{s+1} - \frac{1}{s+2} - \frac{1}{(s+2)^2}$$

$$= \frac{\omega_0}{1+\omega_0^2} \left(\frac{1}{s+1} - \frac{s}{s^2+\omega_0^2} + \frac{1}{s^2+\omega_0^2} \right) + \frac{1}{s+1} - \frac{1}{s+2} - \frac{1}{(s+2)^2}$$

Inverse Laplace transformation of $Y(s)$:

$$y(t) = \frac{\omega_0}{1+\omega_0^2} \left[e^{-t}\varepsilon(t) + \cos(\omega_0 t)\varepsilon(-t) - \frac{1}{\omega_0}\sin(\omega_0 t)\varepsilon(-t) \right] +$$

$$+ (e^{-t} - e^{-2t} - t e^{-2t})\varepsilon(t)$$

$$= \left[\left(1 + \frac{\omega_0}{1+\omega_0^2}\right) e^{-t} - (t+1)e^{-2t} \right] \varepsilon(t) +$$

$$+ \frac{\omega_0}{1+\omega_0^2} \left[\cos(\omega_0 t) - \frac{1}{\omega_0}\sin(\omega_0 t) \right] \varepsilon(-t)$$

Observe that for the inverse Laplace transform, the region of converge is the intersection of ROC$\{H\}$ und ROC$\{X\}$, i.e., ROC$\{Y\} = \{s| -1 < \mathrm{Re}\{s\} < 0\}$.

Solution 7.2

a) We wish to find the solution to the homogenous differential equation $\dot{y}_h(t) + 3y_h(t) = 0$. Using the basic approach $y_h(t) = ce^{at}$ yields:

$$a \cdot ce^{at} + 3 \cdot ce^{at} = 0 \qquad \rightarrow a = -3$$

$$y_h(t) = ce^{-3t}, \quad t > 0, \quad \forall c$$

b) The particular solution is the response of the system to

$$x(t) = 10\cos(4t) = 5e^{j4t} + 5e^{-j4t} = x_1(t) + x_2(t)$$

with $\quad x_{1/2}(t) = 5e^{\pm j4t}, \quad$ for $t > 0$

Because of linearity: $\quad y_s(t) = y_1(t) + y_2(t) \quad$ with $y_{1/2}(t) = S\{x_{1/2}(t)\}$
Using $y_1(t) = Y_1 e^{j4t}, \quad Y_1 \in C$ we obtain:

$$\dot{y}_1(t) + 3y_1(t) = x_1(t)$$

$$(j4+3) Y_1 e^{j4t} = 5e^{j4t}$$

$$Y_1 = \frac{5}{3+j4} = 1 \cdot e^{-j\Theta} \quad \text{with} \quad \Theta = \arctan\frac{4}{3} \approx 53°$$

Similarly, with $y_2(t) = Y_2 e^{-j4t}$ one obtains:

$$Y_2 = \frac{5}{3-j4} = 1 \cdot e^{j\Theta} = Y_1^*$$

Putting in Y_1 and Y_2,

$y_s(t) = e^{j(4t-\Theta)} + e^{-j(4t-\Theta)} = 2\cos(4t - \Theta)$ is obtained.

c) Complete solution: $y(t) = y_h(t) + y_s(t)$

Determining c from the initial conditions:

$y(0) = y_0 = y_h(0) + y_s(0) = c + \underbrace{2\cos(\Theta)}_{= 1.2}$

$c = y_0 - 1.2$

Now the complete solution can be given:

$y(t) \quad = \quad \underbrace{(y_0 e^{-3t}}_{\substack{\text{internal} \\ \text{part}}} \quad + \quad \underbrace{2\cos(4t - \Theta) - 1{,}2\,e^{-3t}}_{\text{external part}})\,\varepsilon(t)$

Solution 7.3

a) $H(s) = \dfrac{1}{s+3}$

b) $X(s) \quad = \quad 10 \cdot \dfrac{s}{s^2 + 16}$

$\begin{aligned} Y(s) \quad &= \quad H(s) \cdot X(s) + \dfrac{1}{s+3}[y_0 - 0] \\ &= \quad \dfrac{10}{s+3} \cdot \dfrac{s}{s^2 + 16} + \dfrac{1}{s+3}[y_0 - 0] \end{aligned}$

(using Eq. 7.16)

c) Partial fraction decomposition:

$H(s)X(s) = \dfrac{A}{s+3} + \dfrac{B}{s+j4} + \dfrac{B^*}{s-j4}$

$A \quad = \quad [(s+3)H(s)X(s)]\big|_{s=-3} \quad = \quad \dfrac{-3 \cdot 10}{25} = -1.2$

$\begin{aligned} B \quad &= \quad [(s+j4)H(s)X(s)]\big|_{s=-j4} \quad = \quad \dfrac{-j4 \cdot 10}{(-j4+3)(-j4-j4)} \\ &= \quad \dfrac{10}{2(3-j4)} = 10 \cdot \dfrac{3+j4}{50} = 0.6 + j0.8 \end{aligned}$

Inverse transformation:

$\begin{aligned} y(t) \quad &= \quad \mathcal{L}^{-1}\{H(s)X(s)\} + \mathcal{L}^{-1}\left\{\dfrac{1}{s+3}y_0\right\} \\ &= \quad \underbrace{\left[A\,e^{-3t} + B\,e^{-j4t} + B^*\,e^{j4t}\right]}\,\varepsilon(t) + y_0\,e^{-3t}\,\varepsilon(t) \end{aligned}$

Conversion to cosines

$\Rightarrow B$ and B^* in magnitude and phase

$$B = 1\,e^{j\Theta}, \quad \Theta = \arctan\left(\frac{4}{3}\right) \approx 0.3\pi \approx 53°$$

$$y(t) \quad = \quad \underbrace{(y_0\,e^{-3t}}_{\substack{\text{internal} \\ \text{part}}} \quad + \quad \underbrace{2\cos(4t - 53°) - 1.2\,e^{-3t}}_{\text{external part}})\,\varepsilon(t)$$

Solution 7.4

Eq. (7.15): $\alpha_1 \dot{y}(t) + \alpha_0 y(t) = \beta_1 \dot{x}(t) + \beta_0 x(t)$

For unilateral signals that start at $t = 0$, with (4.34):

$$\dot{y}(t) \quad \circ\!\!-\!\!\bullet \quad sY(s) - y(0)$$
$$y(t) \quad \circ\!\!-\!\!\bullet \quad Y(s)$$
$$\dot{x}(t) \quad \circ\!\!-\!\!\bullet \quad sX(s) - x(0)$$
$$x(t) \quad \circ\!\!-\!\!\bullet \quad X(s)$$

Putting these pairs into (7.15), yields (7.16)

$$\alpha_1[sY(s) - y(0)] + \alpha_0 Y(s) = \beta_1[sX(s) - x(0)] + \beta_0 X(s)$$

Solution 7.5

a) Initial conditions from the block diagram:

$$y(0-) = z(0)\cdot(-2)\cdot 0.5\cdot 4 = -4z_0$$
$$y(0+) = y(0-) + x(0+)\cdot 0.5 + 4 = -4z_0 + 2$$

b) Since the initial state has been given, (7.21) is suitable for the solution, and the initial conditions are not needed.

$$H(s) = \frac{2\,s}{s+1}, \quad \mathrm{Re}\{s\} > -1 \text{ from the block diagram, direct form II}$$

$$G(s) = \frac{-4}{s+1}$$

$$X(s) = \mathcal{L}\{\varepsilon(t) - t\varepsilon(t) + (t-1)\varepsilon(t-1)\}$$
$$= \frac{1}{s} - \frac{1}{s^2} + \frac{1}{s^2}e^{-s}, \quad \mathrm{Re}\{s\} > 0$$

$$Y(s) = H(s)\,X(s) + G(s)\,z(0) \quad = \quad \frac{2}{s+1} + \frac{2}{s(s+1)}(e^{-s}-1) - \frac{4z_0}{s+1}$$

$$= \frac{2 - 4x_0}{s+1} + \left[\frac{2}{s} + \frac{-2}{s+1}\right](e^{-s}-1)$$

$$= 4\frac{1 - z_0}{s+1} - \frac{2}{s} + \frac{2}{s}e^{-s} - \frac{2}{s+1}e^{-s}$$

$$y(t) = \left[4(1 - z_0)e^{-t} - 2\right]\varepsilon(t) + 2\left[1 - e^{-(t-1)}\right]\varepsilon(t-1)$$

Solution 7.6

a) $X(s) \;=\; \mathcal{L}\{\varepsilon(t) - \varepsilon(t-2)\} \;=\; \dfrac{1}{s} - \dfrac{1}{s}e^{-2s}, \quad \mathrm{Re}\{s\} > 0$

$H(s) \;=\; \dfrac{2s}{s^2 + 4s + 3} \;=\; \dfrac{2s}{(s+1)(s+3)}$

$Y(s) = H(s) \cdot X(s) + \dfrac{2(as+b)}{(s+1)(s+3)} \;=$

$= (1 - e^{-2s})\, \dfrac{2}{(s+1)(s+3)} + \dfrac{2(as+b)}{(s+1)(s+3)} \;=$

$= (1 - e^{-2s}) \left(\dfrac{1}{s+1} - \dfrac{1}{s+3} \right) + \dfrac{-a+b}{s+1} + \dfrac{3a-b}{s+3}$

$Y(s) = \dfrac{1-a+b}{s+1} - \dfrac{1-3a+b}{s+3} - e^{-2s}\left(\dfrac{1}{s+1} - \dfrac{1}{s-3} \right)$

$\bullet$
$\circ$

$y(t) = \big[(1-a+b)\, e^{-t} - (1-3a+b)\, e^{-3t} \big]\, \varepsilon(t) -$
$\quad - \big[e^{-(t-2)} - e^{-3(t-2)} \big]\, \varepsilon(t-2)$

b) No, because the states are not the same: (7.58) only holds for direct form III. The relations (7.56) and (7.57) are different for each choice of state and must be determined beforehand.

Solution 8.1

The response of and RC low-pass filter to a square pulse (8.7)

$y(t) = \begin{cases} \dfrac{1}{T_0}\left[e^{\frac{T_0}{T}} - 1 \right] e^{-\frac{t}{T}}, & t > 0 \\ 0 & \text{otherwise} \end{cases} \;=\; \dfrac{e^{\frac{T_0}{T}} - 1}{T_0} e^{-\frac{t}{T}}\, \varepsilon(t)$

For $T_0 \to 0$ according to L'Hospital:

$\lim_{T_0 \to 0} y(t) = e^{-\frac{t}{T}}\, \varepsilon(t) \lim_{T_0 \to 0} \dfrac{e^{\frac{T_0}{T}} - 1}{T_0} = e^{-\frac{t}{T}}\, \varepsilon(t) \lim_{T_0 \to 0} \dfrac{\frac{1}{T} e^{\frac{T_0}{T}}}{1} = \dfrac{1}{T} e^{-\frac{t}{T}}\, \varepsilon(t)$

The impulse response (8.4) with $a = \dfrac{1}{T}$ is:

$h(t) = \dfrac{1}{T} e^{-\frac{t}{T}}\, \varepsilon(t)$

Solution 8.2

a) $f_a = e^0 = 1$

b) $f_b = e^{-\tau}$

c) $f_c = \dfrac{1}{3} \cdot (0^2 - 2) = -\dfrac{2}{3}$

d) $f_d = \dfrac{1}{|-2|} \cdot 2 \cdot e^{-2} = e^{-2}$

Solution 8.3

a) $x_a(t) \;=\; \varepsilon(t) - \varepsilon(t-1) + \varepsilon(t-2) - \cdots = \displaystyle\sum_{k=0}^{\infty} (-1)^k \varepsilon(t-k)$

$\dot{x}_a(t) \;=\; \displaystyle\sum_{k=0}^{\infty} (-1)^k \dot{\varepsilon}(t-k) = \sum_{k=0}^{\infty} (-1)^k \delta(t-k)$

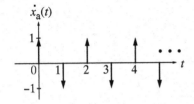

b) $x_b(t) \;=\; \varepsilon(t) \cdot \dfrac{t}{2} - \varepsilon(t-2) - \varepsilon(t-4) - \cdots = \varepsilon(t) \cdot \dfrac{t}{2} - \displaystyle\sum_{k=1}^{\infty} \varepsilon(t-2k)$

$\dot{x}_b(t) \;=\; \dot{\varepsilon}(t) \cdot \dfrac{t}{2} + \varepsilon(t) \cdot \dfrac{1}{2} - \displaystyle\sum_{k=1}^{\infty} \dot{\varepsilon}(t-2k) =$

$\qquad\;=\; \underbrace{\delta(t) \cdot \dfrac{t}{2}}_{=\,0} + \varepsilon(t) \cdot \dfrac{1}{2} - \displaystyle\sum_{k=1}^{\infty} \delta(t-2k)$

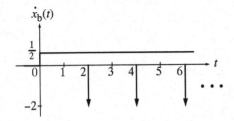

Solution 8.4

$\dot{f}(t) = \dfrac{d}{dt}\,\varepsilon(-t) = \dfrac{d}{dt}\,(1 - \varepsilon(t)) = -\dot{\varepsilon}(t) = -\delta(t)$

Solution 8.5

$$\dot{f}(t) = \frac{d}{dt}\, \varepsilon(at)$$

Substitution: $\tau = at$, $\quad \dfrac{d\tau}{dt} = a \quad \Rightarrow \quad \dfrac{d}{dt} = a\,\dfrac{d}{d\tau}$

$$\dot{f}\left(\frac{\tau}{a}\right) = a\,\frac{d}{d\tau}\varepsilon(\tau) = a\,\delta(\tau)$$

Inverse substitution: $\dot{f}(t) = a\,\delta(at) = \dfrac{a}{|a|}\,\delta(t) = \mathrm{sign}(a)\cdot\delta(t)$

Solution 8.6

$$y(t) = \int_{-\infty}^{\infty} f(t-\tau)g(\tau)\,d\tau = \int_0^4 f(t-\tau)\,d\tau = \int_0^4 (t-\tau)\,\varepsilon(t-\tau)\,d\tau$$

case $t < 0$: $y(t) = 0$

case $0 \le t < 4$: $y(t) = \displaystyle\int_0^t (t-\tau)\,d\tau = \left[t\tau - \frac{\tau^2}{2}\right]_0^t = \frac{t^2}{2}$

case $4 \le t$: $y(t) = \displaystyle\int_0^4 (t-\tau)\,d\tau = 4t - 8$

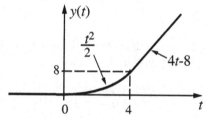

Solution 8.7

$$
\begin{aligned}
y_1(t) &= x_2(t) * x_3(t)\\
y_2(t) &= x_1(t) * x_2(t)\\
y_3(t) &= x_4(t) * x_6(t)\\
y_4(t) &= x_1(t) * x_3(t)\\
y_5(t) &= x_3(t) * x_6(t)\\
y_6(t) &= x_5(t) * x_6(t)\\
y_7(t) &= x_1(t) * x_5(t)\\
y_8(t) &= x_3(t) * x_5(t)\\
y_9(t) &= x_1(t) * x_6(t)
\end{aligned}
$$

Solution 8.8

The calculation is performed as in Sec. 8.4.3, except that $x(\tau)$ is mirrored and shifted. The result is, of course, identical to (8.49).

Solution 8.9

a) $H(s) = \dfrac{I(s)}{U(s)} = \dfrac{1}{Z(s)} = \dfrac{1}{R + \dfrac{1}{\frac{1}{sL} + sC}} = \dfrac{LC \cdot s^2 + 1}{RLC \cdot s^2 + L \cdot s + R}$

Normalizing to 1V, 1mA, 1ms gives the values $R = 1; C = 100; L = 0.1$.

$$H(s) = \frac{10s^2 + 1}{10s^2 + 0.1s + 1} = \frac{s^2 + 0.1}{s^2 + 0.01s + 0.1}$$

b) $H(s) = 1 + \dfrac{-0.01s}{s^2 + 0.01s + 0.1} = 1 + \dfrac{-0.01s}{(s + \sigma_1)^2 + \omega_1^2}$

$\overset{\bullet}{\underset{\circ}{|}}$

$h(t) = \delta(t) - 0.01\varepsilon(t)e^{-\sigma_1 t}\cos\omega_1 t$

with $\sigma_1 = 0.005; \omega_1 = \sqrt{0.1 - 0.005^2} \approx 1$

and $A = -0.01; B = \dfrac{A\sigma_1}{\omega_1} \approx 1.58 \cdot 10^{-4}$

or

$h(t) = \delta(t) + e^{-\sigma_1 t}\varepsilon(t) A_1 \cos(\omega_1 t + \varphi_1)$

with $A_1 \approx 0.01$ and $\varphi_1 \approx -89°$

c) $ROC\{H\} : Re\{s\} > -\sigma_1$. Right-sided since $h(t)$ is the impulse response of a real system and must therefore be causal.

d) The system response $i(t)$ converges if $ROC\{I(s)\}$ contains the imaginary axis. Since $ROC\{I(s)\} = ROC\{H(s)\} \cap ROC\{U(s)\}$, $ROC\{U(s)\}$ must also contain the imaginary axis. So $\sigma_0 > 0$ and ω_0 can be freely chosen. The input signal can have any amplitude, but must decay over time.

Solution 9.1

a) $\mathcal{F}\{x(t)\} = \displaystyle\int_{-\infty}^{\infty} e^{-j\omega_0 t}\varepsilon(t)\, e^{-j\omega t}\, dt = \int_{0}^{\infty} e^{-j(\omega + \omega_0)t}\, dt$ does not converge *

$\mathcal{L}\{x(t)\} = \dfrac{1}{s + j\omega_0}$; $Re\{s\} > 0$

b) $\mathcal{F}\{x(t)\} = \displaystyle\int_{-5}^{5} e^{-j\omega t}\, dt = \left[-\frac{1}{j\omega}e^{-j\omega t}\right]_{-5}^{5} = \dfrac{e^{j5\omega} - e^{-j5\omega}}{j\omega} = \dfrac{j2\sin(5\omega)}{j\omega} =$

$= 10\,\mathrm{si}(5\omega)$

$$\mathcal{L}\{x(t)\} = \mathcal{L}\{\varepsilon(t+5) - \varepsilon(t-5)\} = \frac{e^{+5s}}{s} - \frac{e^{-5s}}{s} \quad , \quad s \in \mathbb{C}, \text{ since } x(t) \text{ has}$$

finite duration.

c) $\mathcal{F}\{x(t)\} = \displaystyle\int_{-\infty}^{\infty} \delta(4t) e^{-j\omega t} \, dt = \frac{1}{|-4|} e^{-j\omega \cdot 0} = \frac{1}{4}$

$$\mathcal{L}\{x(t)\} = \mathcal{L}\{\frac{1}{4}\delta(t)\} = \frac{1}{4} \cdot 1 \quad , \quad s \in \mathbb{C}$$

d) $\mathcal{F}\{x(t)\} = \displaystyle\int_{-\infty}^{\infty} \varepsilon(-t) e^{-j\omega t} \, dt = \displaystyle\int_{-\infty}^{0} e^{-j\omega t} \, dt \quad$ does not converge *

$$\mathcal{L}\{x(t)\} = -\frac{1}{s} \quad , \quad \text{Re}\{s\} < 0$$

e) $\mathcal{F}\{x(t)\}$ does not converge, see a) *

    $\mathcal{L}\{x(t)\}$ does not converge

*Calculating the Fourier integral does not provide the solution because the integral will not converge. Nevertheless, the Fourier transform does exist in the form of a distribution.

Solution 9.2

b) and c), since the region of convergence of the Laplace transform contains the imaginary axis. (Lying on the border is not sufficient!)

Solution 9.3

The Fourier integrals from Exercise 9.1a, d and e do not converge to a function.

for a) Use the modulation theorem on pair (9.7):
$$\mathcal{F}\{\varepsilon(t) e^{-j\omega_0 t}\} = \pi\delta(\omega+\omega_0) + \frac{1}{j(\omega+\omega_0)} .$$

for d) Use the similarity theorem on pair (9.7):
$$\mathcal{F}\{\varepsilon(-t)\} = \pi\delta(\omega) - \frac{1}{j\omega} .$$

for e) Use the principle of duality on pair (9.17):
$$\mathcal{F}\{e^{-j\omega_0 t}\} = 2\pi\delta(\omega+\omega_0) .$$

The duality princple is used as follows:

$$\delta(t-\tau) \quad \circ\!\!-\!\!\bullet \quad e^{-j\omega\tau}$$
$$e^{-j\tau t} \quad \circ\!\!-\!\!\bullet \quad 2\pi\delta(-\omega-\tau) = 2\pi\delta(\omega+\tau)$$

Where τ is any constant that can be appropriately replaced in the result by ω_0.

Solution 9.4

$$\mathcal{F}\left\{\frac{1}{t-a}\right\} = \int_{-\infty}^{\infty} \frac{1}{t-a} e^{-j\omega t}\, dt \stackrel{\tau=t-a}{=} \int_{-\infty}^{\infty} \frac{1}{\tau} e^{-j\omega(\tau+a)}\, d\tau =$$

$$= \lim_{\substack{\varepsilon\to 0 \\ T\to\infty}} \left[\int_{-T}^{-\varepsilon} \frac{1}{\tau} e^{-j\omega(\tau+a)}\, d\tau + \int_{\varepsilon}^{T} \frac{1}{\tau} e^{-j\omega(\tau+a)}\, d\tau\right] =$$

$$= \lim_{\substack{\varepsilon\to 0 \\ T\to\infty}} e^{-j\omega a} \int_{\varepsilon}^{T} \frac{1}{\tau} \left(e^{-j\omega\tau} - e^{j\omega\tau}\right)\, d\tau = \lim_{\substack{\varepsilon\to 0 \\ T\to\infty}} -2j\cdot e^{-j\omega a} \int_{\varepsilon}^{T} \frac{\sin(\omega\tau)}{\tau}\, d\tau =$$

$$= e^{-j\omega a}\cdot \begin{cases} -j\pi & \text{for } \omega > 0 \\ 0 & \text{for } \omega = 0 \\ j\pi & \text{for } \omega < 0 \end{cases} = -j\pi\operatorname{sign}(\omega)e^{-j\omega a}$$

Solution 9.5

a) Zeros of si(x) at $x = n\pi$, $n \in \mathbb{Z}\setminus\{0\}$,

here: $\omega_0 n\cdot 4\pi \stackrel{!}{=} n\pi \Rightarrow \omega_0 = \frac{1}{4}$

b) Only calculate the area of the triangle in Fig. 9.6:

$$\int_{-\infty}^{\infty} x(t)\, dt = \frac{1}{2}\cdot 1\cdot 8\pi = 4\pi$$

c)

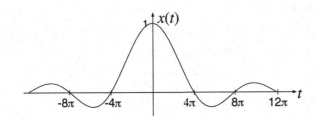

Solution 9.6

si$(10\pi t)$ o—• $\dfrac{\pi}{|10\pi|}\cdot \operatorname{rect}\left(\dfrac{\omega}{20\pi}\right)$

si$(10\pi(t+T))$ o—• $\dfrac{1}{10}\operatorname{rect}\left(\dfrac{\omega}{20\pi}\right)\cdot e^{j\omega T} = X(j\omega)$

a)

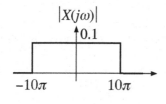

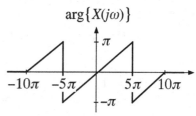

If $|X| = 0$, no phase has been defined; write null.

b)

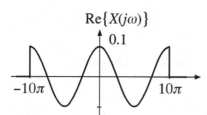

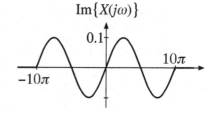

Solution 9.7

$$X_1(s) = \frac{5s + 5}{s^2 + 2s + 17} = \frac{5(s+1)}{(s+1)^2 + 4^2} \quad \text{with} \quad \text{Re}\{s\} > -1$$

$\mathcal{L}\{X_1(s)\} = 5e^{-t}\cos(4t)\epsilon(t)$

Since the ROC of $X_1(s)$ contains the imaginary axis:

$\mathcal{F}^{-1}\{x_1(t)\} = \mathcal{L}^{-1}\{x_1(t)\}|_{s=j\omega}$, so $x_1(t) = 5e^{-t}\cos(4t)\varepsilon(t)$.

$X_2(j\omega) = \text{si}(2\omega)$

$$x_2(t) = \frac{1}{4}\text{rect}\left(\frac{t}{4}\right)$$

$X_3(j\omega) = \text{si}^2(2\omega) = \text{si}(2\omega) \cdot \text{si}(2\omega)$

$$x_3(t) = \frac{1}{4}\text{rect}\left(\frac{t}{4}\right) * \frac{1}{4}\text{rect}\left(\frac{t}{4}\right) = \frac{1}{16} \cdot \begin{cases} 4+t & \text{for } -4 < t \leq 0 \\ 4-t & \text{for } 0 < t < 4 \\ 0 & \text{otherwise} \end{cases}$$

Solution 9.8

a) Conjugate symmetry, i.e. real part even, imaginary part odd:
$X(-j\omega) = X^*(j\omega)$

b) Real part odd, imaginary part even: $X(-j\omega) = -X^*(j\omega)$

Solution 9.9

From $x(t) = \text{Re}\{x_g(t)\} + \text{Re}\{x_u(t)\} + j\text{Im}\{x_g(t)\} + j\text{Im}\{x_u(t)\}$ it follows

a)

$$y_a(t) = \text{Re}\{x_g(t)\} - \text{Re}\{x_u(t)\} + j\text{Im}\{x_g(t)\} - j\text{Im}\{x_u(t)\}$$

$$Y_a(j\omega) = \text{Re}\{X_g(j\omega)\} - \text{Re}\{X_u(j\omega)\} + j\text{Im}\{X_g(j\omega)\} - j\text{Im}\{X_u(j\omega)\}.$$

Both odd parts of $X(j\omega)$ change sign; this corresponds to changing the sign of the argument: $Y_a(j\omega) = X(-j\omega)$.

b) It is sufficient to write down the sign of the four parts.

$$y_b(t) \quad = \quad + \quad \ldots \quad + \quad \ldots \quad - \quad \ldots \quad - \quad \ldots$$

$$Y_b(j\omega) \quad = \quad + \quad \ldots \quad - \quad \ldots \quad - \quad \ldots \quad + \quad \ldots$$

To produce the desired change of sign $(+ - -+)$, the sign of the argument must be reversed $(+ - + -)$ as well forming the complex conjugate $(+ + - -)$: $Y_b(j\omega) = X^*(-j\omega)$.

c)

$$y_c(t) \quad = \quad + \quad \ldots \quad - \quad \ldots \quad - \quad \ldots \quad + \quad \ldots$$

$$Y_c(j\omega) \quad = \quad + \quad \ldots \quad + \quad \ldots \quad - \quad \ldots \quad - \quad \ldots$$

Both imaginary parts change sign: $Y_c(j\omega) = X^*(j\omega)$.

Solution 9.10

$$y(t) = [x(t) + m]\frac{1}{j2}\left(e^{j\omega_T t} - e^{-j\omega_T t}\right)$$

with $[x(t) + m] \circ\!\!-\!\!\bullet X(j\omega) + 2\pi m\, \delta(\omega)$ it holds that:

$$Y(j\omega) = \frac{1}{j2}[X(j(\omega - \omega_T)) - X(j(\omega + \omega_T)) + 2\pi m(\delta(\omega - \omega_T) - \delta(\omega + \omega_T))] =$$

$$= \frac{j}{2}[X(j(\omega + \omega_T)) - X(j(\omega - \omega_T))] + j\pi m[\delta(\omega + \omega_T) - \delta(\omega - \omega_T)]$$

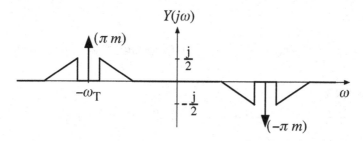

Solution 9.11

$$\sin(\omega_T t) \quad \circ\!\!-\!\!\bullet \quad j\pi[\delta(\omega + \omega_T) - \delta(\omega - \omega_T)]$$

$$Y(j\omega) = \frac{1}{2\pi}[\ldots] * [X(j\omega) + 2\pi m\delta(\omega)] =$$

$$= \frac{j}{2}[X(j(\omega + \omega_T)) - X(j(\omega - \omega_T))] + j\pi m[\delta(\omega + \omega_T) - \delta(\omega - \omega_T)]$$

Solution 9.12

a) from sec. 9.2.2: $\varepsilon(t) \;\circ\!\!-\!\!\bullet\; \pi\delta(\omega) + \dfrac{1}{j\omega}$

   Duality: $\pi\delta(t) + \dfrac{1}{jt} \;\circ\!\!-\!\!\bullet\; 2\pi\varepsilon(-\omega)$

b) from sec. 9.3: $\varepsilon(t)e^{-at} \;\circ\!\!-\!\!\bullet\; \dfrac{1}{j\omega + a}$

   Duality: $\dfrac{1}{jt + a} \quad \circ\!\!-\!\!\bullet \quad 2\pi\varepsilon(-\omega)e^{a\omega}$

   $\dfrac{1}{t - ja} \quad \circ\!\!-\!\!\bullet \quad -2\pi j\varepsilon(-\omega)e^{a\omega}$

c) from sec. 9.4.4: $\dfrac{1}{t} \;\circ\!\!-\!\!\bullet\; -j\pi\mathrm{sign}(\omega)$

   Duality: $-j\pi\mathrm{sign}(t) \quad \circ\!\!-\!\!\bullet \quad 2\pi \cdot \dfrac{1}{-\omega}$

   $\mathrm{sign}(t) \quad \circ\!\!-\!\!\bullet \quad \dfrac{2}{j\omega}$

Solution 9.13

$$\mathcal{F}\{\dot{\delta}(t)\} = \int\limits_{-\infty}^{\infty} \dot{\delta}(t)e^{-j\omega t} = -\frac{d}{dt}\left(e^{-j\omega t}\right)\Big|_{t=0} = j\omega$$

To calculate the integral, the rules for derivated delta impulses (8.23) are used.

Solution 9.14

$$\frac{dX(j\omega)}{d(j\omega)} = \frac{d}{d(j\omega)} \int_{-\infty}^{\infty} x(t)e^{-j\omega t}\, dt = \int_{-\infty}^{\infty} x(t) \cdot (-t)e^{-j\omega t}\, dt = \mathcal{F}\{-tx(t)\}$$

Solution 9.15

a) $\quad \dfrac{dx(t)}{dt} \quad = \quad \dfrac{1}{T}[\varepsilon(t+T) \quad - \quad 2\varepsilon(t) \quad + \quad \varepsilon(t-T)]$

$\quad\quad \dfrac{d^2x(t)}{dt^2} \quad = \quad \dfrac{1}{T}[\delta(t+T) \quad - \quad 2\delta(t) \quad + \quad \delta(t-T)]$

$$\circ\!-\!\bullet \qquad\qquad \circ\!-\!\bullet \qquad \circ\!-\!\bullet \qquad \circ\!-\!\bullet$$

$\quad -\omega^2 X(j\omega) \quad = \quad \dfrac{1}{T}\left[e^{j\omega T} \quad - \quad 2e^0 \quad + \quad e^{-j\omega T}\right]$

In doing so, the differentiation theorem in the time-domain (9.86) and the transform pair (9.17) are used.

b) $\quad x(t) \quad = \quad \dfrac{1}{T}\left[\text{rect}\left(\dfrac{t}{T}\right) \;*\; \text{rect}\left(\dfrac{t}{T}\right)\right]$

$$\circ\!-\!\bullet$$

$\quad\quad X(j\omega) \quad = \quad \dfrac{1}{T}\left[\text{Tsi}\left(\dfrac{\omega T}{2}\right) \cdot \text{Tsi}\left(\dfrac{\omega T}{2}\right)\right] = T\text{si}^2\left(\dfrac{\omega T}{2}\right)$

Solution 10.1

a) $20 \log 10 = 20$ dB
b) 80 dB
c) 3 dB
d) -34 dB
e) 6 dB

Solution 10.2

With $\dfrac{P_{\text{out}}}{P_{\text{in}}} = \dfrac{U_{\text{out}}^2}{U_{\text{in}}^2}$ we obtain:

a) $\dfrac{U_{\text{out}}}{U_{\text{in}}} = 8 \hat{=} 18$ dB $\qquad\qquad$ b) $\dfrac{U_{\text{out}}}{U_{\text{in}}} = \sqrt{2} \hat{=} 3$ dB

Note: The formula $V = 10 \log \dfrac{P_{\text{out}}}{P_{\text{in}}}$ can also be used to calculate the amplification directly from the power ratio. It gives the same result as above.

Solution 10.3

Magnitude $V = 20 \log |H(j\omega)| = 60$ dB $-20 \log \sqrt{\omega^2 + 100^2}$ dB

Phase $\varphi = \arg\{H(j\omega)\} = -\arctan \dfrac{\omega}{100}$

ω	1	10	100	1k	10k	100k
$H(s)$	10	$9.9 - j$	$5 - j5$	$0.1 - j$	$-j0.1$	$-j0.01$
$V[dB]$	20	20	17	0	-20	-40
φ	0	$-6°$	$-45°$	$-84°$	$-90°$	$-90°$

Solution 10.4

$$|H(j10)| = \left|\frac{j10 + 1}{(j10)(j10 + 100)}\right| = \frac{10}{10 \cdot 100} = 10^{-2}$$

$20 \log 10^{-2} = -40$ dB

Amplitude sketch: Proceed as with real poles and zeros. (Sec. 10.4.1).

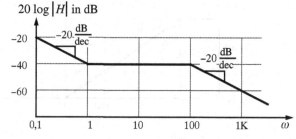

Phase sketch: Proceed as in Sec. 10.4.2.

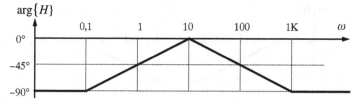

Solution 10.5

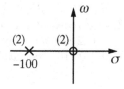

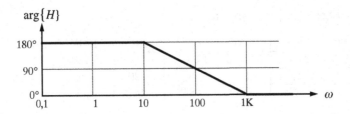

Solution 10.6

a) $h(t) \circ\!\!-\!\!\bullet H(s) = \dfrac{Y(s)}{X(s)} = -\dfrac{1}{500}\left(1 + \dfrac{999}{s+1}\right) = -\dfrac{1}{500}\left(\dfrac{s+1000}{s+1}\right)$

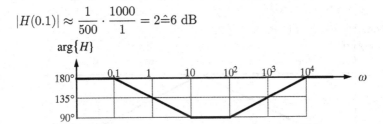

$$|H(0.1)| \approx \dfrac{1}{500} \cdot \dfrac{1000}{1} = 2 \hat{=} 6 \text{ dB}$$

b) It is: $y(t) = |H(j\omega_0)| \cos(\omega_0 t + arg\{H(j\omega_0)\})$, where the magnitude and phase of $H(j\omega)$ can be read from the Bode plot:

$$\begin{aligned}
\omega_0 &= 0.01 \text{ Hz} &\rightarrow\quad & y(t) = -2\cos(\omega_0 t)\\
\omega_0 &= 1 \text{ Hz} &\rightarrow\quad & y(t) = \sqrt{2}\cos(\omega_0 t + 135°)\\
\omega_0 &= 10 \text{ Hz} &\rightarrow\quad & y(t) = 0,2\cos(\omega_0 t + 96°)\\
\omega_0 &= 0.1 \text{ MHz} &\rightarrow\quad & y(t) = -2 \cdot 10^{-3}\cos(\omega_0 t)
\end{aligned}$$

Solution 10.7

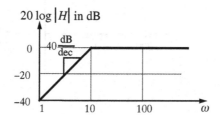

a) $|H(j100)| \approx \dfrac{100^2}{100^2} = 1 \hat{=} 0$ dB

b) $|H(j100)| = \left| \dfrac{(j100)^2}{(j100 + 10)^2} \right| = 0.9901 \hat{=} -0,0864$ dB

Deviation from Bode plot: 0.99%

$|H(j1000)| = \left| \dfrac{(j1000)^2}{(j1000 + 10)^2} \right| = 0.9999 \hat{=} -0.00087$ dB

Deviation from Bode plot: 0,01%

c) 3 dB per pole, so 6 dB in this case.

Exact calculation: $|H(j10)| = \left| \dfrac{(j10)^2}{(j10 + 10)^2} \right| = \dfrac{1}{2} \hat{=} -6.0206$ dB

This value is rounded in the Bode plot to 60 dB leading to a deviation of 0.3433%.

Solution 10.8

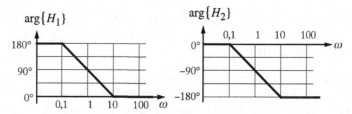

The amplitude sketches are the same, because $|H_1| = |H_2|$.

Solution 10.9

a) $H(s) = 10^4 \cdot \dfrac{(s+1)(s+0.1)}{(s+1)(s+10)(s+1000)} = 10^4 \cdot \dfrac{(s+0.1)}{(s+10)(s+1000)}$

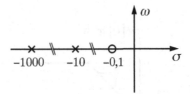

b) Amplitude sketch:

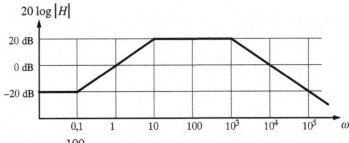

$$|H(j100)| \approx 10^4 \cdot \frac{100}{100 \cdot 1000} = 10 \hat{=} 20 \text{ dB}$$

Phase sketch:

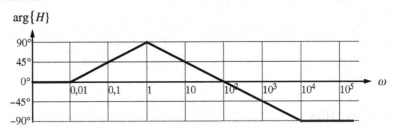

The system has a band-pass behaviour. The maximum amplification is 10. At $\omega = 1$, and at $\omega = 10^4$ it has descended to $\frac{1}{10}$ of the maximum (equivalent to 20 dB).

c) The response to $\varepsilon(t)$ for $t \to \infty$ corresponds to the dc voltage gain, i.e., $|H(\omega = 0)|$. The frequency $\omega = 0$ is not in the diagram, but for $\omega \ll 0.1$, $|H|$ remains constant. The dc voltage gains are 0.1 and -20 dB respectively.

Solution 10.10

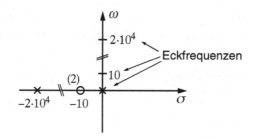

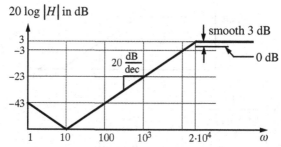

$$H(s) = K \cdot \frac{(s+10)^2}{s \cdot (s+2 \cdot 10^4)}$$

$\omega = 2 \cdot 10^4$ is the corner frequency. The magintude of the transfer function for $\omega \gg 2 \cdot 10^4$ is therefore by a factor of $\sqrt{2}$ greater than at $\omega = 2 \cdot 10^4$.

$$|H(j10^6)| = |K| \cdot \frac{10^{12}}{10^6 \cdot 10^6} = |K| \stackrel{!}{=} 10^{\frac{3}{20}} = \sqrt{2}$$

$$K = \sqrt{2} \text{ or } K = -\sqrt{2}$$

Solution 10.11

a) $-20 \dfrac{\text{dB}}{\text{dec}}$ b) $20 \dfrac{\text{dB}}{\text{dec}}$ c) $40 \dfrac{\text{dB}}{\text{dec}}$ d)$n \cdot 20 \dfrac{\text{dB}}{\text{dec}}$

Solution 10.12

The amplification increases exponentially at $40\frac{\text{dB}}{\text{dec}}$, when $\omega \ll 100$. Doubling the frequency quadruples the amplification, increasing it by $20 \log 4 = 12$ dB.

ω	1	2	4	8
V [dB]	-80	-68	-56	-44

$|H|$ is constant when $\omega \gg 100$, i.e., $V = 20 \log |H| = 0$ dB.

Solution 10.13

$20\frac{dB}{dec}$ means that $|H(j\omega)|$ grows linearly with ω, i.e. doubling the frequency (one octave) results in doubling the amplification corresponding to 6 dB. Thus $20\frac{dB}{dec} \hat{=} 6\frac{dB}{oct}$.

Accordingly, $40\frac{dB}{dec} \hat{=} 12\frac{dB}{oct}$ and $60\frac{dB}{dec} \hat{=} 18\frac{dB}{oct}$.

Solution 10.14

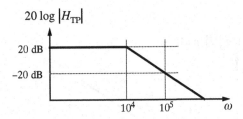

a) $H_{TP}(s) = K \cdot \dfrac{1}{(s + \omega_c)^2}$

$|H_{TP}(0)| = |K| \cdot \dfrac{1}{10^8} \overset{!}{=} 10 \quad \rightarrow \quad |K| = 10^9$

$H_{TP}(s) = \pm \dfrac{10^9}{(s + 10^4)^2}$

b)

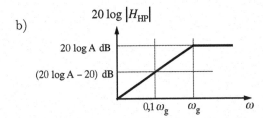

$$H_{HP} = A \cdot \frac{s}{s + \omega_g}$$

Target system: at $\omega = 10$, gain $0.5 \hat{=} - 6\,dB$.

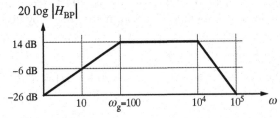

The amplitude sketch indicates $\omega_g = 100$,

and with $(20\log|H_{\text{TP}} \cdot H_{\text{HP}}|)_{\text{max}} = 14$ dB, we obtain $|A| = \dfrac{1}{2}$.

Solution 10.15

a) In this question only the maximum amplitude can be stated. The values at $\omega = 1$ and $\omega = 100$ belong to c).

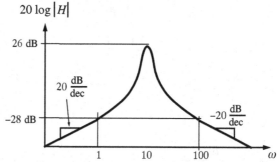

b) shape of $H(s)$ from

 - Increase by $20\frac{\text{dB}}{\text{dec}}$ at $\omega \ll \omega_0$ $\rightarrow$ exactly 1 zero at $s = 0$.

 - Decrease by $20\frac{\text{dB}}{\text{dec}}$ for $\omega \gg \omega_0$ $\rightarrow$ denominator degree= numerator degree $+1 = 2$.

$$\Rightarrow \quad H(s) = \frac{Ks}{s^2 + 2\alpha s + \omega_0^2}$$

 - $Q = 50 \approx \dfrac{\omega_0}{2\alpha}$ $\rightarrow$ $\alpha \approx 0.1$

 - $|H(j\omega_0)| = \left| \dfrac{K \cdot j10}{(j10)^2 + 2\alpha \cdot j10 + 10^2} \right| = \dfrac{K}{2\alpha} \overset{!}{=} 20 \hat{=} 26$ dB

$$\Rightarrow \quad K = 2\alpha \cdot 20 = 0.2 \cdot 20 = 4$$

c) $|H(1)| \approx \dfrac{1 \cdot 4}{10^2} \hat{=} 12\,\text{dB} - 40\,\text{dB} = -28\,\text{dB}$

$|H(100)| \approx \dfrac{100 \cdot 4}{100^2} \hat{=} 52\,\text{dB} - 80\,\text{dB} = -28\,\text{dB}$

d)

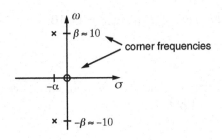

e) $\dfrac{1}{Q_0} = \dfrac{\Delta\omega}{\omega_0} \Rightarrow \Delta\omega = \dfrac{\omega_0}{Q_0} = \dfrac{10}{50} = \dfrac{1}{5}$

$\omega_1 = \omega_0 - \dfrac{\Delta\omega}{2} = 9.9$

$\omega_2 = \omega_0 + \dfrac{\Delta\omega}{2} = 10.1$

f) $H_2(s) = \dfrac{Ks}{s^2 + 2\alpha s + \omega_0{}^2}$

The condition is $\alpha = 0$. In the real world this corresponds to the a loss-free circuit.

Solution 11.1

$x(t) = \displaystyle\sum_{k=-\infty}^{\infty} \delta(t - 5k) = \sum_k \dfrac{1}{5}\delta(\dfrac{t}{5} - k) = \dfrac{1}{5}\, \text{⊥⊥⊥}\left(\dfrac{t}{5}\right)$

Solution 11.2

Insert $T = \dfrac{1}{a}$ into (11.12) ein $\Rightarrow$ $a\,\text{⊥⊥⊥}(at) \circ\!\!-\!\!\bullet\, \text{⊥⊥⊥}\left(\dfrac{\omega}{2\pi a}\right)$

$X(j\omega) = \dfrac{1}{a}\, \text{⊥⊥⊥}\left(\dfrac{\omega}{2\pi a}\right)$

$a = \tfrac{1}{2}$ $a = 1$ $a = 3$

Solution 11.3

a) $x(t) = ⊥⊥⊥(t - t_0)$ $○\!\!-\!\!\bullet$ $X(j\omega) = ⊥⊥⊥\left(\dfrac{\omega}{2\pi}\right) e^{-j\omega t_0}$

b) $t_0 = 0,\ t_0 = \dfrac{1}{2}$: $x(t)$ real + even $\Rightarrow X(j\omega)$ real + even

$t_0 = \dfrac{1}{4}$: $x(t)$ real, asymmetrical $\Rightarrow X(j\omega)$ conjugate symmetry

c) $t_0 = \dfrac{1}{4}$: $X(j\omega) = 2\pi \sum\limits_{\mu} \delta(\omega - 2\pi\mu)\, e^{-j\omega \cdot \frac{1}{4}} = 2\pi \sum\limits_{\mu} \delta(\omega - 2\pi\mu)\, e^{-j\frac{\pi}{2}\mu}$

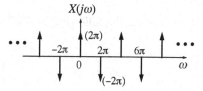

Re$\{X(j\omega)\}$ with arrows at (2π), axis $-2\pi\ 0\ 2\pi\ 6\pi\ 8\pi\ \omega$, (-2π).

Im$\{X(j\omega)\}$ with arrows at (2π), axis $-4\pi\ -2\pi\ 0\ 4\pi\ 6\pi\ 8\pi\ \omega$, (-2π).

$t_0 = \dfrac{1}{2}$: $X(j\omega) = 2\pi \sum\limits_{\mu} \delta(\omega - 2\pi\mu)\, (-1)^\mu$

$X(j\omega)$ with arrows at (2π), axis $-2\pi\ 0\ 2\pi\ 6\pi\ \omega$, (-2π).

Solution 11.4

$$X_1(j\omega) = \frac{2}{\pi} ⊥⊥⊥\left(\frac{\omega}{\pi}\right) + \frac{1}{\pi} ⊥⊥⊥\left(\frac{1}{\pi}\left(\omega + \frac{\pi}{2}\right)\right)$$

$$\begin{aligned}
x_1(t) &= \frac{2}{\pi} \cdot \frac{1}{2} ⊥⊥⊥\left(\frac{t}{2}\right) + \frac{1}{\pi} \cdot \frac{1}{2} ⊥⊥⊥\left(\frac{t}{2}\right) \cdot e^{-j\frac{\pi}{2}t} \\
&= \frac{2}{\pi} \sum_k \delta(t - 2k) + \frac{1}{\pi} \sum_k \delta(t - 2k) \underbrace{e^{-j\pi k}}_{(-1)^k}
\end{aligned}$$

$$X_2(j\omega) = ⊥⊥⊥\left(\frac{\omega}{3}\right) + \frac{1}{2} ⊥⊥⊥\left(\frac{1}{2}(\omega + 1)\right)$$

$$\begin{aligned}
x_2(t) &= \frac{3}{2\pi} ⊥⊥⊥\left(\frac{3t}{2\pi}\right) + \frac{1}{2} \cdot \frac{1}{\pi} ⊥⊥⊥\left(\frac{t}{\pi}\right) \cdot e^{-jt} \\
&= \sum_k \delta\left(t - \frac{2\pi}{3}k\right) + \frac{1}{2} \sum_k \delta(t - \pi k) \underbrace{e^{-j\pi k}}_{(-1)^k}
\end{aligned}$$

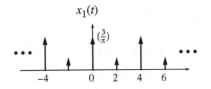

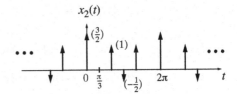

Solution 11.5

a) $x_a(t) = \dfrac{1}{2}\left(e^{j3\omega_0 t} + e^{-j3\omega_0 t}\right) \cdot \left(\dfrac{1}{j2}\right)^2 \left(e^{j2\omega_0 t} - e^{-j2\omega_0 t}\right)^2$

$ = -\dfrac{1}{8}\left(e^{j7\omega_0 t} + e^{j\omega_0 t}\right) + \dfrac{1}{4}\left(e^{j3\omega_0 t} + e^{-j3\omega_0 t}\right) - \dfrac{1}{8}\left(e^{-j\omega_0 t} + e^{-j7\omega_0 t}\right)$

$ = \displaystyle\sum_{\mu} A_\mu e^{j\omega_0 \mu t} \text{ mit } A_\mu = \begin{cases} -\frac{1}{8} & \text{for } \mu \in \{-7; -1; 1; 7\} \\ \frac{1}{4} & \text{for } \mu \in \{-3; 3\} \\ 0 & \text{otherwise} \end{cases}$

b) $x_b(t) = \displaystyle\sum_{\mu} B_\mu e^{j\omega_0 t}$ with the following Fourier coefficients:

μ	$-3; 1$	$-1; 3$	$-5; 7$	$-7; 5$	$-9; 11$	$-11; 9$	otherwise
B_μ	$\frac{j5}{32}$	$-\frac{j5}{32}$	$-\frac{j5}{64}$	$\frac{j5}{64}$	$-\frac{j}{64}$	$\frac{j}{64}$	0

c) $x_c(t) = \displaystyle\sum_{\mu} C_\mu e^{j\omega_0 t}$ with the following Fourier coefficients:

μ	0	$-2; 2$	$-4; 4$	$-6; 6$	$-8; 8$	$-10; 10$	$-12; 12$	$-14; 14$	otherwise
C_μ	$\frac{3}{16}$	$\frac{3}{32}$	$-\frac{1}{16}$	$-\frac{1}{8}$	$-\frac{1}{16}$	$\frac{1}{64}$	$\frac{1}{32}$	$\frac{1}{64}$	0

Solution 11.6

a) In general, the sum of periodic functions is only periodic if each period is a rational proportion of the others. The period of the summed signal T is then the smallest common multiple of the individual periods, and the fundamental frequency of the signal $f = \frac{1}{T}$ is the largest common divisor of the individual fundamental frequencies.

$x_1(t)$: fundamental frequency of circuit $3\omega_0$, period $T_1 = \frac{2\pi}{3\omega_0}$

$x_2(t) = \sin(\omega_0 t)\cos(\sqrt{2}\omega_0 t) = 0,5\left[\sin((1-\sqrt{2})\omega_0 t) + \sin((1+\sqrt{2})\omega_0 t)\right]$

$\dfrac{\omega_1}{\omega_2} = \dfrac{1-\sqrt{2}}{1+\sqrt{2}} \notin \mathbb{Q} \quad \Rightarrow \quad \dfrac{T_1}{T_2} \notin \mathbb{Q} \quad \Rightarrow \quad \text{not periodic}$

$x_3(t) = x_2(t) + \cos(\omega_0 t)\sin(\sqrt{2}\omega_0 t) = \sin((1+\sqrt{2})\omega_0 t)$

Period: $T_2 = \dfrac{2\pi}{(1+\sqrt{2})\omega_0}$

$x_4(t)$: not all ω_ν are proportional, e.g., $\dfrac{\omega_2}{\omega_3} = \sqrt{\dfrac{2}{3}} \notin \mathbb{Q} \quad \Rightarrow \quad$ not periodic

$x_5(t)$: Periods of the summands are $\frac{2\pi}{\pi/2} = 4$ and $\frac{2\pi}{\pi/3} = 6$

$\Rightarrow$ period $T_5 = 12$

b)
$$
\begin{aligned}
X_1(j\omega) &= \pi[\delta(\omega - 6\omega_0) + \delta(\omega + 6\omega_0) + \delta(\omega - 9\omega_0) + \delta(\omega + 9\omega_0)]\\
X_2(j\omega) &= \tfrac{\pi}{2j}[\delta(\omega - (1+\sqrt{2})\omega_0) - \delta(\omega + (1+\sqrt{2})\omega_0)\\
&\quad + \delta(\omega - (1-\sqrt{2})\omega_0) - \delta(\omega + (1-\sqrt{2})\omega_0)]\\
X_3(j\omega) &= \tfrac{\pi}{j}[\delta(\omega - (1+\sqrt{2})\omega_0) - \delta(\omega + (1+\sqrt{2})\omega_0)]\\
X_4(j\omega) &= \tfrac{\pi}{j}[\delta(\omega - \omega_0) - \delta(\omega + \omega_0) + \delta(\omega - \sqrt{2}\omega_0) - \delta(\omega + \sqrt{2}\omega_0)\\
&\quad + \delta(\omega - \sqrt{3}\omega_0) - \delta(\omega + \sqrt{3}\omega_0) + \delta(\omega - \sqrt{4}\omega_0) - \delta(\omega + \sqrt{4}\omega_0)\\
&\quad + \delta(\omega - \sqrt{5}\omega_0) - \delta(\omega + \sqrt{5}\omega_0)]\\
X_5(j\omega) &= j\pi[\delta(\omega + \tfrac{\pi}{2}) - \delta(\omega - \tfrac{\pi}{2})] + \pi[\delta(\omega + \tfrac{\pi}{3}) + \delta(\omega - \tfrac{\pi}{3})]
\end{aligned}
$$

Solution 11.7

a) Squaring the sine function doubles the fundamental frequency, so it becomes $\omega_0 = 2\omega_1$:

$$
A_\mu = \frac{1}{T} \int_0^T \left[\frac{1}{2j}\left(e^{j\omega_1 t} - e^{-j\omega_1 t}\right)\right]^2 e^{-j\omega_0 \mu t}\, dt =
$$

$$
-\frac{1}{4T} \int_0^T \left(e^{j\omega_0(1-\mu)t} - 2\,e^{-j\omega_0 \mu t} + e^{-j\omega_0(1+\mu)t}\right) dt =
$$

$$
-\frac{1}{4T}\left[\underbrace{\int_0^T e^{j\omega_0(1-\mu)t}\,dt}_{= 0 \text{ for } \mu \neq 1} - 2\underbrace{\int_0^T e^{-j\omega_0 \mu t}\,dt}_{= 0 \text{ for } \mu \neq 0} + \underbrace{\int_0^T e^{-j\omega_0(1+\mu)t}\,dt}_{= 0 \text{ for } \mu \neq -1} \right] =
$$

since integration over n periods of $e^{j\omega_0 t}$ equals zero

$$
= -\frac{1}{4T} \begin{cases} -2T & \text{for } \mu = 0 \\ T & \text{for } \mu \pm 1 \\ 0 & \text{otherwise} \end{cases} = \begin{cases} \tfrac{1}{2} & \text{for } \mu = 0 \\ -\tfrac{1}{4} & \text{for } \mu \pm 1 \\ 0 & \text{otherwise} \end{cases}
$$

b) The fundamental has frequency $\omega_0 = \dfrac{2\pi}{T}$:

$$A_\mu = \frac{1}{T} \int\limits_0^{\frac{T}{4}} 1 \cdot e^{-j\frac{2\pi}{T}\mu t}\, dt = \frac{1}{T} \cdot \frac{1}{-j\frac{2\pi}{T}\mu} \left[e^{-j\frac{2\pi}{T}\mu t}\right]_0^{\frac{T}{4}} = \frac{j}{2\pi\mu}\left[(-j)^\mu - 1\right]$$

c) The fundamental has frequency $\omega_0 = 2\pi$:

$$A_\mu = 1 \cdot \int\limits_{-\frac{1}{2}}^{\frac{1}{2}} x_c(t)\, e^{-j2\pi\mu t}\, dt = 2 \cdot \int\limits_{-\frac{1}{2}}^{0} -t\, e^{-j2\pi\mu t}\, dt + 2 \cdot \int\limits_{0}^{\frac{1}{2}} t\, e^{-j2\pi\mu t}\, dt =$$

$$-2 \cdot \int\limits_{\frac{1}{2}}^{0} t\, e^{j2\pi\mu t}\, dt + 2 \cdot \int\limits_{0}^{\frac{1}{2}} t\, e^{-j2\pi\mu t}\, dt = 2 \cdot \int\limits_{0}^{\frac{1}{2}} t\left(e^{j2\pi\mu t} + e^{-j2\pi\mu t}\right)\, dt =$$

$$2 \cdot 2 \int\limits_{0}^{\frac{1}{2}} t\cos(2\pi\mu t)\, dt = 4 \left[\frac{\cos(2\pi\mu t)}{(2\pi\mu)^2} + \frac{t\sin(2\pi\mu t)}{2\pi\mu}\right]_0^{\frac{1}{2}} =$$

$$4\left[\frac{(-1)^\mu}{(2\pi\mu)^2} - \frac{1}{(2\pi\mu)^2}\right] = \begin{cases} -\dfrac{2}{(\pi\mu)^2} & \text{for } \mu \text{ ungerade} \\ 0 & \text{otherwise} \end{cases}$$

Solution 11.8

$$X(j\omega) = \text{Ш}\left(\frac{\omega}{2\pi}\right) = 2\pi \sum_k \delta(\omega - 2\pi k)$$

Fourier series: $X(j\omega) = \sum_\mu A_\mu e^{jT_0\mu\omega}$

with $A_\mu = \dfrac{1}{2\pi} \int\limits_{-\pi}^{\pi} 2\pi \sum_k \delta(\omega - 2\pi k)e^{-jT_0\mu\omega}\, d\omega = 1$

(because of the selective property of the impulse at $\omega = 0$)

With $T_0 = \dfrac{2\pi}{\omega_0} = 1$, $X(j\omega) = \sum_\mu e^{-j\mu\omega}$

Solution 11.9

a) $\tilde{x}(t) \quad = \quad \text{si}(\pi t) \cdot \text{si}(\pi t)$

$$\tilde{X}(j\omega) \quad = \quad \frac{1}{2\pi}\text{rect}\left(\frac{\omega}{2\pi}\right) * \text{rect}\left(\frac{\omega}{2\pi}\right)$$

$$X(j\omega) \quad = \quad \tilde{X}(j\omega) \cdot \text{Ш}\left(\frac{\omega \cdot 4}{2\pi}\right) \quad = \quad \tilde{X}(j\omega) \cdot \frac{\pi}{2}\sum_\mu \delta\left(\omega - \frac{\pi}{2}\mu\right)$$

$$\quad = \quad \frac{\pi}{2}\sum_\mu \tilde{X}\left(j\frac{\pi}{2}\mu\right)\delta\left(\omega - \frac{\pi}{2}\mu\right)$$

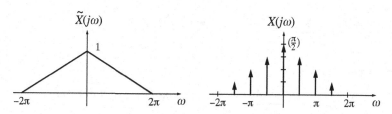

The weights of the delta impulses corresponds to the sample values of $\tilde{X}(j\omega)$ at frequencies $\omega = \frac{\pi}{2}\mu$, multiplied by a factor $\frac{2\pi}{T} = \frac{\pi}{2}$.

b) $A_\mu = \frac{1}{T}\tilde{X}\left(j\frac{2\pi}{T}\mu\right) = \frac{1}{4}\tilde{X}\left(j\frac{\pi}{2}\mu\right)$

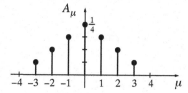

Solution 11.10

$x(t) = \text{rect}\left(\frac{t}{4}\right)$ ○—● $X(j\omega) = 4\,\text{si}(2\omega)$

a) $x_p(t) = \text{rect}\left(\frac{t}{4}\right) * 2 \cdot \sum_k \delta(t - 2k) = 4$, because 2 rectangles are placed over each other at every point in time.

$x_p(t) = 4$ ○—● $X_p(j\omega) = 8\pi\delta(\omega)$

b) $X_p(j\omega) = 4\,\text{si}(2\omega) \cdot 2\,\text{⊥⊥⊥}\left(\frac{\omega}{\pi}\right) = 8\,\text{si}(2\omega) \cdot \pi\sum_k \delta(\omega - \pi k) = 8\pi\delta(\omega)$, because all delta impulses that do not lie at $\omega = 0$ occur at zeros of the si-function.

Solution 11.11

a) Converting (11.14) yields

$$X(j\omega) = X_0(j\omega) \cdot \frac{2\pi}{T}\sum_\mu \delta\left(\omega - \frac{2\pi}{T}\mu\right) = \frac{2\pi}{T}\sum_\mu \delta\left(\omega - \frac{2\pi}{T}\mu\right) X_0\left(j\frac{2\pi}{T}\mu\right),$$

where $X_0(j\omega)$ is the spectrum of one period of $x(t)$. With (11.18) the required relationship follows

$$X(j\omega) = 2\pi\sum_\mu A_\mu\delta\left(\omega - \frac{2\pi}{T}\mu\right)$$

b) $x(t) = \sum\limits_{\nu=-\infty}^{\infty} A_\nu e^{j\frac{2\pi\nu t}{T}}$ ○——● $2\pi \sum\limits_\mu A_\mu \delta\left(\omega - \frac{2\pi}{T}\mu\right) = X(j\omega)$

Solution 11.12

a) $X_a(j\omega) = 2\pi \sum\limits_\mu A_\mu \delta(\omega - \omega_0\mu) = \pi\delta(\omega) - \frac{\pi}{2}\delta(\omega - 2\omega_1) - \frac{\pi}{2}\delta(\omega + 2\omega_1)$

b) $X_b(j\omega) = \sum\limits_k \frac{j}{k}\left[(-j)^k - 1\right]\delta(\omega - k\omega_0)$

c) $X_c(j\omega) = -\frac{4}{\pi}\sum\limits_k \frac{1}{(2k+1)^2}\delta(\omega - (2k+1)\omega_0) =$

$\qquad = -\frac{4}{\pi}\sum\limits_k \frac{1}{(2k+1)^2}\delta(\omega - 2\pi(2k+1))$

Solution 11.13

$h(t) = \frac{1}{2j}\left(e^{jt} - e^{-jt}\right)e^{-0,1t} = \left[\frac{1}{2j}e^{(-0,1+j)t} - \frac{1}{2j}e^{(-0,1-j)t}\right]\varepsilon(t)$

$$H(s) = \mathcal{L}\{h(t)\} = \frac{1}{2j}\left[\frac{1}{s+0,1-j} - \frac{1}{s+0,1+j}\right]$$

$$= \frac{1}{s^2 + 0,2s + 1,01}, \quad \mathrm{Re}\{s\} > -0,1$$

$x(t) = \sum\limits_\mu A_\mu e^{j\omega_0\mu t}, \qquad \omega_0 = \frac{2\pi}{4T}, \qquad \text{period } 4T$

$$A_\mu = \frac{1}{4T}\int\limits_{-2T}^{2T} x(t)e^{-j\omega_0\mu t}\,dt = \frac{1}{4T}\int\limits_{-T}^{T} e^{-j\frac{\pi}{2T}\mu t}\,dt$$

$$= \frac{1}{-j\mu 2\pi}\cdot\left[e^{-j\frac{\pi}{2}\mu} - e^{j\frac{\pi}{2}\mu}\right] = \frac{1}{2}\mathrm{si}\left(\frac{\pi}{2}\mu\right)$$

$y(t) = \sum\limits_\mu B_\mu e^{j\omega_0\mu t}$

with $B_\mu = A_\mu \cdot H\left(j\frac{\pi}{2T}\mu\right) = \frac{1}{2}\mathrm{si}\left(\frac{\pi}{2}\mu\right)\cdot\dfrac{1}{-(\frac{\mu\pi}{2T})^2 + j0,1\frac{\pi\mu}{T} + 1,01}$

Solution 11.14

For a cyclic convolution to exist, both signals must have the same period:

$\left.\begin{array}{lll} f(t) & : & \text{period } 4T \\ g(t) & : & \text{period } \dfrac{2\pi}{\omega_0} \end{array}\right\}$ $4T = \dfrac{2\pi}{\omega_0}; \quad T = \dfrac{\pi}{2\cdot\omega_0}$

$$f(t) = \sum_{\mu} F_{\mu} \, e^{j\mu\omega_0 t}$$

Using the Fourier series for $x(t)$, from Ex. 11.13:

$$f(t) = 2x(t) - 1 \quad \Rightarrow \quad F_{\mu} = \begin{cases} \text{si}\left(\dfrac{\pi\mu}{2}\right) & \text{for} \quad \mu \neq 0 \\ \text{si}(0) - 1 = 0 & \text{for} \quad \mu = 0 \end{cases}$$

$$g(t) = \sum_{\mu} G_{\mu} \, e^{j\mu\omega_0 t} \quad ; \quad G_{\mu} = \begin{cases} \dfrac{1}{2j} & \text{for} \quad \mu = 1 \\ -\dfrac{1}{2j} & \text{for} \quad \mu = -1 \\ 0 & \text{otherwise} \end{cases}$$

$$\begin{aligned} y(t) &= \sum_{\mu} \frac{2\pi}{\omega_0} F_{\mu} G_{\mu} \, e^{j\omega_0 \mu t} = \frac{2\pi}{\omega_0} \text{si}\left(\frac{\pi}{2}\right) \frac{1}{2j} \left(-e^{-j\omega_0 t} + e^{j\omega_0 t}\right) \\ &= \frac{4}{\omega_0} \sin(\omega_0 t) \end{aligned}$$

Solution 11.15

$$\begin{aligned} X_a(j\omega) &= \frac{1}{2\pi} X(j\omega) * \text{⊔⊔⊔}\left(\frac{\omega T}{2\pi}\right) = \frac{1}{2\pi} \cdot X(j\omega) * \frac{2\pi}{T} \sum_{\mu} \delta\left(\omega - \frac{2\pi\mu}{T}\right) \\ &= \frac{1}{T} \sum_{\mu} X(j(\omega - \mu\omega_a)), \quad \omega_a = \frac{2\pi}{T} \end{aligned}$$

case 1: $\quad \omega_{a1} = \dfrac{2\pi}{T_1} = \dfrac{2\pi \cdot 2\omega_g}{\pi} = 4\omega_g$

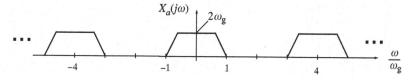

case 2: $\quad \omega_{a2} = \dfrac{2\pi}{T_2} = 2\omega_g$

case 3: $\quad \omega_{a3} = \dfrac{2\pi}{T_3} = \omega_g$

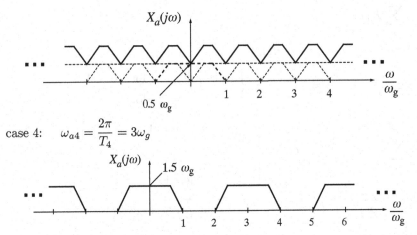

case 4: $\omega_{a4} = \dfrac{2\pi}{T_4} = 3\omega_g$

Aliasing in case 3. Critical sampling in case 2.

Solution 11.16

a) $x(t) = \dfrac{\omega_g}{2\pi}\mathrm{si}^2\left(\dfrac{\omega_g t}{2}\right)$

$X(j\omega) = \dfrac{1}{2\pi}\cdot\dfrac{\omega_g}{2\pi}\cdot\dfrac{2\pi}{\omega_g}\mathrm{rect}\left(\dfrac{\omega}{\omega_g}\right)*\dfrac{2\pi}{\omega_g}\mathrm{rect}\left(\dfrac{\omega}{\omega_g}\right) =$

$$= \begin{cases} 1+\dfrac{\omega}{\omega_g} & \text{for } -\omega_g \le \omega \le 0 \\[2mm] 1-\dfrac{\omega}{\omega_g} & \text{for } 0 < \omega \le \omega_g \\[2mm] \quad 0 & \text{otherwise} \end{cases}$$

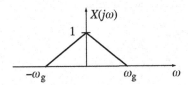

b) $x_a(t) = x(t)\cdot\dfrac{1}{T}\mathbf{III}\left(\dfrac{t}{T}\right)$

$X_a(j\omega) = \dfrac{1}{2\pi}X(j\omega)*\mathbf{III}\left(\dfrac{\omega T}{2\pi}\right) = \dfrac{1}{2\pi}X(j\omega)*\mathbf{III}\left(\dfrac{\omega}{3\omega_g}\right)$

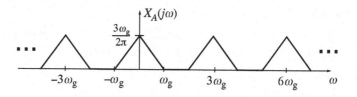

Solution 11.17

a) $r(t) = a \operatorname{rect}\left(\dfrac{t}{T}\right) \circ\!\!-\!\!\bullet \; R(j\omega) = aT\operatorname{si}\left(\dfrac{\omega T}{2}\right)$

$|R(j\omega)| = \left|aT\operatorname{si}\left(\dfrac{\omega T}{2}\right)\right|,$

$\arg\{R(j\omega)\} = \begin{cases} 0 & 4\pi n \leq |\omega T| \leq 2\pi(2n+1) \\ \pi & 2\pi(2n-1) \leq |\omega T| \leq 4\pi n \end{cases} \quad n \in \mathbb{N}_0$

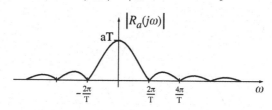

b)

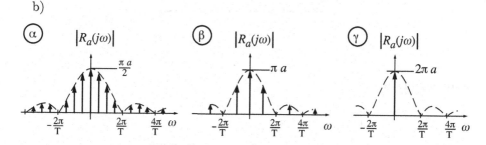

c) periodic repetition of the time-domain signal with:

$\alpha)\; T_p = \dfrac{2\pi}{\omega_0}, \quad \beta)\; T_P = 2T, \quad \gamma)\; T_P = T$

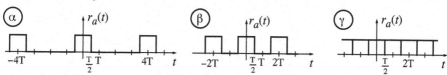

Solution 11.18

a) $X(j\omega)$: real   o—• $x(t)$: conjugate symmetry
    $X(j\omega)$: asymmetrical   o—• $x(t)$: complex

b) Bandwidth: $\Delta\omega = 2\omega_g$

Critical sampling: $\omega_A = \Delta\omega \Rightarrow f_A = \dfrac{1}{2\pi}\cdot 2\omega_g = \dfrac{\omega_g}{\pi}, T_A = \dfrac{1}{f_A} = \dfrac{\pi}{\omega_g}$

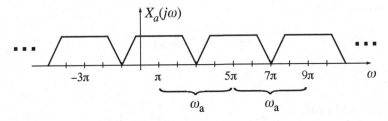

c) $\omega_A = 3\omega_g$

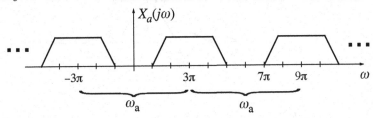

Solution 11.19

a)

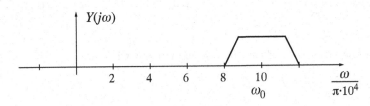

b)

When choosing w_A, 2 criteria must be satisfied:

- An image of the spectrum $Y(j\omega)$ must lie exactly in the baseband, so $n \cdot w_A = w_0$, $\quad n \in \mathbb{N}$

- No aliasing (overlapping images) may occur:
$$w_A \geq \Delta\omega = \frac{w_0}{2.5}$$

$\Rightarrow$ Two possible solutions: $w_A = w_0$ and $w_A = \dfrac{w_0}{2}$. The reconstruction filter must remove all images outside of the baseband, and let the baseband pass unchanged.

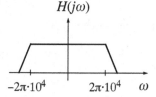

$H(j\omega)$

$-2\pi \cdot 10^4 \qquad 2\pi \cdot 10^4 \qquad \omega$

Solution 11.20

a) real band-pass signal

b) Yes, because the allocated frequency band is a multiple of the bandwidth, so (11.42) is satisfied.

c) $w_{a1} = \dfrac{2\pi}{T_1} = 2w_0$

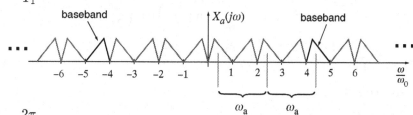

$w_{a2} = \dfrac{2\pi}{T_2} = 3w_0$

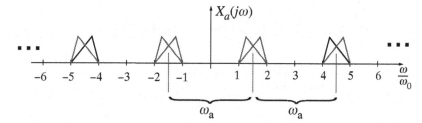

$$\omega_{a3} = \frac{2\pi}{T_3} = 10\omega_0$$

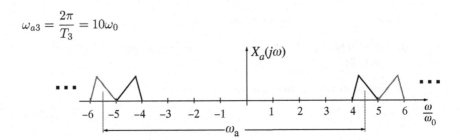

Solution 11.21

a) to d) are real band-pass signals $\Rightarrow$ check (11.42)

a) no; $\omega_{a,min} = 3.8\omega_0$
b) yes; $\omega_{a,min} = 2\Delta\omega = 3\omega_0$
c) yes; $\omega_{a,min} = 2\Delta\omega = 2\omega_0$
d) no; $\omega_{a,min} = 2.2\omega_0$

e) and f) are complex band-pass signals $\Rightarrow$ critical sampling is always possible. In both cases, $\omega_{a,min} = \Delta\omega = 2\omega_0$.

Solution 11.22

According to Table 11.1, the following holds for critical sampling of real band-pass signals:
$$\omega_a = \frac{2\pi}{T} = 2\Delta\omega.$$ With $\omega_0 = 0$ and the above condition, (11.44) becomes:

$$h(t) = \text{si}\left(\frac{\pi t}{2T}\right)\cos\left(\frac{\pi t}{2T}\right) = \frac{1}{\left(\frac{\pi t}{2T}\right)}\sin\left(\frac{\pi t}{2T}\right)\cos\left(\frac{\pi t}{2T}\right)$$

$$= \frac{1}{\left(\frac{\pi t}{2T}\right)}\frac{1}{2}\sin\left(\frac{\pi t}{T}\right) = \text{si}\left(\frac{\pi t}{T}\right)$$

Solution 11.23

a) $a(t) = \frac{1}{\tau}\text{rect}\left(\frac{t}{\tau}\right) * \text{rect}\left(\frac{t}{\tau}\right)$

$$A(j\omega) = \frac{1}{\tau}\left[\tau\text{si}\left(\frac{\omega\tau}{2}\right)\right]^2 = \tau\text{si}^2\left(\frac{\omega\tau}{2}\right) = \frac{\pi}{\omega_a}\text{si}^2\left(\frac{\pi\omega}{2\omega_a}\right)$$

with $\tau = \frac{1}{2f_a} = \frac{\pi}{\omega_a}$, $f_a = 40$ kHz

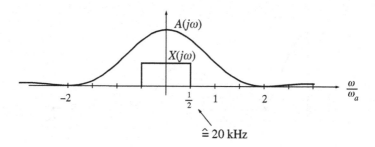

$$\triangleq 20\,\mathrm{kHz}$$

b) $A(j0) = \dfrac{\pi}{\omega_a}$

$$A\left(j\frac{\omega_a}{2}\right) = \frac{\pi}{\omega_a} \cdot \mathrm{si}^2\left(\frac{\pi\,0.5\omega_a}{2\omega_a}\right) = \frac{\pi}{\omega_a}\mathrm{si}^2\left(\frac{\pi}{4}\right)$$

Amplification decays to $\dfrac{A\left(j\frac{\omega_a}{2}\right)}{A(j0)} = \mathrm{si}^2\left(\dfrac{\pi}{4}\right) = 0.81 \triangleq -1.8\ \mathrm{dB}$

c) The first zero of the aperture function should be at half the sample frequency,
   i.e. at $\dfrac{\omega_a}{2}$ the argument of the si function should be equal to π:

$$\frac{\omega_a \tau}{4} = \pi \quad \Rightarrow \quad \tau = \frac{4\pi}{\omega_a} = \frac{2}{f_a} = \frac{2}{10\mathrm{kHz}} = 0.2\ \mathrm{ms}.$$

Solution 11.24

$$\begin{aligned}
h(t) &= \mathrm{rect}\left(\frac{t}{T_0}\right) * [\delta(t - 0.5T_0) + \delta(t - 2.5T_0) + \delta(t - 4.5T_0)] \\
&= \mathrm{rect}\left(\frac{t - 2.5T_0}{T_0}\right) * [\delta(t + 2T_0) + \delta(t) + \delta(t - 2T_0)]
\end{aligned}$$

$$\circ\!\!-\!\!\bullet$$

$$H(j\omega) = T_0\mathrm{si}\left(\frac{\omega T_0}{2}\right) \cdot e^{-j2.5T_0\omega} \cdot [1 + 2\cos(2T_0\omega)]$$

with $T = \dfrac{1}{24}\mathrm{s}$ and $T_0 = \dfrac{T}{6} = \dfrac{1}{144}\mathrm{s}$

$$|H(j\omega)| = T_0\left|\mathrm{si}\left(\frac{\omega T_0}{2}\right) \cdot [1 + 2\cos(2T_0\omega)]\right|$$

The first zero of the si term is at

$$\omega = \frac{2\pi}{T_0} \quad \Rightarrow \quad f = \frac{\omega}{2\pi} = \frac{1}{T_0} = 144\,\mathrm{Hz}.$$

The cos term has period $\omega_0 = \dfrac{\pi}{T_0} \rightarrow f_0 = 72$ Hz. Accordingly, the zeros of the term in square brackets are at 24 Hz and 48 Hz.

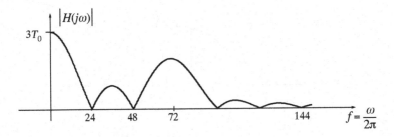

Solution 12.1

a) $x_R(t) = e^{\ln \frac{1}{4}t} \cos(2\pi t) = \left(\frac{1}{4}\right)^t \cos(2\pi t)$

$x_I(t) = \left(\frac{1}{4}\right)^t \sin(2\pi t)$

$\Sigma = \frac{1}{4} \ln \frac{1}{4}, \quad \Omega = \frac{\pi}{2}$

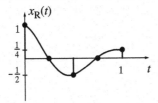
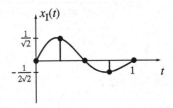

$x_R[k] = \left(\frac{1}{\sqrt{2}}\right)^k \cos\left(\frac{\pi}{2}k\right) = \begin{cases} \left(-\frac{1}{2}\right)^{\frac{k}{2}} & \text{for even } k \\ 0 & \text{for odd } k \end{cases}$

$x_I[k] = \left(\frac{1}{\sqrt{2}}\right)^k \sin\left(\frac{\pi}{2}k\right) = \frac{1}{\sqrt{2}} \cdot \begin{cases} \left(-\frac{1}{2}\right)^{\frac{k-1}{2}} & \text{for odd } k \\ 0 & \text{for even } k \end{cases}$

b) $x_R(t) = \left(\frac{1}{4}\right)^t \cos(10\pi t)$

$x_I(t) = \left(\frac{1}{4}\right)^t \sin(10\pi t)$

$\Sigma = \frac{1}{4} \ln \frac{1}{4}, \quad \Omega = \frac{5}{2}\pi$

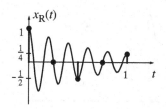
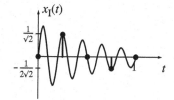

$$x_R[k] = \left(\frac{1}{\sqrt{2}}\right)^k \cos\left(\frac{5\pi}{2}k\right) = \left(\frac{1}{\sqrt{2}}\right)^k \cos\left(\frac{\pi}{2}k\right)$$

$$x_I[k] = \left(\frac{1}{\sqrt{2}}\right)^k \sin\left(\frac{5\pi}{2}k\right) = \left(\frac{1}{\sqrt{2}}\right)^k \sin\left(\frac{\pi}{2}k\right)$$

Solution 12.2

a) $\Sigma = -2$; $\quad \Omega = 0$

b) $\Sigma = \ln 0.9$; $\quad \Omega = 0$

c) $\Sigma = \ln 0.9$; $\quad \Omega = \pi$

d) $\Sigma = 0$; $\quad \Omega = \dfrac{\pi}{2}$

e) $\Sigma = \ln \dfrac{1}{\sqrt{2}} = -\dfrac{1}{2}\ln 2$; $\quad \Omega = \dfrac{\pi}{4}$

f) $\Sigma = 0$; $\quad \Omega = \dfrac{3}{2} \cdot 3\pi = \dfrac{9}{2}\pi \,\hat{=}\, \dfrac{\pi}{2}$

g) $\Sigma = 0$; $\quad \Omega = \dfrac{5}{2}\pi \,\hat{=}\, \dfrac{\pi}{2}$

h) $\Sigma = 0$; $\quad \Omega = \dfrac{9}{2}\pi \,\hat{=}\, \dfrac{\pi}{2}$

d), f), g) and h) are the same.

Solution 12.3

$$
\begin{aligned}
x[k] &= \frac{1}{2\pi} \int_{-\pi}^{\pi} X(e^{j\Omega}) e^{jk\Omega}\, d\Omega \\[2mm]
&= \frac{1}{2\pi} \int_{-\pi}^{\pi} \sum_{\mu} x[\mu] e^{-j\Omega\mu} e^{jk\Omega}\, d\Omega \\[2mm]
&= \frac{1}{2\pi} \int_{-\pi}^{\pi} \sum_{\mu} x[\mu] e^{j\Omega(k-\mu)}\, d\Omega
\end{aligned}
$$

$$= \frac{1}{2\pi} \sum_{\mu} x[\mu] \int_{-\pi}^{\pi} e^{j\Omega(k-\mu)} \, d\Omega$$

$$= \frac{1}{2\pi} \sum_{\mu} x[\mu] \cdot \underbrace{\left\{ \begin{array}{ll} 2\pi & \text{for } k = \mu \\ 0 & \text{otherwise} \end{array} \right.}_{x[k] \cdot 2\pi}$$

$$= x[k]$$

Solution 12.4

a) $X(e^{j\Omega}) = \sum_{k=-\infty}^{\infty} \text{si}\left(\frac{\pi}{2}k\right) e^{-j\Omega k} = 2 \sum_{k=1}^{\infty} \text{si}\left(\frac{\pi}{2}k\right) \cos(\Omega k) + 1 \cdot \cos(\Omega \cdot 0)$

$\phantom{a) X(e^{j\Omega})} = 1 + \frac{4}{\pi} \cos\Omega - \frac{4}{3\pi} \cos 3\Omega + \frac{4}{5\pi} \cos 5\Omega - \ldots$

b) when $k = \dfrac{t}{T}$:

$$x_a(t) = \sum_k \delta(t - kT)\text{si}\left(\frac{\pi}{2} \cdot \frac{t}{T}\right) = \text{si}\left(\frac{\pi t}{2T}\right) \cdot \frac{1}{T}\underset{\bot\bot\bot}{}\left(\frac{t}{T}\right)$$

$$\circ$$
$$\bullet$$

$$X_a(j\omega) = \frac{1}{2\pi} 2T \, \text{rect}\left(\frac{\omega T}{\pi}\right) * \underset{\bot\bot\bot}{}\left(\frac{\omega T}{2\pi}\right) = 2 \sum_{\mu} \text{rect}\left(\frac{T}{\pi}\left(\omega - \frac{2\pi\mu}{T}\right)\right)$$

$$X(e^{j\Omega}) = X_a(j\omega) \quad \text{mit} \quad \Omega = \omega T$$

$$X(e^{j\Omega}) = 2 \sum_{\mu} \text{rect}\left(\frac{1}{\pi}(\Omega - 2\pi\mu)\right)$$

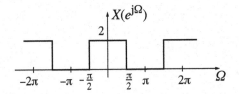

Note: Even when $e^{j\Omega}$ terms do not appear in the result, the spectrum of a series usually is written $X(e^{j\Omega})$.

c) Fourier series $X(e^{j\Omega}) = \sum_{\mu} A_{\mu} e^{jT_0\mu\Omega}$ with fundamental period

$$T_0 = \frac{2\pi}{\Omega_{per}} = 1.$$

$$A_\mu = \frac{1}{2\pi} \int\limits_{-\pi}^{\pi} X(e^{j\Omega}) e^{-jT_0\mu\Omega}\, d\Omega$$

$$= \frac{1}{2\pi} \int\limits_{-\frac{\pi}{2}}^{\frac{\pi}{2}} 2\cdot e^{-j\mu\Omega}\, d\Omega = \frac{1}{-j\mu\pi}\left[e^{-j\mu\Omega}\right]_{-\frac{\pi}{2}}^{\frac{\pi}{2}}$$

$$= \frac{1}{-j\mu\pi}\underbrace{\left[e^{-j\mu\frac{\pi}{2}} - e^{j\mu\frac{\pi}{2}}\right]}_{-j2\sin\left(\mu\frac{\pi}{2}\right)} = \frac{2}{\mu\pi}\sin\left(\frac{\mu\pi}{2}\right) = \mathrm{si}\left(\frac{\mu\pi}{2}\right)$$

This Fourier series corresponds to the first step of a).

d) A discrete low-pass with cut-off frequency $\dfrac{\pi}{2}$, called a half-band low-pass.

Solution 12.5

$$H_2(e^{j\Omega}) = \sum x_1[k]\cdot(-1)^k\, e^{-j\Omega k} = \sum x_1[k] e^{j\pi k}\, e^{-j\Omega k}$$
$$= \sum x_1[k] e^{-j(\Omega-\pi)k} = H_1(e^{j(\Omega-\pi)})$$

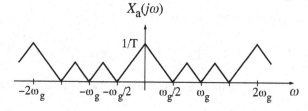

H_2 is a high-pass filter.

Solution 12.6

a) It is evident from spectrum $X(j\omega)$ that $X(j\omega) = 0$ for $|\omega| > \omega_g$. To sample at the nyquist rate, $T = \dfrac{\pi}{\omega_g}$ must be chosen.

$$X_a(j\omega) = \frac{1}{2\pi}\cdot X(j\omega) * ⊥⊥⊥\left(\frac{\omega T}{2\pi}\right) = \frac{1}{T}\sum_\mu X\left(j\left(\omega - \frac{2\pi\mu}{T}\right)\right)$$

b) According to (12.43), the $\mathcal{F}$-transform of a sampled signal corresponds to the $\mathcal{F}_*$-transform of the equivalent discrete series for $\Omega = \omega T$.

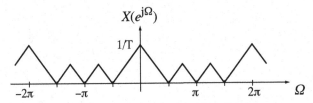

c) $H_1(e^{j\Omega}) = T\mathrm{rect}\left(\dfrac{\Omega}{\pi}\right)$ for $-\pi \leq \Omega \leq \pi$

As the $\mathcal{F}_*$-transform is 2π periodic, the spectrum of $H_1(e^{j\Omega})$ and $Y(e^{j\Omega})$ are also periodic with 2π.

According to: b) $Y_a(j\omega) = Y(e^{j\Omega})$ with $\Omega = \omega T$

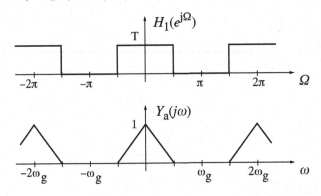

d) $H_2(j\omega) = K \cdot \left(\dfrac{1}{j\omega + 0{,}75\omega_g}\right)^N$

$H_2(\omega \to 0) = K \cdot \dfrac{1}{(0{,}75\omega_g)^N} = 1 \quad \Rightarrow \quad K = \left(\dfrac{3\omega_g}{4}\right)^N$

First spectral image at $1.5\omega_g$, i.e. $18\frac{\text{dB}}{\text{oct.}} \hat{=} 60\frac{\text{dB}}{\text{dec.}}$ damping needed $\Rightarrow$

$N = 3$

Solution 12.7

Without loss of generality we choose $\Omega_0 \in [-\pi; \pi]$

$$x[k] = \frac{1}{2\pi} \int\limits_{-\pi}^{\pi} \underset{}{\text{⊥⊥⊥}}\left(\frac{\Omega - \Omega_0}{2\pi}\right) e^{jk\Omega}\, d\Omega = \frac{1}{2\pi} \int\limits_{-\pi}^{\pi} 2\pi\delta(\Omega - \Omega_0)\, e^{jk\Omega}\, d\Omega = e^{jk\Omega_0}$$

Only one of the delta impulses in the impulse train falls with the limits of the integration.

Solution 12.8

a) $\displaystyle\sum_k x[k-\kappa]\,e^{-j\Omega k} \overset{\overset{\mu=k-\kappa}{\downarrow}}{=} \sum_\mu x[\mu]e^{-j\Omega[\mu+\kappa]} = \underbrace{\sum_\mu x[\mu]\,e^{-j\Omega\mu}}_{X(e^{j\Omega})}\cdot e^{-j\Omega\kappa}$

b) $\displaystyle\sum_k e^{j\Omega_0 k}x[k]\,e^{-j\Omega k} = \sum_k x[k]e^{j(\Omega-\Omega_0)k}$

Solution 12.9

$$\mathcal{F}_*\{f[k]*g[k]\} = \sum_k\left[\sum_\kappa f[\kappa]g[k-\kappa]\right]e^{-jk\Omega} = \sum_\kappa\sum_k f[\kappa]g[k-\kappa]e^{-jk\Omega}$$

$$= \sum_\kappa f[\kappa]\underbrace{\sum_k g[k-\kappa]e^{-jk\Omega}}_{\text{Shift theorem}} = \sum_\kappa f[\kappa]e^{-j\Omega\kappa}G(e^{j\Omega})$$

$$= F(e^{j\Omega})\cdot G(e^{j\Omega})$$

Solution 12.10

$$\frac{1}{2\pi}F(e^{j\Omega})\circledast G(e^{j\Omega}) = \frac{1}{2\pi}\int_{-\pi}^{\pi} F(e^{j\eta})G(e^{j(\Omega-\eta)})d\eta$$

Inverse transform according to (12.16):

$$\frac{1}{2\pi}\int_{-\pi}^{\pi}\left[\frac{1}{2\pi}\int_{-\pi}^{\pi} F(e^{j\eta})G(e^{j(\Omega-\eta)})d\eta\right]e^{j\Omega k}d\Omega$$

$$= \frac{1}{2\pi}\int_{-\pi}^{\pi} F(e^{j\eta})\underbrace{\left[\frac{1}{2\pi}\int_{-\pi}^{\pi} G(e^{j(\Omega-\eta)})e^{j\Omega k}d\Omega\right]}_{\text{Modulation theorem}} d\eta$$

$$= \frac{1}{2\pi}\int_{-\pi}^{\pi} F(e^{j\eta})e^{j\eta k}g[k]d\eta = g[k]\cdot\frac{1}{2\pi}\int_{-\pi}^{\pi} F(e^{j\eta})e^{j\eta k}d\eta$$

$$= g[k]\cdot f[k]$$

Solution 13.1

$$X_1(z) = z^{-3} - 4z^{-2} + 6z^{-1} - 4 + z = \frac{(z-1)^4}{z^3}$$

$$X_2(z) = \sum_{k=2}^{\infty} e^{-ak}z^{-k} = \sum_{\kappa=0}^{\infty} e^{-a(\kappa+2)}z^{-(\kappa+2)}$$

$$= e^{-2a}z^{-2}\sum_{\kappa=0}^{\infty}\left(\frac{e^{-a}}{z}\right)^\kappa = e^{-2a}\frac{1}{z(z-e^{-a})}, \quad |z| > |e^{-a}|$$

$$X_3(z) = \sum_{k=-10}^{0} (-0.8)^{-k} z^{-k} + \sum_{k=0}^{10} (-0.8)^k z^{-k} - (-0.8^0 \cdot z^0)$$

$$= \sum_{\kappa=0}^{10} (-0.8z)^\kappa + \sum_{k=0}^{10} \left(\frac{-0.8}{z} \right)^k - 1$$

$$= \frac{(-0.8z)^{11} - 1}{-1 - 0.8z} + \frac{\left(-\frac{0.8}{z} \right)^{11} - 1}{-\frac{0.8}{z} - 1} - 1, \quad 0.8 < |z| < 1.25$$

Solution 13.2

$$x_1[k] = \left(\frac{1}{a} \right)^k \varepsilon[k] \circ\!\!-\!\!\bullet X_1(z) = \frac{z}{z - \frac{1}{a}}, \quad |z| > \left| \frac{1}{a} \right|$$

$$x_2[k] = -\left(\frac{1}{a} \right)^k \varepsilon[-k] \circ\!\!-\!\!\bullet X_2(z) = \frac{1}{a} \cdot \frac{1}{z - \frac{1}{a}}, \quad |z| < \left| \frac{1}{a} \right|$$

$$X_3(z) = \frac{z}{z - 0,5} - \frac{0.8}{z - 0.8}, \quad 0.5 < |z| < 0.8$$

$X_4(z)$ does not exist because the regions of convergence do not overlap.

$$x_5[k] = a^k \varepsilon[k-1] + a^{-k} \varepsilon[-k] \circ\!\!-\!\!\bullet X_5(z) = \frac{a}{z - a} - \frac{1}{a} \cdot \frac{1}{z - \frac{1}{a}}, \quad a < |z| < \frac{1}{a}$$

only exists when $a < 1$

Solution 13.3

a)

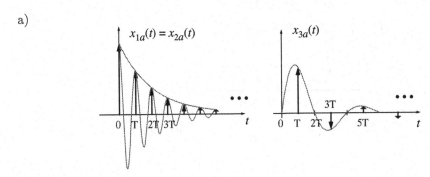

b) If $x_\nu[k]$ is chosen according to (13.15), then:

$$\mathcal{L}\{x_{\nu a}(t)\} = \mathcal{Z}\{x_\nu[k]\}, \quad \text{with } z = e^{sT}$$

$$x_1[k] = e^{-0.5k} \varepsilon[k] \circ\!\!-\!\!\bullet X_1(z) = \frac{z}{z - e^{-0.5}} \quad \Rightarrow X_{1a}(s) = \frac{e^{sT}}{e^{sT} - e^{-0.5}}$$

$$X_{2a}(s) = \frac{e^{sT}}{e^{sT} - e^{-0.5}}, \quad \text{da } x_{1a}(t) = x_{2a}(t).$$

$$x_3[k] = e^{-2k} \sin\left(\frac{\pi k}{2}\right) \cdot \varepsilon[k] = \frac{1}{2j}\left[e^{(-2+j0,5\pi)k} - e^{(-2-j0,5\pi)k}\right]\varepsilon[k]$$

$$= \frac{1}{2j}\left[(je^{-2})^k - (-je^{-2})^k\right]\varepsilon[k]$$

$$X_3(z) = \frac{1}{2j}\left(\frac{z}{z-je^{-2}} - \frac{z}{z+je^{-2}}\right) = \frac{ze^{-2}}{z^2+e^{-4}}$$

$$X_{3a}(s) = \frac{e^{-2+sT}}{e^{2sT}+e^{-4}}$$

Solution 13.4

a) $X_1(z) = 0.5z^{-1} + 1 + 0.5z = \dfrac{(z+1)^2}{2z}$

$X_2(z) = z^{-1} + 1 + z = \dfrac{z^2+z+1}{z}$

$X_3(z) = z^{-2} + z^{-1} + 1 = \dfrac{z^2+z+1}{z^2}$

$X_4(z) = -\dfrac{2}{3\pi}z^{-3} + \dfrac{2}{\pi}z^{-1} + 1 + \dfrac{2}{\pi}z - \dfrac{2}{3\pi}z^3$

b) $X_1(e^{j\Omega}) = 0.5e^{-j\Omega} + 1 + 0.5e^{j\Omega} = 1 + \cos\Omega$

$X_2(e^{j\Omega}) = 1 + 2\cos\Omega$

$X_3(e^{j\Omega}) = (1 + 2\cos\Omega)\,e^{-j\Omega}$

$X_4(e^{j\Omega}) = 1 + \dfrac{4}{\pi}\cos\Omega - \dfrac{4}{3\pi}\cos3\Omega$

All filters are low-pass filters.

Solution 13.5

a) $\mathcal{Z}\{x[k-\kappa]\} \;=\; \displaystyle\sum_{k=-\infty}^{\infty} x[k-\kappa]\,z^{-k} = \sum_{\mu=-\infty}^{\infty} x[\mu]z^{-(\mu+\kappa)}$

$\qquad\qquad\;\; = \; z^{-\kappa}\displaystyle\sum_{\mu} x[\mu]z^{-\mu} = z^{-\kappa}X(z)$

b) $\mathcal{Z}\{a^k x[k]\} = \displaystyle\sum_{k} x[k]\left(\dfrac{z}{a}\right)^{-k} = X\left(\dfrac{z}{a}\right)$

c) $-z\dfrac{dX(z)}{dz} \;=\; -z\dfrac{d}{dz}\displaystyle\sum_{k} x[k]z^{-k} = -z\sum_{k} x[k]\dfrac{d}{dz}z^{-k}$

$\qquad\qquad\;\; = \; -z\displaystyle\sum_{k} x[k](-k)z^{-k-1} = \sum_{k} k\,x[k]z^{-k} = \mathcal{Z}\{k\,x[k]\}$

d) $\mathcal{Z}\{x[-k]\} = \displaystyle\sum_{k} x[-k]z^{-k} = \sum_{\kappa} x[\kappa]z^{\kappa} = \sum_{\kappa} x[\kappa]\left(\dfrac{1}{z}\right)^{-\kappa} = X\left(\dfrac{1}{z}\right)$

Solution 13.6

$X(z) = z^{-1} + 2 + z = \dfrac{(z+1)^2}{z}$

$X_m(z) = X(z\,e^{-j\Omega_0}) = (z\,e^{-j\Omega_0})^{-1} + 2 + z\,e^{-j\Omega_0}$

$\Omega_0 = 0$: $X_m(z) = \dfrac{(z+1)^2}{z}$

$\Omega_0 = \dfrac{\pi}{2}$: $X_m(z) = jz^{-1} + 2 - jz = \dfrac{(z+j)^2}{jz}$

$\Omega_0 = \pi$: $X_m(z) = -z^{-1} + 2 - z = -\dfrac{(z-1)^2}{z}$

Spectrum: $X_m(e^{j\Omega}) = \left(e^{j(\Omega-\Omega_0)}\right)^{-1} + 2 + e^{j(\Omega-\Omega_0)} = 2 + 2\cos(\Omega - \Omega_0)$

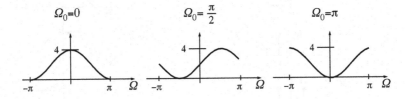

Solution 13.7

$X(z) = \mathcal{Z}\{x[k]\}, \quad \text{ROC}: |z| > 0.5$

Then it follows with

linearity:
$$X_1(z) = k_0 X(z), \quad \text{ROC}: |z| > 0.5$$

shift theorem:
$$X_2(z) = z^{-k_0} X(z), \quad \text{ROC}: |z| > 0.5$$

modulation theorem:
$$X_3(z) = X\left(\frac{z}{(-e)^\alpha}\right), \quad \text{ROC}: \left|\frac{z}{(-e)^\alpha}\right| > 0.5$$
$$\Rightarrow |z| > 0.5 \cdot e^{\text{Re}\{\alpha\}}$$

Solution 13.8

The location of the poles is the deciding factor (see Sec. 13.5).

1. No poles except at infinity $\Rightarrow$ finite, $\text{ROC}: |z| < \infty$

2.,3. No poles except at the origin $\Rightarrow$ finite, $\text{ROC}: 0 < |z|$

4. $H_4(z) = \dfrac{2z + 0.5}{z - 0.5}$ $\Rightarrow$ infinite, $\text{ROC}: |z| > 0.5$

5. $H_5(z) = \dfrac{2(z-1)}{z}$ $\Rightarrow$ finite, $\text{ROC}: 0 < |z|$

Solution 13.9

Because the series are infinite, the inverse transform is carried out by splitting partial fractions of $\dfrac{X_\nu(z)}{z}$:

$$\frac{X_1(z)}{z} = \frac{6}{z} - \frac{2}{z^2} - \frac{6}{z + 0.5}$$

$$X_1(z) = 6 - \frac{2}{z} - \frac{6z}{z + 0.5}$$

$$x_1[k] = 6\,\delta[k] - 2\,\delta[k-1] - 6(-0.5)^k \varepsilon[k]$$

$$\frac{X_2(z)}{z} = -\frac{4}{z} + \frac{4}{z + 0.5} + \frac{3}{(z + 0.5)^2}$$

$$X_2(z) = -4 + \frac{4z}{z + 0.5} + \frac{3z}{(z + 0.5)^2}$$

$$x_2[k] = -4\,\delta[k] + (4 - 6k)(-0.5)^k \varepsilon[k]$$

Solution 13.10

a) $H_a(z) = (z-1)(z^2 + z + 1) = (z-1)\left(z + \dfrac{1}{2} + j\dfrac{\sqrt{3}}{2}\right)\left(z + \dfrac{1}{2} - j\dfrac{\sqrt{3}}{2}\right)$

b) $H_b(z) = \dfrac{1 - z^3}{z^3} = -\dfrac{H_a(z)}{z^3}$

c) $H_c(z) = \dfrac{z^6 + 1}{z^3}$

Zeros: $z = \sqrt[6]{-1}$; $z_\nu = e^{j\left(\frac{\pi}{6} + \frac{\pi}{3}\nu\right)}$, $\nu \in [0; 5]$

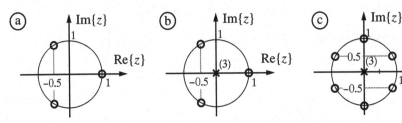

Solution 13.11

$$H_1(z) = A\left(z - e^{j\frac{5\pi}{6}}\right)\left(z - e^{-j\frac{5\pi}{6}}\right) = A(z^2 + \sqrt{3}z + 1)$$

$$h_1[k] = A\delta[k+2] + \sqrt{3}A\delta[k+1] + A\delta[k]$$

$$H_2(z) = \frac{H_1(z)}{z} \quad \Rightarrow \quad h_2[k] = A(\delta[k+1] + \sqrt{3}\delta[k] + \delta[k-1])$$

$$H_3(z) = zH_1(z) \quad \Rightarrow \quad h_3[k] = A(\delta[k+3] + \sqrt{3}\delta[k+2] + \delta[k+1])$$

The difference is a shift and a constant factor

Solution 13.12

a) $x[k] = \varepsilon[k] - 2\varepsilon[k - r - 1] + \varepsilon[k - 2r - 2]$

with the shift theorem:

$$X(z) = \mathcal{Z}\{x[k]\} = \frac{z}{z-1}(1 - 2z^{-r-1} + z^{-2r-2}) = \frac{(z^{r+1} - 1)^2}{z^{2r+1}(z - 1)}$$

b) double zeros at $z^{r+1} = 1 \Rightarrow z_n = e^{\frac{2\pi n}{r+1}}$, $n = 0, 1, 2, \ldots, r$

The single pole at $z = 1$ is nullified by a zero at $z = 1$.

pole with multiplicity of $2r + 1$ at $z = 0$

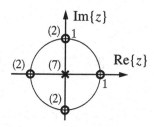

c)

Solution 14.1

a) LTI

b) LTI

c) TI, not L, since
$$S\{x_1[k] + x_2[k]\} = a + x_1[k] + x_2[k] \neq$$
$$S\{x_1[k]\} + S\{x_2[k]\} = 2a + x_1[k] + x_2[k]$$

d) L, because $S\{x_1[k] + x_2[k]\} = a^k x_1[k] + a^k x_2[k] = S\{x_1[k]\} + S\{x_2[k]\}$
not TI, because
$$S\{x[k - N]\} = a^k x[k - N] \neq$$
$$y[k - N] = a^{k-N} x[k - N]$$

e) LTI

f) L, not TI, since
$$S\{x[k - N]\} = \sum_{\mu=0}^{k} x[\mu - N] \neq$$
$$y[k - N] = \sum_{\mu=0}^{k-N} x[\mu] = \sum_{\tilde{\mu}=N}^{k} x[\tilde{\mu} - N]$$

g) L, TI, because
$$S\{x[k - N]\} = \sum_{\mu=-\infty}^{k} x[\mu - N] =$$
$$y[k - N] = \sum_{\mu=-\infty}^{k-N} x[\mu] = \sum_{\tilde{\mu}=-\infty}^{k} x[\tilde{\mu} - N]$$

h) LTI, because difference equations describe LTI systems.

i) L, not TI, see d)

j) TI, not L, because
$$S\{x_1[k] + x_2[k]\} = a^{x_1[k]+x_2[k]} \neq$$
$$S\{x_1[k]\} + S\{x_2[k]\} = a^{x_1[k]} + a^{x_2[k]}$$

Solution 14.2

a) shift-variant, because

$$S\{x[k-N]\} = x[2(k-N)] \neq y[k-N] = x[2k-N]$$

b) likewise shift-variant, see a).

c) The system sets to zero every sample with odd k:

$$y[k] = \begin{cases} x[k], & k \text{ even} \\ 0, & k \text{ odd} \end{cases}$$

$$S\{x[k-N]\} = \begin{cases} x[k-N], & k \text{ even} \\ 0, & k \text{ odd} \end{cases}$$

$$y[k-N] = \begin{cases} x[k-N], & k-N \text{ even} \\ 0, & k-N \text{ odd} \end{cases}$$

Example for $N = 1$: shift-variant

$$S\{x[k-1]\} = \begin{cases} x[k-1], & k \text{ even} \\ 0, & k \text{ odd} \end{cases}$$

$$y[k-1] = \begin{cases} 0, & k \text{ even} \\ x[k-1], & k \text{ odd} \end{cases}$$

Example for $N = 2$: shift-*invariant*

$$S\{x[k-2]\} = \begin{cases} x[k-2], & k \text{ even} \\ 0, & k \text{ odd} \end{cases}$$

$$y[k-2] = \begin{cases} x[k-2], & k \text{ even} \\ 0, & k \text{ odd} \end{cases}$$

The system is invariant when shifted by an even number of cycles, but shifting $x[k]$ by an odd number of cycles does not deliver a shifted version of $y[k]$. Note: Systems that are invariant for a certain shift and its multiples are called *periodic shift-invariant*.

Solution 14.3

a) $y[-1] = 0$

$y[0] = x[0] = 1$
$y[1] = x[1] - 2y[k-1] = 0 - 2 \cdot 1 = -2$
$y[2] = x[2] - 2y[k-1] - y[k-2] = 0 - 2 \cdot (-2) - 1 = 3$

$$y[3] = 0 - 2 \cdot 3 + (-2) = -4$$
$$y[4] = 0 - 2 \cdot (-4) + 3 = 5$$
$$\vdots$$

$$y[k] = (-1)^k (k+1)\varepsilon[k]$$

b) $y[k] = 1, -1, 2, -2, 3, -3, \ldots$

Solution 14.4

$$y[k] = x[k] - y[k-2]$$

$$y[2] = x[2] - y[0] = 1$$
$$y[3] = x[3] - y[1] = -6$$
$$y[4] = x[4] - y[2] = 0$$
$$y[5] = x[5] - y[3] = 5$$

Solution 14.5

- Internal part: leave out input signal

$$y_{\text{int}}[0] = 0.5 \cdot 2 = 1$$
$$y_{\text{int}}[1] = 0.5 \cdot y_{\text{int}}[0] = 0.5$$
$$y_{\text{int}}[2] = 0.5 \cdot y_{\text{int}}[1] = 0.25$$
$$y_{\text{int}}[3] = 0.125$$

- External part: set initial state to zero

$$y_{\text{ext}}[0] = 2 \cdot 1 = 2$$
$$y_{\text{ext}}[1] = 2 \cdot 1 + 0.5 \cdot (1 + y_{\text{ext}}[0]) = 3.5$$
$$y_{\text{ext}}[2] = 2 \cdot 1 + 0.5 \cdot (1 + 3.5) = 4.25$$
$$y_{\text{ext}}[3] = 2 \cdot 1 + 0.5 \cdot (1 + 4.25) = 4.625$$

- Initial condition

$$y[0] = y_{\text{int}}[0] + y_{\text{ext}}[0] = 1 + 2 = 3$$

Solution 14.6

a) A unique assignment of initial states is only possible with canonical forms.

- DF I: Begin with the feedback branch.

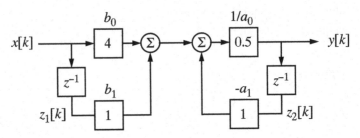

The initial condition is fulfilled by $0.5z_1[0] + 0.5z_2[0] + 4 \cdot 0.5x[0] = 3$, and
$0.5z_1[0] + 0.5z_2[0] = 1 = y_{int}[0]$.

Note: The internal part alone is sufficient to determine the initial states.

- DF II: as shown in Fig. 14.7 and Fig. 14.9.

  Initial condition: $y_{int}[0] = \left(b_1 + b_0 \cdot \dfrac{-a_1}{a_0} \right) z[0] = 3z[0] \quad \Rightarrow \quad z[0] = \frac{1}{3}$

- DF III: see also Fig. 2.5

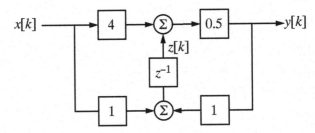

$y_{int}[0] = 0.5z[0]] \quad \Rightarrow \quad z[0] = 2$

b)   (I) : $z[k+1] = x[k] + y[k]$

            (II) : $y[k] = 2x[k] + 0.5z[k]$

     (I) in (II) : $y[k] = 2x[k] + 0.5(x[k-1] + y[k-1])$

Difference equation: $y[k] - 0,5y[k-1] = 2x[k] + 0,5x[k-1]$

Multiplying the difference equation by 2 yields the coefficients from a).

Solution 14.7

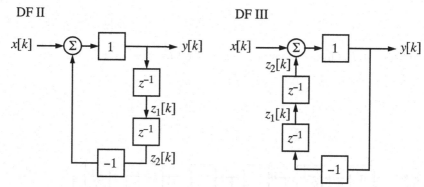

DF II   DF III

The initial states realise $y_{int}[0] = y[0] - y_{ext}[0] = -1$ and $y_{int}[1] = $
$= y[1] - y_{ext}[1] = 6$. Only then will the auxiliary conditions be fulfilled, see
Exercise 14.4.

DF II: $z_2[0] = 1$; $z_1[0] = -6$
DF III: $z_2[0] = -1$; $z_1[0] = 6$

Solution 14.8

$$x[k] \circ\!\!-\!\!\bullet \frac{z}{z - \frac{1}{2}} \quad |z| > \frac{1}{2}$$

Partial fractions: $\dfrac{Y(z)}{z} = \dfrac{1}{z - \frac{1}{2}} + \dfrac{j\frac{1}{\sqrt{3}}}{z - \frac{1}{4} + j\frac{\sqrt{3}}{4}} - \dfrac{j\frac{1}{\sqrt{3}}}{z - \frac{1}{4} - j\frac{\sqrt{3}}{4}}$

$$y[k] = \left(\frac{1}{2}\right)^k \varepsilon[k] + \frac{j}{\sqrt{3}} \varepsilon[k] \left[\left(0.5 e^{-j\frac{2\pi}{3}}\right)^k - \left(0.5 e^{j\frac{2\pi}{3}}\right)^k \right]$$

$$= \left(\frac{1}{2}\right)^k \left[\frac{2}{\sqrt{3}} \sin(\frac{2\pi}{3}k) + 1 \right] \varepsilon[k]$$

Solution 14.9

a) IIR system, because $H_1(z)$ has poles outside the origin, which means that
the output signal is fed back into the system.

$$\frac{H_1(z)}{z} = \frac{az}{(z - z_p)(z - z_p^*)} = \frac{A}{z - z_p} + \frac{A^*}{z - z_p^*}$$

with $z_p = \dfrac{1 + j}{2}$, $A = \dfrac{a(1 - j)}{2}$

$$h_1[k] = \left[Az_p^k + A^* z_p^{*k}\right]\varepsilon[k] = |A| \cdot |z_p|^k \cdot \left[e^{\arg\{A\}}e^{j\frac{\pi}{4}k}\right.$$

$$\left. + e^{\arg\{A^*\}}e^{-j\frac{\pi}{4}k}\right]\varepsilon[k] = a\sqrt{2}\left(\frac{1}{\sqrt{2}}\right)^k \cos\left[\frac{\pi}{4}(k-1)\right]\varepsilon[k]$$

b) Calculate the first five values of $h_1[k]$ and use them for $h_2[k]$:

$$\Rightarrow \quad h_2[k] = a, a, \frac{a}{2}, 0, -\frac{a}{4}, -\frac{a}{4}, 0, \dots$$

$$H_2(z) = a \cdot (1 + z^{-1} + 0.5z^{-2} - 0.25z^{-4} - 0.25z^{-5})$$

$$= a\frac{z^5 + z^4 + 0.5z^3 - 0.25z - 0.25}{z^5}$$

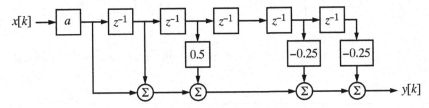

Solution 14.10

a) $H(z) = \dfrac{2 + 0.5z^{-1}}{1 - 0.5z^{-1}} = \dfrac{2z + 0.5}{z - 0.5}, \quad |z| > 0.5$

$X(z) = \dfrac{z}{z - 1}, \quad |z| > 1$

$$\dfrac{Y_{\text{ext}}(z)}{z} = \dfrac{2z + 0.5}{(z-1)(z-0.5)} = \dfrac{5}{z-1} - \dfrac{3}{z-0.5}$$

$$y_{\text{ext}}[k] = 5\varepsilon[k] - 3 \cdot 0.5^k \varepsilon[k]$$

b) $Y_{\text{int}}(z) = A\dfrac{z}{z - 0.5}, \quad |z| > 0.5$

$$y_{\text{int}}[k] = A \cdot 0.5^k \varepsilon[k]$$

$$y_{\text{int}}[0] = y[0] - y_{\text{ext}}[0] = 3 - 2 = 1 \quad \Rightarrow \quad A = 1$$

c) $y[k] = y_{\text{ext}}[k] + y_{\text{int}}[k] = (5 - 2 \cdot 0.5^k)\varepsilon[k]$

Solution 14.11

External part:

$$H(z) = \dfrac{1}{1 + z^{-2}} = \dfrac{z^2}{z^2 + 1} = \dfrac{z^2}{(z + j)(z - j)}, \quad |z| > 1$$

$$X(z) = \frac{z}{z+1}, \quad |z| > 1$$

$$\frac{Y_{\text{ext}}(z)}{z} = \frac{z^2}{(z+1)(z+j)(z-j)} = \frac{0.5}{z+1} + \frac{0.25(1-j)}{z+j} + \frac{0.25(1+j)}{z-j}$$

$$
\begin{aligned}
y_{\text{ext}}[k] &= \left[0.5 \cdot (-1)^k + 0.25\left((-j)^k + j^k\right) - j0.25\left((-j)^k - j^k\right)\right]\varepsilon[k] \\
&= 0.5\varepsilon[k]\left[(-1)^k + \cos\left(\frac{\pi}{2}k\right) - \sin\left(\frac{\pi}{2}k\right)\right]
\end{aligned}
$$

Internal part:

$$Y_{\text{int}}(z) = A\frac{z}{z-j} + A^*\frac{z}{z+j}, \quad |z| > 1$$

$$y_{\text{int}}[k] = \left[Aj^k + A^*(-j)^k\right]\varepsilon[k]$$

From Exercise 14.4: $y_{\text{int}}[0] = -1, \quad y_{\text{int}}[1] = 6$

$k = 0: \quad A + A^* = -1 \quad \Rightarrow \quad \text{Re}\{A\} = -0.5$

$k = 1: \quad jA - jA^* = 6 \quad \Rightarrow \quad \text{Im}\{A\} = -3$

$$y_{\text{int}}[k] = \left[-\cos\left(\frac{\pi}{2}k\right) + 6\sin\left(\frac{\pi}{2}k\right)\right]\varepsilon[k]$$
complete solution

$$y[k] = 0.5\varepsilon[k] \cdot \left[(-1)^k - \cos\left(\frac{\pi}{2}k\right) + 11\sin\left(\frac{\pi}{2}k\right)\right]$$

Solution 14.12

$$
\begin{aligned}
\text{a)} \quad c[k] &= \sum_{\kappa=-\infty}^{\infty} a[\kappa]b[k-\kappa] = \sum_{\kappa=0}^{2} b[k-\kappa] \\
&= \delta[k] + 2\delta[k-1] - \delta[k-2] + \delta[k-1] + 2\delta[k-2] - \delta[k-3] \\
&\quad + \delta[k-2] + 2\delta[k-3] - \delta[k-4] \\
&= \delta[k] + 3\delta[k-1] + 2\delta[k-2] + \delta[k-3] - \delta[k-4]
\end{aligned}
$$

$$
\begin{aligned}
\text{b)} \quad c[k] &= a[k+2] + 0.8a[k+1] = 0.8^{k+2}\varepsilon[k+2] + 0.8 \cdot 0.8^{k+1}\varepsilon[k+1] \\
&= 0.8^{k+2}\delta[k+2] + 0.8^{k+2}\varepsilon[k+1] + 0.8 \cdot 0.8^{k+1}\varepsilon[k+1] \\
&= 0.8^{k+2}\delta[k+2] + 2 \cdot 0.8^{k+2}\varepsilon[k+1]
\end{aligned}
$$

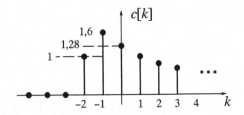

Solution 14.13

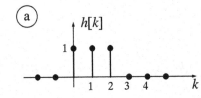

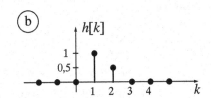

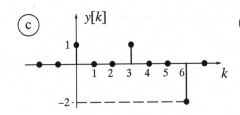

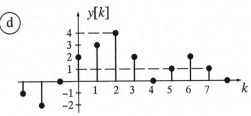

Solution 14.14

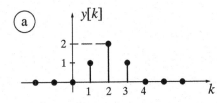

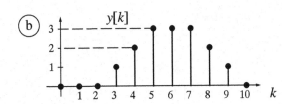

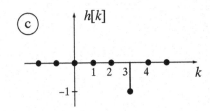

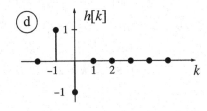

Solution 15.1

a) not causal

b) causal

c) causal

d) causal

e) not causal, "reacts" to future $x[k]$, when $k > 0$ e.g., $y[1] = x[2]$

Solution 15.2

a) $\mathcal{H}\{e^{j\omega_0 t}\} = e^{j\omega_0 t} * \dfrac{1}{\pi t} \circ\!\!-\!\!\bullet\ 2\pi\delta(\omega - \omega_0) \cdot (-j\,\mathrm{sign}(\omega)) = -2\pi j \delta(\omega - \omega_0)$

$\mathcal{H}\{e^{j\omega_0 t}\} = \mathcal{F}^{-1}\{-j2\pi\delta(\omega - \omega_0)\} = -je^{j\omega_0 t} = e^{j(\omega_0 t - \frac{\pi}{2})}$

b) $\mathcal{H}\{\sin\omega_0 t\} = \dfrac{1}{2j}\left[e^{j\omega_0 t} - e^{-j\omega_0 t}\right] * \dfrac{1}{\pi t}$

$\begin{matrix}\circ \\ \shortmid \\ \bullet\end{matrix}$

$\dfrac{1}{2j} \cdot 2\pi[\delta(\omega - \omega_0) - \delta(\omega + \omega_0)] \cdot (-j\,\mathrm{sign}(\omega)) = -\pi[\delta(\omega - \omega_0) + \delta(\omega + \omega_0)]$

$\mathcal{H}\{\sin\omega_0 t\} = -\cos\omega_0 t = \sin\left(\omega_0 t - \dfrac{\pi}{2}\right)$

c) $\mathcal{H}\{\cos\omega_0 t\} = \dfrac{1}{2}\left[e^{j\omega_0 t} + e^{-j\omega_0 t}\right] * \dfrac{1}{\pi t}$

$\begin{matrix}\circ \\ \shortmid \\ \bullet\end{matrix}$

$\dfrac{1}{2} \cdot 2\pi[\delta(\omega - \omega_0) + \delta(\omega + \omega_0)] \cdot (-j\,\mathrm{sign}(\omega)) = -j\pi[\delta(\omega - \omega_0) - \delta(\omega + \omega_0)]$

$\mathcal{H}\{\cos\omega_0 t\} = \sin\omega_0 t = \cos\left(\omega_0 t - \dfrac{\pi}{2}\right)$

d) $\mathcal{H}\{\cos 2\omega_0 t\} \circ\!\!-\!\!\bullet\ -j\pi[\delta(\omega - 2\omega_0) - \delta(\omega + 2\omega_0)]$

$\mathcal{H}\{\cos 2\omega_0 t\} = \sin 2\omega_0 t = \cos\left(2\omega_0 t - \dfrac{\pi}{2}\right)$

In all cases, the Hilbert transform causes a phase shift in the time-domain by $\dfrac{\pi}{2}$.

Solution 15.3

a) $h(t) = \dfrac{1}{\pi t} \circ\!\!-\!\!\bullet\ H(j\omega) = -j\,\mathrm{sign}(\omega)$

$$|H(j\omega)| = \begin{cases} 1 & |\omega| \neq 0 \\ 0 & \text{otherwise} \end{cases} \qquad \arg\{H(j\omega)\} = \begin{cases} -\dfrac{\pi}{2} & \omega > 0 \\ 0 & \omega = 0 \\ \dfrac{\pi}{2} & \omega < 0 \end{cases}$$

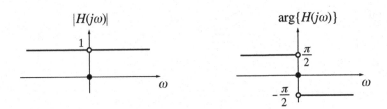

The magnitude of $X(j\omega)$ stays unchanged until $\omega = 0$:

$$|X_h(j\omega)| = \begin{cases} 0 & \omega = 0 \\ |X(j\omega)| & \text{otherwise} \end{cases}$$

for the argument:

$$\arg\{X_h(j\omega)\} = \begin{cases} \arg\{X(j\omega)\} - \dfrac{\pi}{2} & \omega > 0 \\ 0 & \omega = 0 \\ \arg\{X(j\omega)\} + \dfrac{\pi}{2} & \omega < 0 \end{cases}$$

b) $y(t)$ ist real, since both $x(t)$ and $h(t)$ are real. Using the symmetry scheme 9.61 and $Y(j\omega) = -j\text{sign}(\omega)X(j\omega)$, we obtain:

$$x_{g,\text{real}} \circlearrowright\!\!\!\!\bullet\; X_{g,\text{real}} \xrightarrow{\mathcal{H}} Y_{u,\text{imag}} \bullet\!\!\!\!\circlearrowright y_{u,\text{real}}$$

und

$$x_{u,\text{real}} \circlearrowright\!\!\!\!\bullet\; X_{u,\text{imag}} \xrightarrow{\mathcal{H}} Y_{g,\text{real}} \bullet\!\!\!\!\circlearrowright y_{g,\text{real}}$$

The even part of $x(t)$ becomes odd by the Hilbert transform and vice versa.

c) Since is $y(t)$ real, $Y(j\omega)$ has conjugate symmetry, according to (9.94):

$$\int_{-\infty}^{\infty} x(t)y(t)\,dt = \frac{1}{2\pi}\int_{-\infty}^{\infty} X(j\omega)Y(-j\omega)\,d\omega = \frac{1}{2\pi}\int_{-\infty}^{\infty} X(j\omega)Y^*(j\omega)$$

$$= \frac{1}{2\pi}\int_{-\infty}^{\infty} X(j\omega)j\text{sign}(\omega)X^*(j\omega)\,d\omega$$

$$= \frac{j}{2\pi}\int_{-\infty}^{\infty} |X(j\omega)|^2\text{sign}(\omega)\,d\omega = 0,$$

$X(j\omega) = X^*(-j\omega)$, and therefore $|X(j\omega)|^2$ is an even function. The Hilbert transform of a real signal creates an *orthogonal* signal.

d) With the result from Exercise 15.2 and $\text{sign}(0) = 0$, we obtain:

$$\mathcal{H}\{x_F(t)\} = \sum_{\nu=1}^{\infty} [a_\nu \sin(\omega_0\nu t) - b_\nu \cos(\omega_0\nu t)]$$

Solution 15.4

a) $x(t)$ $= \dfrac{1}{\pi} \cdot \dfrac{1}{t} - \dfrac{1}{\pi} \cos \omega_g t \cdot \dfrac{1}{t}$

$\circ$
$\bullet$

$X(j\omega)$ $= \dfrac{1}{\pi}[-j\pi \mathrm{sign}(\omega)] - \dfrac{1}{\pi} \cdot \dfrac{1}{2\pi} \cdot \pi[\delta(\omega - \omega_g) + \delta(\omega + \omega_g)] * [-j\pi \mathrm{sign}(\omega)]$

$= -j\mathrm{sign}(\omega) + \dfrac{j}{2} [\mathrm{sign}(\omega - \omega_g) + \mathrm{sign}(\omega + \omega_g)]$

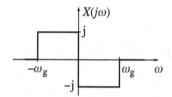

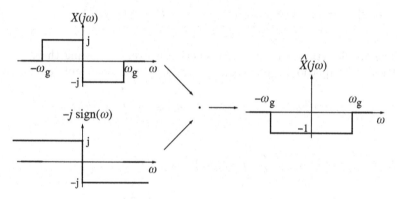

b) $\hat{X}(j\omega)$ $= -\mathrm{rect}\left(\dfrac{\omega}{2\omega_g}\right)$

$\bullet$
$\circ$

$\hat{x}(t)$ $= -\dfrac{\omega_g}{\pi} \mathrm{si}(\omega_g t)$

c) $X_a(j\omega) = 0$ for $\omega < 0$, causal

d) $X_a(j\omega) = X(j\omega) + j\hat{X}(j\omega) = X(j\omega)(1 + j[-j\mathrm{sign}(\omega)])$

$= X(j\omega)(1 + \mathrm{sign}(\omega)) = \begin{cases} 2X(j\omega) & \text{for} \quad \omega > 0 \\ X(j\omega) & \text{for} \quad \omega = 0 \\ 0 & \text{for} \quad \omega < 0 \end{cases}$

Solution 15.5

$$\hat{x}(t) \quad = \quad \mathcal{H}\{x(t)\}$$

$$\circ\!\!\!-\!\!\!\bullet$$

$$\hat{X}(j\omega) \quad = \quad -j\text{sign}(\omega) \cdot X(j\omega) \quad \Rightarrow \quad |\hat{X}(j\omega)| = |X(j\omega)| \quad \forall \omega \neq 0$$

According to Parseval, signals with the same magnitude spectrum have the same energy. The exception at $\omega = 0$ does not make any difference after integration.

Solution 15.6

a) Es gilt:

$$X_2(j\omega) \quad = \quad [1 + j(-j\text{sign}(\omega))]X_1(j\omega)$$

$$\bullet\!\!\!-\!\!\!\circ$$

$$x_2(t) \quad = \quad x_1(t) + j \cdot x_1(t) * \frac{1}{\pi t} = x_1(t) + j\mathcal{H}\{x_1(t)\}$$

b) The Fourier transform of a real function is even. Knowing that, and using Parseval's equation, the energy of $x_1(t)$ and $x_2(t)$ be calculated as follows:

$$E_1 = \int_{-\infty}^{\infty} |x_1(t)|^2 \, dt = \frac{1}{2\pi} \int_{-\infty}^{\infty} |X_1(\omega)|^2 \, d\omega = \frac{2}{2\pi} \int_{0}^{\infty} |X_1(\omega)|^2 \, d\omega$$

$$E_2 = \int_{-\infty}^{\infty} |x_2(t)|^2 \, dt = \frac{1}{2\pi} \int_{-\infty}^{\infty} |X_2(\omega)|^2 \, d\omega$$

$$= \frac{1}{2\pi} \int_{-\infty}^{\infty} |(1 + \text{sign}(\omega))(X_1(\omega))|^2 \, d\omega = \frac{1}{2\pi} \int_{0}^{\infty} 4|X_1(\omega)|^2 \, d\omega$$

$$= \frac{4}{2\pi} \int_{0}^{\infty} |X_1(j\omega)|^2 \, d\omega = 2E_1$$

c) $$E_2 = \int_{-\infty}^{\infty} |x_2(t)|^2 \, dt = \int_{-\infty}^{\infty} |x_1(t) + j\mathcal{H}\{x_1(t)\}|^2 \, dt$$

$$= \int_{-\infty}^{\infty} x_1(t)^2 + \mathcal{H}\{x_1(t)\}^2 \, dt$$

$$= \int_{-\infty}^{\infty} |x_1(t)|^2 \, dt + \underbrace{\int_{-\infty}^{\infty} |\mathcal{H}\{x_1(t)\}|^2 \, dt}_{E_1} = 2E_1$$

$$\underbrace{\phantom{\int_{-\infty}^{\infty} |x_1(t)|^2 \, dt}}_{E_1}$$

Solution 15.7

a) For a causal system :

$$Q(j\omega) = -\frac{1}{\pi} P(j\omega) * \frac{1}{\omega} = -\mathcal{H}\{P(j\omega)\}$$

This yields:

$$H(j\omega) = P(j\omega) - j\frac{1}{\pi} P(j\omega) * \frac{1}{\omega}$$

$$\bullet \atop \circ$$

$$h(t) = p(t) - j2p(t) \cdot \mathcal{F}^{-1}\left\{\frac{1}{\omega}\right\} = p(t) + p(t)\mathcal{F}^{-1}\left\{\frac{2}{j\omega}\right\}$$

$$= p(t) + p(t)\mathrm{sign}(t) = 2p(t)\varepsilon(t)$$

In particular, $P(j\omega) = \dfrac{1}{\omega_g} \cdot \mathrm{rect}\left(\dfrac{\omega}{\omega_g}\right) * \mathrm{rect}\left(\dfrac{\omega}{\omega_g}\right) \;\bullet\!\!-\!\!\circ\; \dfrac{\omega_g}{2\pi}\mathrm{si}^2\left(\dfrac{\omega_g t}{2}\right) = p(t)$

yields: $\quad h(t) = \dfrac{\omega_g}{\pi}\mathrm{si}^2\left(\dfrac{\omega_g t}{2}\right)\varepsilon(t)$

b) $Q(j\omega) = -\mathcal{H}\{P(j\omega)\} = -\dfrac{1}{\pi}\displaystyle\int_{-\infty}^{\infty} \dfrac{P(j\eta)}{\omega - \eta}\, d\eta$

$$P(j\omega) = \left[1 + \frac{\omega}{\omega_g}\right][\varepsilon(\omega + \omega_g) - \varepsilon(\omega)] + \left[1 - \frac{\omega}{\omega_g}\right][\varepsilon(\omega) - \varepsilon(\omega - \omega_g)]$$

$$Q(j\omega) = -\frac{1}{\pi}\int_{-\infty}^{\infty} \frac{\left[1 + \frac{\eta}{\omega_g}\right][\varepsilon(\eta + \omega_g) - \varepsilon(\eta)]}{\omega - \eta}\, d\eta$$

$$-\frac{1}{\pi}\int_{-\infty}^{\infty} \frac{\left[1 - \frac{\eta}{\omega_g}\right][\varepsilon(\eta) - \varepsilon(\eta - \omega_g)]}{\omega - \eta}\, d\eta$$

$$Q(j\omega) = -\frac{1}{\pi\omega_g}\left[(\omega_g + \omega)\ln\left|\frac{\omega_g + \omega}{\omega}\right| + (\omega - \omega_g)\ln\left|\frac{\omega - \omega_g}{\omega}\right|\right]$$

c) $H(j\omega) = P(j\omega) + jQ(j\omega)$

$$\bullet \atop \circ$$

$$h(t) = p(t) + jq(t)$$

Since $P(j\omega)$ is real and even, the symmetry properties (9.61) show that:

$p(t)$ is real and even

$q(t)$ is imaginary and odd

$Q(j\omega)$ is real and odd

Solution 15.8

a) $s_M(t) = s(t)\cos(\omega_0 t) \; \circ\!\!-\!\!\bullet \; S_M(j\omega) = \dfrac{1}{2\pi} S(j\omega) * \pi[\delta(\omega - \omega_0) + \delta(\omega + \omega_0)]$

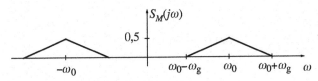

b) $S(j\omega) : 2 \cdot \omega_g = 2\pi \cdot 8 \text{ kHz}$

   $S_M(j\omega) : 2 \cdot 2 \cdot \omega_g = 2\pi \cdot 16 \text{ kHz} \qquad \Rightarrow \qquad \dfrac{\text{bandwidth } S_M(j\omega)}{\text{bandwidth } S(j\omega)} = 2$

c) $\quad s_{EM}(t) \quad = \quad s(t)\cos(\omega_0 t) + \left[s(t) * \dfrac{1}{\pi t}\right]\sin(\omega_0 t)$

$\quad\quad\quad \circ \atop \bullet$

$\quad S_{EM}(j\omega) \quad = \quad \dfrac{1}{2\pi}\Big\{ S(j\omega) * \pi[\delta(\omega - \omega_0) + \delta(\omega + \omega_0)] +$

$\quad\quad\quad\quad\quad\quad\quad +[S(j\omega)\cdot(-j)\text{sign}(\omega)] * \tfrac{\pi}{j}[\delta(\omega - \omega_0) - \delta(\omega + \omega_0)]\Big\}$

$\quad\quad\quad\quad\quad = \quad \dfrac{1}{2}\Big\{ S(j(\omega - \omega_0)) + S(j(\omega + \omega_0)) + S(j(\omega + \omega_0))\cdot\text{sign}(\omega + \omega_0) -$

$\quad\quad\quad\quad\quad\quad\quad -S(j(\omega - \omega_0))\text{sign}(\omega - \omega_0)\Big\}$

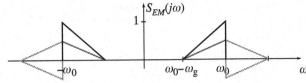

d) $S(j\omega) : 2 \cdot \omega_g = 2\pi \cdot 8 \text{ kHz}$

   $S_{EM}(j\omega) : 2 \cdot \omega_g = 2\pi \cdot 8 \text{ kHz} \qquad \Rightarrow \qquad \dfrac{\text{bandwidth } S_{EM}(j\omega)}{\text{bandwidth } S(j\omega)} = 1$

Solution 16.1

a) $\displaystyle\int\limits_{-\infty}^{\infty} |h(t)|\,dt = \int\limits_{0}^{\infty} e^{-0.1t}|\sin(5\pi t)|\,dt < \int\limits_{0}^{\infty} e^{-0.1t}\cdot 1\,dt = \frac{1}{-0.1}\left[e^{-0.1t}\right]_{0}^{\infty} =$

$= 10 \quad\Rightarrow\quad$ stable

b) with Table 4.1 and the modulation theorem of the Laplace transform, we obtain

$$H(s) = \frac{5\pi}{(s+0.1)^2 + (5\pi)^2} = \frac{5\pi}{(s-s_p)(s-s_p^*)} \;;\; s_p = -0,1 \pm j5\pi$$

Poles lie in the left half-plane $\Rightarrow$ stable

Solution 16.2

Bounded input: $\displaystyle\sum_{k=-\infty}^{\infty} |x[k]| < M_1\,,\; M_1 < \infty$

Sufficient for bounded output:

$$|y[k]| = \left|\sum_{\kappa=-\infty}^{\infty} x[\kappa]h[k-\kappa]\right| \le \sum_{\kappa=-\infty}^{\infty} |x[\kappa]|\cdot|h[k-\kappa]| <$$

$$< M_1 \sum_{\kappa=-\infty}^{\infty} h[k-\kappa] < M_1 M_4 \quad,\; M_4 < \infty$$

Necessary condition: choose input signal: $x[k] = \dfrac{h^*[-k]}{|h[-k]|}$

$$\Rightarrow y[0] = \sum_{\kappa=-\infty}^{\infty} x[\kappa]h[-\kappa] = \sum_{\kappa=-\infty}^{\infty} |h[\kappa]|$$

Solution 16.3

a) $Y(z) = X(z) - a^6 z^{-6} X(z) \;\Rightarrow\; H_1(z) = \dfrac{z^6 - a^6}{z^6}$

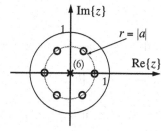

Region of convergence: $0 < |z| < \infty$

$H_1(z)$ is stable for all a.

b) $H_2(z)H_1(z) \overset{!}{=} 1 \Rightarrow H_2(z) = \dfrac{1}{H_1(z)} = \dfrac{z^6}{z^6 - a^6}$

Possible regions of convergence:

1. $|z| < |a| \Rightarrow H_2(z)$ is not causal and stable for $|a| > 1$, because the poles of a left-sided sequence have to lie outside of the unit circle.

2. $|z| > |a| \Rightarrow H_2(z)$ is causal and stable for $|a| < 1$.

Solution 16.4

a) $H(z) = \dfrac{1}{1 - \frac{1}{2}z^{-1} + \frac{1}{4}z^{-2}} = \dfrac{z^2}{z^2 - \frac{1}{2}z + \frac{1}{4}}$

b) α)

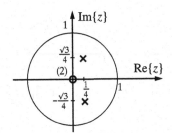

Zeros: $z_{1/2} = 0$

Poles: $z_{1/2} = \frac{1}{4}(1 \pm j\sqrt{3})$; $|z_{1/2}| = \frac{1}{2} < 1 \Rightarrow$ stable.

β) 1. $H(z = 1)$ is finite

2. $z^2 - \dfrac{1}{2}z + \dfrac{1}{4}\bigg|_{z=\frac{s+1}{s-1}} = \dfrac{\frac{3}{4}s^2 + \frac{3}{2}s + \frac{7}{4}}{(s-1)^2}$; Numerator is a Hurwitz poly-

nomial $\Rightarrow$ stable.

c) All zeros inside the unit sample are $\Rightarrow$ minimum phase.

d) Yes, evident from the difference equation.

Solution 16.5

The internal part with order i can be split into partial fractions:

$$Y_{\text{int}}(z) = \sum_{i=1}^{N} \dfrac{A_i z}{z - z_i} \quad \bullet\!\!-\!\!\circ \quad y_{\text{int}}[k] = \sum_{i=1}^{N} A_i z_i^k \varepsilon[k]$$

The series z_i^k decays where $k \to \infty$, if $|z_i| < 1$, i.e., if the pole lies within the unit circle.

Solution 16.6

a) Denominator polynomial $N(s) = -2.5[s^4 + 4.5s^3 + 8s^2 + 7s + 2]$

all coefficients are positive and exist $\Rightarrow$ possibly stable.

Hurwitz test:

$\Delta_1 = a_1 = 4.5 > 0$

$\Delta_2 = a_1 a_2 - a_3 = 4.5 \cdot 8 - 7 > 0$

$\Delta_3 = a_1 a_2 a_3 - a_1^2 a_4 - a_3^2 = 162.5 > 0$

$\Delta_4 = a_1 a_2 a_3 a_4 - a_4^2 a_1^2 - a_3^2 a_4 = 325 > 0$

$\Rightarrow N(s)$ is a Hurwitz polynomial, so $H(s)$ is stable.

$H(s)$ does not have minimum phase, because zeros lie in the right half-plane.

b) Not all coefficients in $N(s) = s^4 - 4.5s^3 + 8s^2 - 7s + 2$ are positive $\Rightarrow$ so it cannot be a Hurwitz polynomial, and $H(s)$ is not stable.

$H(s)$ does not have minimum phase, because zeros lie in the right half-plane.

c) Not all coefficients in $N(s) = s^3 + s + 2$ are positive since there is no term with $s^2 \Rightarrow$ not a Hurwitz polynomial, so $H(s)$ is not stable.

$H(s)$ has minimum phase, because no zeros lie in the right half-plane.

Solution 16.7

Since the transform is the same in both directions:

- Inside the unit circle of the z-plane $\rightarrow$ left half of the s-plane

- left half of the z-plane $\rightarrow$ inside the unit circle of the der s-plane

- Intersection $\hat{=}$ left half of the unit circle on the z-plane $\rightarrow$
  Intersection $\hat{=}$ left half of the unit circle on the s-plane.

Solution 16.8

a) $H_0(s) = 2V_0(s^2 + 1)$

$$H(s) = \frac{E(s)}{1 + E(s)G(s)} = \frac{E(s)}{1 + H_0(s)} = \frac{s}{1 + 2V_0(s^2 + 1)} = \frac{s}{2V_0 s^2 + 2V_0 + 1}$$

b) $H(s)$ is unstable, because the denominator of $H(s)$ is not a Hurwitz polynomial.

Poles: $s_{1,2} = \pm j \sqrt{1 + \frac{1}{2V_0}}$

c) No, denominator polynomial is not a Hurwitz polynomial for all $V_0 > 0$.

d) When $s = j\omega$: $j\omega = \pm j\sqrt{1 + \frac{1}{2V_0}}$ $\rightarrow$ $\omega = \pm 6$

e) $U(j\omega) = \frac{\pi}{j}[\delta(\omega - 6) - \delta(\omega + 6)]$

$\quad R(j\omega) \quad = \quad H_0(j\omega)U(j\omega) = -U(j\omega)$

$\quad\quad \bullet\!\!\downarrow\!\!\circ$

$\quad r(t) \quad = \quad -\sin(6t) = -u(t)$

Solution 16.9

a) $H(s)$ has poles at $s = \pm j \Rightarrow$ unstable

b) $H_r(s) = \dfrac{H(s)}{1 - H(s)Ks} = \dfrac{1}{s^2 + 1 - Ks}$

c) $s^2 - Ks + 1$ must be a Hurwitz polynomial $\Rightarrow K < 0$

d) $H_r(s) \quad = \quad \frac{1}{(s - \frac{K}{2})^2 + a^2}$; $a = \sqrt{1 - \frac{K^2}{4}}$

$\quad\quad \bullet\!\!\downarrow\!\!\circ$

$\quad h_r(t) \quad = \quad \frac{1}{a} e^{\frac{K}{2}t} \sin(at)\varepsilon(t)$

$\displaystyle\int_{-\infty}^{\infty} |h_r(t)|\, dt = \frac{1}{|a|} \int_0^{\infty} |e^{\frac{K}{2}t} \sin(at)|\, dt < \frac{1}{|a|} \int_0^{\infty} e^{\frac{K}{2}t}\, dt < \infty$

$h_r(t)$ is integrable $\Longleftrightarrow$ stable

Solution 16.10

a)

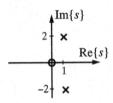

Unstable, because of poles in the right half-plane

b) $G(s) = K$; $H(s) = \dfrac{F(s)}{1 + K \cdot F(s)} = \dfrac{s}{s^2 + (K - 2)s + 5}$

points for root locus:

$K = 0$:	$s_p = 1 \pm j2$
$K = 2$:	$s_p = \pm j\sqrt{5}$
$K = 4$:	$s_p = -1 \pm j2$
double pole $K = 2 + 2\sqrt{5}$:	$s_p = -\sqrt{5}$
$K \to \infty$:	$s_{p1} \to -\infty \; ; \; s_{p2} \to 0-$

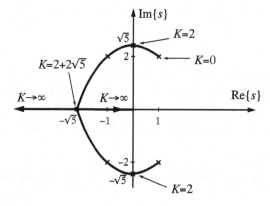

Stablisation succeeds if $2 < K < \infty$.

Solution 16.11

a) $H(s) = \dfrac{F(s)}{1 + K \cdot F(s)} = \dfrac{1}{s^2 - 2s + 5 + K}$

points for root locus:

$K \to -\infty$:	$s_{p1} \to -\infty \; ; \; s_{p2} \to \infty$
$K = -5$:	$s_{p1} = 0 \; ; \; s_{p2} = 2$
$K = -4$:	$s_{p1} = s_{p2} = 1$
$K = 0$:	$s_p = 1 \pm j2$
$K = 5$:	$s_p = 1 + \pm j3$
$K = \to \infty$:	$s_p = 1 \pm j\infty$

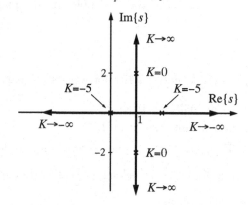

Cannot be stabilised because both poles do not lie in the left half-plane for any k.

b) $H(s) = \dfrac{1}{s^2 + (K - 2)s + 5}$

Hurwitz: necessary + sufficient that all coefficients are positive
$\Rightarrow$ stabilisation for $K > 2$ (compare Exercise 16.10)

Solution 17.1

a) The linear ensemble mean $E\{x(t)\}$ is equal to the linear time-average $\overline{x_i(t)}$ of each sample function.

Ensembles 1 and 2: Yes
Ensemble 3: No, because the time-averages are different. $(\overline{x_i(t)} \neq \overline{x_j(t)}, \ i \neq j)$
Ensemble 4: Yes
Ensemble 5: No, because the ensemble mean is time-dependent, $E\{x(t)\} \neq$ constant.

b) The ensemble mean squared $E\{x^2(t)\}$ is equal to the time-average squared $\overline{x_i^2(t)}$ for each sample function.

Ensemble 1: Yes
Ensemble 2: No, because the ensemble mean squared $E\{x^2(t)\}$ is time-dependent.
Ensemble 3: No, because the time-averages squared $\overline{x_i^2(t)}$ of each sample function are different.
Ensemble 4: No, in this case the time-averages squared $\overline{x_i^2(t)}$ are also different. The sample functions have the same mean, but different amplitudes and therefore different mean powers $\overline{x_i^2(t)}$.
Ensemble 5: no, because the ensemble mean squared $E\{x^2(t)\}$ is time-dependent.

Solution 17.2

a) According to (17.14) und (17.15), for weak stationary random processes:

  - $\mu_x(t) = $ constant: except at 5,
  - $\sigma_x^2(t) = $ constant: except at 2,
  - the ACF $\varphi_{xx}(t, t-\tau)$ does not depend on t: except at 1.

Only ensembles 3 and 4 can belong to weak stationary processes.

b) For weak ergodic processes:

- $E\{x(t)\} = \overline{x_i(t)}$, because of (17.21). Applies to ensembles 1, 2 and 4, see Exercise 17.1a.

- $E\{x(t_1) \cdot x(t_2)\} = \overline{x_i(t_1) \cdot x_i(t_2)}$, because of (17.20). In Exercise 17.1b the special case $t_1 = t_2$ is investigated. The condition only applies to ensemble 1. Because this random process is not weak stationary, however, it cannot be weak ergodic.

$\Rightarrow$ none of the random processes are weak ergodic.

Solution 17.3

a) Linear time-average $\overline{x_i(t)} = 0$, because there is no dc component.
The expected value at a certain point t_0, $E\{x(t_0)\}$, is spread across many points of the CD signal (at most by as many samples as there are on the CD), because in our experiment, each sample function comes from a random time. Since music signals in general do not have dc components, $E\{x(t)\} = 0$.

b) With first order expected values we can only discuss the condition for stationarity given in (17.14): In a) we acertained that $\mu_x = E\{x(t)\} = $ constant. With the same notation we can assume $E\{x^2(t)\} = $ constant, because it is determined from many values of the output signal $x^2(t)$.
$\Rightarrow$ the random process could be stationary.

c) For ergodicity, the first and second order ensemble means must agree with the corresponding time-averages of any sample function, e.g., $E\{x^2(t)\} = \overline{x_i^2(t)}$. This does not apply because to form the time-average squared $\overline{x_i^2(t)}$ only a ten second section of the CD is considered, which for different sample functions i, is taken from different points on the CD. The averages are generally different, following to the loudness of the music.
$\Rightarrow$ the random process is not ergodic.

Solution 17.4

$\mu_y = \mu_{x_1} + \mu_{x_2} = 2$

$E\{y^2(t)\} = E\{(x_1(t) + x_2(t))^2\} = E\{x_1^2(t)\} + \underbrace{2\,E\{x_1(t)x_2(t)\}}_{=0} + E\{x_2^2(t)\} = 7$

$\sigma_y^2 = E\{y^2(t)\} - \mu_y^2 = 3$

Solution 17.5

$\sigma_v^2 = E\{(v(t) - \mu_v(t))^2\} = E\{[x(t) + y(t) - (\mu_x(t) + \mu_y(t))]^2\}$

Since $y(t)$ is deterministic, $\mu_y(t) = y(t)$, and therefore:

$\sigma_v^2 = E\{(x(t) - \mu_x(t))^2\} = \sigma_x^2 = 10$

Adding a deterministic signal does not change the variance.

Solution 17.6

a) $\mu_y(t) = 1 + K$

$\sigma_y^2(t) = E\{(y(t) - \mu_y)^2\} = E\{(x(t) + K - (1 + K))^2\} = E\{(x(t) - \mu_x)^2\} =$
$= \sigma_x^2 = 4$

$E\{y^2(t)\} = \sigma_y^2 + \mu_y^2 = K^2 + 2K + 5$

$\overline{y_i(t)} = 1 + K$

Ergodic.

b) $\mu_y(t) = 1 + \sin(t) \neq$ constant. $\Rightarrow$ not stationary

$\sigma_y^2(t) = E\{(x(t) + \sin(t) - 1 - \sin(t))^2\} = \sigma_x^2 = 4$

$E\{y^2(t)\} = 5 + 2\sin(t) + \sin^2(t)$

$\overline{y_i(t)} = 1$

Not ergodic because it is not stationary.

c) $\mu_y(t) = 1 + \varepsilon(t) \neq$ constant $\Rightarrow$ not stationary

$\sigma_y^2(t) = \sigma_x^2$, see b)

$E\{y^2(t)\} = 4 + (1 + \varepsilon(t))^2 = 5 + 3\varepsilon(t)$

$\overline{y_i(t)} = 1.5$

Not ergodic.

d) $\mu_y(t) = 5\varepsilon(t) \neq$ constant $\Rightarrow$ not stationary

$\sigma_y^2(t) = 25\varepsilon(t) \cdot \sigma_x^2 = 100\varepsilon(t)$

$E\{y^2(t)\} = 125\varepsilon(t)$

$\overline{y_i(t)} = 2.5$

Not ergodic.

Solution 17.7

$\mu_x(t) = x(t)$

$E\{x^2(t)\} = x^2(t) = e^{-0.2t}\varepsilon(t)$

$\sigma_x^2(t) = E\{x^2(t)\} - \mu_x^2(t) = x^2(t) - x^2(t) = 0$

$$\overline{x_i(t)} = \lim_{T \to \infty} \frac{1}{2T} \int_0^T e^{-0,1t}\, dt = \lim_{T \to \infty} \left[\frac{1}{2T} \cdot \frac{1}{-0,1}\left(e^{-0,1T} - 1\right)\right] = 0$$

Solution 17.8

a) Ergodic, because the properties of a die do not change with time.

$\mu_x = 3.5$ (see Example 17.1)

$$E\{x^2[k]\} = \frac{1+4+9+16+25+36}{6} = \frac{91}{6} = 15.17$$

$$\sigma_x^2 = E\{x^2[k]\} - \mu_x^2 = \frac{35}{12} = 2.92$$

b) Likewise ergodic, because the properties of a die do not change with time.

$\mu_x = 15.17$

$$E\{x^2[k]\} = \frac{1}{6}\sum_{k=1}^{6} k^4 = 379.17$$

$$\sigma_x^2 = 149.14$$

Solution 17.9

To calculate the time-averages, we use the results from Ex. 17.8a), where $x(t)$ is the number on a "normal" die, $\mu_x = \frac{7}{2}$ und $E\{x^2(t)\} = \frac{91}{6}$.

$$\overline{y_i[k]} = \frac{6 + 2\mu_x}{3} = \frac{13}{3}$$

$$\overline{y_i^2[k]} = \frac{1}{3}(2 \cdot E\{x^2[k]\} + 6^2) = \frac{199}{9}$$

The ensemble means of $y[k]$ for $k \neq 3N$, $N \in \mathbb{Z}$, agree with those of $x[k]$ from Exercise 17.8a). For $k = 3N$, $N \in \mathbb{Z}$:

$\mu_y[k] = 6$

$E\{y^2[k]\} = 36$

$\sigma_y{}^2[k] = E\{y^2[k]\} - \mu_y{}^2 = 0$

Neither stationary nor ergodic, because the ensemble means are not constant.

Solution 17.10

$$\varphi_{xx}(\tau) = E\{x(t+\tau)x^*(t)\} = E\{K \cdot K^*\} = |K|^2$$

Solution 17.11

As every sample function has a period of 10s, the ACF $\varphi_{xx}(t_0, t_0 + \tau)$ is also period in τ with a period of 10s. A good approximation is given by $\varphi_{xx}(t_0, t_0 + \tau)$, independent of time t_0. The justification is similar to Exercise 17.3: The correlation properties of a sigal mixture from the whole CD are virtually unchanged within 10s:

$$\varphi_{xx}(t_0, t_0 + 10s) = \varphi_{xx}(t_0, t_0) = E\{x^2(t_0)\} \approx \text{constant}.$$

Solution 17.12

a) $E\{(x(t) + y(t))^2\} = E\{x^2(t) + 2x(t)y(t) + y^2(t)\} =$
$= E\{x^2(t)\} + 2E\{x(t)y(t)\} + E\{y^2(t)\} = E\{x^2(t)\} + E\{y^2(t)\}$
To satisfy the condition, $E\{x(t)y(t)\} = 0$.

b) The special case of $\tau = 0$ in (17.48), $\varphi_{xx} = E\{x(t+\tau)y(t)\}$ (17.46) and $\mu_x = E\{x(t)\}$ yield:
$E\{x(t)y(t)\} = E\{x(t)\} \cdot E\{y(t)\}$

c) Either $E\{x(t)\}$ or $E\{y(t)\}$, or both expected expected values must be zero (at least one of the random signal must have zero mean).

Solution 17.13

With deterministic signal, the quantities in question must be calculated with time-averages, because the ensemble means correspond to the signal itself.

Power: $\displaystyle \lim_{T \to \infty} \frac{1}{2T} \int_{-T}^{T} d^2(t)\, dt = \overline{d^2(t)}$

dc component: $\displaystyle \lim_{T \to \infty} \frac{1}{2T} \int_{-T}^{T} d(t)\, dt = \overline{d(t)}$

Effective value: $\sqrt{\overline{d^2(t)}}$

Power of the ac component: $\overline{(d(t) - \overline{d(t)})^2}$

Solution 17.14

Forming the time-average according to (17.18) is linear. With $\mu = \overline{d(t)}$, therefore:
$\overline{(d(t) - \mu)^2} = \overline{d^2(t)} - 2\mu\overline{d(t)} + \mu^2 = \overline{d^2(t)} - 2\mu\overline{d(t)} + \mu^2 = \overline{d^2(t)} - \mu^2$
see derivation of (17.8)

Solution 17.15

$\psi_{xy}(\tau) = E\{(x(t) - \mu_x)(y(t-\tau) - \mu_y)\} =$
$= E\{x(t)\, y(t-\tau)\} - \mu_x\, E\{y(t-\tau)\} - \mu_y\, E\{x(t)\} + \mu_x\mu_y = \varphi_{xy}(\tau) - \mu_x\mu_y$

Solution 17.16

For $\tau \to \infty$ any chosen random processes are generally uncorrelated:
$\varphi_{xy}(\tau \to \infty) = \mu_x\mu_y \;\Rightarrow\; \mu_y = \displaystyle \lim_{\tau \to \infty} \frac{4\tau^2 + 10}{1 + \tau^2} = 4$

$\varphi_{yx}(\tau) = \varphi_{xy}(-\tau) = \varphi_{xy}(\tau)$, da $\varphi_{yx}(\tau)$ is even.

Taking the solution of Ex. 17.15 we obtain:
$\psi_{yx}(\tau) = \varphi_{yx}(\tau) - \mu_x\mu_y = \dfrac{4\tau^2 + 10}{1 + \tau^2} - 4 = \dfrac{6}{1 + \tau^2} = \psi_{xy}(\tau).$

Solution 17.17

a) $\mu_v = \varphi_{vv}(\tau \to \infty) = 0$

$\mu_u = 0$, because the delay does not change the dc component.

$\varphi_{uv}(\tau) = E\{u(t)v(t-\tau)\} = E\{v(t-10)v(t-\tau)\} = \varphi_{vv}(\tau-10) = e^{-|\tau-10|}$

$\varphi_{vu}(\tau) = \varphi_{uv}(-\tau) = e^{-|-\tau-10|} = e^{-|\tau+10|}$

$\varphi_{uu}(\tau) = \varphi_{vv}(\tau)$, because a delay does not change the correlation properties of a stationary random process.

b) $\varphi_{vv}(\tau) = e^{-|\tau|} = \varepsilon(\tau)e^{-\tau} + \varepsilon(-\tau)e^{\tau}$

$\Phi_{vv}(j\omega) = \dfrac{1}{j\omega+1} - \dfrac{1}{j\omega-1} = \dfrac{2}{\omega^2+1}$

(with (9.12) and Table 4.1)

$\Phi_{uv}(j\omega) = e^{-j10\omega}\Phi_{vv}(j\omega) = e^{-j10\omega}\dfrac{2}{\omega^2+1}$

$\Phi_{vu}(j\omega) = e^{j10\omega}\dfrac{2}{\omega^2+1}$

c) $\Phi_{vv}(j\omega)$: real + even, since $\varphi_{vv}(\tau)$ real + even

$\Phi_{uv}(j\omega)$ and $\Phi_{vu}(j\omega)$: conjugate symmetry.

$v(t)$ real, since ACF real + even.

Solution 17.18

$\varphi_{xx}[\kappa] = E\{x[k]x[k-\kappa]\}$

$\varphi_{xx}[0] = E\{x^2[k]\} = \dfrac{91}{6}$, see Ex. 17.8.

Since the numbers of dice are uncorrelated with different time k, $\varphi_{xx}[\kappa \neq 0] = \mu_x^2 = \frac{49}{4}$.

$\varphi_{xx}[\kappa] = \dfrac{49}{4} + \dfrac{35}{12}\delta[\kappa]$

$\psi_{xx}[\kappa] = \varphi_{xx}[\kappa] - \mu_x^2 = \dfrac{35}{12}\delta[\kappa]$

$\Phi_{xx}(e^{j\Omega}) = \mathcal{F}_*\left\{\dfrac{49}{4} + \dfrac{35}{12}\delta[k]\right\} = \dfrac{49}{4}\,\underset{2\pi}{\sqcup\!\sqcup\!\sqcup}\left(\dfrac{\Omega}{2\pi}\right) + \dfrac{35}{12}$

Solution 17.19

$\mu_x = \dfrac{0+1+1+1}{4} = \dfrac{3}{4}$

$\sigma_x^2 = E\{x^2[k]\} - \mu_x^2 = \dfrac{0^2+1^2+1^2+1^2}{4} - \dfrac{9}{16} = \dfrac{3}{16}$

$\varphi_{xx}[\kappa \neq 0] = \mu_x^2$, because the outcome of a die does not correlate to different times.

$$\varphi_{xx}[\kappa] = \frac{9}{16} + \frac{3}{16}\delta[\kappa]$$

$$\psi_{xx}[\kappa] = \frac{3}{16}\delta[\kappa]$$

$$\Phi_{xx}(e^{j\Omega}) = \frac{9}{16}\, \text{⊥⊥⊥}\left(\frac{\Omega}{2\pi}\right) + \frac{3}{16}$$

Solution 17.20

$$\mu_y = \frac{5}{2}$$

$$\sigma_y{}^2 = E\{y^2[k]\} - \mu_y{}^2 = \frac{5}{4}$$

$$\varphi_{yy}[\kappa] = \frac{25}{4} + \frac{5}{4}\delta[\kappa]$$

$$\varphi_{xy}[\kappa] = \varphi_{yx}[\kappa] = \mu_x \mu_y = \frac{15}{8}, \text{ because the two processes are uncorrelated.}$$

Solution 17.21

Power: $$E\{x^2(t)\} = \frac{1}{2\pi}\int_{-\infty}^{\infty}\Phi_{xx}(j\omega)d\omega = \frac{1}{\pi}$$

ACF: $$\Phi_{xx}(j\omega) = \frac{1}{2}\text{rect}\left(\frac{\omega}{2}\right) * \text{rect}\left(\frac{\omega}{2}\right)$$

$$\varphi_{xx}(\tau) = \frac{1}{\pi}\text{si}^2(\tau)$$

Solution 18.1

a) $$\begin{aligned}\varphi_{xy}(\tau) &= E\{x(t+\tau)y^*(t)\} = E\{x(t+\tau)(C^* + x^*(t))\} = \\ &= C^*\mu_x + E\{x(t+\tau)x^*(t)\} = C^*\mu_x + \varphi_{xx}(\tau)\end{aligned}$$

$$\begin{aligned}\varphi_{yy}(\tau) &= E\{y(t+\tau)y^*(t)\} = E\{(C + x(t+\tau))(C^* + x^*(t))\} = \\ &= |C|^2 + C^*\mu_x + C\mu_x^* + \varphi_{xx}(\tau) = \\ &= |C|^2 + 2\,\text{Re}\{C^*\mu_x\} + \varphi_{xx}(\tau)\end{aligned}$$

b) We consider $x(t)$ and C as signals of the same kind.

$$\begin{aligned}\varphi_{xy}(\tau) &= \varphi_{xx}(\tau) + \varphi_{xC}(\tau) = \varphi_{xx}(\tau) + E\{x(t+\tau)C^*\} = \varphi_{xx}(\tau) + \mu_x C^*\\ \varphi_{yy}(\tau) &= \varphi_{xx}(\tau) + \varphi_{xC}(\tau) + \varphi_{Cx}(\tau) + \varphi_{CC}(\tau) = \\ &= \varphi_{xx}(\tau) + \mu_x C^* + C\mu_x^* + |C|^2 = \\ &= \varphi_{xx}(\tau) + 2\text{Re}\{\mu_x C^*\} + |C|^2\end{aligned}$$

Solution 18.2

a) With (18.8): $\varphi_{vv}(\tau) = |A|^2 \varphi_{xx}(\tau)$

b) With (18.12): $\begin{aligned}\varphi_{yy}(\tau) &= \varphi_{vv}(\tau) + 2\operatorname{Re}\{\mu_v B^*\} + |B|^2 = \\ &= |A|^2 \varphi_{xx}(\tau) + 2\operatorname{Re}\{A\mu_x B^*\} + |B|^2\end{aligned}$

c) With (18.7): $\varphi_{xv}(\tau) = A^*\varphi_{xx}(\tau)$

Solution 18.3

a) $\begin{aligned}\varphi_{xy}(\tau) &= \mathrm{E}\{x(t+\tau)y^*(\tau)\} = \\ &= \mathrm{E}\{x(t+\tau)(A^*x^*(t) + B^*)\} = \\ &= A^*\mathrm{E}\{x(t+\tau)x^*(t)\} + B^*\mathrm{E}\{x(t+\tau)\} = \\ &= A^*\varphi_{xx}(\tau) + B^*\mu_x\end{aligned}$

b) No, because we are not considering an LTI system. Adding a constant makes the system non-linear (see Ch. 1).

Solution 18.4

a) $\begin{aligned}\varphi_{ww}(\tau) &= \mathrm{E}\{w(t+\tau)w^*(t)\} = \\ &= \mathrm{E}\{A(B + u(t+\tau))(A^*(B^* + u^*(t)))\} = \\ &= |A|^2 \big[|B|^2 + B^*\underbrace{\mathrm{E}\{u(t+\tau)\}}_{=0} + B\underbrace{\mathrm{E}\{u^*(t)\}}_{=0} + \mathrm{E}\{u(t-\tau)u^*(t)\}\big] = \\ &= |A|^2[|B|^2 + \varphi_{uu}(\tau)]\end{aligned}$

b) $\begin{aligned}\varphi_{vw}(\tau) &= \mathrm{E}\{v(t+\tau)w^*(t)\} = \\ &= \mathrm{E}\{v(t+\tau)A^*(B^* + u^*(t))\} = \\ &= A^*B^*\underbrace{\mathrm{E}\{v(t+\tau)\}}_{=0} + A^*\mathrm{E}\{v(t+\tau)u^*(t)\} = \\ &= A^*\varphi_{vu}(\tau)\end{aligned}$

c) With (17.54): $\varphi_{wv}(\tau) = \varphi_{vw}^*(-\tau) = A\varphi_{vu}^*(-\tau)$

d) $\begin{aligned}\varphi_{uw}(\tau) &= \mathrm{E}\{u(t+\tau)w^*(t)\} = \\ &= \mathrm{E}\{u(t+\tau)A^*(B^* + u*(t))\} = \\ &= A^*B^*\underbrace{\mathrm{E}\{u(t+\tau)\}}_{=0} + A^*\mathrm{E}\{u(t+\tau)u^*(t)\} = \\ &= A^*\varphi_{uu}(\tau)\end{aligned}$

e) With (17.54): $\varphi_{wu}(\tau) = \varphi_{uw}^*(-\tau) = A\varphi_{uu}^*(-\tau) = A\varphi_{uu}(\tau)$

Solution 18.5

Using (18.21) and (18.7), for uncorrelated signals x and y holds

$$\varphi_{xz}(\tau) = A^*\varphi_{xx}(\tau)$$

$$\varphi_{yz}(\tau) = B^*\varphi_{yy}(\tau)$$

With (18.8) and (18.17), we obtain:

$$\varphi_{zz}(\tau) = |A|^2\varphi_{xx}(\tau) + |B|^2\varphi_{yy}(\tau)$$

Solution 18.6

a) $H(j\omega) = \dfrac{-\omega^2 - 2j\omega + 2}{(j\omega + 2)(-\omega^2 + 2j\omega + 2)}$

$\quad |H(j\omega)|^2 = \dfrac{(2 - \omega^2)^2 + 4\omega^2}{(4 + \omega^2)((2 - \omega^2)^2 + 4\omega^2)} = \dfrac{1}{4 + \omega^2}$

$\quad |H(j\omega)|^2 \; \bullet\!\!-\!\!\circ \; \varphi_{hh}(\tau) = \dfrac{1}{4}e^{-2|\tau|}$

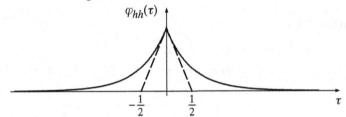

b) $\varphi_{yy}(\tau) = \varphi_{hh}(\tau) * \varphi_{xx}(\tau)$

$\quad \varphi_{xx}(\tau) = \delta(t) \rightarrow \varphi_{yy}(\tau) = \varphi_{hh}(\tau)$

$\quad P_y = \varphi_{yy}(0) = \varphi_{hh}(0) = \dfrac{1}{4}$

Solution 18.7

a) $\Phi_{xx}(j\omega) = \mathcal{F}\{\varphi_{xx}(\tau)\} = \mathcal{F}\{\delta(\tau)\} = 1$

b) $\mu_x = 0$ (x has zero mean)

$\quad \mu_y = H(0) \cdot \mu_x = 0$

c) Simple calculation of $\varphi_{hh}(\tau)$ in the frequency-domain

$$
\begin{array}{ccccc}
\varphi_{hh}(\tau) & = & h(\tau) & * & h^*(-\tau) \\
\circ\!\!\!\uparrow\!\!\!\bullet & & \circ\!\!\!\uparrow\!\!\!\bullet & & \circ\!\!\!\uparrow\!\!\!\bullet \\
\Phi_{hh}(j\omega) & = & H(j\omega) & \cdot & H^*(j\omega) & = & |H(j\omega)|^2
\end{array}
$$

$$H(j\omega) = \mathcal{F}\{\text{si}(t)\} = \pi\,\text{rect}(\frac{\omega}{2})$$

$$\varphi_{hh}(\tau) = \mathcal{F}^{-1}\{\pi^2\text{rect}(\frac{\omega}{2})\} = \pi\,\text{si}(\tau)$$

d) $\Phi_{yy}(j\omega) = |H(j\omega)|^2\Phi_{xx}(j\omega) = \pi^2\text{rect}(\frac{\omega}{2})$

$$\varphi_{yy}(\tau) \quad = \pi\,\text{si}(\tau)$$

e) $\varphi_{xy}(\tau) = \varphi_{xx}(\tau) * h^*(-\tau)$

$\varphi_{xy}(\tau) = \varphi_{xx}(\tau) * h^*(-\tau) = \delta(\tau) * h^*(-\tau) = h^*(-\tau) = \text{si}^*(-\tau) = si(\tau)$

f) Power = mean squared= $E\{|x(t)|^2\} = \varphi_{xx}(0)$

$\varphi_{xx}(0) \to \infty$ (white noise has infinite power)

$\varphi_{xx}(0) = \sigma_x^2 + \mu_x^2$

$\Rightarrow$ Varianz $\sigma_x^2 = \varphi_{xx}(0) \to \infty$

$\varphi_{yy}(0) = \pi$, $\sigma_y^2 = \pi$ (band-limited white noise has finite power)

Solution 18.8

a) Inverse Fourier transform with the shift theorem:

$$H(j\omega) \quad = \quad \frac{1}{2}\text{rect}(\frac{\omega}{2\omega_g})(e^{j\frac{\pi\omega}{2\omega_g}} + e^{-j\frac{\pi\omega}{2\omega_g}})$$

$$h(t) \quad = \quad \frac{\omega_g}{2\pi}\left[\text{si}\left(\omega_g(t + \frac{\pi}{2\omega_g})\right) + \text{si}\left(\omega_g(t - \frac{\pi}{2\omega_g})\right)\right]$$

b) Likewise, calculation of $\varphi_{hh}(\tau)$ can be achieved with the shift theorem:

$$|H(j\omega)|^2 \quad = \quad \cos^2(\frac{\pi}{2\omega_g}\cdot\omega)\text{rect}(\frac{\omega}{2\omega_g}) =$$

$$\frac{1}{4}\left[e^{j\frac{\pi}{2\omega_g}\cdot\omega} + e^{-j\frac{\pi}{2\omega_g}\cdot\omega}\right]^2\text{rect}(\frac{\omega}{2\omega_g}) =$$

$$\left[\frac{1}{4}e^{j\frac{\pi}{\omega_g}\omega} + \frac{1}{2} + \frac{1}{4}e^{-j\frac{\pi}{\omega_g}\omega}\right]\text{rect}(\frac{\omega}{2\omega_g})$$

$$\varphi_{hh}(\tau) \quad = \quad \frac{1}{4}\frac{\omega_g}{\pi}\text{si}(\omega_g\tau - \pi) + \frac{1}{2}\frac{\omega_g}{\pi}\text{si}(\omega_g\tau) + \frac{1}{4}\frac{\omega_g}{\pi}\text{si}(\omega_gt + \pi)$$

c) $\varphi_{xx}(\tau) = \mathcal{F}^{-1}\{\Phi_{xx}(j\omega)\} = N_0\cdot\delta(\tau) + \frac{m}{2\pi}$

$\mu_x^2 = \lim\limits_{\tau\to\infty}\varphi_{xx}(\tau) = \frac{m}{2\pi}$

d) $\mu_y = \mu_x \cdot H(0) = \sqrt{\dfrac{m}{2\pi}} \cdot 1 = \sqrt{\dfrac{m}{2\pi}}$

$$\Phi_{yy}(j\omega) = \Phi_{xx}(j\omega) \cdot |H(j\omega)|^2 = N_0 \cos^2\left(\frac{\pi\omega}{2\omega_g}\right)\mathrm{rect}\left(\frac{\omega}{2\omega_g}\right) + m \underbrace{|H(j\omega)|^2}_{=1} \delta(\omega)$$

$$\varphi_{yy}(\tau) = \varphi_{xx}(\tau) * \varphi_{hh}(\tau) = \frac{N_0}{2\pi} \frac{\sin(\omega_g t)}{t[1 - (\frac{\omega_g t}{\pi})^2]} + \frac{m}{2\pi}$$

$$P_y = \varphi_{yy}(0) = \frac{\omega_g N_0}{2\pi} + \frac{m}{2\pi}$$

Solution 18.9

ACF:

$$|H(j\omega)|^2 = \frac{1 + \omega^2}{(2 - \omega^2)^2 + 9\omega^2} = \frac{1}{\omega^2 + 4}$$

$$\Phi_{yy}(j\omega) = |H(j\omega)|^2 \Phi_{xx}(j\omega) = N_0 \frac{1}{\omega^2 + 4} \quad \bullet\!\!-\!\!\circ \quad \varphi_{yy}(\tau) = \frac{N_0}{4} e^{-2|\tau|}$$

Mean: $\mu_y = \varphi_{yy}(\tau \to \infty) = 0$

Variance: $\sigma_y^2 = \varphi_{yy}(0) - \mu_y^2 = \dfrac{N_0}{4}$

Solution 18.10

a) $\varphi_{yy}(\tau) = N_0 \dfrac{\alpha}{2} e^{-\alpha|\tau|} (\alpha > 0) \circ\!\!-\!\!\bullet N_0 \dfrac{\alpha}{2} \dfrac{2\alpha}{\omega^2 + \alpha^2} = \Phi_{yy}(j\omega)$

$\Phi_{yy}(j\omega) = |H(j\omega)|^2 \Phi_{xx}(j\omega) \rightarrow |H(j\omega)|^2 = \dfrac{\alpha^2}{\omega^2 + \alpha^2} \rightarrow H(s) = \dfrac{\alpha}{\alpha + s}$

b) No, any number of all-pass stages will still not change $|H(j\omega)|^2$.

Solution 18.11

a) With (18.38) and (18.39):

$\varphi_{y_1 y_1}(\tau) = \varphi_{h_1 h_1}(\tau) * \varphi_{xx}(\tau) = h_1(\tau) * h_1^*(-\tau) * \varphi_{xx}(\tau)$

$\varphi_{y_2 y_2}(\tau) = \varphi_{h_2 h_2}(\tau) * \varphi_{xx}(\tau) = h_2(\tau) * h_2^*(-\tau) * \varphi_{xx}(\tau)$

b) With (18.52):

$\varphi_{y_1 x}(\tau) = \varphi_{xx}(\tau) * h_1(\tau)$

$\varphi_{y_2 x}(\tau) = \varphi_{xx}(\tau) * h_2(\tau)$

c) put $r = y_2$ and $y = y_1$ in (18.65):

$\varphi_{y_1 y_2}(\tau) = \varphi_{xy_2}(\tau) * h_1(\tau) = \varphi_{y_2 x}^*(-\tau) * h_1(\tau) = \varphi_{xx}^*(-\tau) * h_2^*(-\tau) * h_1(\tau) =$
$= \varphi_{xx}(\tau) * h_2^*(-\tau) * h_1(\tau)$

We used (17.54) and (17.58) to rearrange the equations.

Solution 18.12

Proceed as in Sec. 18.2.4:

$$\varphi_{xy}[\kappa] \;=\; \mathrm{E}\{x[k+\kappa] \sum_\mu h^*[\mu] x^*[k-\mu]\} = \sum_\mu h^*[\mu] \mathrm{E}\{x[k+\kappa]x^*[k-\mu]\} =$$
$$=\; \sum_\mu h^*[\mu] \varphi_{xx}[k+\mu]$$

Substitution $\nu = -\mu$

$$\varphi_{xy}[\kappa] = \sum_\nu h^*[-\nu] \varphi_{xx}[\kappa - \nu] = \varphi_{xx}[\kappa] * h^*[-\kappa]$$

Solution 18.13

Proceed as in Sec. 18.2.3:

$$\varphi_{yy}[\kappa] \;=\; \mathrm{E}\left\{ \sum_\mu h[\mu]x[k+\kappa - \mu] \cdot \sum_\nu h^*[\nu]x^*[k-\nu] \right\} =$$
$$=\; \sum_\mu \sum_\nu h[\mu]h^*[\nu] \cdot \mathrm{E}\{x[k+\kappa-\mu]x^*[k-\nu]\} =$$
$$=\; \sum_\mu \sum_\nu h[\mu]h^*[\nu] \cdot \varphi_{xx}[\kappa - \mu + \nu]$$

Substituting $\vartheta = \mu - \nu$ and swapping the sums yields:

$$
\begin{aligned}
\varphi_{yy}[\kappa] &= \sum_{\vartheta} \sum_{\mu} h[\mu] h^*[\mu - \vartheta] \cdot \varphi_{xx}[\kappa - \vartheta] = \\
&= \sum_{\vartheta} h[\vartheta] * h^*[-\vartheta] \cdot \varphi_{xx}[\kappa - \vartheta] = \\
&= \sum_{\vartheta} \varphi_{hh}[\vartheta] \cdot \varphi_{xx}[\kappa - \vartheta] = \varphi_{hh}[\kappa] * \varphi_{xx}[\kappa]
\end{aligned}
$$

Solution 18.14

a) Difference equation: $s[k] = n[k] + 0.9s[k-1]$

in the frequency-domain: $S(z) = N(z) + 0.9z^{-1}N(z)$

$$H(z) = \frac{S(z)}{N(z)} = \frac{1}{1 - 0.9z^{-1}} = \frac{z}{z - 0.9}$$

$$H(e^{j\Omega}) = \frac{e^{j\Omega}}{e^{j\Omega} - 0.9}$$

b) $\Phi_{ss}(e^{j\Omega}) = |H(e^{j\Omega})|^2 \Phi_{nn}(e^{j\Omega}) = \dfrac{|e^{j\Omega}|^2}{|e^{j\Omega} - 0.9|^2} \cdot 1 = \dfrac{1}{|e^{j\Omega} - 0.9|^2}$

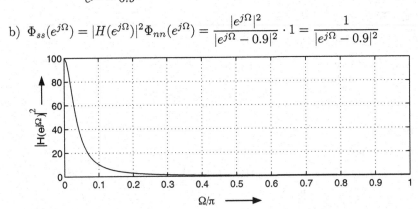

Solution 18.15

a) To determine the Wiener filter we need to know the power density spectrum $\Phi_{yy}(j\omega)$ of the received signal $y(t)$ and the cross-spectrum $\Phi_{xy}(j\omega)$ between $y(t)$ and the transmitted signal $x(t)$. The optimal solution to this exercise is:

$$H(j\omega) = \frac{\Phi_{yx}^*(j\omega)}{\Phi_{yy}(j\omega)} = \frac{\Phi_{xy}(j\omega)}{\Phi_{yy}(j\omega)}.$$

b) Read the linear distortion without noise $\Rightarrow H(j\omega) = \dfrac{1}{G(j\omega)}$

from the block diagramm: $G(s) = b\dfrac{1}{s} + a = \dfrac{as + b}{s}$

$$H(j\omega) = \frac{j\omega}{j\omega \cdot a + b} = \frac{j\omega}{j\omega + 100}$$

$$\Phi_{yy}(j\omega) = |G(j\omega)|^2 \cdot \Phi_{xx}(j\omega) = \left|\frac{100}{j\omega} + 1\right|^2 \cdot 1 = \frac{10^4 + \omega^2}{\omega^2}$$

$$\Phi_{\tilde{x}\tilde{x}}(j\omega) = |H(j\omega)|^2 \cdot |G(j\omega)|^2 \cdot \Phi_{xx}(j\omega) = \Phi_{xx}(j\omega)$$

$\Phi_{xx}(j\omega) = \Phi_{\tilde{x}\tilde{x}}(j\omega)$ $\qquad\qquad$ $\Phi_{yy}(j\omega) = |G(j\omega)|^2$ $\qquad\qquad$ $|H(j\omega)|$

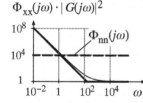 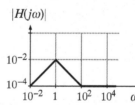 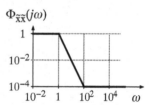

Before sketching the Bode plots, the zeros and poles of $|G(s)|^2$ are determined: $s_n = \pm 10^2$ and $s_p = 0$. The zeros have the combined effect of a double zero at -10^2.

c) Linear distortion and additive noise

$$\Rightarrow H(j\omega) = \frac{\Phi_{xx}(j\omega) G^*(j\omega)}{\Phi_{xx}(j\omega) |G(j\omega)|^2 + \Phi_{nn}(j\omega)}$$

$$H(j\omega) = \frac{G^*(j\omega)}{|G(j\omega)|^2 + N_0} = \frac{\omega^2 + 100j\omega}{10^4(\omega^2 + 1)}$$

$$\begin{aligned}\Phi_{\tilde{x}\tilde{x}}(j\omega) &= \left[\Phi_{xx}(j\omega) \cdot |G(j\omega)|^2 + N_0\right] \cdot |H(j\omega)|^2 \\ &= \left[\frac{\omega^2 + 10^4}{\omega^2} + 9999\right] \cdot \frac{\omega^4 + 10^4\omega^2}{10^8(\omega^2+1)^2} = \frac{\omega^2 + 10^4}{10^4(\omega^2+1)}\end{aligned}$$

$\Phi_{xx}(j\omega) \cdot |G(j\omega)|^2$ $\qquad\qquad$ $|H(j\omega)|$ $\qquad\qquad$ $\Phi_{\tilde{x}\tilde{x}}(j\omega)$

Solution 18.16

Calculation of $\Phi_{sx}(j\omega)$:

According to (18.22), $\Phi_{sv}(j\omega) = \Phi_{ss}(j\omega)$, since $s(t)$ and $n(t)$ are uncorrelated.

$\Phi_{sx}(j\omega) = \Phi_{sv}(j\omega) \cdot G^*(j\omega) = \Phi_{ss}(j\omega) \cdot G^*(j\omega)$, according to (18.67)

Calcuation of $\Phi_{xx}(j\omega)$:

$\Phi_{xx}(j\omega) = \Phi_{vv}(j\omega) \cdot |G(j\omega)|^2 = \left(\Phi_{nn}(j\omega) + \Phi_{ss}(j\omega)\right)|G(j\omega)|^2$

Wiener filter:

$$H(j\omega) \quad = \quad \frac{\Phi_{sx}(j\omega)}{\Phi_{yx}(j\omega)} = \frac{\Phi_{ss}(j\omega)\,G^*(j\omega)}{\left(\Phi_{nn}(j\omega) + \Phi_{ss}(j\omega)\right)|G(j\omega)|^2}$$

$$= \quad \begin{cases} \dfrac{j\omega + 10}{2j\omega} & \text{for } |\omega| < \omega_g \\[2mm] 0 & \text{otherwise} \end{cases}$$

Solution 18.17

The noise-free signal at the output is $\tilde{y}(t)$.

$$\varphi_{yx}(\tau) = \mathrm{E}\{y(t+\tau)\,x(t)\} = \mathrm{E}\{[n(t+\tau) + \tilde{y}(t+\tau)]\,x(t)\} =$$
$$= \mathrm{E}\{n(t+\tau)\,x(t)\} + \mathrm{E}\{\tilde{y}(t+\tau)\,x(t)\} = 0 + \varphi_{\tilde{y}x}(\tau)$$

With $\varphi_{xx}(\tau) = \delta(\tau)$ gilt analog zu (18.77), (18.78) $\varphi_{\tilde{y}x}(\tau) = h(\tau)$ and therefore $\Phi_{yx}(j\omega) = \Phi_{\tilde{y}x}(j\omega) = H(j\omega)$.

Appendix B Tables of Transformations

Appendix B.1 Bilateral Laplace Transform Pairs

$x(t)$	$X(s) = \mathcal{L}\{x(t)\}$	ROC
$\delta(t)$	1	$s \in \mathbb{C}$
$\varepsilon(t)$	$\dfrac{1}{s}$	$\mathrm{Re}\{s\} > 0$
$e^{-at}\varepsilon(t)$	$\dfrac{1}{s+a}$	$\mathrm{Re}\{s\} > \mathrm{Re}\{-a\}$
$-e^{-at}\varepsilon(-t)$	$\dfrac{1}{s+a}$	$\mathrm{Re}\{s\} < \mathrm{Re}\{-a\}$
$t\varepsilon(t)$	$\dfrac{1}{s^2}$	$\mathrm{Re}\{s\} > 0$
$t^n\varepsilon(t)$	$\dfrac{n!}{s^{n+1}}$	$\mathrm{Re}\{s\} > 0$
$te^{-at}\varepsilon(t)$	$\dfrac{1}{(s+a)^2}$	$\mathrm{Re}\{s\} > \mathrm{Re}\{-a\}$
$t^n e^{-at}\varepsilon(t)$	$\dfrac{n!}{(s+a)^{n+1}}$	$\mathrm{Re}\{s\} > \mathrm{Re}\{-a\}$
$\sin(\omega_0 t)\varepsilon(t)$	$\dfrac{\omega_0}{s^2 + \omega_0^2}$	$\mathrm{Re}\{s\} > 0$
$\cos(\omega_0 t)\varepsilon(t)$	$\dfrac{s}{s^2 + \omega_0^2}$	$\mathrm{Re}\{s\} > 0$
$e^{-at}\cos(\omega_0 t)\varepsilon(t)$	$\dfrac{s+a}{(s+a)^2 + \omega_0^2}$	$\mathrm{Re}\{s\} > \mathrm{Re}\{-a\}$
$e^{-at}\sin(\omega_0 t)\varepsilon(t)$	$\dfrac{\omega_0}{(s+a)^2 + \omega_0^2}$	$\mathrm{Re}\{s\} > \mathrm{Re}\{-a\}$
$t\cos(\omega_0 t)\varepsilon(t)$	$\dfrac{s^2 - \omega_0^2}{(s^2 + \omega_0^2)^2}$	$\mathrm{Re}\{s\} > 0$
$t\sin(\omega_0 t)\varepsilon(t)$	$\dfrac{2\omega_0 s}{(s^2 + \omega_0^2)^2}$	$\mathrm{Re}\{s\} > 0$

Appendix B.2   Properties of the Bilateral Laplace Transform

$x(t)$	$X(s) = \mathcal{L}\{x(t)\}$	ROC
Linearity $Ax_1(t) + Bx_2(t)$	$AX_1(s) + BX_2(s)$	ROC $\supseteq$ ROC$\{X_1\}$ $\cap$ROC$\{X_2\}$
Delay $x(t - \tau)$	$e^{-s\tau}X(s)$	not affected
Modulation $e^{at}x(t)$	$X(s - a)$	Re$\{a\}$ shifted by Re$\{a\}$ to the right
'Multiplication by t', Differentiation in the frequency domain $tx(t)$	$-\dfrac{d}{ds}X(s)$	not affected
Differentiation in the time domain $\dfrac{d}{dt}x(t)$	$sX(s)$	ROC $\supseteq$ ROC$\{X\}$
Integration $\displaystyle\int_{-\infty}^{t} x(\tau)d\tau$	$\dfrac{1}{s}X(s)$	ROC $\supseteq$ ROC$\{X\}$ $\cap\{s : \text{Re}\{s\} > 0\}$
Scaling $x(at)$	$\dfrac{1}{|a|}X\left(\dfrac{s}{a}\right)$	ROC scaled by a factor of a

Appendix B.3   Fourier Transform Pairs

$x(t)$	$X(j\omega) = \mathcal{F}\{x(t)\}$
$\delta(t)$	1
1	$2\pi\delta(\omega)$
$\dot{\delta}(t)$	$j\omega$
$\dfrac{1}{T}\,\text{Ш}\left(\dfrac{t}{T}\right)$	$\text{Ш}\left(\dfrac{\omega T}{2\pi}\right)$
$\varepsilon(t)$	$\pi\delta(\omega) + \dfrac{1}{j\omega}$
$\text{rect}(at)$	$\dfrac{1}{\lvert a\rvert}\,\text{si}\left(\dfrac{\omega}{2a}\right)$
$\text{si}(at)$	$\dfrac{\pi}{\lvert a\rvert}\,\text{rect}\left(\dfrac{\omega}{2a}\right)$
$\dfrac{1}{t}$	$-j\pi\text{sign}(\omega)$
$\text{sign}(t)$	$\dfrac{2}{j\omega}$
$e^{j\omega_0 t}$	$2\pi\delta(\omega - \omega_0)$
$\cos(\omega_0 t)$	$\pi[\delta(\omega + \omega_0) + \delta(\omega - \omega_0)]$
$\sin(\omega_0 t)$	$j\pi[\delta(\omega + \omega_0) - \delta(\omega - \omega_0)]$
$e^{-\alpha\lvert t\rvert},\ \alpha > 0$	$\dfrac{2\alpha}{\alpha^2 + \omega^2}$
$e^{-a^2 t^2}$	$\dfrac{\sqrt{\pi}}{a}e^{-\frac{\omega^2}{4a^2}}$

Appendix B.4 Properties of the Fourier Transform

	$x(t)$	$X(j\omega) = \mathcal{F}\{x(t)\}$
Linearity	$Ax_1(t) + Bx_2(t)$	$AX_1(j\omega) + BX_2(j\omega)$
Delay	$x(t - \tau)$	$e^{-j\omega\tau}X(j\omega)$
Modulation	$e^{j\omega_0 t}x(t)$	$X(j(\omega - \omega_0))$
'Multiplication by t' Differentiation in the frequency domain	$tx(t)$	$-\dfrac{dX(j\omega)}{d(j\omega)}$
Differentiation in the time domain	$\dfrac{dx(t)}{dt}$	$j\omega X(j\omega)$
Integration	$\displaystyle\int_{-\infty}^{t} x(\tau)d\tau$	$X(j\omega)\left[\pi\delta(\omega) + \dfrac{1}{j\omega}\right]$ $= \dfrac{1}{j\omega}X(j\omega) + \pi X(0)\delta(\omega)$
Scaling	$x(at)$	$\dfrac{1}{\|a\|}X\left(\dfrac{j\omega}{a}\right), \quad a \in \mathbb{R}\backslash\{0\}$
Convolution	$x_1(t) * x_2(t)$	$X_1(j\omega) \cdot X_2(j\omega)$
Multiplication	$x_1(t) \cdot x_2(t)$	$\dfrac{1}{2\pi}X_1(j\omega) * X_2(j\omega)$
Duality	$x_1(t)$ $x_2(jt)$	$x_2(j\omega)$ $2\pi x_1(-\omega)$
Symmetry relations	$x(-t)$ $x^*(t)$ $x^*(-t)$	$X(-j\omega)$ $X^*(-j\omega)$ $X^*(j\omega)$
Parseval theorem	$\displaystyle\int_{-\infty}^{\infty} \|x(t)\|^2\, dt$	$\dfrac{1}{2\pi}\displaystyle\int_{-\infty}^{\infty} \|X(j\omega)\|^2 d\omega$

Appendix B.5   Two-sided z-Transform Pairs

$x[k]$	$X(z) = \mathcal{Z}\{x[k]\}$	ROC
$\delta[k]$	1	$z \in \mathbb{C}$
$\varepsilon[k]$	$\dfrac{z}{z-1}$	$|z| > 1$
$a^k \varepsilon[k]$	$\dfrac{z}{z-a}$	$|z| > |a|$
$-a^k \varepsilon[-k-1]$	$\dfrac{z}{z-a}$	$|z| < |a|$
$k\varepsilon[k]$	$\dfrac{z}{(z-1)^2}$	$|z| > 1$
$ka^k \varepsilon[k]$	$\dfrac{az}{(z-a)^2}$	$|z| > |a|$
$\sin(\Omega_0 k)\varepsilon[k]$	$\dfrac{z \sin \Omega_0}{z^2 - 2z \cos \Omega_0 + 1}$	$|z| > 1$
$\cos(\Omega_0 k)\varepsilon[k]$	$\dfrac{z(z - \cos \Omega_0)}{z^2 - 2z \cos \Omega_0 + 1}$	$|z| > 1$

Appendix B.6   Properties of the z-Transform

Property	$x[k]$	$X(z)$	ROC
Linearity	$ax_1[k]+bx_2[k]$	$aX_1(z) + bX_2(z)$	ROC $\supseteq$ ROC$\{X_1\}\cap$ROC$\{X_2\}$
Delay	$x[k - \kappa]$	$z^{-\kappa}X(z)$	ROC$\{x\}$; separate consideration of $z = 0$ and $z \to \infty$
Modulation	$a^k x[k]$	$X\left(\dfrac{z}{a}\right)$	ROC$= \left\{ z \left| \dfrac{z}{a} \in$ROC$\{x\}\right.\right\}$
Multiplication by k	$kx[k]$	$-z\dfrac{dX(z)}{dz}$	ROC$\{x\}$; separate consideration of $z = 0$
Time inversion	$x[-k]$	$X(z^{-1})$	ROC$=\{z\,|\,z^{-1}\in$ROC$\{x\}\}$
Convolution	$x_1[k] * x_2[k]$	$X_1(z) \cdot X_2(z)$	ROC $\supseteq$ ROC$\{x_1\}\cap$ROC$\{x_2\}$
Multiplication	$x_1[k] \cdot x_2[k]$	$\dfrac{1}{2\pi j}\oint X_1(\zeta)X_2\left(\dfrac{z}{\zeta}\right)\dfrac{1}{\zeta}d\zeta$	multiply the limits of the ROC

Appendix B.7   Discrete-Time Fourier Transform Pairs

$x[k]$	$X(e^{j\Omega}) = \mathcal{F}_*\{x[k]\}$
$\delta[k]$	1
$\varepsilon[k]$	$\dfrac{1}{1-e^{-j\Omega}} + \dfrac{1}{2}\,⊥⊥⊥\left(\dfrac{\Omega}{2\pi}\right)$
1	$⊥⊥⊥\left(\dfrac{\Omega}{2\pi}\right)$
$e^{j\Omega_0 k}$	$⊥⊥⊥\left(\dfrac{\Omega-\Omega_0}{2\pi}\right)$
$\cos\Omega_0 k$	$\dfrac{1}{2}\left[⊥⊥⊥\left(\dfrac{\Omega+\Omega_0}{2\pi}\right) + ⊥⊥⊥\left(\dfrac{\Omega-\Omega_0}{2\pi}\right)\right]$
$\sin\Omega_0 k$	$\dfrac{j}{2}\left[⊥⊥⊥\left(\dfrac{\Omega+\Omega_0}{2\pi}\right) - ⊥⊥⊥\left(\dfrac{\Omega-\Omega_0}{2\pi}\right)\right)\right]$
$x[k] = \begin{cases} 1 & \text{for} \quad 0 \le k < N \\ 0 & \text{otherwise} \end{cases}$	$e^{-j\Omega\frac{N-1}{2}} \cdot \dfrac{\sin\left(\frac{N\Omega}{2}\right)}{\sin\left(\frac{\Omega}{2}\right)}$

Appendix B.8   Properties of the Discrete-Time Fourier Transform

Property	$x[k]$	$X(e^{j\Omega}) = \mathcal{F}_*\{x[k]\}$
Linearity	$ax_1[k] + bx_2[k]$	$aX_1(e^{j\Omega}) + bX_2(e^{j\Omega})$
Delay	$x[k-\kappa]$	$e^{-j\Omega\kappa}X(e^{j\Omega})$, $\kappa \in \mathbb{Z}$
Modulation	$e^{j\Omega_0 k}x[k]$	$X(e^{j(\Omega-\Omega_0)})$, $\Omega_0 \in \mathbb{R}$
Convolution	$x_1[k] * x_2[k]$	$X_1(e^{j\Omega})\,X_2(e^{j\Omega})$
Multiplication	$x_1[k]\,x_2[k]$	$\dfrac{1}{2\pi}X_1(e^{j\Omega})\circledast X_2(e^{j\Omega})$
Parseval theorem	$\displaystyle\sum_{k=-\infty}^{\infty}|x[k]|^2$	$\dfrac{1}{2\pi}\displaystyle\int_{-\pi}^{\pi}|X(e^{j\Omega})|^2 d\Omega$ one period!

Bibliography

[1] J. S. Bay. *Fundamentals of Linear State Space Systems*. WCB/McGraw-Hill, Boston, 1999.

[2] Jr. C.H. Edwards and D.E. Penney. *Elementary Differential Equations*. Prentice-Hall, Englewood Cliffs, 3rd edition, 1993.

[3] C.-T. Chen. *Linear System Theory and Design*. Oxford University Press, New York, 1999.

[4] R.V. Churchill. *Operational Mathematics*. McGraw–Hill, New-York, 3rd edition, 1972.

[5] A. Fettweis. *Elemente nachrichtentechnischer Systeme*. B.G.Teubner, Stuttgart, 1990.

[6] N. Fliege. *Systemtheorie*. B.G.Teubner, Stuttgart, 1991.

[7] E. Hänsler. *Statistische Signale*. Springer Verlag, Berlin, 2 edition, 1997.

[8] S. Haykin and B. Van Veen. *Signals and Systems*. John Wiley & Sons, New York, 1999.

[9] W. Kamen and B. S. Heck. *Fundamentals of Signals and Systems Using the Web and Matlab*. Prentice-Hall, Englewood Cliffs, 2nd edition, 1997.

[10] Z.Z. Karu. *Signals and Systems Made Ridiculously Simple*. ZiZiPress, Cambridge, MA, USA, 1995.

[11] H. Kwakernaak and R. Sivan. *Modern Signals and Systems*. Prentice-Hall, Englewood Cliffs, 1991.

[12] B.P. Lathi. *Signal Processing and Linear Systems*. Berkeley Cambridge Press, Carmichael, 1998.

[13] D. K. Lindner. *Introduction to Signals and Systems*. WCB/McGraw-Hill, Boston, 1999.

[14] F. Oberhettinger and L. Badii. *Tables of Laplace Transforms*. Springer-Verlag, Berlin, 1973.

[15] A. Oppenheim and A. Willsky. *Signale und Systeme*. VCH Verlagsgesellschaft, Basel, 1989.

[16] J.G. Reid. *Linear System Fundamentals*. McGraw–Hill, New-York, 1983.

[17] R. Sauer and I. Szabó. *Mathematische Hilfsmittel des Ingenieurs, Teil I*. Springer-Verlag, Berlin, 1967.

[18] H.W. Schüßler. *Netzwerke, Signale und Systeme 1*. Springer-Verlag, Berlin, 2 edition, 1990.

[19] H.W. Schüßler. *Netzwerke, Signale und Systeme 2*. Springer-Verlag, Berlin, 3 edition, 1991.

[20] W.McC. Siebert. *Circuits, Signals, and Systems*. The MIT Press, Cambridge, MA, USA, 1986.

[21] F. Szidarovsky and A.T. Bahill. *Linear Systems Theory*. CRC Press, Boca Raton, Florida, 1992.

[22] R. Unbehauen. *Grundlagen der Elektrotechnik*. Springer Verlag, Berlin, 4 edition, 1994.

[23] R. Unbehauen. *Systemtheorie*. Oldenbourg Verlag, München, 7 edition, 1997.

[24] G. Wunsch. *Geschichte der Systemtheorie*. Oldenbourg Verlag, München, 1985.

Index